Fracture Mechanics of Ceramics

Volume 8
Microstructure, Methods, Design, and Fatigue

Volume 1 Concepts, Flaws, and Fractography
Volume 2 Microstructure, Materials, and Applications
Volume 3 Flaws and Testing
Volume 4 Crack Growth and Microstructure
Volume 5 Surface Flaws, Statistics, and Microcracking
Volume 6 Measurements, Transformations, and High-Temperature Fracture
Volume 7 Composites, Impact, Statistics, and High-Temperature Phenomena
Volume 8 Microstructure, Methods, Design, and Fatigue

Fracture Mechanics of Ceramics

Volume 8
Microstructure, Methods, Design, and Fatigue

Edited by R. C. Bradt

College of Engineering
University of Washington
Seattle, Washington

A. G. Evans

Department of Materials and Mineral Engineering
University of California
Berkeley, California

D. P. H. Hasselman

Department of Materials Engineering
Virginia Polytechnic Institute and State University
Blacksburg, Virginia

and F. F. Lange

Rockwell International
Science Center
Thousand Oaks, California

PLENUM PRESS • NEW YORK-LONDON

ISBN 0-306-42273-5
Library of Congress Catalog Card Number 83-641076

Second part of the proceedings of the Fourth International Symposium on the Fracture Mechanics of Ceramics, held June 19–21, 1985, at the Virginia Polytechnic Institute and State University, Blacksburg, Virginia

A Division of Plenum Publishing Corporation
233 Spring Street, New York, N.Y. 10013

Printed in the United States of America

PREFACE

These volumes, 7 and 8, of Fracture Mechanics of Ceramics constitute the proceedings of an international symposium on the fracture mechanics of ceramic materials held at Virginia Polytechnic Institute and State University, Blacksburg, Virginia on June 19, 20 and 21, 1985. These proceedings constitute the fourth pair of volumes of a continuing series of conferences.

The theme of this conference, as the previous three, focused on the mechanical behavior of ceramic materials in terms of the characteristics of cracks, particularly the roles which they assume in the fracture process. The 78 contributed papers by over 100 authors and co-authors represent the current state of the field. They address many of the theoretical and practical problems of interest to those concerned with brittle fracture.

The program chairmen gratefully acknowledge the financial assistance for the Symposium provided by the EXXON Foundation, the Army Research Office, the National Science Foundation, and the Office of Naval Research. Without their support, this conference simply would not have been possible. The suggestions of Drs. J. C. Hurt, R. C. Pohanka, and L. Toth were particularly helpful in assuring the success of this symposium.

Special appreciation is extended to Professor J. I. Robertson, C. P. Miles Professor of History, whose presentation following the banquet on the American Civil War was very well received by the audience.

Finally, we wish to also thank our joint secretaries, especially Karen Snider, for their patience and help in finally bringing these proceedings to press.

R.C. Bradt
Seattle, Washington

A.G. Evans
Berkeley, California

D.P.H. Hasselman
Blacksburg, Virginia

F.F. Lange
Tou

June, 1985

CONTENTS

FRACTURE MECHANICS AND MICROSTRUCTURES

M.G. Gee and R. Morrell

National Physical Laboratory
Teddington
Middlesex, UK

ABSTRACT

The influence of microstructure on defects in ceramics, and the consequences of their presence for the application of fracture mechanics theories are reviewed. The complexities of microstructures, especially the multiphase nature, the crystallographic anisotropy and the resultant anisotropic physical properties, and the variation of microstructure and surface finish from point to point in real components, all lead to considerable uncertainties in the actual performance of any particular component. It is concluded that although the concepts of fracture mechanics have been and will continue to be most useful for the qualitative explanation of fracture phenomena, the usefulness as a predictive tool with respect to most existing types of material is limited by the interrelation between material microstructure and mechanical properties. At present, the only method of eliminating components with unsatisfactory mechanical properties is to proof-test them, despite the fact that proof-testing itself is limited in ability to cope with changes to the component in service. The aim of the manufacturer must be to improve quality and consistency within individual components, from component to component, and from batch to batch. The aim of the fracture specialist must be to study longer-term properties to improve the accuracy of behaviour predictions with a stronger data base. Materials development needs to concentrate on obtaining defect-free materials that can be translated into more-reliable products, using our present understanding of the influence of microstructure on strength and toughness.

INTRODUCTION

Over the last 10-15 years, our understanding of the mechanical behaviour of ceramics has improved considerably as a direct result of the application of fracture mechanics to ceramics, in a sense extending the basic Griffith concepts established as long ago as 1920. These concepts have been applied successfully to glasses, principally because the microstructure is almost featureless down to the atomic scale. This means that the fracture process can be modelled by examining an atomically sharp crack in an isotropic continuum. Furthermore, since the structure is usually defect-free, cracks can be nucleated far more readily at the surface than within the bulk. The flaw population tends to be controlled almost exclusively by the mechanical state of the surface, fatigue effects are controlled by the chemistry of the external environment, and strength and lifetime parameters tend to be fairly predictable for a given bulk composition.

In direct contrast, polycrystalline ceramic materials are not microstructurally homogeneous, and while many display fatigue effects in a manner similar to glasses, they seem to be much more variable in behaviour, and thus it is far more difficult to make lifetime predictions with the same level of assurance as for glasses.

This paper reviews the relationship between microstructure and fracture mechanics, and examines the levels of understanding of materials properties and performance that will be required in order to progress with the engineering use of ceramic components in a non-empirical manner.

THE CURRENT POSITION

When designing with metallic materials, the problem of stress concentrations at points of contact is ameliorated considerably because the plasticity of the metal automatically limits the stresses that can be applied, allowing a considerable degree of local stress tolerance. Only as the strain-to-failure becomes more limited in the harder, more brittle alloys is it necessary to take more care to avoid excessive stress concentration. At the extreme of the scale, ceramics are essentially totally brittle, and do not conveniently relax contact stresses except at elevated temperatures. The design engineer is therefore confronted with the need to consider carefully and to minimise not only the effects of geometrical stress concentrations due to real component shape, but also localised stresses at the points of loading. Finite element analysis and other techniques for stress analysis can cope with such situations, and as computing time becomes cheaper, these techniques become more widely available. However, the potential errors can be large due to a lack of knowledge of relevant material parameters, surface conditions, precise geometry (out-of-tolerance), and service conditions. Thus although it is possible _in principle_ to estimate failure probabilities and lifetimes, even under complex conditions [1-3], in practice the benefits are limited. As a result there remains a gulf between drawing-board design and engineering success, so much so that frequently many design iterations are required before prototype engineering trials begin to show sufficient promise. The lack of appropriate design methodology seems a serious one for the future of engineering ceramics, especially in developments of heat engines, and can lead to poor design or inappropriate use.

In order to assess the lifetime of a component, many material-based factors have to be taken into account.

(1) Basic materials properties - elastic moduli, thermal properties, chemistry and microstructure.
(2) Variations in properties as a consequence of fabrication method, defect distributions, anisotropy.
(3) Surface finish and surface properties, particularly residual stresses, friction, strength and fatigue effects.
(4) Service conditions, and the progressive change in factors (1) to (3) that might result.
(5) Suitability of the methods of prediction of stresses, and of extrapolation of relevant property data.

The difficulties in defining material properties and service conditions which lead to design uncertainties are discussed below.

STRENGTH AND FRACTURE MECHANICS THEORIES

The well-established Griffith concept of atomically sharp flaws works demonstrably well for annealed glasses with featureless microstructures. Many fracture mechanics theories, notably that of subcritical crack growth (e.g.[4]), have been well tested, and good correlations obtained between delayed failure lifetimes and predictions made on short-term properties and slow crack growth exponents. Many of the basic features of fracture behaviour can be demonstrated. These theories are well-known, and are available in a number of reviews. However, they are based on a number of simplifying assumptions that may not always be generally valid for polycrystalline materials. These are now examined.

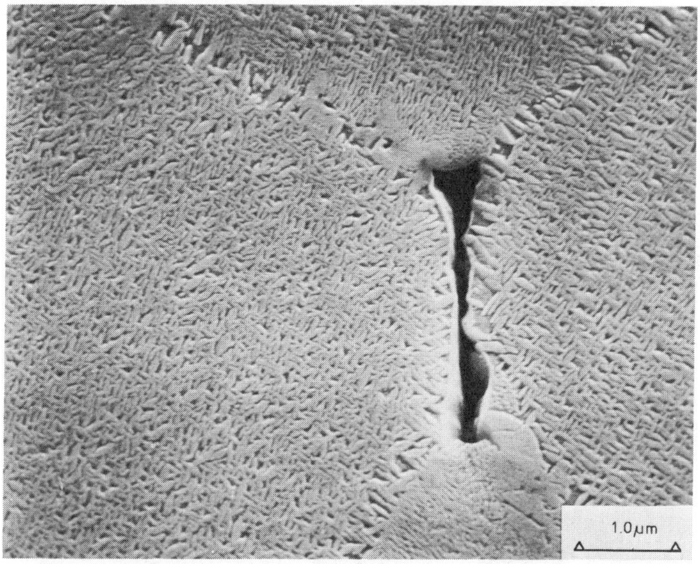

Fig.1 A grain-boundary pore in commercial toughened partially stabilised zirconia. Note also the different microstructure in the boundaries compared with the bulk.

The nature of flaws

Strength-controlling flaws are traditionally conceived as being pre-existing sharp cracks. This may be true in many materials, notably those in which the density of defects in the bulk microstructure are substantially eliminated and strength is controlled by surface flaws such as machining damage. In others, the situation is less clear. Internal defects of various types can exist, and these may not appear sharp. Examples are pores and groups of pores (Fig.1,2), large grains (Fig.3) with associated thermal expansion mismatch stresses, foreign particles, and regions of contamination (Fig.4). These are discussed later in more detail, but the common feature is the difficulty of deciding which defect governs the limitation in strength in any particular component. This arises because of the indeterminate ability of each defect to initiate a sharp crack, a step that must precede macroscopic failure. It has been proposed[5] that both sharp and blunt "original" defects can propagate subcritically during a strength test, with a progressive increase in apparent fracture toughness, reaching a K_{Ic} level at a size rather larger than the "original" size, and with a rather different microstructure at the tip. In this respect it is therefore difficult to define an effective "flaw size" for the purpose of fracture

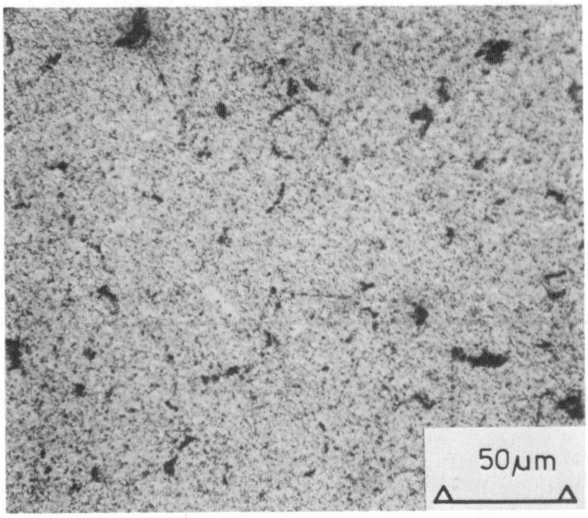

Fig.2 Porosity distribution in dry-pressed spray-dried β-alumina, showing the traces of inter-particle porosity not removed by particle collapse in pressing.

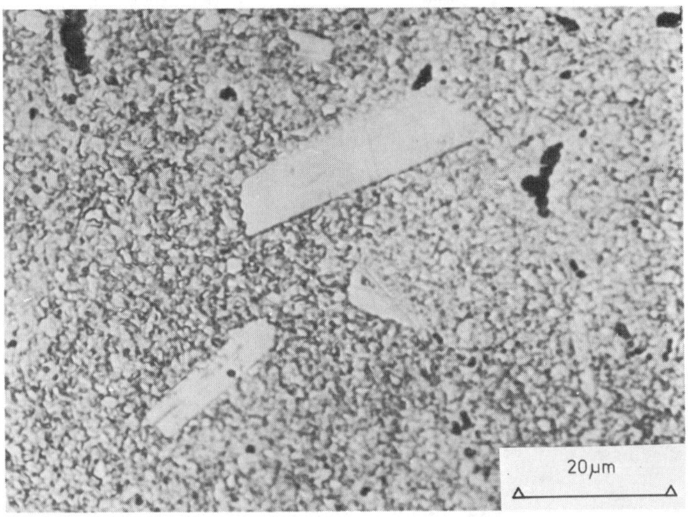

Fig.3 Large grains in a fine-grained matrix of β-alumina control the strength attainable.

mechanics. The sharp crack may not appear until stress is applied. Examples include failure from pores in glasses and ceramics[6], and the evidence for the opening and closing of cracks on applying and removing stress found using surface acoustic waves[7,8]. The original defects can be much smaller than the critical flaw size, which is often calculated from strength and fracture toughness to be larger than any microstructural feature observable. Conversely, the fracture toughness that has been estimated from the apparent size of small flaws may be smaller than that for large flaws or pre-cracks in fracture toughness tests[9], because the scale of the microstructure involved and its homogeneity (as in Fig.1 for example) are different. In addition, the exact fracture criterion will depend on the shape of the flaw and its orientation relative to the applied stress field. Different materials have different types of microstructure and different

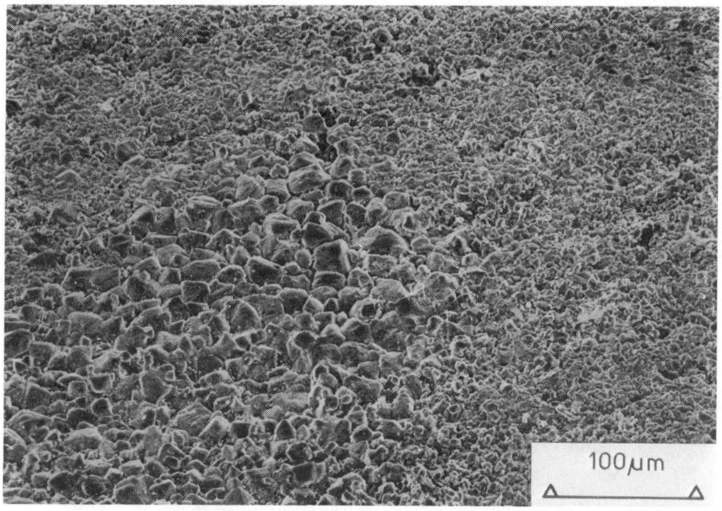

Fig.4 A region of contamination in a commercial alumina ceramic caused by a speck of iron on an as-fired surface.

types of defects that require consideration.

The logical continuation of this argument is that it will be extremely difficult to determine which original defect in a component will be strength-limiting, unless of course it is an obvious gross manufacturing defect. Non-destructive testing techniques which can detect small defects will thus have as their fundamental limitation, irrespective of resolution, the inability to predict whether sharp crack nucleation and propagation during subsequent stressing can occur from any particular defect at a given stress level.

Defect distributions

The scatter in strengths obtained from a nominally identical set of components or test-pieces is usually attributed to the variable size of the "worst" flaw caused by the randomness of the microstructure. Commonly, the distribution of strengths is represented as a two-parameter Weibull distribution, principally because this function is particularly flexible. The probability of failure, P_f, is given by:

$$P_f = 1 - \exp\{-U(\sigma)\} \qquad \text{where } U(\sigma) = A\sigma_{max}^m \int_V c^m \, dV$$

where A and m are constants, σ_{max} is the maximum stress in the body, and c is the ratio of actual to maximum stress throughout volume V. It allows a straight-line fit to be made to the cumulative distribution, the slope being the Weibull modulus, m, which is a measure of strength consistency. It also allows convenient separation of the stress magnitude and spatial (over volume V) parts of the probability of failure. With most other distributions, the volume integral is not separable from the stress integral[10]. This factor makes a Weibull distribution particularly convenient for scaling purposes, but it is purely empirical and is used merely for convenience. However, it can be shown that it is related to the extreme values of an inverse power law defect size distribution[11]:

$$Q(x) = Qx^{-p}$$

where Q(x)dx is the number of defects in the size range x to x + dx in unit volume, and Q and p are constants.

Other distributions may fit the data from a limited number of tests equally well[12]. However, when the distributions are extrapolated to larger or smaller volumes and/or low probabilities of failure, predictions become widely different. In addition there may be more than one defect population with different size dependencies (e.g. where both surface and bulk defects need to be taken into account)[13], and these become dominant in different size or stress regimes. They are difficult to analyse, and so attempts to study size effects often do not produce simple relationships as are predicted by Weibull theory, e.g.[14]. In this respect, it has to be appreciated that the flaw distribution in a component will be controlled by the method of manufacture and the surface finishing procedures it receives. Flexural tests on batches of bar specimens that have had various surface finishes, such as vibro-milling, grinding or lapping, have strength distributions with different Weibull moduli[15], and changing the size of the specimen or the test environment can also change "m"[14]. This may reflect changes not only in surface flaw or edge flaw distributions, but also different amounts of subcritical crack growth before failure. Add to these distributions those due to internal defects, and there becomes a considerable degree of uncertainty in predicting how different the net flaw distribution will be for different components, or for changes in manufacturing specification, or even for changes of grinding machine and operator[6]. This points to the obvious need for consistency of product, not only in the basic shape but also in surface finishing.

Stress-state effects

Most studies examining strength and fracture prediction theories for ceramics have employed uniaxial strength tests to check the agreement of theory with observed strength. In service, components experience complex multiaxial stresses which vary with position and time. The simplest theories, e.g. with a maximum tensile stress criterion, which are adequate for uniaxial stressing, may produce misleading results when applied to components under multiaxial stresses. Thus whereas Giovan and Sines[16] found that for an alumina, the biaxial strength was less than the strength in four-point bending, Petrovic and Stout[17] found for another alumina that the strength under combined tension and torsion loading was greater than the uniaxial strength.

An important element of the predictive theories is the crack extension (fracture) criterion. In the classical Weibull formulation, a normal stress criterion is used, which is likely to be correct when the dominant crack is normal to a uniaxial tensile stress. However, in a multiaxial stress field with flaws orientated at arbitrary angles to the applied stresses, both normal tensile stresses and in-plane shear stresses will influence the deformation and fracture processes[18,19], and lead to a different fracture response to that of the uniaxial case. Several possible shear-sensitive criteria have been proposed, and a strain energy release rate criterion has been used quite effectively to evaluate multiaxial strength test results under some conditions[18,20,21].

Different types of flaw are likely to give different changes in fracture response as the multiaxiality of the stress field changes. Shetty et al.[22] suggested that for sharp cracks associated with a blunt defect, a biaxial strengthening would increase as the crack length to flaw radius ratio increased. They later[18] reported theoretical work by Babel and Sines[23] which suggested that failure from blunt flaws such as pores would lead to a biaxial strengthening, but that failure from crack-like defects would result in biaxial weakening. They use this result to explain the biaxial strengthening they observed in alumina. Petrovic and Stout[17] suggested that a material's response to fracture may change in different stress states because of possible influences of the stress state on microcracking and process zone formation. It is also quite probable that a multiaxial stress state will lead to changes in the rate of slow crack growth, and hence in fatigue rates[18,24].

It is clear that the influence of multiaxial stresses on the strength and fracture properties of ceramics is not clearly understood, and depends on the interaction of the detailed microstructure of the ceramic with the fracture and deformation processes. These will change from one material to another, probably also if the component shape is changed, so that great care is needed in the interpretation of multiaxial strength tests and their application to the prediction of service life.

Flaw extension characteristics

Many ceramics, most notably those based on oxides, or containing oxide-based glassy layers at grain boundaries, suffer from progressive loss of remaining strength with time under load, variously termed static fatigue or delayed failure. The phenomenon is usually attributed to the subcritical growth of flaws to the point where a critical length is achieved and fracture ensues. The lifetime under load of any particular component or test-piece depends on three main factors:
(1) the initial crack length;
(2) the environment at the crack tip;
(3) the stress applied.
The external environment is usually more severe than that within a dense ceramic, so it is more likely that the flaw growth process occurs for a surface flaw than for an internal flaw of the same size. On the other hand, an internal flaw can determine life if it is sufficiently large. Consider the case where internal and external flaws in an isotropic continuum are subject to identical stress conditions. Assuming an empirical crack growth law of the form:

$$v = AK_I^{\,n}$$

where v is the crack velocity, K_I is the stress intensity factor, and A and n are constants, holds for the particular material, the time to failure, t_f, from a flaw of length c is:

$$t_f \simeq \frac{2}{AY^2\sigma^2(n-2)K_{Ii}^{\,n-2}}$$

where K_{Ii} refers to the initial conditions of crack geometry, and where Y is a flaw shape parameter relating the applied stress, σ, to the stress intensity factor K_I:

$$K_I = Y\sigma c^{1/2}$$

The complexity of this equation indicates that the detailed propagation characteristics of both internal and external flaws have to be known before it can be predicted whether the external environment, which usually causes a low value of n and hence greater susceptibility to crack growth, controls lifetime, rather than an internal flaw, perhaps of different geometry.

Flexure tests tend to bias failure from surface or edge flaws rather than bulk flaws, and the flaw extension characteristics that might be determined may not be applicable to larger components subject to multiaxial stressing. Unless the occurrence of significant internal flaws can be minimised by good control of microstructure, there will always be a risk in using flaw extension characteristics determined on flexure specimens for predicting behaviour of other shapes.

As noted earlier, the usual power law for crack growth is an empirical one which is often presumed to fit all circumstances, for example the calculation of n from dynamic fatigue tests, without proof that the law is valid. However, there is a fundamental difficulty in obtaining proof. This arises because the production of K - v curves in various environments is usually done by observing crack growth over long distances in large plate specimens. In such tests, the result pertains strictly to average rates of growth, and treats the microstructure as uniform. In contrast, flaws that limit strength in components are usually much smaller, and have some element of residual stress associated with them. Their propagation is controlled by the local microstructure. For these reasons, there is seldom agreement between values of n obtained from plate specimens with large cracks, and those derived from delayed failure or dynamic fatigue tests. It can therefore be argued that a K - v relationship derived from one situation will not hold for the other. Furthermore, the relationship used to fit experimental data over the range of K_I used may not be the correct choice if data were to be available over a wider range. It can be shown that whereas errors in lifetime estimation over the same range of K_I as for the experimental data are small if alternative fitting equations, such as[4]:

$$v = A'\exp\{K_I n'\}$$

are used, the errors increase rapidly when such equations are extrapolated beyond the range of the data. This arises because the time to failure is very sensitive to K_I. In practice, the errors might still be small compared with the observed scatter in component lifetimes, but nevertheless this suggests that in order to minimise errors, the material's characteristics need to be carefully examined at crack growth rates relevant to practical lifetimes. Laboratory data usually do not extend beyond a few hours duration. There are numerous cases in the literature where uncertain evidence has been presented for the existence or non-existence of fatigue limits of crack growth[25,26]. Most early data have been obtained from large plate specimens with through-cracks, but because of the microstructural questions raised earlier, the trend is currently to use so-called dynamic fatigue or constant loading rate tests, e.g. [27]. The range of times to failure examined by this method is limited to the lowest loading rates available on testing machines, giving equivalent times to failure of a few seconds only. This goes counter to the requirement to study the existence of fatigue limits. In addition, chemical changes at the tip of the crack occur in almost all practical environments. These may not assume signifi- cance unless the rate of propagation is extremely low. The suggestions obtained from work on high-alumina ceramics are that it may be necessary to consider not only enhanced rates of slow crack propagation in certain environments, but also the risk of chemical dissolution [28] giving an effective lower limit to crack propagation rate. The oxidation of non-oxides at high temperatures poses a similar type of uncertainty, with lifetime controlled by a mixture of crack growth due to oxidation, phase changes of oxide layers (e.g. cristobalite), and creep cavitation, or perhaps blunting by viscous effects. Factors such as these suggest that predictions of lifetime based on short-term properties are at best order-of-magnitude estimates, and the possible errors accumulate rapidly as extrapolations are made beyond the time-scale in which laboratory tests are made. Although tedious to undertake, this conclusion suggests that more effort be placed on delayed failure tests to acquire data for use in critical situations.

Service conditions

The imposed service conditions control the stress levels applied to the flaws in a component, and these are usually not easy to define with any certainty. From the predictive point of view, the simplest solution is to choose the highest stress seen in practice and to assume it is constantly applied. This has the advantage of underestimating the time to failure, even though it might lead to inefficient design. In addition, there are a number of other service features that can lead to problems.

(1) Vibration, however caused, may play an overriding role in determining life. Simple theories of cyclic fatigue for vibration superimposed on steady externally applied stress show that the effect of stresses above the mean level is critical. Predicting the magnitude of vibrational stresses in service is a complex problem, exacerbated by the very limited damping available in ceramics. Furthermore, it is not clear what effect cyclic fatigue has in practice. The simple cumulative damage theories (e.g. [29]) assume it has the same effect as steady loading, but under varying stress conditions. However, there is evidence in the literature that this cannot be assumed in cases where the stress variations allow the crack faces to come into contact during parts of the stress cycle. Depending on the type of crack and the microstructure through which it is propagating, the crack tip could remain highly stressed because closure is inhibited by a small plastic zone, elastic relaxations, debris, or even phase transformations. The data of Krohn and Hasselman[30] for cyclic fatigue of an alumina ceramic may be explained by such a phenomenon. Here, a decrease in fatigue life was obtained when the stress amplitude was increased while keeping the maximum tensile stress constant, despite a reduction in the effective time under high load near the peak of the stress cycle. Such a phenomenon, which is quite common in metals, suggests that different types of material will behave in different ways. There are insufficient examples available at present to be certain as to what type of allowance has to be made for cyclic stress conditions. It could be rather more than simple crack extension theories suggest.

(2) Similar difficulties are posed with respect to defining the surface stresses being applied. Since most engineering ceramics are stiffer than metals, elastic strains due to contact loads, and hence areas of contact, are smaller. In the absence of yield, local stresses can become high and readily cause Hertzian-type cracking. In addition, surface stresses are influenced by the roughness and friction characteristics of the contacting surfaces, and these are generally unknown, or not readily quantified. In this respect, probably the best approach is to design to avoid potential problems by use of soft seatings for ceramic components, although when intermittent high-temperature use is involved, there is a considerable problem of designing to tolerate expansion coefficient mismatches, especially between ceramic and metal components.

(3) The problem also arises in predicting resistance to surface fatigue, whether caused by vibration, by impact or by wear. The literature on practical measurements under such conditions is small. An example of repeated flexural impact on a 95% alumina ceramic[31] suggests a very rapid fatiguing effect, much more significant than obtained under static conditions. The difficulties arise not only in predicting what the impact stresses are and how they are dynamically distributed, but also in defining the cumulative effect on the material, as described above.

(4) Thermal stresses are particularly difficult to model in the sense that the heat transfer coefficient is usually not known with any certainty, especially at elevated temperatures where radiation and convection both contribute. The thermal stress distribution can be modelled with

precision only if the heat transfer distribution is known, but this cannot always be measured. Perhaps the best solution is a practical one: prove the component by thermal fatigue testing and correlate the result with predicted behaviour based on estimated thermal conditions.

(5) In practical applications, the environment is usually not constant. Temperatures and humidities can vary. Corrosive chemicals can vary in composition and concentration. Partial pressures of reaction or decomposition products may change. To predict accurately the effects of environment on stressed components requires extensive practical data under all relevant permutations of environment. Simple exponential temperature dependencies through a constant activation energy cannot always be assumed. For example, condensation of a corrosive vapour can produce more corrosion than the same species as a vapour at higher temperatures. The same is possibly true for the effects of water. The cumulative effect on fatigue properties of production and removal of protective coatings on non-oxides is difficult to model, especially when thermal cycling is involved.

At best, the effect of service conditions on life can be modelled only as well as the practical data allow. This poses problems of total effort required, even for one application.

Summary

The theories of fracture mechanics have allowed considerable progress in understanding the behaviour of well-characterised materials under stress. Provided conditions of applied stress and environment are well-defined and constant, good correlations can be obtained between practical and estimated lifetimes. However, the theory requires many assumptions about the interaction of cracks with ceramic microstructures and the effect of environment, many of which may have only a limited range of validity. Lifetime estimates, the principal purpose of the theories, are subject to considerable errors of extrapolation for the variety of reasons described above. It is probably safest to assume the worst effects are always present. This may underestimate life, but until practical trials on actual components prove it is an underestimate, there will always be uncertainties.

To extend the applicability of theories further, we are probably at the stage where we need more effort placed in practical demonstrations of long-term properties, and of the ways in which microstructures influence the detailed fracture mechanics on a scale related to the defects that limit strength.

REAL MATERIALS

Ceramic microstructures

Unlike glasses, ceramics are strictly not elastic continua on the scale at which fracture processes occur. Most basic ceramic phases have non-cubic crystallography, with the major consequence that there is anisotropy of thermal and elastic properties in each grain. In some cases, such as SiC, the effects are minimal, but in others, e.g. Al_2O_3, they are of considerable significance. Thus a polycrystalline alumina ceramic comprising randomly orientated grains is a highly strained material, the thermal stresses being developed simply by cooling from the firing temperature. The energy associated with a grain boundary is sufficient to cause it to crack when the grain size exceeds about 60 µm[32], as demonstrated by hysteresis of Young's modulus during thermal cycling[33]. For smaller grain sizes the stresses remain locked in.

Most ceramics are composed of more than one phase as a consequence of the development of fabrication routes to achieve dense structures. The major phase is usually bonded by an intergranular phase which may be glassy or crystalline, and the composition and distribution of this phase controls mechanical properties at all temperatures. In sialons and sintered silicon nitrides it plays a major role in determining high-temperature strength[34,35]. In many aluminas, the bonding phase is based on calcium silicate additions, and the expansion coefficient of the non-equilibrium phase mixture produced is broadly similar to that of alumina itself. However, this phase mixture is not sufficiently resistant to hot acids in applications requiring this property, and so it has been replaced by magnesium silicate additions[36]. These tend to produce phases of relatively low thermal expansion, with the result that there is a more pronounced mismatch of expansion with alumina. This substitution causes a major reduction of K_{Ic}, from typically 4 - 5 $MNm^{-3/2}$ to 1 - 2 $MNm^{-3/2}$, correlating with an increased tendency to suffer edge chipping during machining or handling[28].

In addition, microstructures can exhibit considerable variation from point to point in terms of distributions of grain size, phases, and porosity, as well as overall anisotropy of grain orientation. These factors all originate from the preparation of the starting powder batch and the method of shaping it into a blank or a component (Fig.5,6). Thus the microstructural homogeneity of a particular type of material must be expected to vary considerably with shape and source unless very stringent controls are exercised in manufacture. For most products, the controls are adequate to ensure manufacture to size and shape specifications, but the resulting properties may not be reproducible to the level that would be desirable for lifetime estimation procedures. There is a considerable risk that properties in some regions of a component are not as good as in others, which in turn are rather different to the data generally quoted for the material. Particular problems are likely to arise in near-net-shaping procedures, such as injection moulding, where the homogeneity of the highly plasticised powder batch and its flow characteristics in the mould determine the homogeneity of the moulded shape, particularly the porosity distribution.

As further examples, alumina ceramics are frequently made from Bayer process alumina, which has a plate-like habit. Wet processing methods such as slip-casting, tape-casting and extrusion can produce varying degrees of alignment of the alumina particles (Fig.7,8) which is retained during subsequent firing, even though the grain structure appears equiaxed as a result of grain growth[37]. The consequence of alignment of grains is a reduction in average levels of internal stress due to thermal expansion anisotropy, a general increase in mean strength, and a reduction in risk of failure associated with large grains. However, properties such as thermal expansion, elastic moduli, fracture toughness and permittivity become anisotropic. A similar situation can arise in many clay-based or talc-based structural ceramics, and in hot-pressed silicon nitride where the anisotropy in flexural strength is around 20 - 30%.

More generally, pore distributions in all ceramics present obvious defects[37], even if it is difficult to envisage the real dimensions of the sharp defect required to estimate strength. Pores arise because the non-uniformity of packing of the starting powder, which is often in the form of spray-dried granules, is not eliminated in firing (Fig.2). Pore distributions can therefore reflect the inhomogeneity of packing, and can be localised in such a way as to form areas of weakness which thus constitute defects far larger than the individual pore size. The strength of RBSN is critically determined by the homogeneity of the porosity (Fig.9); in particular, higher strengths are achieved by the removal of larger extended pores[39].

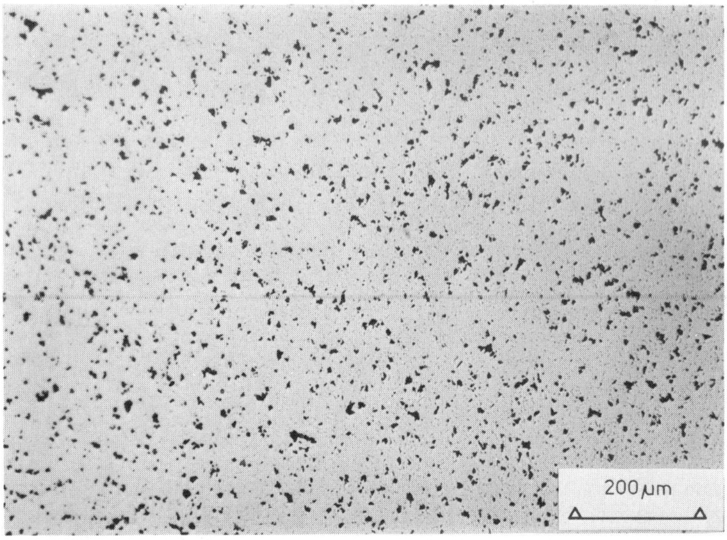

Fig.5 Porosity distribution in an isostatically pressed 95% alumina.

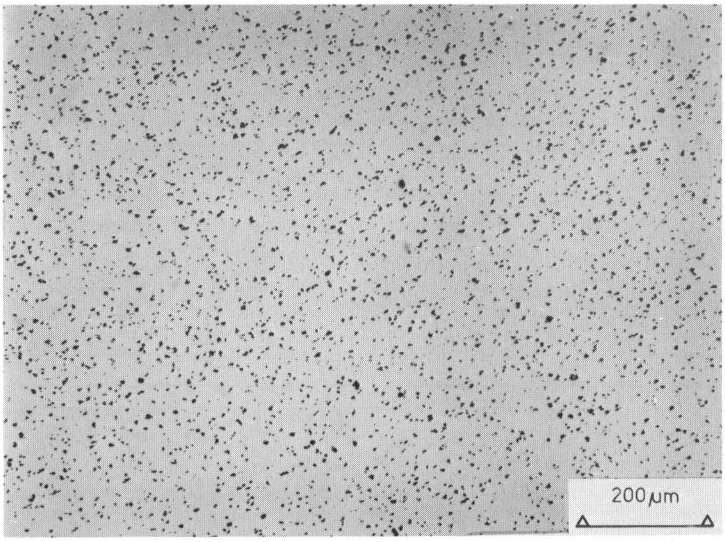

Fig.6 Porosity distribution in a die-pressed 95% alumina of the same commercial code as that in Fig.5.

The green density, and hence the final fired density, may vary thoughout a component, often being greater near the original surface than in the bulk if the transmission of pressing pressure to all parts of the component is restricted by inter-particle friction. Properties that are sensitive to small changes in the levels of porosity, such as strength, toughness, and hardness, will then be spatially variable.

It is clear from this discussion that the basic microstructure of a ceramic, aside from gross manufacturing defects such as cracks, delaminations or foreign particles, is a source of considerable variation in

Fig.7 Microstructure of an extruded 88% alumina with the extrusion axis horizontal, showing the alignment of plate-like grains.

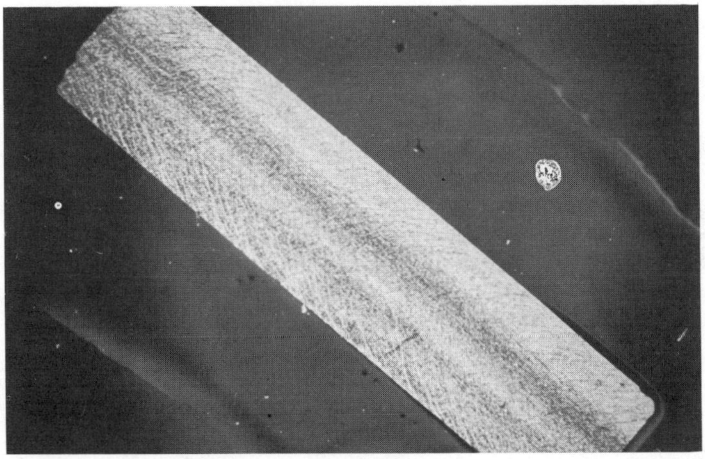

Fig.8 Thin section of a 4mm extruded 95% alumina rod at 45°C to the axis of light polarisation showing the complex texture introduced by the auger screw of the extruder.

detailed properties. Attention has to be paid to the cleanliness and reproducibility of manufacture in order to improve the reproducibility of component strength. However, microstructural variations and strength-limiting defects exist in all types of ceramic. The estimation of lifetime parameters needs to take such variations into account, but generally speaking, the detailed information does not exist, or is not relevant to the component shape in question.

Basic material properties

The chemistry, phase assemblage and microstructure of a ceramic essentially defines a range of properties which are intrinsic to the product.

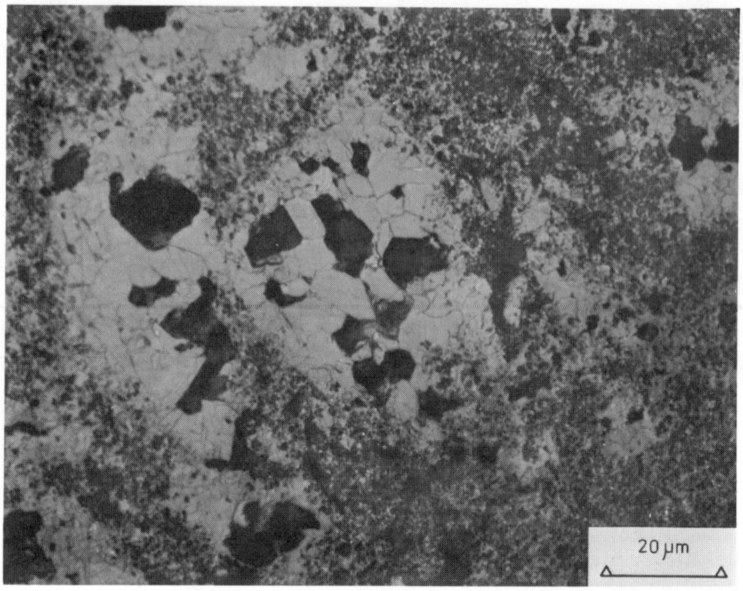

Fig.9 Groups of pores associated with dense regions in a commercial reaction-bonded silicon nitride.

Often it requires substantial changes in phase assemblage to change the intrinsic properties by significant amounts, and thus such properties do not represent critical data for stress estimation purposes. Thermal properties such as thermal expansion, thermal conductivity and specific heat are typical cases. For example, the thermal expansion of silicon nitride based ceramics is only a weak function of composition, tending to increase slowly with increasing second-phase content or alloying content (as in sialon materials[40]). Exceptions are materials in which phase changes occur during thermal cycling, or in which crystallographic anisotropy leads to hysteresis[41], but these effects occur mostly in non-structural materials.

With thermal conductivity, a similar argument can used. In this case, the variation of true thermal conductivity with temperature can be the principal source of uncertainty with data, which for many ceramics show a steep decline over the temperature range 0 - 300°C. Thermal conductivity measurements are normally made employing a temperature gradient along the test-piece, while thermal diffusivity measurements are frequently made employing unspecified temperature excursions (e.g. in the laser-flash method). In both cases, there is uncertainty as to what temperature the mean conductivity or diffusivity refers.

Elastic moduli are also subject more to uncertainty of measurement than to great real variation with microstructure. This leads to even greater uncertainty in the calculation of Poisson's ratio, with different literature sources frequently showing a scatter of results of up to 50% for the same basic material type. Errors here can lead to very significant errors in stress analysis.

These basic properties, if accurately measured, are not major sources of error when used to calculate stresses in components, and are therefore mostly only of peripheral concern to the stress analyst or fracture mechanics specialist. However, it is felt that the use of inaccurate or inappropriate intrinsic data, particularly when affected by microstructural anisotropy, can lead to significant errors in calculating performance requirements.

Surface quality

The basic microstructure of a material, including the manufacturing defects that may be present, is not the only strength-limiting feature of importance. The exterior surface represents a further dimension, and the way in which that surface is prepared in manufacture or is modified in service can have a profound effect on mechanical properties, particularly in circumstances where the internal microstructure is of high quality with no large strength-limiting features.

Component surfaces are usually not in the as-fired state for several reasons: dimensions may not be sufficiently accurate and machining is required; as-fired skins may be of relatively poor quality and require removal; adhering furnace debris, moulding flash or green machining marks may have to be removed. In all cases, the surface of the component is subjected to mechanical abrasion or impact which can introduce damage. The damage will interact with the microstructure to produce a surface flaw distribution which will vary according to the treatment received[15,42]. For example, many automated grinding procedures do not adjust for removal of larger than average amounts of excess material caused by shape inaccuracy. The grinding forces imposed, and hence the damage introduced will vary from piece to piece, and from area to area in the same piece[16]. This aspect of manufacture needs to be very carefully controlled to avoid subsequent problems.

Generally the machining procedure will place the surface layer into a degree of directional compression[8,15]. The surface flaws may be trapped within this layer, or may extend through it. Strength will be determined by the combined effect of all these factors, and it is for this reason that variable mean strengths and strength distributions (e.g. as characterised by Weibull parameters) can be obtained from the same material.

The strength can be changed subsequently by any condition of use which modifies the surface in any way. Raising the temperature substantially can anneal out the surface stress field and blunt flaws, but this can also remove surface compression disadvantageously if it was inhibiting the growth of flaws[15]. Several reports in the literature demonstrate the complete change of strength distribution by annealing and/or oxidation (of non-oxides), completely eliminating any truncation of the distribution resulting from proof-testing prior to raising the temperature[43,44].

Surfaces can be extensively modified by chemical or mechanical action in service. Chemical attack in ceramics frequently takes place preferentially in grain boundaries, opening fissures into the material, modifying surface stresses and weakening the surface. Impact and abrasion can introduce considerable addition damage into a surface, most noticeably on carefully prepared surfaces. Even sliding wear can generate wide surface cracks, probably by progressive plastic deformation and/or thermal fatigue. Table 1 gives some results of strength tests conducted on alumina and tungsten carbide hardmetal specimens which have been subjected to abrasive wear or scratching. It can be seen that both processes produce marked degradation in strength in all materials except the high cobalt content hardmetal because of the associated lateral cracking.

Table 1. Changes in strength resulting from abrading or scratching a surface.

Material type	Wear test applied	Fracture strength ± std.error, MNm^{-2}	No.of tests
94% alumina, 10 μm grain size	As lapped *	245 ± 27	9
	ASTM B611 **	197 ± 9	3
	Scratched ***	152 ± 23	5
99.5% alumina, 18 μm grain size	As lapped *	241 ± 17	5
	ASTM B611 **	192 ± 17	3
	Scratched ***	162 ± 20	4
WC/6% Co, < 1 μm grain size	As lapped *	1446 ± 251	6
	ASTM B611 **	861 ± 161	4
WC/15% Co, 3.5 μm grain size	As lapped *	1404 ± 262	9
	ASTM B611 **	1593 ± 208	4

All tests were performed on specimens 3 x 2 x 50 mm over a span of 30 mm in four-point bending at a machine cross-head rate of 0.02 mm min^{-1}.

* Diamond lapped with 1 μm grit.

** Specimens cut along the grinding direction from tiles subjected to the steel wheel/abrasive grit (30 mesh, 630 μm brown alumina) test ASTM B611 for 500 revolutions (1000 for the hardmetals) under 200 N force.

*** Repeated single traverses with a diamond indenter of 0.2 mm tip radius made at 0.2 mm separation under 90 N load at 0.5 mms^{-1} traverse speed. Specimens cut parallel to the tracks.

In view of all the factors described above, it is clear that the external surface of a ceramic is of at least equal if not greater importance than the bulk microstructure in determining strength performance. Preparation techniques will need to be carefully standardised and controlled in order to maintain strength levels. Inadvertent heavy grinding must be avoided. Degradation of strength as a result of abrasion, sliding wear or chemical attack needs to be well characterised and allowed for, but, so often, insufficient systematic testing has been carried out to determine behaviour. This should perhaps be the area of greatest effort in properties research. The investigation of surface coatings to reduce the effects of surface damage should also be pursued.

Test procedures, data acquisition and interpretation

As shown above, from a theoretical point of view it is possible to carry the extrapolation of laboratory strength data on small test bars to any desired extent, based on determination of a small number of parameters:

- m   the Weibull modulus, or alternatively some other measure of strength distribution;
- n   the slow crack growth exponent;
- $\bar{\sigma}_i$ an "inert" mean strength;
- $\bar{\sigma}$ a mean strength under the usual environment.

The parameter K_{Ic} is not normally required for such extrapolation exercises, but may be needed for describing the propagation of cracks under thermal or mechanical fatigue conditions, or for estimating the effect of damage due to impact or wear.

Mostly, strength data are acquired on small uniaxially stressed bend specimens, testing a sufficient number to obtain some degree of statistical assurance of the result. The specimens are tested at fixed loading rates or, more properly, fixed machine cross-head displacement rates, and batches of specimens may be tested at a variety of rates to determine the slow crack growth exponent n, the so-called dynamic fatigue method. Alternatively, delayed failure tests under static loading are used, but since this is a time-consuming method, dynamic fatigue testing is more commonly adopted.

Material consistency is often judged by using strength tests but, more especially for higher strength materials, the results are determined by precise specimen finishing procedures and test conditions. It is thus possible, in the absence of uniquely specified procedures and test conditions, to control the result to any desired level. Careful polishing and lapping of tensile test faces, and rounding and polishing of edges to reduce the risk of edge failure, often raise apparent strengths. Such a finish is usually not realistic in practice where lapping and polishing is not employed, and where the surfaces and edges are subject to damage by external agencies. In addition, Weibull parameters are often surface-finish and specimen-size dependent, and there are reports in the literature where different environments produce different values of m, presumably by influencing the significance levels of subcritical growth in different types of flaw[14].

To avoid the edge-finish problem, test geometries such as the four-ball test (ASTM F394) are becoming increasingly popular. Nevertheless, it is still true to say that in order to employ strength data for design purposes a great deal of care must be taken to ensure that the test data are for the appropriate surface condition and that the risks of failure from edges are sufficiently small that they can be ignored. Recent studies[27,45] show that even the slow crack growth parameter, n, can be influenced by the shape and degree of prestressing of strength-limiting flaws.

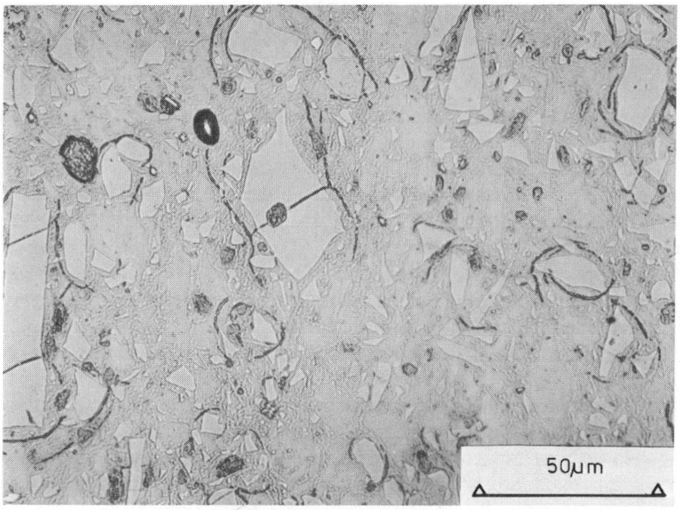

Fig.10 Internal cracks in a siliceous porcelain associated with the quartz grains.

For determining slow crack growth parameters, the use of large cracks is falling from favour because it is now realised that these do not necessarily model strength-limiting flaws adequately. As perhaps an extreme example, a porcelain tested by the double-torsion stress-relaxation method[24] gave consistent n-values well in excess of 100, while the same material subjected to dynamic fatigue tests gave values of around 30, more appropriate to glasses which comprise the matrix phase[46]. In the former case, the large crack has to propagate long distances through the extremely inhomogeneous microstructure and probably becomes pinned at quartz grains by their circumferential ring cracks (Fig.10). In the latter, at much higher stress levels, the critical crack is much smaller, and delayed failure characteristics are controlled by its propagation for short distances between the quartz particles.

It can be argued that the dynamic fatigue or delayed failure test is more realistic for determining n, because it tests real flaws, not large cracks. However, to determine n accurately requires a large number of strength tests over as wide a range of loading rates as possible. This is necessary in order to reduce the possible error in mean strength at any one loading rate. There remains a risk of error in subsequent use of such data because the effective time under load at the slowest practical test machine loading rate is short compared with normally expected component lifetimes in service, as described earlier.

To reduce the scatter in strengths obtained from test bars, some authors advocate the use of controlled flaws inserted by means of a diamond indenter. Failure then occurs from a predetermined point. Some scatter still remains because of variations in the size of cracks produced around indentations. However, the residual stresses associated with the indentation can modify crack propagation behaviour[45], and natural flaws may be preferred.

To extend the duration of tests in order to take into account longer-term factors it is necessary to undertake statically-loaded, delayed-failure tests. The range of times to failure is normally large and uncertain, and the resulting test programme is long and tedious. It can be reduced by truncating the strength distribution by prior proof-loading. An alternative approach is to provide a delayed failure check on dynamic fatigue tests. A series of delayed failure tests at a relevant stress level can be performed on the basis of stepwise increments in stress until failure occurs. Using simple crack growth theory, all the results can be corrected back to a single short survival time based on a dynamic fatigue value of n. The strength distribution obtained can be compared with the similarly corrected dynamic fatigue distribution. If the agreement between the two distributions is good, the value of n used is sufficiently accurate to extrapolate dynamic fatigue data to times to failure at least as long as those of the delayed failure tests, which can be continued as long as is desired at any stress level. This approach to checking the validity of short-term data is probably the most useful for prediction purposes.

Techniques to measure K_{Ic} are extremely varied, and most can be criticised on several counts. The problems vary from calibration difficulties due to uncertainties associated with complex stress fields, to the method's relevance to real flaws in various practical situations. In many cases the microstructure of the material in question introduces further uncertainties. The scale of the inhomogeneity of microstructure along the crack tip at the point of rapid failure will determine the appropriate K_c figure. In some materials, the distance the crack propagates can determine K_c. This is due to the increasing tendency to bifurcate as the crack front moves, giving rise to the well-known R-curve effect seen in fracture toughness tests (e.g.[47]). It is therefore clear that measurements of K_{Ic} on large cracks may not be completely relevant to small cracks. The degree to which cracks are prestressed by their means of production, such as machining, indentation, thermal shock or wear, can also influence the apparent K_{Ic} figure. In consequence, K_{Ic} data need to be obtained and used with caution, particularly with reference to the prediction of damage introduced by impact or by thermal shock.

PROGRESS FROM THE CURRENT POSITION

Improved materials

Fracture mechanics applied to ceramics has taught us how important microstructure is in determining strength and fracture properties. Recent advances in tough materials (e.g. those containing dispersed zirconia, Fig.1) have come as an indirect result of fracture studies, and there remains much to be done to optimise systems to obtain high reliable levels of strength on a commercially effective scale. Perhaps this will be achieved in ceramic composites or by further manipulation of monolithic ceramic microstructures. In either case, the key to success will be obtaining microstructures that not only have useful strengthening/toughening mechanisms, but ones that can be made reliably with few defects and can withstand engineering service. This problem needs to be addressed mainly by materials scientists and chemists. Manufacturers need to be aware of the sources of strength-limiting defects in their products, and to seek to eliminate them by process control, for example by routine hot isostatic pressing before or after final machining[48], because it is unlikely that non-destructive testing will be able adequately to screen components for small flaws, while it can wasteful to have to eliminate defective components later on.

Will we ever be in a position to expect materials always to be consistent? To some extent this may become a real possibility as the important factors in processing are identified and controlled. But, as pointed out by Ritter and Davidge[38], there will always be a limitation to the degree of perfection of microstructure that will be manufacturable, and so some variability of strength will always be obtained. It is therefore likely that the only technique that will be able to guarantee each component is some form of mechanical proof test.

Design reliability

The variability of materials on the one hand, and uncertainties in predictive theories on the other, clearly lead to severe problems in many situations of design. Theories make many assumptions about materials behaviour, and the material data required to justify the assumptions are either not available, or are subject to considerable variation depending on precise manufacturing conditions. How should the design engineer approach this problem and obtain answers which are reliable? It requires very close attention to detail, coupled with good understanding of materials behaviour. Stress analysis is becoming easier and more reliable, and does not pose the problem that it once did. The real need now seems to be for better data measured by standardised techniques, and better understanding of long-term properties and the role of surface conditions.

Until it can be demonstrated that theory and practice give the same answer reliably, it is unlikely that engineers will rely on theory. We must therefore expect them to continue to use an expensive iterative practical approach of trial and error for the foreseeable future to test materials and components.

Proof testing

Much research into proof testing has been carried out, and this should be continued, especially with respect to establishing the degree of reliability that can be achieved with the surviving fraction. It is known that proof testing is not a water-tight guarantee of future life[44], but presently it is the only method available for eliminating weak components from a batch, whatever the cause of their weakness.

Many authors have developed proof-test diagrams for various types of material. These have usually been obtained by manipulation of simple test-bar data, and give an indication of the proof-test ratio (peak test stress/service stress) required to guarantee a minimum life in service. However, like all the predictive theories they suffer from errors in the data used, and so at best can give only a rough guide to the ratio required to test components. In view of this, choice of correct conditions of test is subject to some possible large error. However, this does not negate the value of proof-testing. It can still be used in an empirical way to remove by trial and error the low-strength tail of the distribution of strengths if it is found that the gulf between diagram and practice is too large for the diagram to be useful.

Proof testing cannot guarantee a component against later changes to the flaw distribution, such as damage by contact stresses, impact, abrasion, corrosion, oxidation or creep. The effect of such factors has to allowed for separately by experiment. Only by building up an extensive bank of data on materials response to conditions of use will it be possible to predict performance to reliability levels the engineer needs.

CONCLUSIONS

In the application of fracture mechanics to real materials, many simplifying assumptions are usually made. The extent to which these assumptions are justified depends on the detailed microstructure of the material in question, including the manufacturing and finishing defects that are introduced. In addition the interpretation of test data and its extrapolation to long times to failure or to different shapes of component may be undertaken only with extreme caution. The predictive power of fracture mechanics is therefore limited by the variability of materials, from point to point, and from component to component. At present, good predictions can be made only under circumstances in which careful control of all variables is maintained, but this is seldom achieved on a practical scale.

The rate-controlling steps in advancing predictive design based on fracture mechanics appear to be:

(1) improving the consistency of materials by improved process control, eliminating bulk defects and producing microstructures with uniform defect sizes which are tolerant to variable surface machining conditions;

(2) establishing longer-term properties in relevant environments, and providing proof that extrapolations of short-term data are reasonably accurate;

(3) widening the understanding of multiaxial stressing and the effects it may have on longer-term properties;

(4) widening the understanding of how to allow for stress cycling, especially in vibration, impact, fretting and thermal shock.

It seems unlikely that properties will ever be sufficiently consistent that practical testing can be eliminated and total reliance placed on prediction in applications where high stresses and/or temperatures are involved.

Advances in materials preparation will undoubtedly be made, perhaps with ceramic matrix composites, but successful application of the materials relies on convincing the engineer of the level of reliability. It is felt that proof testing plus a thorough understanding of material behaviour in service is the only practical route to success.

REFERENCES

1. E.A.Almond, M.G.Gee and L.N.McCartney, in "Towards improved tool materials", Metals Soc., London, (1983).
2. A.Paluszny, in "Ceramics for high-performance applications - III, reliability", E.M.Lenoe, R.N.Katz, J.J.Burke, eds., Plenum Press, New York (1983).
3. C.A.Rau, P.M.Besuner and K.G.Sorensen, Met.Sci. 14:463 (1980).
4. S.M.Wiederhorn, in "Fracture mechanics of ceramics - 2", R.C.Bradt, D.P.H.Hasselman, F.F.Lange, eds., Plenum Press, New York (1974).
5. R.W.Davidge, Met.Sci. 14:459 (1980).
6. R.W.Rice, J.Mater.Sci. 19:895 (1984).
7. J.W.Tien, B.T.Khuri-Yakub, G.S.Kino, D.B.Marshall and A.G.Evans, J.Nondestr.Eval. 2:219 (1981).
8. D.B.Marshall, A.G.Evans, B.T.Yakub-Khuri, J.W.Tien and G.S.Kino, Proc.Roy.Soc.Lond. A385:461 (1983).
9. T.Hoside, H.Furuya, Y.Nagase and T.Yamada, Int.J.Fracture 26:229 (1984).
10. P.Kittl and G.Camilo, Res.Mech.Lett. 1:115 (1981).
11. R.A.Hunt and L.N.McCartney, Int.J.Fracture 15:365 (1979).
12. M.G.Gee, Int.J.Mech.Sci. 26:85 (1984).
13. T.E.Easler, R.C.Bradt and R.E.Tressler, Comm.Am.Ceram.Soc. 64:C53 (1981).
14. G.K.Bansal, W.H.Duckworth and D.E.Niesz, J.Am.Ceram.Soc. 59:472 (1976).
15. M.J.Hanney and R.Morrell, Proc.Brit.Ceram.Soc. 32:277 (1982).
16. M.N.Giovan and G.Sines, J.Am.Ceram.Soc. 62:510 (1979).
17. J.J.Petrovic and M.G.Stout, J.Am.Ceram.Soc. 64:656 (1981).
18. D.K.Shetty, A.R.Rosenfield and W.H.Duckworth, in "Structural reliability of brittle materials", ASTM STP 844, S.W.Freiman and C.M.Hudson, eds., ASTM, Philadelphia (1984).
19. S.B.Batdorf and H.L.Heinisch, J.Am.Ceram.Soc. 61:355 (1978).
20. J.Lamon and A.G.Evans, J.Am.Ceram.Soc. 66:177 (1983).
21. K.Ikeda and H.Igakio, J.Am.Ceram.Soc. 67:538 (1984).
22. D.K.Shetty, A.R.Rosenfield, G.K.Bansal and W.H.Duckworth, J.Am.Ceram.Soc. 64:1 (1981).
23: H.W.Babel and G.Sines, J.Basic Eng. 90:285 (1968).
24. B.J.Pletka and S.M.Wiederhorn, J.Mater.Sci. 17:1247 (1982).
25. E.J.Minford, J.A.Costello, I.S.T.Tsong and R.E.Tressler, in "Fracture mechanics of ceramics - 6", R.C.Bradt, A.G.Evans, D.P.H.Hasselman, F.F.Lange, eds., Plenum Press, New York (1983).
26. S.M.Wiederhorn and L.H.Bolz, J.Am.Ceram.Soc. 53:543 (1970).
27. A.C.Gonzalez, H.Multhopp, R.F.Cook, B.R.Lawn and S.W.Freiman, in "Methods of assessing the structural reliability of brittle materials", ASTM STP 844, S.W.Freiman and C.M.Hudson, eds., ASTM, Philadelphia (1984).
28. W.P.Byrne, M.J.Hanney and R.Morrell, Proc.Brit.Ceram.Soc. 32:303 (1982).
29. A.G.Evans and T.G.Langdon, Prog.Mater.Sci. 21:171 (1976).
30. D.A.Krohn and D.P.H.Hasselman, J.Am.Ceram.Soc. 55:208 (1972).
31. B.K.Sarkar and T.K.Glinn, Trans.J.Brit.Ceram.Soc. 69:199 (1970).
32. R.W.Davidge, Proc.Brit.Ceram.Soc. 32:199 (1982).
33. E.D.Case, J.R.Smyth and O.Hunter, in "Fracture mechanics of ceramics - 6", R.C.Bradt, A.G.Evans, D.P.H.Hasselman, F.F.Lange, eds., Plenum Press, New York (1983).

34. S.M.Winder and M.H.Lewis, J.Mater.Sci.Lett. 4:241 (1985).
35. J.Y.Laval, C.Delamare, M.C.Amamra and D.Broussaud, J.Mater.Sci. 20:381 (1985).
36. G.Richards, Trans.J.Brit.Ceram.Soc. 80:120 (1981).
37. M.McNamee and R.Morrell, Sci.Ceram. 12:629 (1984).
38. J.E.Ritter and R.W.Davidge, J.Am.Ceram.Soc. 67:432 (1984).
39. D.C.Larsen, Report AFML-TR-79-4188 (1979).
40. S.P.Howlett and R.Morrell, J.Mater.Sci.Lett. 4:227 (1985).
41. J.A.Kuszyk and R.C.Bradt, J.Amer.Ceram.Soc. 65:420 (1973).
42. M.W.Hawman, P.H.Cohen, J.C.Conway and R.N.Pangborn, J.Mater.Sci. 20:482 (1985).
43. S.M.Wiederhorn and N.J.Tighe, J.Mater.Sci. 13:1781 (1978).
44. J.E.Ritter, P.B.Oates, E.R.Fuller and S.M.Wiederhorn, J.Mater.Sci. 15:2275 (part 1), and 15:2282 (part 2) (1980).
45. E.R.Fuller, B.R.Lawn and R.F.Cook, J.Am.Ceram.Soc. 66:314 (1983).
46. D.Taylor, private communication.
47. R.Steinbrech, H.Blanke, R.Knehans and W.Schaarwächter, Sci.Ceram. 12:655 (1984).
48. T.J.Whalen, R.M.Williams and B.N.Juterbock, in "Proceedings of the 11th Plansee Seminar", H.Bildstein and H.M.Ortner, eds., Metallwerk Plansee GmbH, Reutte, Austria, 2:783 (1985).

MICROSTRUCTURE AND THE STRENGTH OF CERAMICS

C.J. Fairbanks, B.R. Lawn, R.F. Cook, and Y.W. Mai*

Center for Materials Science
National Bureau of Standards
Gaithersburg, MD 20899

ABSTRACT

Microstructural influences on ceramic strength become significant at small flaw sizes. These influences are readily quantified by strength testing with controlled indentation flaws. Data are presented here for alumina and glass-ceramic specimens broken under both inert and fatigue conditions. As the flaw size is systematically reduced there is a tendency to a reduction in strength relative to that predicted from macroscopic toughness measurements, reflecting R-curve behavior. This tendency is critically dependent on the microstructural detail, e.g. presence of glassy phases at the grain boundaries in the aluminas. However, the fatigue susceptibility is found to be relatively insensitive to the microstructural influence over the same flaw-size range.

A fracture mechanics framework for incorporating a "microstructural stress intensity factor" is outlined. The description establishes a proper basis for extrapolating fracture data from traditional large-scale crack specimens into the domain of naturally occurring flaws. Direct observations of the indentation crack response during actual strength testing indicates that the principal mechanism of the microstructural "toughening" effect is crack restraint by grain-localized ligamentary "bridging" behind the advancing front.

1. INTRODUCTION

It is now well known that the fracture properties of brittle ceramics can be strongly influenced by the material microstructure.[1] It becomes pertinent to ask whether toughness and crack velocity data on well-developed, macroscopic fracture specimens can be extrapolated to the domain of strength-controlling microscopic flaws. At issue is the long-standing Griffith strength formalism on which all modern-day fracture mechanics design is based.

---

*On leave from the Department of Mechanical Engineering, University of Sydney, N.S.W. 2006, Australia

In confronting this issue we are led to the phenomenon of crack-resistance, or R-curve, behavior. At very small crack sizes relative to the scale of the microstructure the toughness is expected to be determined simply by a bulk cleavage or grain boundary interface energy, depending on whether the fracture mode is transgranular or intergranular. At large crack sizes the toughness can be considerably greater than these intrinsic levels, representative of the polycrystalline aggregate. Hence the toughness is no longer a constant material parameter, but becomes some function of crack size - the R-curve. How, then, may we determine such toughness characteristics experimentally, and thence incorporate these characteristics into a formal fracture mechanics description of strength? What effect does R-curve behavior have on fatigue properties, in particular on the effective crack velocity exponent? At the more fundamental level, what are the underlying physical processes responsible for the R-curve, and how do these processes relate to the material microstructure? The future success in developing ultra-strength ceramics surely rests with the answers to questions like these.

In this paper we survey the results of recent studies in these laboratories[2-5] aimed at elucidating the nature of R-curve behavior in selected aluminas and glass-ceramics, materials in which the elements of the microstructure are reasonably well understood. In our experiments indentation flaws are introduced into strength bars, so that the size of the crack which leads to failure may be varied systematically. This conveniently allows us to bridge the gap in scale of fracture from the traditional macroscopic test specimen down to the microstructurally-determined flaw. Moreover, the origin of failure is now predetermined, so we can follow the crack evolution directly during an actual strength test. As we shall see, these experiments provide new insights into the fracture mechanisms. A stress intensity factor for the microstructural influence on the net crack driving force is thereby determined, at two levels of rigor: first, semi-empirically (i.e. without recourse to a specific R-curve mechanism), sufficient to account for the trends in both inert and dynamic-fatigue strength data; second, in terms of a specific physical model of distributed restraining forces behind the crack front. The restraining forces in the latter instance are associated with the formation of microstructurally-localized ligaments across the separating fracture interface.

2. EXPERIMENTAL

We have conducted a test program on several ceramics.[2,3] Here we present results on three representative aluminas - single crystal sapphire, polycrystal nominally "pure" (Coors Vistal), polycrystal with 1% grain boundary glassy phase (Freidrichsfeld F99) - and two glass-ceramics - low expansion (Corning C9606, Pyroceram) and machinable (Corning C9658, Macor). Specimens were prepared in disk or bar form suitable for strength testing.

Most of the specimens were indented with a Vickers diamond at the centers of the prospective tensile faces. Some specimens were left unindented as controls. The range of contact loads covered was 1 to 300 N. The strengths of the specimens were determined in biaxial or four-point flexure. These flexure tests were run in either inert environment (silicone oil), to determine the toughness characteristics,[2] or fatigue environment (water), to determine crack velocity characteristics.[3]

Optical microscopy was used to examine the fracture patterns, both after and during the testing to failure. The post-indentation examination was needed to confirm the indentation flaw as the source of failure. It also provided useful information on the crack morphology. However, the most revealing observations were those made in situ, as the flexural loading was being applied. For this latter purpose, a special loading facility was attached directly to the stage of an inverted microscope, with the indented specimen surface facing downward, allowing the entire crack evolution to failure to be monitored. A video recording attachment was extremely beneficial in helping to interpret some of the seemingly complex features of the crack growth behavior enroute to failure.

3. INTERPRETATION AND DISCUSSION OF RESULTS

3.1 Inert Strengths and R-Curves

Inert strength data are shown as a function of indentation load, $\sigma_m(P)$, for the aluminas and glass-ceramics in Figs. 1(a,b,c) and 2(a,b), respectively. Individual data points represent means and standard deviations (evaluated in logarithmic coordinates) for breaks at indentation flaws at each prescribed load. Hatched regions at left of each plot represent strengths of the unindented controls. The solid curves are theoretical fits to the data (see below).

It is immediately apparent, even without a detailed fracture mechanics analysis, that there is disparate behavior in the strength responses within each material class. This is seen most clearly in the alumina data of Fig. 1. The sapphire data plot more or less in accordance with the simple power-law relation $\sigma_m \propto P^{-1/3}$ from "ideal" indentation theory[6,7] (i.e. with microstructure terms excluded). Insofar as the proportionality factor in this relation involves the material

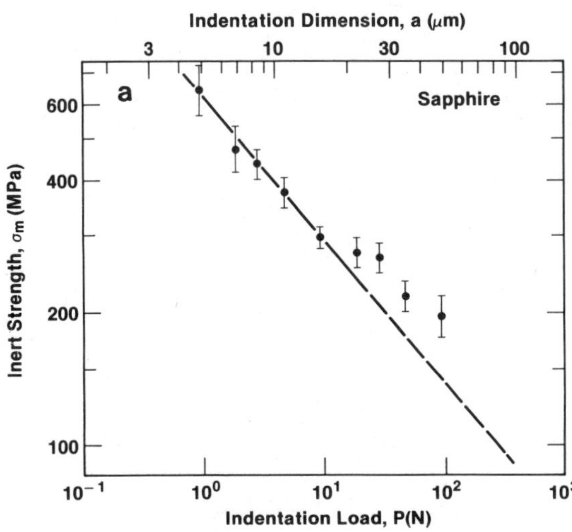

Figure 1. Inert strength as function of indentation load for three aluminas. Upper horizontal axis shows equivalent flaw size, plotted as indentation half-diagonal. After Ref. 2.

(b and c continued on next page)

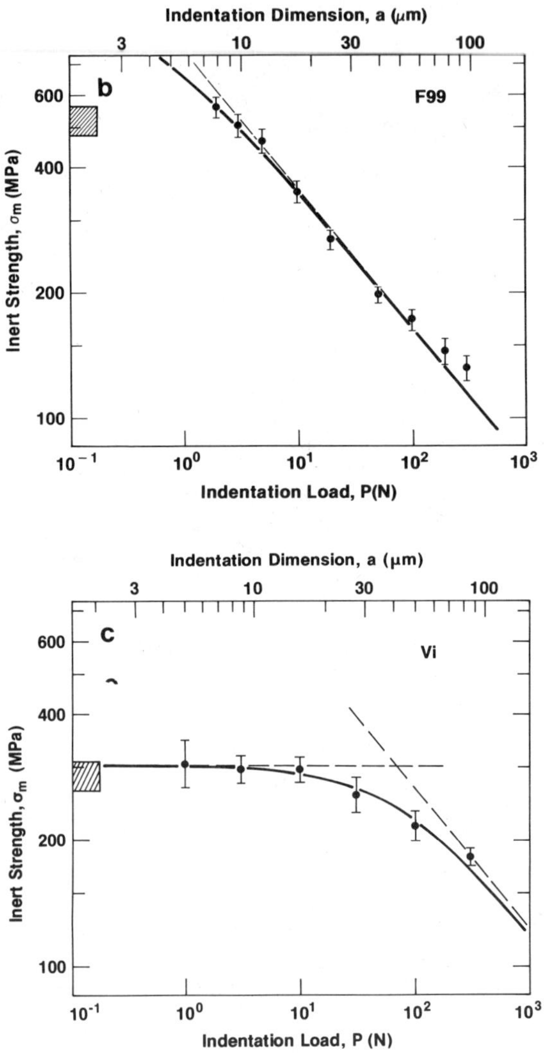

Figure 1. (Continued)

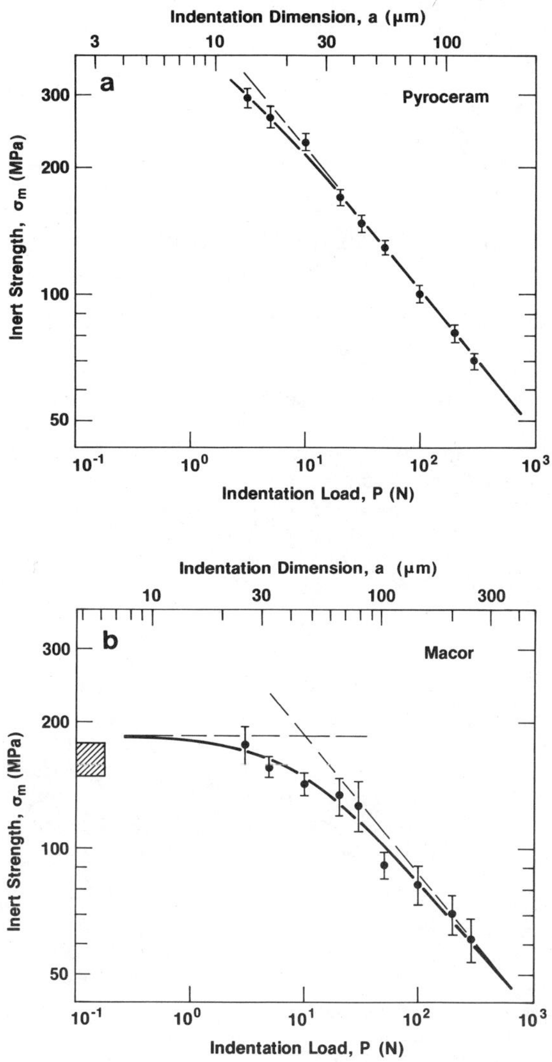

Figure 2. Inert strength as function of indentation load for two glass-ceramics.

toughness (see Eq. 3 below), we may regard the linear fit in this case as an extrapolation from large-scale crack resistance data. The data for the polycrystalline aluminas show indications of a low-load "saturation" in strength levels, much more so in the Vi than in the F99 material. In these cases the extrapolations from macroscopic toughness evaluations, represented as the inclined broken line asymptotes, overestimate the strengths in the small-flaw region. The need to include a microstructural term into the strength analysis is evident.

The starting point for such an analysis is an appropriate stress intensity factor.[2] For an indentation crack of size c, formed at contact load P and subsequently subjected to a bending stress σ_a, we may write

$$K = K_a + K_r + K_m$$
$$= \psi\sigma_a c^{1/2} + \chi P/c^{3/2} + K_m(c,d,.....) = T \qquad (1)$$

where the equality with intrinsic toughness T (referred to elsewhere as K_c) denotes a state of equilibrium (pertinent to inert testing conditions). Here ψ is a crack geometry constant ($\cong 2/\pi^{1/2}$), χ is a residual contact field parameter, and d is some characteristic dimension of the crack-microstructure interaction. The object of the exercise is to determine K_m (c,d,....) explicitly for given microstructural systems. Then we may solve Eq. 1 for the critical stress σ_m at failure (i.e. the inert strength) by imposing the instability requirement dK/dc = 0.

In the absence of any knowledge as to the specific mechanism of crack-microstructure interaction we can only proceed by semi-empirical representation of K_m. This was effectively the approach adopted in Ref. 2 (see also Ref. 8). In terms of the current terminology, we may write the microstructural functional from that earlier work in the form, for $c \geq d$,

$$K_m = -(\mu Q/d^{3/2})[1 - (d/c)^{3/2}] \qquad (2)$$

where Q is some microstructurally associated "force" (real or effective) and μ is a constant. We note that $K_m = 0$ at c = d, designating a cutoff crack size below which the toughness is determined exclusively by the intrinsic value T. At c > d, K_m becomes increasingly negative so that the toughness, insofar as it may be considered to be augmented by this term (i.e. $T-K_m$ in Eq., 1), appears as a rising function of crack size. It is Eq. 2 which provides the basis for the curve fits in Figs. 1 and 2.

Once we have an explicit expression for Eq. 1, we should be able to invert the solution to determine the effective R-curve (or perhaps more strictly, in our present notation, T^*-curve) characteristics from the $\sigma_m(P)$ data. Indeed, we can carry out such an inversion for individual data points without a full solution, by calculating an "apparent" toughness from a base formula representative of the microstructure-free state (i.e. with $K_m = 0$ in Eq. 1). Thus we can show that

$$T^* = \eta(\sigma_m P^{1/3})^{3/4} \qquad (3)$$

where $\eta = \eta(\psi,\chi) = 0.88$ is determined by experimental calibration,[9] and where the asterisk is to signify that we intend to use Eq. 3 outside the strict realms of its applicability (which is along the linear asymptotes of slope -1/3 in Figs. 1 and 2). In this way, we plot the apparent toughness vs load data for the aluminas and glass-ceramics in Figs. 3 and 4, respectively. The R-curve tendencies are clearly illustrated in these plots. Particularly noteworthy is the way in which the data for Vi alumina in Fig. 3 cross those for the other two aluminas, indicating

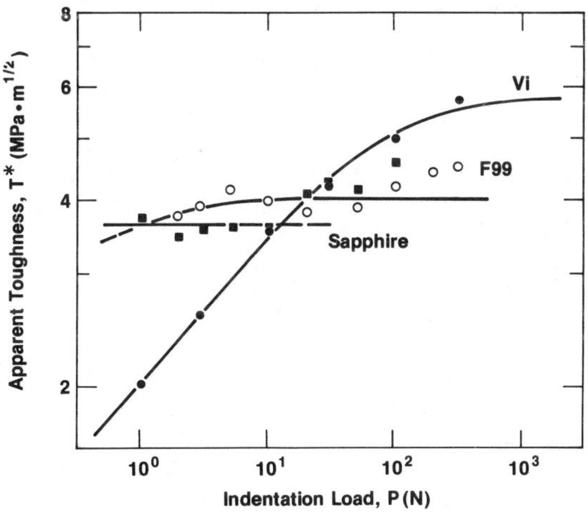

Figure 3. Apparent toughness (evaluated from Eq. 3) as function of indentation load for the aluminas represented in Fig. 1. Note tendency for polycrystal data (especially Vi) to cross above the sapphire data at high loads.

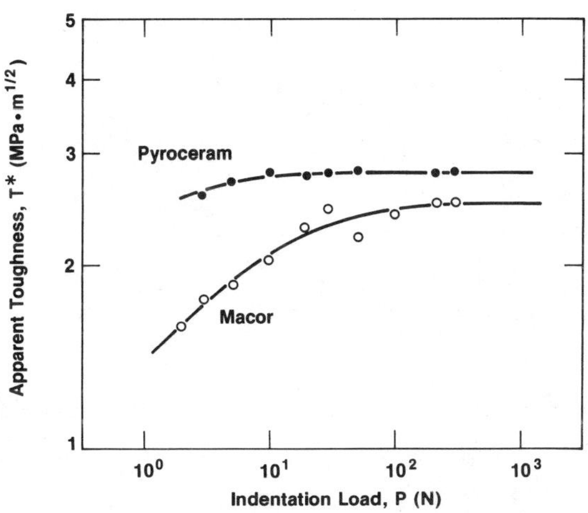

Figure 4. Apparent toughness as function of indentation load for the glass-ceramics represented in Fig. 2.

that toughness evaluations made in the macroscopic region can be totally misleading in assessing the relative merits of different materials at the microstructural flaw level. The solid curves in these plots are again fits using the empirical function Eq. 2 to obtain analytical solutions to $T^*(P)$ from Eq. 1.[2]

3.2 Fatigue Strengths and Crack Velocity Exponents

Figures 5 and 6 show dynamic fatigue strength as a function of stressing rate, $\sigma_f(\dot{\sigma}_a)$, for those materials which displayed the strongest R-curve behavior in the preceding subsection, viz. Vi alumina and Macor glass-ceramic. The results are obtained for two indentation loads, towards opposite ends of the R-curves, for each material. Data points represent means and standard deviations (log coordinates) at each stressing rate. Horizontal broken lines at right are inert strength limits at the appropriate loads. The solid lines are linear fits in accordance with theoretical prediction (below), notwithstanding some apparent curvature in the plots.

We see that, despite the extreme R-curve behavior shown by these two materials, there is no significant sensitivity of the fatigue susceptibility to crack size; the curves for different values of P in Figs. 5 and 6 are closely parallel.

An analytical basis for explaining this insensitivity can be established by combining the stress intensity factor K(c) in Eq. 1 (in the subcritical region $K < T$) with a crack velocity function v(K),[3] here assumed to be of standard power-law form#

$$v = AK^n. \tag{4}$$

The resultant formulation is a differential equation which, for constant $\dot{\sigma}_a$ and empirical K_m function in Eq. 2, has a solution of the same form as that for Griffith-like flaws,

$$\sigma_f = (\lambda'\dot{\sigma}_a)^{1/(n'+1)} , \tag{5}$$

differing only in the values of the exponent and intercept terms (hence the prime notation). It turns out that the exponent n' is related to the true crack velocity exponent n via the simple transformation equation

$$n' = 3n/4 + 1/2 \tag{6}$$

totally independently of P. This accounts for the parallelism between the curves in Figs. 5 and 6. The intercept λ' does turn out to be dependent on P, however, and this needs to be allowed for in any complete description of fatigue properties.[3]

3.3 Direct Observations of Crack Growth in Strength Specimens

Observations of the crack response to the applied loading provide valuable clues to the mechanism responsible for the rising R-curve behavior described earlier.[4] We have looked at all the materials represented in the strength data plots, but have focussed particularly on the Vi alumina, the material with the strongest effect. Optimal conditions for microscopy were obtained by polishing the surfaces down to

---

#There are several subtleties in writing the crack velocity function this way for materials with rising R-curves. These subtleties will be explored elsewhere.

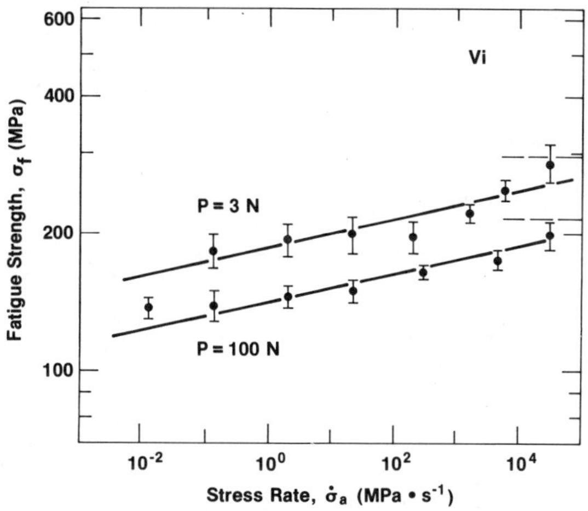

Figure 5. Dynamic fatigue plots for Vi alumina at two indentation loads. After Ref. 3.

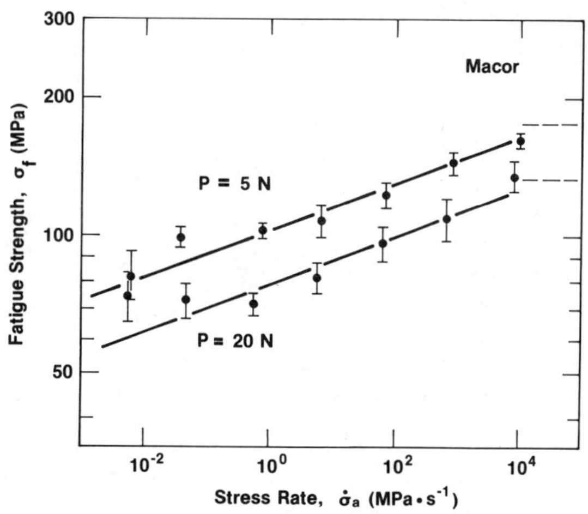

Figure 6. Dynamic fatigue plots for Macor glass-ceramic at two indentation loads.

0.3 μm Al_2O_3 paste, thermally etching to reveal the grain boundaries, and metal coating to improve the reflectivity.

The first microscopy examinations made were of the final failure patterns. Some of the general features of these patterns are evident in Fig. 7, which shows the immediate region around the indentation in a fractured Vi alumina disk. It is seen that the fracture is almost exclusively intergranular, once the initial radial cracks traverse the first grains containing the indentation impression. This intergranular mode was evident in all the aluminas we tested, regardless of degree of R-curve behavior. This is not unexpected, of course, in view of the distinct tendency for the $T^*(P)$ polycrystal data to cross below the single crystal data at low P in fig. 3. An analogous "intergranular" mode was observed in the glass-ceramics, with matrix cracks propagating around rather than straight through the larger second-phase crystallites.[10,11] In all cases well-defined primary cracks were observed, with no obvious subsidiary bifurcation or microcracking.

Post-test inspections in Vi revealed one further, telling piece of information. At failure (marked by an abrupt drop in applied stress to zero) the radial cracks propagated to the edges of the specimens. Continuous traces of these cracks could be followed along both top and bottom surfaces. Yet the specimens did not fragment - some additional applied force was necessary to separate the pieces. It appeared that there must be remnant links of some kind across the "broken" interfaces.

It was in this connection that the in situ observations were most revealing, especially on the Vi alumina. Tests were run at slow stressing rates, in air, to allow each step in the crack growth sequence to be followed carefully. In the initial stage of loading the cracks remained quite stationary. Then one arm of the radial crack pattern extended abruptly, typically over 5 to 10 grain diameters. Further loading produced more such jumps, either in the same or adjacent arms, with increasing frequency, until failure occurred. Similar initiations were occasionally observed from prominent "natural" flaws.[12] The picture we conjure up is one of highly stable, but erratic, growth prior to failure.

The stable growth comes as no surprise, since it is well established that the residual stress term K_r in Eq. 1 (by virtue of its inverse dependence on crack size) manifests itself in precisely this way.[6] But what causes the discrete jump-arrest sequences? Closer inspection of the separating crack behind the main advancing front provides important clues. There are points where the crack appears to be disjointed on the microscale, as though the main front has had some difficulty in deciding which way it might propagate around a certain grain (or second-phase particle). An extreme example is shown in Fig. 8. Rarely did the distance between the adjacent crack segments exceed more than one or two grain diameters (including in the subsurface direction, as indicated by interrupting the test and viewing in transmitted light). Independent observations of such regions by Swanson in these laboratories[13] have demonstrated the continued evolution of the crack segments at distances greater than 1 mm behind the main front in Vi alumina. This evolution ceases only when one of these segments runs back into the plane of the other, at which point the opposing crack walls completely separate. In "favorable" cases the two segments link up simultaneously, causing grain "pop-outs" (e.g. in upper left of crack trace, Fig. 7).

We are led to conclude that the crack is "hung up" by grain-localized, restraining ligaments distributed over the separating crack interface. This conclusion is consistent with the "memory effect"

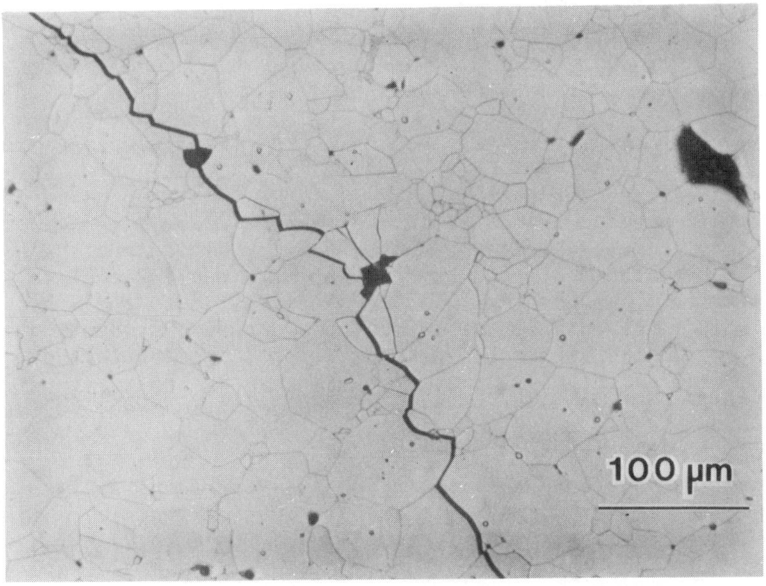

Figure 7. Optical micrograph of Vickers indentation site in a fractured Vi alumina disk. Note well-defined primary, intergranular crack patterns.

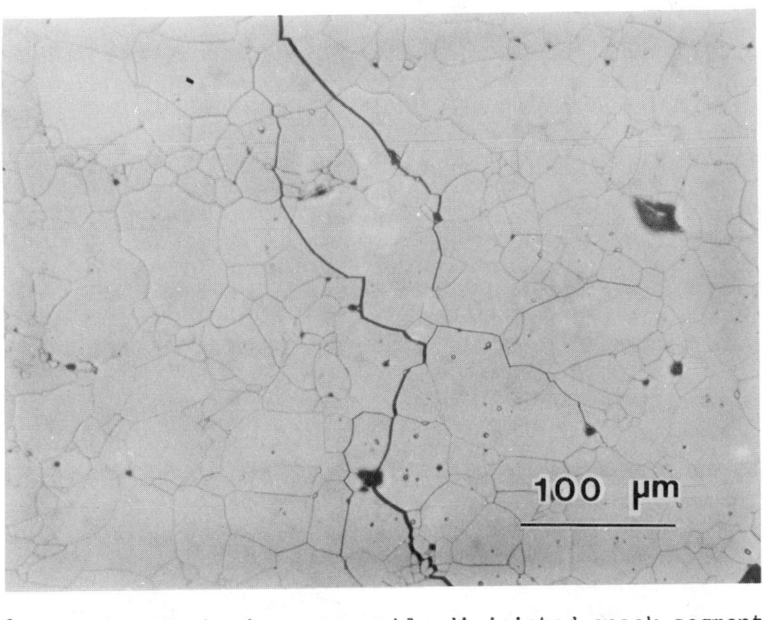

Figure 8. Micrograph showing apparently disjointed crack segments along main fracture path in Vi alumina.

reported by Knehans and Steinbrech;[14] that any previous progression up an R-curve can be negated by cutting away material from the walls of the extended crack (e.g. by sawcutting). We now have an experimental basis for explaining the R-curves of Fig. 3 more rigorously.

4. BRIDGING MODEL OF R-CURVE CHARACTERISTICS

We have indicated in our discussion of Eq. 1 the need for explicit knowledge of the microstructure-associated term $K_m(c,....)$ in any quantitative fracture mechanics analysis of the R-curve phenomenon in ceramics. This need is paramount in the context of design: we have already alluded (notably in our consideration of Fig. 3) to the potential dangers of extrapolating toughness data into the domain of strength-controlling flaws. We have thus far been restricted to the use of a semi-empirical representation, Eq. 2, which, although accounting well enough for the trends in the strength data, provides little physical insight into the material element in the toughening process. Clearly, we have to incorporate basic information on the specific crack resistance mechanism into our microstructural modelling.

The direct crack observations in Sect. 3.3 suggest that we should consider a model based on the restraining effect of grain-localized "bridges" across the primary crack interface. There is well-established precedent in the concrete[15-17] and ceramic composites[18] literature for this kind of modelling. Our aim in this section is to outline how one sets up the fracture mechanics, leaving the details of the calculations to a separate paper.[5]

Before proceeding with our bridging model, however, it is well that we should give reasons for discounting other mechanisms which have been proposed to account for apparent toughening effects.[1] The most vaunted of the R-curve mechanisms, transformation toughening, does not operate in the materials studied (certainly not in the near-monophase aluminas). Models based on crack geometry alone, e.g. deflection,[19] appear incapable of accounting for the long range of the R-curve effect, e.g. as manifested by the gradual transition in strength behavior of Figs. 1 and 2. Moreover, materials which show widely different R-curve characteristics (e.g. F99 and Vi aluminas) have remarkably similar (intergranular) fracture morphologies. Explanations invoking internal stresses (such as those due to thermal expansion mismatch)[20] appear to be able to account for a lower toughness at small indentation loads, where the microscale cracks are presumed to feel the grain boundary tensile forces to full effect. But in the large-load limit the crack should average out the alternate local tensile and compressive stresses, tending ultimately to the grain boundary toughness; thus, since the grain boundary is weaker than the matrix single crystal (for otherwise the fracture would be transgranular), we are unable to account for the crossing of the Vi $T^*(P)$ curve above the sapphire curve in Fig. 3. Finally, there is the possibility of a microcracking process zone which grows steadily with crack size;[21] as indicated in Sect. 3.3, we saw no evidence for any such zone.

Consider now the model shown schematically in Fig. 9. We suppose that there are bridging ligaments which exert interfacial closure forces $F_i(x,z)$ behind the advancing crack tip. It is assumed, as in the case of fiber-reinforced ceramic composites,[18] that these forces can be considered to be continuously distributed, $F_i(x,z) = \sigma_i(x)\ d^2$, once the crack extends beyond the first bridges, where d is the mean spacing between bridging sites. This assumption represents a compromise between retention of an element of discreteness and mathematical tractability.

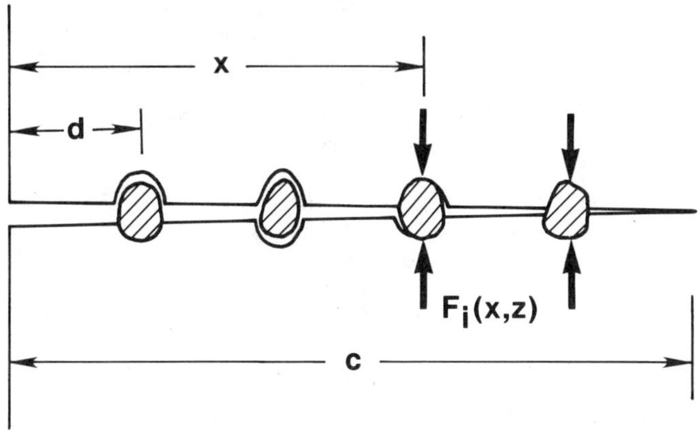

Figure 9. Model of grain-localized crack bridging mechanism. Grains behind main crack front remain partially attached to both sides of separating interface, thereby exerting closure forces $F_i(u)$ on the crack. When crack segment (see Fig. 8) runs around one side of grain ligament the main crack separates completely, $F_i(u)$ goes to zero. If segment runs around both sides simultaneously, grains are detached from surface.

Then we may resort to standard Greens function solutions for the stress intensity factor.[22] In the case of penny-like cracks pertinent to the indentation geometry we have (treating closure forces as positive here)

$$K_m = -(\psi/c^{1/2}) \int_d^c \sigma_i(x)\ x\ dx/(c^2-x^2)^{1/2}. \tag{7}$$

Unfortunately, we do not know the functional dependence $\sigma_i(x)$ a priori, so we have to make some approximations. A convenient ploy is to make use of Sneddon's equation for the wall opening displacements of a traction-free, uniformly loaded crack (plane stress)[23]

$$u(x,c) = (\psi K/Ec^{1/2})(c^2-x^2)^{1/2}\ . \tag{8}$$

None of the conditions for validity of this expression are strictly met here, but this is not expected to reflect strongly in the material dependencies.[18] Then for equilibrium cracks (K = T), Eqs. 7 and 8 combine to give

$$K_m = -(E/T) \int_o^{u(d,c)} \sigma_i(u)\ du\ . \tag{9}$$

The problem is therefore reduced to determining the force-separation function for ligament rupture, which of course is dependent on the details of the rupture process.

Once $\sigma_i(u)$ has been determined, the requisite stress intensity factor $K_m(c,d,....)$ may be evaluated. We note that the crack-size dependence enters via the upper limit in Eq. 9 (in conjunction with Eq. 8). To this point our evaluations of Eq. 9 have been restricted to trial functions $\sigma_i(u)$, of the kind proposed in the analysis of concrete fracture.[15,16] There are some encouraging trends emerging from our preliminary calculations, however. Thus we find that the $K_m(c)$ function, although dependent on the choice of the force-separation curve, is in all cases monotonically increasing (indicating that Eq. 2 does at least contain the essence of the crack-size influence). Also, by virtue of the low limit cutoff in Eq. 7 (below which the toughening mechanism is not in operation), the calculations lead to the prediction of the same kind of radial crack pop-in behavior as observed in the indentation-strength experiments earlier (Sect. 3.3). On the other hand, a full understanding of the role of material properties, apart from E and T in Eq. 9, awaits further development of the micromechanics of ligament rupture (via $\sigma_i(u)$). Only then can we expect to account for such factors as grain boundary phase (as is so clearly manifest in the contrasting R-curves for Vi and F99 aluminas in Fig. 3, Sect. 3.1), grain size, etc.

5. CONCLUSIONS

(i) Indentation-strength techniques provide a powerful methodology for analyzing the dependence of toughness and fatigue properties of brittle ceramics on crack size.

(ii) Microstructural influences are strongly felt in the inert strength data, reflecting R-curve behavior. They have relatively no effect on the slopes of the dynamic fatigue plots (at least in the crack-size range covered here).

(iii) The experimental R-curves depend sensitively on the microstructural detail, e.g. grain boundary phase.

(iv) Direct observations of the crack evolution to failure suggest that "toughening" in the aluminas and glass-ceramics is due to the closure effect of unruptured grain-scale ligaments behind the main crack front. The basis of a fracture mechanics description of this process is outlined.

ACKNOWLEDGEMENTS

The authors wish to thank P.L. Swanson and B.J. Hockey for many useful discussions on parts of this work, particularly in relation to the observations in Sect. 3.3. Funding was provided by the U.S. Air Force Office of Scientific Research.

REFERENCES

1. S. M. Wiederhorn, Ann. Rev. Mater. Sci. 14; 373 (1984).
2. R. F. Cook, B. R. Lawn and C. J. Fairbanks, J. Amer. Ceram. Soc., in press.
3. R. F. Cook, B. R. Lawn and C. J. Fairbanks, J. Amer. Ceram. Soc., in press.
4. P. L. Swanson, C. J. Fairbanks, B. J. Hockey, B. R. Lawn and Y. W. Mai, in preparation.
5. Y. W. Mai and B. R. Lawn, in preparation.
6. D. B. Marshall, B. R. Lawn and P. Chantikul, J. Mater. Sci. 14; 2225 (1979).
7. B. R. Lawn, p. 1 in Fracture Mechanics of Ceramics, Vol. 5, R. C. Bradt, A. G. Evans, D. P. H. Hasselman and F. F. Lange, eds., Plenum, New York (1983).

8. A. G. Evans, J. Am. Ceram. Soc. 63; 115 (1980).
9. P. Chantikul, G. R. Anstis, B. R. Lawn and D. B. Marshall, J. Am. Ceram. Soc. 64; 539 (1981).
10. R. Morena, K. Niihara and D. P. H. Hasselman, p. 145 in Fracture Mechanics of Ceramics, Vol. 5, R. C. Bradt, A. G. Evans, D. P. H. Hasselman and F. F. Lange, eds., Plenum, New York (1983).
11. R. Morena, K. Niihara and D. P. H. Hasselman, J. Am. Ceram. Soc. 66; 673 (1983).
12. T. Okada, G. Sines and D. J. Green, J. Am. Ceram. Soc. 65; C-64 (1982).
13. P. L. Swanson, private communication.
14. R. Knehans and R. Steinbrech, J. Mater. Sci. Letters 1; 327 (1982).
15. A Hillerborg, M. Modeer and P. E. Petersson, Cement and Concrete Res. 6; 773 (1976).
16. M. Wecharatana and S. P. Shah, J. Eng. Mechs. Div., ASCE 109; 1231 (1983).
17. Y. W. Mai, in Application of Fracture Mechanics to Cementitious Composites, S. P. Shah, ed., Martinus Nijthoff, Netherlands, in press.
18. D. B. Marshall, B. N. Cox and A. G. Evans, in preparation.
19. K. T. Faber and A. G. Evans, Acta Met. 31; 565, 577 (1983).
20. R. W. Rice, R. C. Pohanka and W.J. McDonough, J. Am. Ceram. Soc. 63; 703 (1980).
21. A. G. Evans and K. T. Faber, J. Am. Ceram. Soc. 67; 255 (1984).
22. G.C. Sih, "Handbook of Stress Intensity Factors," Lehigh University, Bethlehem (1973).
23. I. N. Sneddon and M. Lowengrubb, "Crack Problems in the Classical Theory of Elasticity," Wiley, New York (1969).

MECHANICAL BEHAVIOR OF LIGHTWEIGHT CERAMICS

David J. Green

Dept. of Materials Science and Engineering
The Pennsylvania State University
University Park, PA 16802

ABSTRACT

The current understanding of the mechanical behavior of lightweight ceramics is reviewed. It is shown that these materials may have considerable potential in structural applications. To capitalize on this potential, it is clear that several challenges need to be met. These include establishing a clear understanding of the relationship between microstructure and mechanical behavior in these materials, developing new and innovative techniques for their fabrication and devising appropriate techniques for their engineering design.

INTRODUCTION

The majority of scientific studies into the mechanical behavior of ceramics have invariably emphasized materials close to theoretical density. Indeed, most structural applications of materials, with the exception of wood, utilize dense materials. Conversely, Nature invariably uses fibrous or cellular materials in such applications and presumably these represent a careful evolutionary optimization. As pointed out by Ashby [1], cellular materials permit the simultaneous optimization of stiffness, strength and overall weight. There are a wide range of man-made lightweight materials, with polymeric foams being the most familiar group. Within the field of ceramics there are several examples, including fibrous and foamed materials that are used for thermal insulation and open cell materials, that are used for filtering molten metals. As these materials represent a rather new class of ceramics, it is reasonable to expect they may have a wide range of potential applications that are not yet fully explored. Currently, these materials are not chosen for their structural attributes but for some other special quality. It is, however, still important to maximize the strength of such materials for an optimum design. For example, the tiles used for the thermal protection system (TPS) of the space shuttles were chosen for their high temperature capability, low thermal conductivity, low coefficient of thermal expansion and weight, and yet prior to the first flight their mechanical behavior and its understanding was a critical issue [2-4].

MICROSTRUCTURES OF LIGHTWEIGHT CERAMICS

For the purposes of this paper, let us consider lightweight ceramics as having fractional densities ρ/ρ_s <0.5, where ρ and ρ_s are the densities of the lightweight and solid materials respectively. For thermal insulation purposes, values of ρ/ρ_s are often <0.1 and rigid materials with values as low as 0.03 have been fabricated [5]. There are two broad categories of lightweight ceramics, i.e., fibrous or cellular, which reflects the nature of their microstructure. The fibrous materials generally consist of a tangled network of fibers, whereas the cellular materials consist of a relatively regular array of cells. For the cellular materials, there is generally a further sub-classification into closed or open cells, depending on whether the faces of the cell are present or not. These various classes of lightweight ceramics are illustrated by some examples in Fig. 1.

The microstructures of the fibrous materials are often anisotropic. This is usually a result of the processing, in which fiber slurries are pressed and filtered. As the fibers will prefer to lie perpendicular to the pressing direction and/or gravity, this preferred orientation will persist in the final microstructure. This leads to anisotropy in the properties of these materials. The tiles used on the TPS of the space shuttle are transversely isotropic, i.e., they are isotropic in one principal plane but possess a different set of properties for the direction perpendicular to this plane. Thus, when studying the mechanical behavior of these materials, it is essential to be aware of the anisotropy of strength, fracture toughness and the elastic constants [6,7]. Cellular ceramics tend to be relatively isotropic and are often produced by foaming techniques (closed cell) or by using reticulated polymeric foams as a skeleton (open cell). An alternative approach to producing closed cell ceramics is by the sintering or bonding of hollow spheres [8-10].

In terms of microstructural and engineering design, it is crucial to develop relationships between the microstructural parameters, e.g., cell size, fiber diameter, cell wall thickness, etc., and the mechanical properties, e.g., fracture toughness, strength and the elastic constants, especially with respect to variations in ρ/ρ_s. A powerful scientific approach to accomplish this, is to model the microstructure using an idealized arrangement of micro mechanical unit cells.

MICROMECHANICAL MODELS

There has been a variety of attempts to model cellular and fibrous materials. Let us first consider cellular materials, for which there is an extensive literature, mainly in the field of polymer science [11-18]. In these studies, the first assumption is that the mechanical behavior of the cellular material can be analyzed by considering the deformation of a particular unit cell. There has been a wide variety of cells chosen, including open and closed cubic [12,14] and various dodecahedrons [16-18]. These latter cells are chosen to reflect the morphology of foams, as this is often the approach to fabricating cellular materials. This type of approach has been discussed by Patel and Finnie [17]. During foaming, the pressure difference between adjacent cells is expected to be small due to diffusion. This implies the reduction of surface area should be the overriding factor in establishing the final geometry. This behavior is equivalent to that of soap bubbles, for which surfaces can only meet in two ways [19]. That is, either three surfaces meet along a curve at angles of 120° or six surfaces (four edges) meet at a vertex with angles of 109.47°. There is no regular polyhedron that meets these requirements as well as being able to fill space. Kelvin [20] showed, that if the edges of a truncated octahedron were curved, both the required angles and the compatibility conditions could be satisfied. This polyhedron is called the minimum area tetrakaidecahedron. In actual foams, however, such curvature

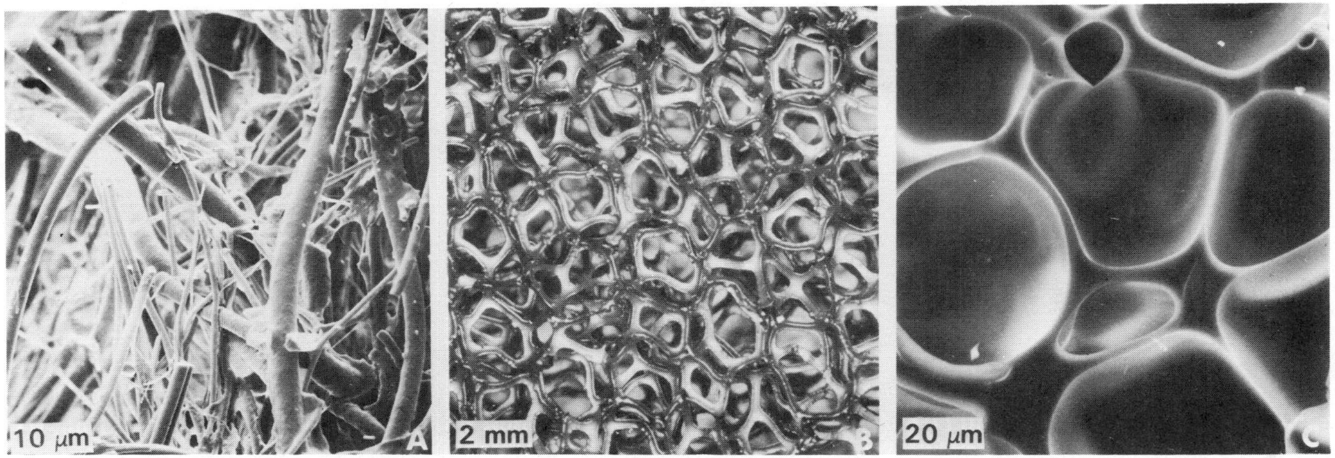

Fig. 1. Microstructures of various lightweight ceramics; (A) fibrous silica, (B) open cell SiC and (C) closed cell sodium borosilicate glass.

of the edges and faces is not usually observed and so plane-faced figures, such as dodecahedrons, represent a reasonable approach, even though they do not fill space.

An alternative approach has been suggested by Ashby and his co-workers [1,21], in which the unit cell is even more idealized than suggested in the previous discussion, but argue that the essential micromechanics can be obtained by considering the deformation of such a cell. Fig. 2 compares this unit cell with an earlier one suggested by Gent and Thomas [12]. An important aspect of the Gibson and Ashby approach [21] is that bending of cell edges, due to stresses applied in a principal direction, is incorporated into the model. Such bending has been seen to occur in both model foams [17,22-24] and in actual cellular materials [21]. Bending of cell edges had been incorporated into some previous analyses [16,17,25] but Ashby and his co-workers have extended this concept to many aspects of the mechanical behavior of cellular materials [1,21,26,27]. The final aspect of the choice of unit cell is the difference between open and closed cells. Ashby [1] argues that closed cell materials behave mechanically in a similar way to the open cell materials because surface tension draws most of the material into the cell edges. Patel and Finnie [17], however, argued previously that the application of a tensile stress to a closed cell material must place some of the cell faces in tension. More recently, the analysis of a simple model that incorporated cell face tension, showed that provided the cell wall face is less than one tenth of the cell edge dimension, a closed cell model should approach that of an open cell [28].

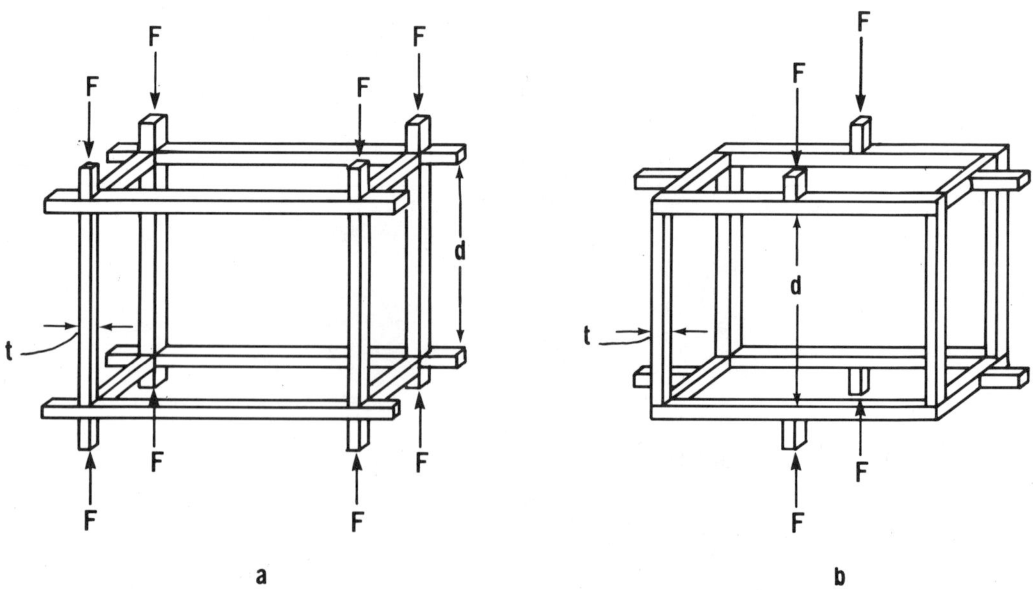

Fig. 2. Comparison of micromechanical unit cells used in refs. 12 and 21.

The models for fibrous materials have been developed along similar grounds to those of cellular materials. As shown previously, the microstructure of these materials is much more inhomogeneous, but the abstraction to analyze the deformation of a fibrous unit cell has been found to be a powerful approach. For work on the space shuttle tiles, Green and Lange [29] chose a model similar to that of Gent and Thomas [12] shown in Fig. 2a. The main difference was that the spacing of fibers in one direction was different from that in the other two directions, to simulate the anisotropic nature of the tiles, and the fibers in the various directions acted primarily independent of each other. In tensile failure of the tiles, there was evidence for failure of the fibers in bending [6] and from microstructural observations, such as Fig. 1, it is clear that some bending must occur. Even still, it appeared that the fibers were breaking at high tensile stresses, as the fibers fragmented during fracture [6]. Models based on fiber bending have been developed in other studies [30,31].

For all the models discussed, it is possible immediately to relate ρ/ρ_s to the microstructural geometry. For example, at low densities, ρ/ρ_s is simply proportional to $(t/d)^2$. This ignores the contribution to density from the cell corners, which becomes important for the higher densities. For these cases, the relationship usually has the form [12,21]

$$(\rho/\rho_s) = \{(t/d)^2 + b(t/d)^3]/(b(1 + t/d)^3) \qquad (1)$$

where b is a constant that depends on the shape of the cell corners. Gent and Thomas [12] disregarding the corner shape use a value of b=0.333, whereas Gibson and Ashby [21] suggest a value of 0.766 is more appropriate for a foamed material.

THEORY OF MECHANICAL BEHAVIOR

The stress-strain behavior of fibrous and cellular materials in compression is very similar and qualitatively independent of the choice of material [1]. The typical behavior is shown in Fig. 3. Initially, the materials deform elastically, which is followed by a plateau of deformation at almost constant stress and finally, there is a region of densification as the material crushes together. The behavior in the plateau region is a result of various mechanisms, in which the cells or fibers collapse, yield or fracture. Fig. 4 shows the compressive behavior of a space shuttle tile (LI 900). For this material, the plateau region is primarily brittle crushing with evidence of shattering only at the higher strains. This is followed by the densification stage. For a brittle cellular material (Fig. 5), the stress in the plateau region varies substantially and involves brittle crushing of the cells but the specimen crumbles before densification occurs. Densification will, however, occur in a cellular ceramic if the specimen is contained [27]. The tensile behavior of lightweight ceramics is usually linearly elastic followed by brittle catastrophic fracture.

Once the various deformation modes and mechanisms have been identified, the next step is to analyze the chosen unit cells in these terms. Aspects of the deformation modes relevant to lightweight ceramics will be discussed below. For simplicity in this paper, the cells shown in Fig. 2 will be used for illustration, with comparisons where necessary for the fibrous ceramics.

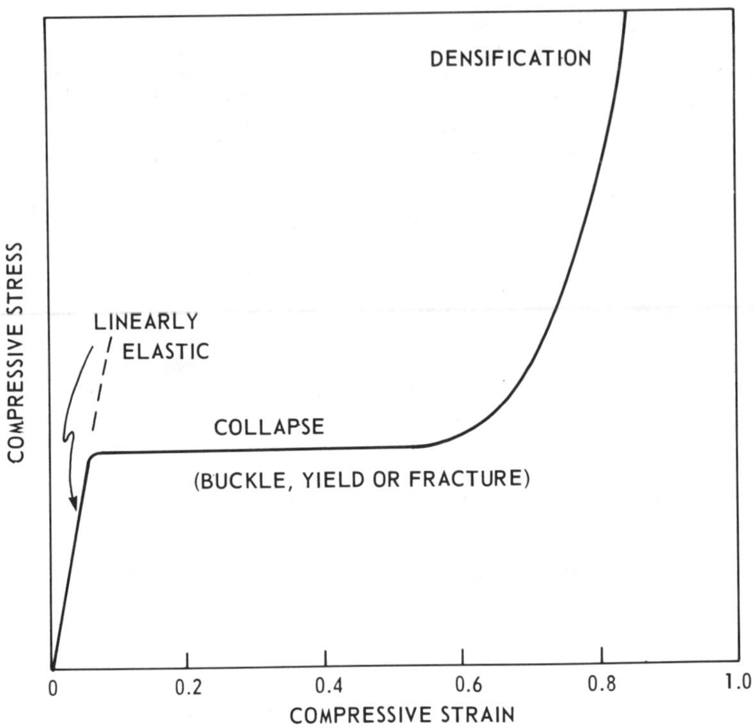

Fig. 3. Idealized compressive stress/strain behavior for fibrous and cellular materials.

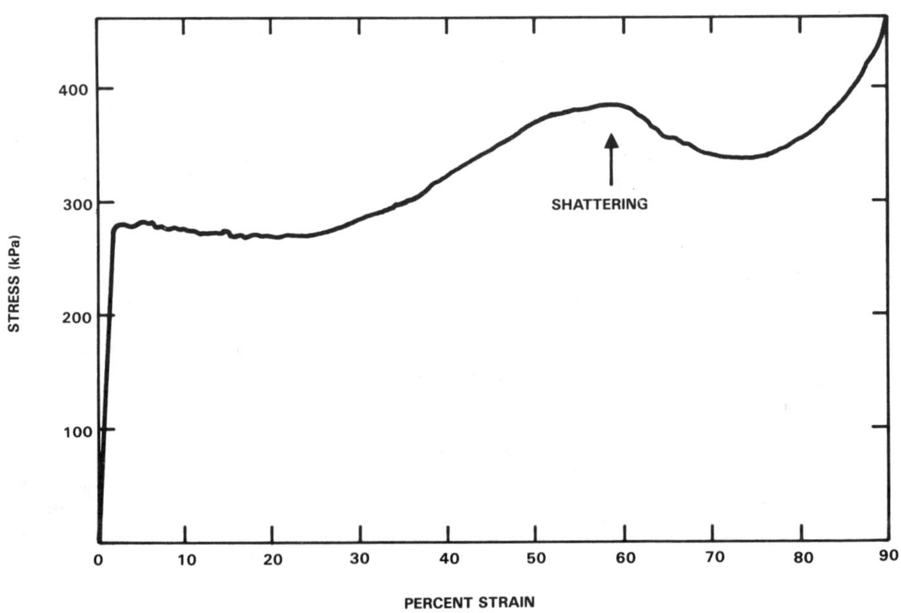

Fig. 4. Compressive stress/strain behavior of fibrous silica tiles used for TPS of space shuttles.

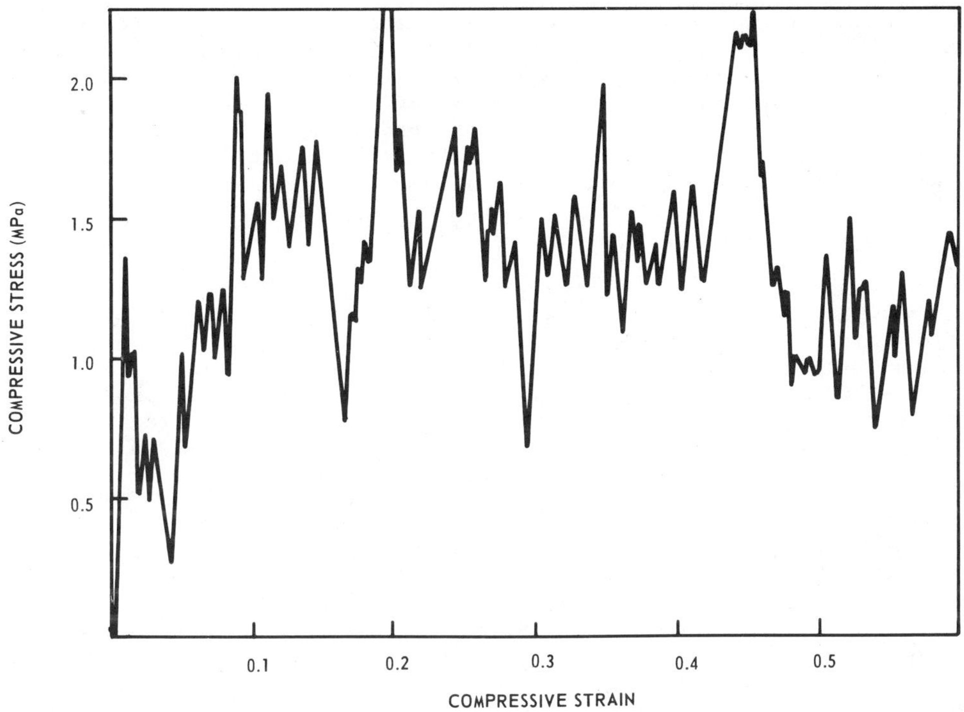

Fig. 5. Compressive stress/strain behavior of an open cell $Al_2O_3(\rho/\rho_s=0.1)$.

Linear Elastic Behavior

For the model in Fig. 2a, a force F applied as shown, causes the vertical members to be placed in compression (or tension). The applied stress is proportional to F/d^2 and the resulting strain is $F/(E_s t^2)$, where E_s is the Young's modulus for the cell edge material. The Young's modulus of the cellular or fibrous material (E) is

$$E/E_s = B_1(t/d)^2 = C_1(\rho/\rho_s) \tag{2}$$

for low densities, where B_1 and C_1 are constants that depend on the exact cell geometry. Conversely, for the model in Fig. 2b, the primary deformation will be the bending of the horizontal members [21]. Using simple beam theory, Gibson and Ashby [21] showed that the strain is proportional to $Fd^2/E_s t^4$ and the equation for E becomes

$$E/E_s = B_2(t/d)^4 = C_2(\rho/\rho_s)^2 \tag{3}$$

The dimensionless constants B_i and C_i where i is an integer, are used in this paper in the same spirit as those used by Ashby [1] and they depend on cell geometry or can be determined experimentally. For this paper, the relationship $\rho/\rho_s = (t/d)^2$ from Eq. 1 at low densities, will be used in deriving the various equations. Gibson and Ashby [21] also developed the relationship for shear modulus and it scales in the same fashion as Eq 3. This implies that Poisson's ratio is independent of density, which agrees approximately with experimental measurements [21]. For anisotropic fibrous materials, Green and Lange [29] developed equations similar to Eq. 2, except there are two values for the Young's moduli in the weak and strong directions. Alers [32] took this work further, to develop the 5 elastic constants that are needed for a transversely isotropic material. A semi-empirical approach has been suggested by Rice [33] for estimating E/E_s for cellular microstructures. An understanding of the wide range of variation of elastic constants with density is important for the structural design of lightweight ceramics.

The Crushing Strength

As indicated earlier, lightweight ceramics collapse by brittle crushing. Ashby [1] has developed an approach to this mechanism by considering the moment acting on the cell wall. The stress due to this moment is proportional to Fd/t^3 and the applied stress acting on the cellular material is proportional to F/d^2. The cell edge will then fail when the stress due to the moment reaches the strength of the cell edge (σ_f). The crushing strength (σ_{fc}) is therefore given by

$$(\sigma_{fc}/\sigma_f) = B_3(t/d)^3 = C_3(\rho/\rho_s)^{3/2} \tag{4}$$

For a cellular material with no defects bigger than the cell size, Ashby [1] suggests that Eq. 4 would also describe the behavior in tension. Thus, the surprising conclusion is that the strength in tension (σ_{ft}) could be the same as that in compression. This is approximately true for foamed glass [34] and Ashby [1] estimates C_3 to be 0.65. This implies that for a material with $\rho/\rho_s = 0.2$ and if σ_f could be 1 GPa, then a strength of 58 MPa should be feasible. Another interesting aspect of lightweight ceramics is their indentation behavior. Unlike most dense solids, lightweight ceramics can change their volume when indented. Work on the space shuttle tiles indicated that the material simply crushed under the indentor and this occurred when the stress due to the indentor was approximately σ_{fc} [35]. Thus, for these materials, the hardness is approximately given by Eq. 4.

Fracture Toughness and Tensile Strength

Unlike compression, the tensile strength of lightweight ceramics is expected to be sensitive to the presence of defects, e.g., broken cell walls, large cells, the amount of interfiber bonding, etc. For the work on the space shuttle tiles [29], the defects were simulated by a series of broken fibers as shown in Fig. 6. This type of geometry had been analyzed previously [36] assuming that the joints could rotate. It was shown that the stress field in front of the broken fibers could be described by fracture mechanics parameters and in particular, the critical stress intensity factor (K_{IC}) would be given by [29,36]

$$K_{IC} = B_4\sigma_f t^{1/2}(t/d)^{3/2} \tag{5}$$

$$= C_4\sigma_f t^{1/2}(\rho/\rho_s)^{3/4} \tag{6}$$

$$= C_4\sigma_f d^{1/2}(\rho/\rho_s)$$

These 3 equations are a useful basis for indicating the critical parameters that need to be controlled in the fabrication of these materials. At a given density, substantial gains can be made by increasing σ_f, thus toughening mechanisms that increase the toughness of the cell wall material and elimination of defects in the cell wall would be beneficial. The equation indicates that increasing t would be useful but as strength generally decreases with increasing size, this may involve optimization.

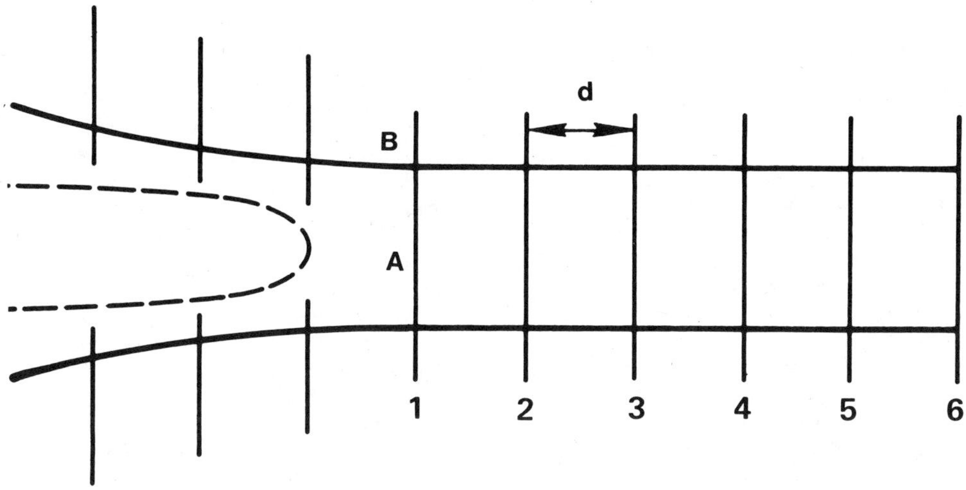

Fig. 6. Configuration of fibers at a crack tip used to describe the fracture process.

For the fibrous space shuttle tiles, it was necessary to use two values of K_{IC} for the weak and strong directions [6].

Maiti et al. [26] used a different approach to obtain K_{IC}. Firstly, they simply assumed the fracture mechanics equations to be valid and they indicated that bending of the cell edge at the crack tip will be predominant. In terms of Fig. 6, this implies it is the bending of cell edges in locations like A and B that control the fracture process. For the tensile stress on the cell edge (A), their expression is equivalent to Eqs. 5 and 6, but when bending is incorporated they obtain

$$K_{IC} = C_5 \sigma_f (\pi d)^{1/2} (\rho / \rho_s)^{3/2} \tag{7}$$

Maiti et al. [26] estimated C_5 to be 0.65 and thus, if σ_f=1 GPa, d=10μm and ρ/ρ_s=0.2 then K_{IC}=0.33 $MPa/m^{1/2}$. This implies that on a weight basis, fracture toughness values of lightweight ceramics can approach those of dense ceramics. As will be seen later this conclusion is supported by experimental results.

The relationship between tensile strength and K_{IC} is well known, i.e., $\sigma_{ft}=K_{IC}/Y c^{1/2}$, where 2c is the crack size and Y is a geometric constant. If one uses $Y=\pi^{1/2}$, as was used in determining Eq. 7, the tensile strength is given by

$$\sigma_{ft}/\sigma_f = C_5 (d/c)^{1/2} (\rho/\rho_s)^{1/2} \tag{8}$$

This equation assumes $c \geq d$ and reduces to Eq. 4 when c=d. To obtain the maximum tensile strength, defects in the material need to be less than the cell size. Further gains can only be produced by increasing K_{IC}. The density exponent would be changed in Eq. 8 if Eq. 6 is used instead of Eq. 7.

In addition to the types of mechanical behavior discussed above, Ashby has reviewed his work on non-linear elestic behavior, plastic yielding, indentation and creep of cellular materials [1]. More recently, Maiti et al. [27] have developed deformation and energy absorption diagrams for cellular solids. For example, Fig. 7 shows a deformation diagram constructed following their analysis for a brittle cellular material. The diagram shows the expected behavior of these materials in compression. The various modes of deformation are shown, with the expected stress-strain contours for various initial densities. For values of $\rho/\rho_s > 1/3$, the cell walls become too stocky to collapse and thus there is no crushing (plateau) region. For the lower densities, crushing follows the linear elastic behavior and then densification occurs when the relative density is ≅1/3. As indicated before, densification in cellular ceramics only occurs if the specimen is contained. Maiti et al. [27] suggest that the plateau region is like a two phase field, in that part of the material will be crushed while the rest of the material is elastic. That is, the crushing occurs in a band that broadens with increasing strain.

The energy absorption diagrams developed by Maiti et al. [27] should also be useful for lightweight ceramics, especially in compressive or impact loading. Their type of analysis indicates that compared to dense ceramics, quite large amounts of energy can be absorbed by brittle crushing.

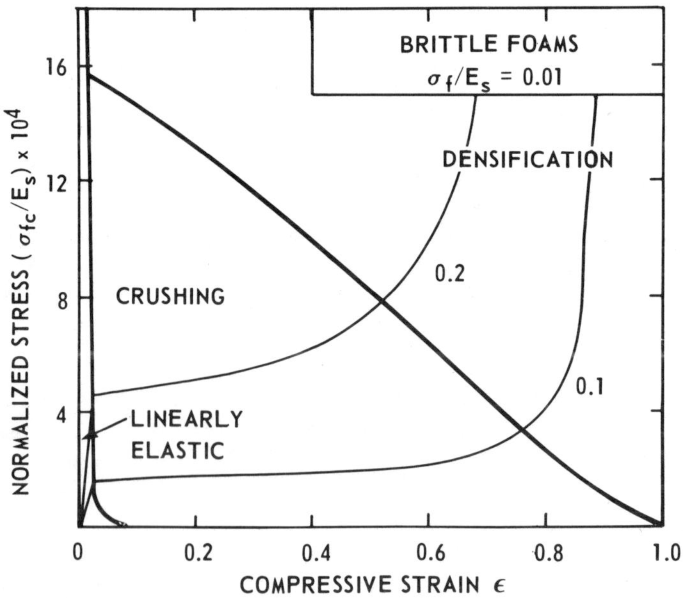

Fig. 7. Theoretical deformation map for the compressive behavior of a brittle cellular material, after ref. [27].

MECHANICAL BEHAVIOR OF SPECIFIC SYSTEMS

Until recently, there has been very few studies on the mechanical properties of lightweight ceramics. Most of the work has considered foamed glass, which is commercially available [18,37,38] but there has also been studies on sintered or bonded hollow sphere ceramics [9,10,39], fibrous silica [4], cellular mullite [27] and foamed cements [40-42]. The mechanical behavior of the first three of these materials will be now discussed in more detail.

Foamed Glass

Values of E for foamed glasses have been measured at several densities [18,37] and are summarized in Fig. 8. The results are compared with the micromechanical models discussed earlier [12,21]. In order to do this, a value of E_s=69 GPa was chosen, which is reasonable for a bulk glass. The uncertainty of this assumption for the 4 compositions present in Fig. 8, complicates the assessment of the two models. Linear regression of the data indicated the exponent was 1.2 (95% confidence $\pm$ 0.2). Thus, unless there is a systematic variation of E_s with density, the experimental data supports the tensile model [12]. The values of K_{IC} and σ_{ft} (flexure) were also measured for most of the glasses in Fig. 8 [18]. Surprisingly, both parameters appeared to vary linearly with density and for K_{IC} the values extrapolated to that of dense glass for ρ/ρ_s=1. If one assumes constant values of σ_f and d, the data appear to be in better agreement with Eq. 6 than Eq. 7 but again the study involved 4 compositions, the cell sizes were not reported for all compositions and it is not clear that σ_f should be a constant. It is clear that further work is needed to understand the variation of these properties with density.

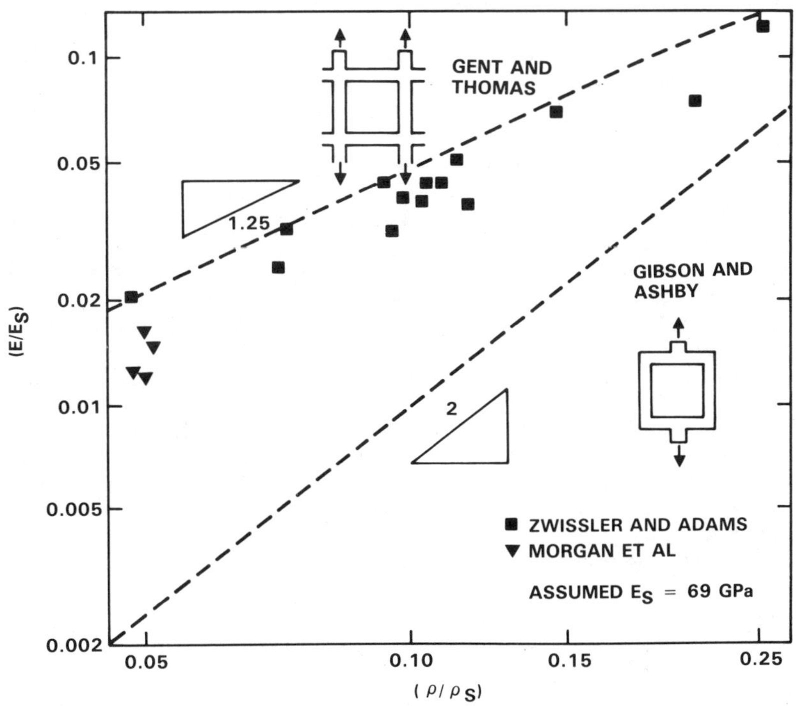

Fig. 8. Comparison of predicted variation of Young's modulus with density for cellular solids based on the tensile and bending micromechanical models [12,27]. The experimental data for foamed glass [18,37] is included assuming E_s=69 GPa.

It is worth considering the size of the critical flaws in these materials. For a commercial foamed glass (Foamglas®, d=1.2 mm, ρ/ρ_s=0.05), Zwissler and Adams [18] found K_{IC}=40 kPa/$m^{1/2}$ and σ_{ft} (flexure) =800 kPa. This allowed them to estimate the critical flaw size and they found it to be in the range 1.5-4.0 mm, which is 1-3 cell diameters. Morgan et al. [37] studying the same glass found that the strengths in tension, flexure and compression were all increased if the cell size were decreased. Using the K_{IC} value given above, their analysis gives a value of the critical flaw size, which is very close to that of Zwissler and Adams [18]. The increase in strength with decreasing cell size is an important observation for the development of lightweight ceramics. Comparison with Eq. 8 indicates at first that this is unexpected but if c is dependent on d or if σ_f increases with decreasing cell size as expected, then the dependence on cell size becomes clearer. In their work, Morgan et al. [37] found that the strengths increased in the order tension, flexure and compression. For example, at a cell size of 1.2 mm, the strengths were approx. 0.5, 0.6 and 1.1 MPa, respectively. Following up on this work, Conley et al. [38], showed the strengths were dependent on stressing rate. This was interpreted in terms of the stress corrosion effect that gives rise to subcritical crack growth in dense glass. They did, however, find that the stress intensity exponent was different in tension, compression and bending. This is unexpected in terms of the micromechanical models, as the failure process is qualitatively similar in all 3 modes. Zwissler and Adams [18] also found substantial differences in the subcritical crack growth behavior of bulk and cellular glasses.

Hollow Sphere Ceramics

As an alternative to foaming, cellular ceramics can be formed by sintering or bonding hollow spheres [8-10]. Fig. 1(C) is an example of a closed cell ceramic that was formed by the sintering of hollow glass spheres [9]. In this study, it was found that the sintering behavior was complex and in particular, local densification could lead to the presence of large, strength-limiting voids. An advantage of this type of approach is that ceramics with relatively small cell sizes can be produced. For the material in Fig. 1(C), the average cell size was ≅32μm. The variation of E/E_s with the degree of sintering is shown in Fig. 9 and is compared to the micromechanical models discussed earlier. At the higher densities, when the cellular microstructure is well-developed, the variation of E/E_s is in good agreement with the model of Gibson and Ashby [21] in terms of the density exponent. At the lower densities, however, there is a rapid change in E, which is not expected from their model. This transition was found to be related to the microstructure produced by the sintering process. Fig. 10 shows a fracture surface of the material in the early stages of sintering. The fracture process has involved failure of the sintered contact areas. This can be explained as follows. In the early stages of sintering, the annular contact areas are very small, so that if the material is stressed, the stresses and strains in the vicinity of the contact areas will be very large and will thus give rise to a low value of E and will make the contact areas preferential fracture sites. As further sintering occurs, however, the size of the contact area will change rapidly until the microstructure approaches that of closed cell, giving a transition in the E behavior.

In order to analyze the behavior of the sintered hollow sphere materials, a different type of model was formulated [39]. This is illustrated in Fig. 11, where a thin hollow sphere is loaded through 2 contact areas. It was then possible to use elastic thin shell theory to determine the stresses and strains in the sphere. By extending this to N contacts and combining the mechanics with a simple densification theory, it

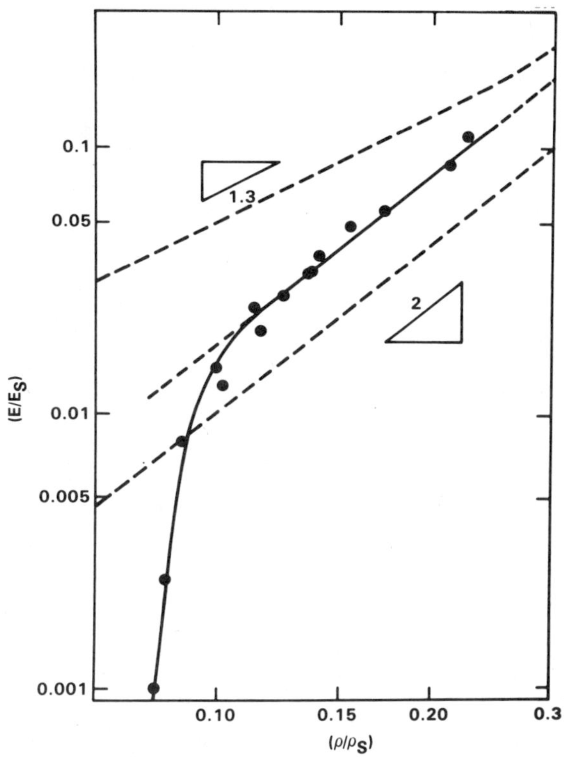

Fig. 9. Variation of Young's modulus with density for sintered hollow glass sphere (SHGS) materials and a comparison with the theoretical predictions.

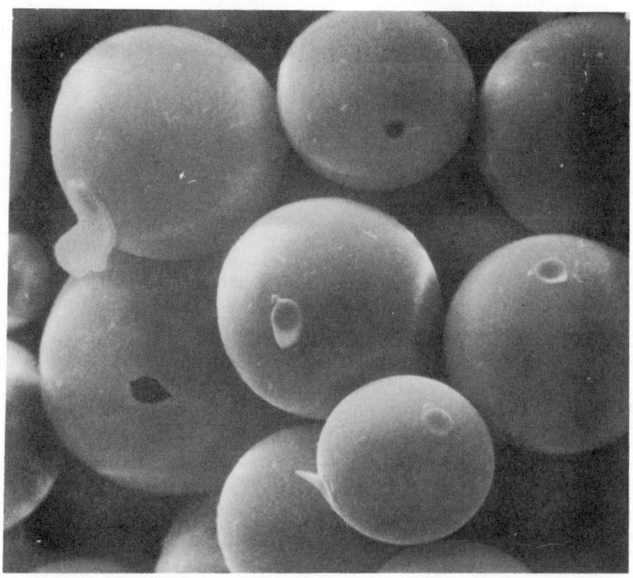

Fig. 10. Fracture surface of SHGS materials showing failure of the sintered contact areas at the lower densities.

was possible to predict the variation of E/E_s and K_{IC} with the degree of sintering. Fig. 12 shows the comparison between the predicted and experimental variation of $K_{IC}r_o[\sigma_f t_o(\pi r_o)^{1/2}]$, where t_o and r_o are the average thickness and radius of the hollow spheres. The only unknown parameter was σ_f, which was used as a fitting parameter. A value of σ_f =430 MPa, which is reasonable for a thin glass wall, gave the excellent agreement shown in Fig. 12. Overall, the thin shell model gave a better understanding of the hollow sphere ceramics and should be useful in optimizing the fabrication and the design of lightweight ceramics produced this way.

Verweij et al. [10] have studied the bonding of silica spheres and produced microstructures similar to Fig. 10. They found that E/E_s values were in the range 0.007 to 0.014, which is similar to the sintered data of Fig. 9 at the lower densities.

Fibrous Silica Bodies

The lightweight tiles used for the TPS of the current space shuttles are based on silica fibers, as shown in Fig. 1(A). These tiles are subjected to a variety of stresses during flight and thus, it is critical to understand their mechanical behavior. In a series of papers [4,6,7,29,43,44], that were recently reviewed [4], it was found that fracture mechanics and micromechanical modelling could be very useful in the design process and in understanding the relationships between microstructure and mechanical properties. This approach acted as a powerful basis for understanding techniques that aid in ensuring reliability, such as proof testing and nondestructive evaluation and in indicating directions for improved materials. Some of these aspects will be briefly discussed.

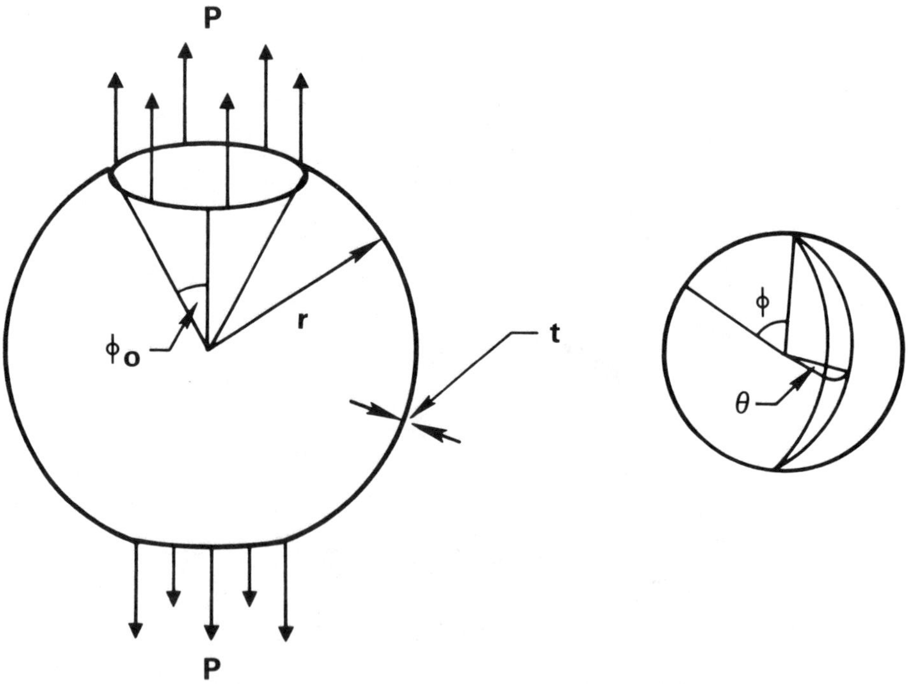

Fig. 11. Thin-walled hollow sphere loaded through annular contact areas. The size of the contact area is described by the angle ϕ_o.

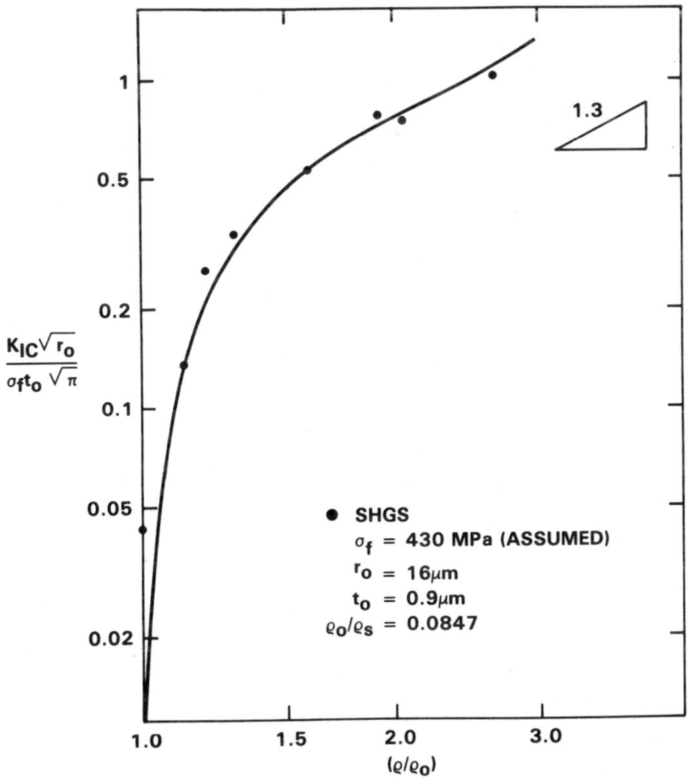

Fig. 12. Comparison of theoretical analysis [39] and experimental data for the variation of K_{IC} with the degree of sintering for the sintered hollow glass sphere materials.

Time Dependent Strength Behavior. As with bulk glasses, the fibrous silica tiles were found to be subject to subcritical crack growth. Using fracture mechanics techniques developed in recent years [45-47], it was possible to characterize the subcritical crack growth [6,43] and then to predict the behavior of the tiles for a variety of time-dependent stresses. The characterization was primarily performed using strength/stress rate data, such as shown in Fig. 13. In the characterization, it was found that the stress intensity factor exponent of crack velocity, which is related to the slope of the strength/stress rate data [45], did not depend on the density of the tiles and was very similar to that of bulk silica glass [4,6,43]. Indeed, by analyzing the model of Fig. 6, it was shown that such behavior would be expected. Included in Fig. 13 is some recent data [48] for a new tile material, known as fibrous refractory composite insulation (FRCI) [49]. It appears that this material, which consists of a combination of silica and alumina borosilicate fibers, is more sensitive to subcritical crack growth.

In terms of the first flight of the space shuttle, it was known that the strength of the tiles was similar to the design stresses and it was decided to proof test tiles in critical areas. It was, however, important to determine the effect of the proof testing on the surviving tiles, to ensure it did not lead to futher weakening by the subcritical crack growth process. Fig. 14 shows the experimental strength distribution for one class of tiles and the predicted strength distribution after a 122 kPa proof test.

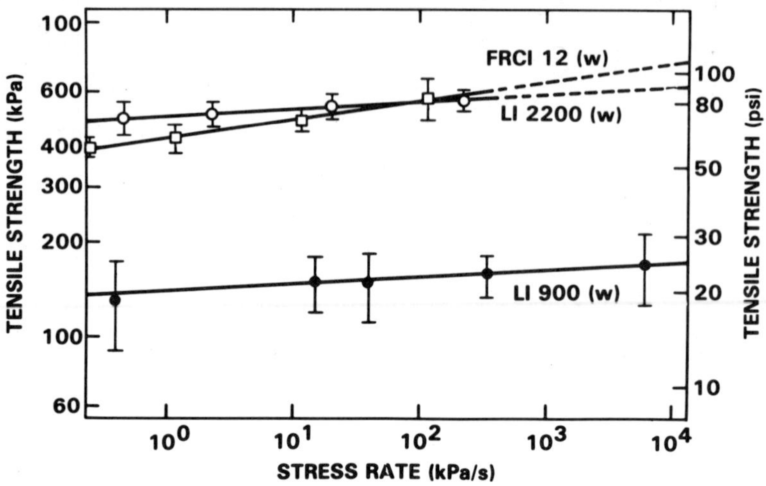

Fig. 13. Tensile strength/stress rate data for a variety of fibrous materials used for the TPS of the space shuttles.

Included in Fig. 14 is the measured after proof test strength distribution. The agreement showed that fracture mechanics was an excellent approach and that, along with other tests on the various tile systems, proof testing should significantly reduce the failure probability at the maximum proof test stress.

Fracture Toughness and Tensile Strength. As pointed out earlier, the mechanical behavior of the fibrous silica tiles were analyzed using a model that was similar to that of Fig. 2a. By combining equations similar to 2 and 6, it was possible to show that K_{IC} for a principal direction should be related to the value of E in that direction and in particular $K_{IC}=AE^{3/4}$, where A is a constant that depends on the fiber radius, strength and Young's modulus. The excellent agreement between this expression and the experimental data is shown in Fig. 15, which summarizes a large amount of data. In reality, there is a large amount of scatter about the various lines but it was found that the dependence of K_{IC} or σ_{ft} on E was the same, regardless of density or testing direction. It is somewhat surprising that σ_{ft} should have a similar dependence on E as K_{IC}, as the strength should also depend on the critical flaw size. For these materials, however, there are large variations in K_{IC} and these may tend to overshadow the flaw size variations. Included in Fig. 15 is some recent data for FRCI and shows the same type of relationship between tensile strength, fracture toughness and Young's modulus as the silica materials. It has been shown that FRCI has tensile strengths ≅3 times the silica materials at a given density [49].

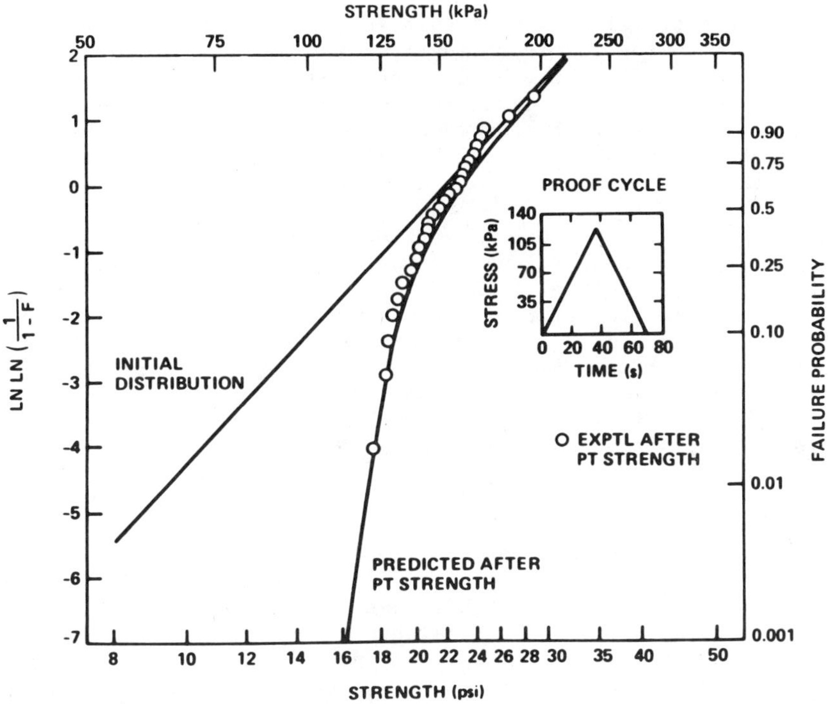

Fig. 14. Comparison of experimental and predicted influence of 122 kPa proof test on tensile strength distribution of LI 900 fibrous material. The initial strength distribution was determined experimentally and fitted to a two-parameter Weibull equation.

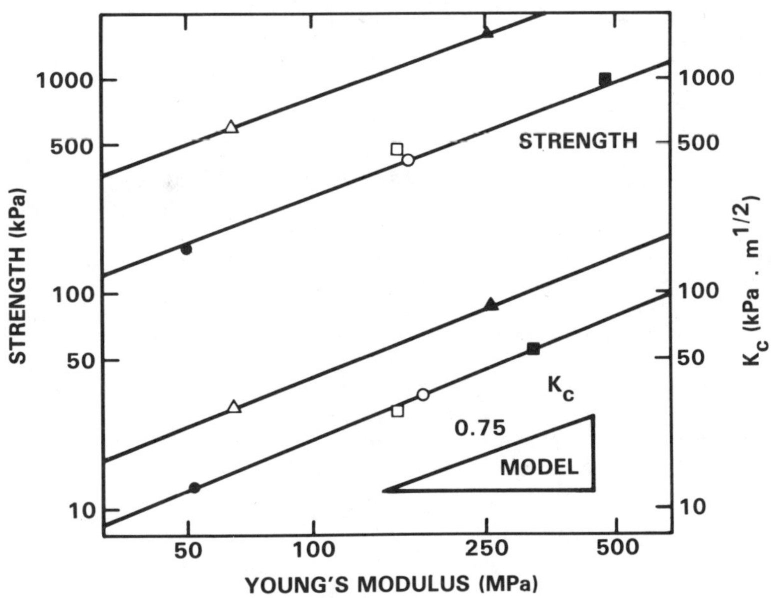

Fig. 15. Experimental correlations between tensile strength or fracture toughness and Young's modulus for a variety of space shuttle TPS fibrous materials and a comparison with the predicted behavior. (LI 900, ● weak direction, o strong; LI 2200, □ weak, ■ strong; FRCI-12, △ weak, ▲ strong).

As pointed out earlier, the theoretical equations are useful for pointing out directions for improved fabrication. Indeed, the improvements found in the FRCI material can be rationalized in these terms. For example, the alumina borosilicate fibers are stronger and have a larger diameter. Moreover, there is found to be improved bonding between fibers [49], which will increase E. Thus, it is expected that the addition of such fibers should increase both K_{IC} and σ_{ft}.

Nondestructive Evaluation. The relationship between K_{IC} and E was found to be useful in understanding a nondestructive evaluation technique that is used to evaluate the silica tiles. This technique was originally based on an empirical correlation between σ_{ft} and the sonic velocity in a given direction [30]. This can be understood as follows. The sonic velocity is related to E and σ_{ft} is related to K_{IC}, so as discussed above, if K_{IC} and E are related then so must σ_{ft} and the sonic velocity. Thus, by non-destructively measuring the sonic velocity, it is possible to identify specimens with low K_{IC} and hence, probably, with low strength [44].

Summary

The K_{IC} and σ_{ft} (flexure) for the 3 systems discussed in this section are shown in Figs. 16 and 17. The cellular materials appear to slightly better but developments, such as FRCI, has made the difference less significant. Overall, the theoretical analysis has indicated that lightweight ceramics have potential for significant improvements over the data discussed here. Another direction, which has not been discussed here, is the use of sandwich structures. For example, lightweight ceramics can be bonded to dense ceramics, giving a composite structure that is strong and stiff with improved resistance to contact damage. The mechanical behavior of sandwich panels is not completely understood and can involve a number of failure mechanisms [1], but it is clear that an understanding of the mechanical behavior of the lightweight core will be an essential part of such an understanding. A sandwich-type structure has been suggested as a solar mirror support and its impact behavior has been studied by Varner et al. [50].

CONCLUSIONS

It is hoped that this review has shown that lightweight ceramics are an important class of materials and an understanding of their mechanical behavior is an essential and exciting new field. In particular, it has been suggested that such materials have an untapped potential in terms of improved properties and possible applications. Although work in recent years has indicated that a micromechanical approach is extremely powerful, many of the equations and concepts need to be verified. In order to capitalize on the potential of lightweight ceramics, several challenges need to be met. These include challenges to the processing ceramist to improve fabrication techniques, to the physical ceramist to understand the relationship between microstructure and mechanical behavior and to the engineer to design with these unusual materials.

ACKNOWLEDGEMENTS

The author would like to acknowledge the large number of people involved in this work over the last 5 years, especially F. F. Lange, J. E. Ritter, Jr., R. G. Hoagland and M. G. Metcalf. The manuscript was prepared with the financial support of a Research Initiation Grant from The Pennsylvania State University.

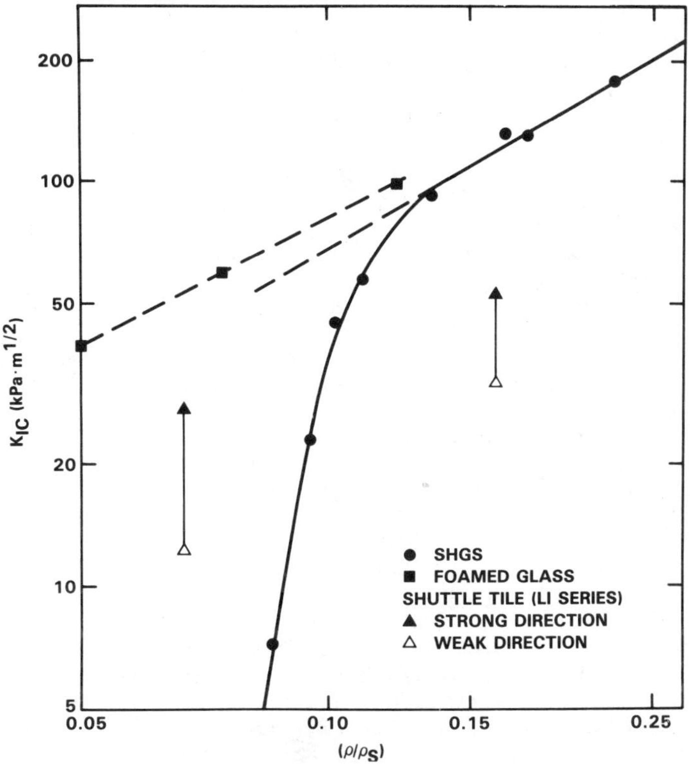

Fig. 16. Variation of K_{IC} with fractional density for various lightweight ceramics.

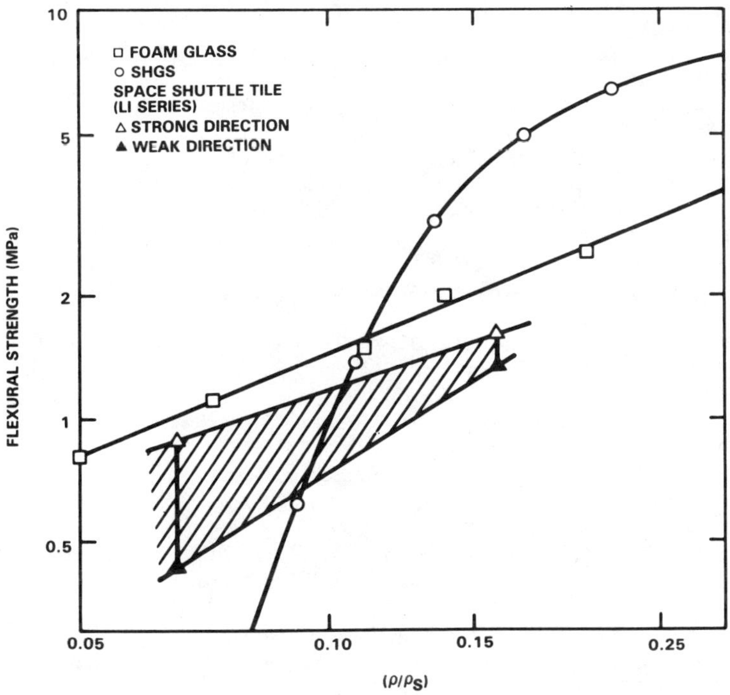

Fig. 17. Comparison of the variation of flexural strength with density for a variety of lightweight ceramics.

REFERENCES

1. M. F. Ashby, Metall. Trans. 14A:1755 (1983).
2. P. A. Cooper and P. F. Holloway, Astronaut, Aeronaut, 19:24 (1980).
3. L. J. Korb and H. M. Clancy, in "Proc. 26th National SAMPE Symposium," SAMPE, Azusa, CA (1981).
4. D. J. Green, in "Industrial Materials Science and Engineering," L. E. Murr, ed., Marcel Dekker, New York (1984).
5. Lockheed Missile and Space Co., Sunnyvale, CA, News Release (1983).
6. D. J. Green, J. E. Ritter Jr., and F. F. Lange, J. Am. Ceram. Soc., 65:141 (1982).
7. D. J. Green, C. A. Hawkins and M. M. Hirlinger, J. Am. Ceram. Soc., 66:46 (1983).
8. A. Briggs, in: "Energy and Ceramics, Proc. 4th Intl. Meeting on Modern Ceramic Technologies," P. Vincenzini, ed., Elsevier Publ. Co., New York (1980).
9. D. J. Green, J. Am. Ceram. Soc., 68: to be published (1985).
10. H. Verweij, G. deWith and D. Veeneman, J. Mater. Sci., 20:1069 (1985).
11. A. N. Gent and A. G. Thomas, J. Appl. Polym. Sci., 1:107 (1959).
12. A. N. Gent and A. G. Thomas, Rubber Chem. Technol., 36:597 (1963).
13. J. M. Lederman, J. Appl. Polym. Sci., 15:693 (1971).
14. V. A. Matonis, SPE J, 1024 (1964).
15. W. L. Ko, J. Cell. Plast., 1:45 (1965).
16. R. Chan and M. Nakamura, J. Cell. Plast., 5:112 (1969).
17. M. R. Patel and I. Finnie, J. Mater., 5:909 (1970).
18. J. G. Zwissler and M. A. Adams, in: "Fracture Mechanics of Ceramics, Vol. 6," R. C. Bradt et al., eds., Plenum Press, New York (1983).
19. F. J. Almgren, Jr., and J. E. Taylor, Sci. Am., 235:82 (1976).
20. W. Thompson, "Collected Mathematical and Physical Papers," Vol. V, Cambridge Univ. Press, England (1911).
21. L. J. Gibson and M. F. Ashby, Proc. R. Soc. London, Ser. A, 382:43 (1982).
22. L. J. Gibson, Ph.D. Thesis, Engineering Department, Cambridge University (1981).
23. L. J. Gibson, M. F. Ashby, G. S. Schajer and C. I. Robertson, Proc. R. Soc. London, Ser. A, 382:25 (1982).
24. F. K. Abd. El Sayed, R. Jones and I. W. Burgess, Composites, 10:209 (1979).
25. G. Menges and F. Knipschild, Polym. Engng. Sci., 15:623 (1975).
26. S. K. Maiti, M. F. Ashby and L. J. Gibson, Scripta Metall., 18:213 (1984).
27. S. K. Maiti, L. J. Gibson and M. F. Ashby, Acta Metall., 32:1963 (1984).
28. D. J. Green, Final Rept., AFOSR Contract No. F49620-83-C0078, June 1984.
29. D. J. Green and F. F. Lange, J. Am. Ceram. Soc., 65:138 (1982).
30. G. A. Alers and R. M. Zimmerman, Proc. of Ultrasonic Symposium, Vol. 1, IEEE, Piscataway, NJ (1980).
31. G. C. Wei and J. M. Robbins, Am. Ceram. Soc. Bull., 64:691 (1985).
32. G. A. Alers, Personal Communication (1981).
33. R. W. Rice, J. Am. Ceram. Soc., 59:536 (1976).
34. Pittsburgh-Corning Inc., Foamglas Data Sheets., 1984.
35. D. J. Green, Unpublished Data.
36. J. Christoffersen, J. Appl. Mech., 47:71 (1980).
37. J. S. Morgan, J. L. Wood and R. C. Bradt, Mater. Sci. Eng., 47:37 (1981).
38. P. H. Conley, H. C. Chandan and R. C. Bradt, in: "Fracture Mechanics of Ceramics, Vol. 4," R. C. Bradt et al., eds., Plenum Press, New York (1978).
39. D. J. Green and R. G. Hoagland, J. Am. Ceram. Soc., 68: to be published (1985)

40. M. D. Meiser and R. E. Tressler, Am. Ceram. Soc. Bull., 60:901 (1981).
41. M. D. Meiser and R. E. Tressler, Cement and Concrete Research, 12:279 (1982).
42. R. R. Hengst and R. E. Tressler, Cement and Concrete Research, 13:127 (1983).
43. D. J. Green, Time-Dependent Failure in Low Density Fibrous Ceramics, Can. Metall.Q., 23:33 (1984).
44. D. J. Green, J. Am. Ceram. Soc., 66:288 (1983).
45. J. E. Ritter, Jr., in: "Fracture Mechanics of Ceramics, Vol. 4," R. C. Bradt et al., eds., Plenum Press, New York (1978).
46. S. M. Wiederhorn and J. E. Ritter, Jr., in: "Fracture Mechanics Applied to Brittle Materials, ASTM Spec. Tech. Publ., No. 678," S. W. Freiman, ed., American Society for Testing and Materials, Philadelphia, PA (1979).
47. J. E. Ritter Jr., S. M. Weiderhorn, N. J., Tighe and E. R. Fuller, Jr., NBSIR 80-2047 (1980).
48. D. J. Green, Unpublished Data.
49. D. B. Leiser, M. Smith and H. E. Goldstein, Am. Ceram. Soc. Bull., 60:1201 (1981).
50. J. R. Varner, M. Akbar Ali and M. A. Adams, in: "Fracture Mechanics of Ceramics, Vol. 5," R. C. Bradt et al., eds., Plenum Press, New York (1983).

PRECIPITATION AND TOUGHNESS IN ALUMINA-RICH SPINEL SINGLE CRYSTALS

D. Lewis, B.A. Bender, R.W. Rice*, J. Homeny+, and T. Garino#

Naval Research Laboratory
Washington, D.C.

ABSTRACT

Results are presented for the effects of heat treatment of Al_2O_3-rich $MgAl_2O_4$ crystals 850, 1000, and 1150C. Results are given as a function of aging time for hardness, fracture toughness and flexural strength. In contrast to Bansal and Hener's previous study [5], the present work shows no evidence of toughening by precipitation in the alumina-rich spinel single crystals heat treated at 850°C. Some effects noted are attributed to preferential precipitation at surfaces and cracks.

INTRODUCTION

Theories for crack-precipitate interaction toughening (CPI) in ceramics suggest that substantial increases in toughness can be produced by dispersions of second phase particles or precipitates with appropriate volume fraction, shape and properties [1]. Studies and use of polycrystalline ceramics toughened by CPI are complicated by the presence of grain boundaries. Fracture through and along grain boundaries limits the degree of toughening achieved, and glassy grain boundaries limit high temperature structural use of CPI-toughened ceramics, regardless of the intra-grain properties. Grain boundary failure also complicates theoretical interpretation of experimental results, as well as greatly complicating theoretical modeling in such 2 materials. Conversely, single crystal ceramics, toughened by a precipitate dispersion, are appealing on several counts. Fracture in such materials can be readily modeled, even to the point of incorporating all of the pertinent material and structural parameters (e.g., precipitate shape, size, volume fraction and anisotropy, matrix anisotropy, fracture plane, etc.) [2] Experimental measurements, both at ambient and elevated temperatures, can be made in a controlled and unambiguous fashion, and compared with theory in a meaningful sense. The single crystal materials offer great potential as high temperature structural materials, through the absence of weak or glassy grain boundaries. The same lack of grain boundaries also suggest use for

---

*Now with W. R. Grace and Company, Columbia, MD.
+University of Illinois, Urbana, IL
#Massachusetts Institute of Technology, Cambridge, MA

precipitate-toughened single crystals as high strength, high toughness optical materials, where the precipitate size can be kept sufficiently small that optical scattering is not a problem.

The present study is part of an ONR-sponsored program investigating the use of second phase toughening in both polycrystalline and single crystal ceramics. The present study is an outgrowth of earlier work on single crystal partially stabilized zirconia (PSZ) [3,4] and studies on TiO_2-doped sapphire. The work on single crystal PSZ showed some of the potential for precipitation toughening in single crystal materials. The work on the Al_2O_3-TiO_2 system suggested an intrinsic limitation with that material (essentially variations of star sapphire) in respect to limited volume fraction of precipitates and limited matrix/precipitate mismatches in thermal and elastic properties. The normal precipitates in star sapphire are TiO_2 (rutile), are coherent with the alumina matrix, produce little or no residual stress field, and can be obtained in only limited amount, <1-2 v/o. Thus, no CPI toughening would be expected, and none appears to occur. The present study, on alumina-rich spinel, is on a system in which much greater quantities of precipitates can be obtained, and for which previous studies have shown evidence of CPI toughening [5].

EXPERIMENTAL PROCEDURE

Single crystals, prepared by a commercial manufacturer* using the Verneuil method, were obtained in the form of boules approx. 2.5 cm in diameter and approx. 5 cm long, of the nominal composition, $MgO{:}3.5Al_2O_3$. Test specimens were sawn and ground from these boules using conventional diamond tooling. Test specimens were aged at temperatures of 850, 1000 and 115°C for times of 0-1 hr for the 1150 C tests, 0-40 hr for the 1000C tests, and 0-150 hr for the tests at 850C. The limits on the aging times were determined during the course of the tests by determining where significant and persistent degradation in mechanical properties occurred. All heat treatments were done in an air environment, in furnaces using molybdenum disilicide heaters, with the specimens placed on coarse alumina setter powder in high purity alumina crucibles. Mechanical properties evaluated, all at 20 -25C and ca. 40% R.H., included flexural strength, fracture toughness by two methods, and Vicker's hardness. The hardness test, performed at a load of 2 kg. was also used to characterize the fracture toughness, by the indent crack length technique [6]. The specimens for indentation hardness and toughness were indented perpendicular to the [100] planes, with the indent diagonals parallel to the <100> directions. Flexural strength was measured on 2 x 3 mm cross-section bars, tested in three-point flexure on a 12 mm span; the bars were surface ground in the lengthwise direction using a 600 grit diamond wheel, with edges radiused on a 15 um diamond lap. The same bars, rotated by 90°, with the 3 mm dimension as the height, were used for fracture toughness measurements via the single edge notched beam (SENB) technique, utilizing a 150 um wide notch, approx. 1-1.5 mm in depth. For both the fracture strength and single edge notched beam fracture toughness tests, the samples were orientated with the [100] planes parallel to the faces of the test bars. In this manner, the strength and toughness measurements were performed on the [100] planes and in the <100> directions. Toughness measurements were made on specimens notched before and after heat treatment to evaluate the possible effects of preferential nucleation of precipitates from machining-induced surface damage.

---

*Hrand Djevahirdjiian Ltd.,

Typically, 5 specimens each were used for flexural strength and notched beam tests, for each set of aging time and temperature, with indent measurements using ten indents for each set of conditions. Transmission electron microscopy (TEM) was used to characterize the precipitate structure in a few of the aged spinel crystals, with selected area electron diffraction (SAD) employed to identify the phases present. Limited scanning electron microscopy was used to examine fracture surfaces for evidence of crack-precipitate interaction, which might be associated with increases in fracture toughness.

RESULTS AND DISCUSSION

The hardness results are presented in Fig. 1, which compares the results of the present study for the three aging temperatures with those of Bansal and Heuer [5] for their single aging temperature of 850C. Note that the results of Bansal and Heuer are for Knoop hardness while those of the present study are Vicker's hardness. While the results of the previous study showed two apparent maxima in hardness, which were correlated with particular events supposedly occurring during the precipitation process, the present study shows only a smooth increase in hardness with aging time, for all aging temperatures. The increase becomes progressively more rapid with increasing aging temperature. This type of hardness increase would be expected in a system where the density and/or size of a second phase dispersion increased smoothly with aging time, leading to an increase in the yield stress (hence hardness) through the dispersion strengthening mechanism. A recent study of precipitation in these crystals [7] provides evidence that the precipitate population does change in this gradual fashion, rather than in the two stage process suggested in the earlier work [5].

The results for fracture toughness, as shown in Fig. 2 and 3, are rather complex. Note that all of the results shown are from the present study; the previous study referenced [5] did not measure fracture toughness. The results for indentation fracture toughness, which samples a surface region, approx. 400-500 um deep, are essentially constant for the 850C heat treatment temperature, out to a time of approx. 2000 min., at which point the toughness decreases rapidly with increasing heat treatment. The indent results for higher heat treatment temperatures show more drastic reductions in toughness, occurring at shorter times. In the case of the 1000C heat treatment, the decline in measured toughness begins at times of approx. 20 min.; for the 1150 soak, the decline begins immediately; in both cases the toughness decreases from an initial value of approx. 1.9 MPa-$m^{1/2}$ to a final value (where aging tests were terminated) of <1.1 MPa-$m^{1/2}$. The time and temperature for the 1000C aging correspond to one of the boundaries in the time-temperature-transsformation diagram of reference [7], where one of the transition phases begins to appear, Type I_T. The t-T-T diagram does not extend to cover the point, 1150C and 1 min, where the toughness declines for the heat treatment temperature. The point at which the toughness decreases for the 850C heat treatments is very close to the boundary on the t-T-T diagram marking the formation of alpha-alumina precipitates.

The results for the SENB toughness are somewhat different, with the toughness for pre-notched specimens appearing to increase with heat treatment at 850C up to a time of approx. 2000 min, at which point the toughness decreases. This toughness increase may well be only an apparent one since th toughness for specimens notched after heat treatment does not show the same increase in toughness with heat treatment. Rather these

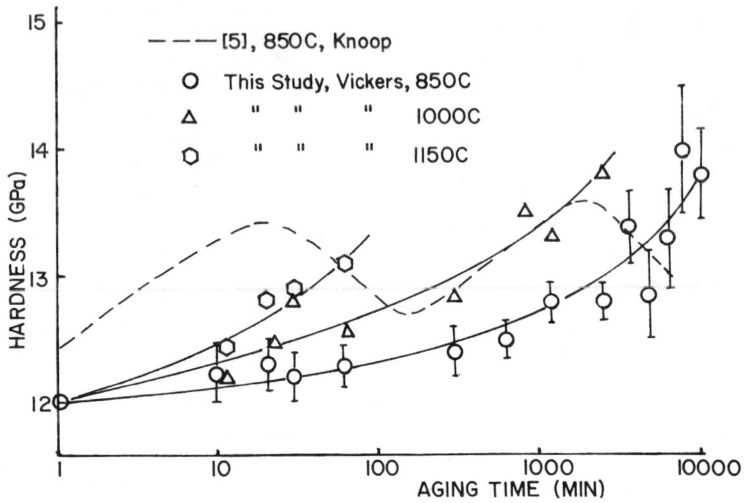

Figure 1. Indentation hardness vs. heat treatment time and temperature-note gradual hardness increase with heat treatment and precipitation

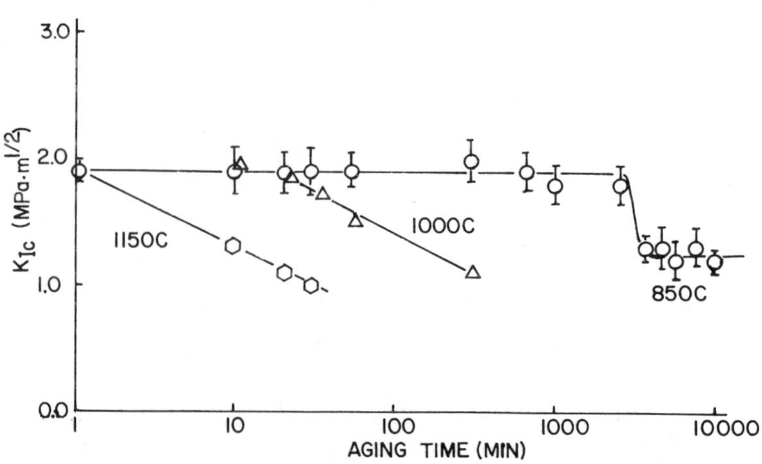

Figure 2. Indentation fracture toughness vs. heat treatment time-note monotonic decrease in fracture toughness with precipitation

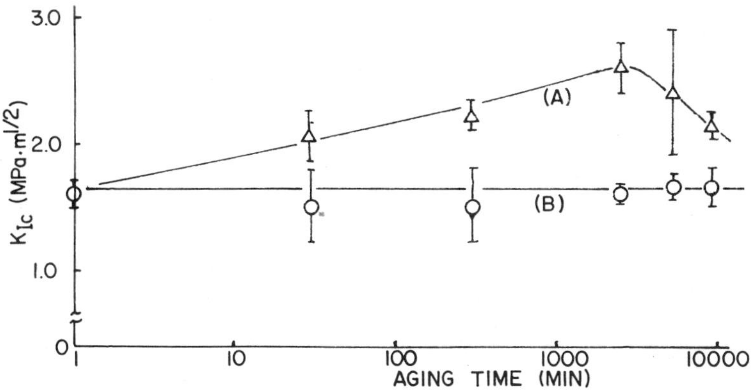

Figure 3. SENB fracture toughness vs. heat treatment time at 850C, (A) specimens notced prior to heat treatment, (B) specimens notched after heat treatment

latter results show an essentially constant value of approx. 1.6 MPa-m$^{1/2}$ out to the longest heat treatment time used, approx. 10,000 min. These latter results appear to show that the toughness of the spinel, at a depth of approx. 1 mm from the surface, is unaffected by the heat treatment, while the results for the pre-notched specimens suggest that either crack blunting or healing is occurring during the heat treatment, increasing the stress intensity required for fracture, or a surface damage-induced precipitation process is increasing the toughness of the near surface region, with the same effect. One previous study [8] has shown this type of effect, with much greater precipitation from a machined surface than from a polished surface of the same material. However blunting, due to precipitate formation along surfaces and at tips of pre-existing cracks is also possible because of surface energy conditions favoring formation of precipitates on a free surface. Past studies have in fact shown preference for precipitation on cracks in other materials.[8] The indent results may be explained by a precipitation process which has no effec in the earlier stages but induces a tensile stress in the surface in the late, stages (with the formation of corundum), or one which induces a compressive stress in the earlier stages of precipitation (explaining the higher toughnes for the indent tests relative to the SENB tests), which disappears with the conversion of the precursor monoclinic precipitates [6] to corundum.

The flexural strength, which is controlled by the fracture toughness of the near surface region of the specimens and the strength-controlling flaws, which are typically due to the machining process, follows quite different behavior from the hardness. Note that strength measurements were only made for the aging temperature of 850C, because of material limitations, and the availability of the earlier results of Bansal and Heuer for this temperature. In this case, as shown in Fig. 4, the strength remains essentially constant, at approx. 180 MPa, to an aging

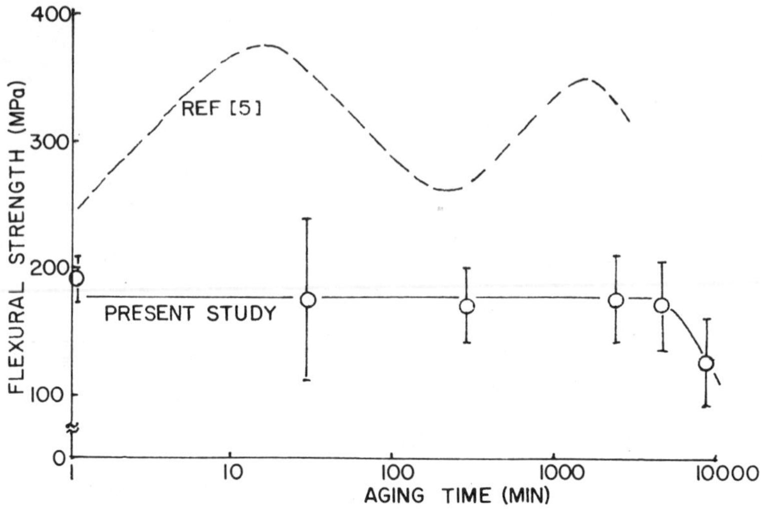

Figure 4. Flexural strength vs. heat treatment time at 850C-note no strength increase with aging-induced precipitation

time of approx. 2000 min, at which point the strength begins to decrease rapidly with further aging. This type of strength behavior could be explained in several ways. One possibility is a relative constancy in the fracture toughness with aging, with the formation of large precipitates, at sufficiently long aging times, which act as strength limiting flaws, thus decreasing the strength. It is also possible, in the intermediate times, that competing effects are acting to keep the strength relatively constant, e.g., crack healing is reducing the severity of the preexistent flaws, while microstructural changes are decreasing the toughness of the material. The other alternative, increasing fracture toughness with aging, associated with increased flaw severity with aging, appears quite unlikely; when the aging process is producing precipitates with dimensions of microns or less which are not likely to act as strength con-trolling flaws. The present results also differ dramatically from those of the previous study cited [5], in which maxima in strength were noted at approx. 20 and 2000 min.

SUMMARY AND CONCLUSIONS

To summarize the preceding results, it appears that the bulk of the precipitation process has little or no effect on the toughness of the spinel single crystals. TEM of the precipitates formed suggests little or no mismatch of the precipitates with the matrix from the absence of strain contours. The absence of significant residual stress fields at the precursor precipitates in the spinel and their coherency would preclude any significant CPI toughening. Calculations of lattice and thermal expansion mismatches between corundum and spinel agree in predicting only small residual stress fields around corundum precipitates in a spinel matrix. The decline in toughness, as measured by the indent method, the constancy of the toughness, as measured by the SENB technique with post-notched specimens, and the initial increase in toughness and subsequent decrease, for pre-notched SENB specimens, all taken together suggest a detrimental effect of the formation of alpha-corundum in the latter stages of precipitation in the spinel. This detrimental effect may result from small tensile stresses arising from the small mismatches between corundum and spinel.

REFERENCES

1. K.T. Faber and A.G. Evans, Acta Metall 31:565 (1983).

2. R.P. Ingel, Ph.D. Thesis, Catholic Univ. (1982).

3. R.P. Ingel, et. al., J. Am. Ceram. Soc. 65: C108 (1982).

4. R.P. Ingel et.al., J. Am. Ceram. Soc. 65: C150 (1982).

5. G. K. Bansal and A. H. Heuer, Phil. Mag. 29:709 (1974).

6. A.G. Evans and E.A. Charles, J. Am. Ceram. Soc. 59: 371 (1976).

7. C.M. Jantzen and H. Herman, "Two Stage Pre-precipitation in Alumina Rich Spinel," accepted for publication in J. Am. Ceram. Soc. (1985).

8. G.R. Pulliam, J. Am. Ceram. Soc. 42: 477 (1959).

MICROFRACTURE BEHAVIOUR OF FINE GRAINED ALUMINA STUDIED BY INDENTATION AND ACOUSTIC EMISSION IN VARIOUS ENVIRONMENTS

T. Sperisen, C. Carry and A. Mocellin

Ecole Polytechnique Fédérale de Lausanne
Laboratoire de céramique, 34 chemin de Bellerive
CH-1007 Lausanne, Switzerland

ABSTRACT

Fracture micromechanisms in fully dense fine grained aluminas have been investigated by carefully controlled indentation and concurrent recording of acoustic emissions. Samples in the form of hot-pressed and polished Al_2O_3 discs, 40 mm diamater by 4 mm thickness, and of varying grain sizes were submitted to a diamond Vickers indenter in a specially designed testing rig allowing for controlled environments. Experimental testing parameters, acoustic activities and microstructural observations were correlated and are presented, allowing different types of mechanical behaviours to be documented, as functions of microstructure and environment. Attempts at rationalizing the results in the light of our present understanding of indentation fracture are discussed.

I. INTRODUCTION

In the past ten years or so, indentation techniques based on classical hardness tests have often been used to study fracture phenomena in ceramics. A number of investigations on glassy and transparent materials have led to a good understanding of the conditions for the formation and development of the various possible crack patterns in continuous, homogeneous, isotropic solids in the absence of microstructural effects (1-6). Thus, on the basis of observations made on glasses models have been developed (1-5) for elasto-plastic interactions between samples and indentors. Semi-empirical relationships involving elastic modulus E, hardness H, maximum aplied load P, crack length C and toughness K_{IC} have been proposed. But it has not been possible yet to rationalize material dependent proportionality constants which also appear in such relationships.

Only recently (7-9) have environmental effects on indentation cracking in glasses begun to be investigated. In polycrystals which are most usually non transparent indentation techniques have been used to introduce controlled-size defects in samples to be subsequently tested in bending (10-12) and/or for direct K_{IC} determination (13,14). In such cases only the length of the crack trace on the exposed surface has been of concern. Some doubt still exists as to its exact geometry in the indenter diagonal plane below the surface, e.g. a single semi-circular shape or two smaller Palmqvist type cracks. The nature of the material and detail of its microstructure are likely to be of major importance in this context. Palmqvist cracking has been claimed to be dominant in cemented carbides (15-17). Also the indentation fracture test itself has been argued not to be valid at all grain sizes (18).

The present work was undertaken in order to generate experimental information on alumina, a model polycrystalline ceramic, under carefully controlled operating conditions and rather unusual compared with common practice. Parameters of particular interest in the experiments have included: programmed loading and unloading schedules, well defined chemical environments (water and silicone oil respectively) and microstructures, and continuous recording of acoustic emissions generated during both loading and unloading.

II. MATERIALS AND PROCEDURES

Samples used in this study were fine grained dense hot-pressed aluminas prepared as previously described (19) from a powder of the same origin and quality except for a residual potassium content of 40 ppm compared to ~140 ppm in the earlier work (19). Indentation experiments were performed on two hot pressed materials: 0.25 hour at 1450°C and 0.5 hour at 1600°C. Average linear grain sizes $\bar{d}$ were 1.1 μm and 6.2 μm respectively, determined as $\bar{d} = 1.38\sqrt{\bar{s}}$, where $\bar{s}$ is the average grain section area obtained by counting on the plane of polish.

Such polycrystals in the form of discs, 40 mm diameter by 7 mm initial thickness were diamond-ground to ~4 mm thick. The final surface finish was obtained with 0.25 μm diamond paste on the face to be indented, and 1 μm diamond paste on the opposite face where the acoustic emission transducer was to be placed. Indentations were made in zones close to the disc centers only in order to minimize spurious effects due to possible structural differences between center and rim.

Figure 1 shows a schematic of the testing rig, specially designed for the present purpose. Features of interest include:

i) constant loading of the transducer onto the bottom face of the sample by means of a spring.
ii) controlled testing environment (either deionized water or diffusion pump oil) by immersing the diamond indentor in a "pool" of the liquid covering the upper face of the sample. The liquid medium systematically was brought only after the complete sample containing rig had been dried in an oven at 60 °C for 30 minutes.

Loading was effected by a standard (136° angle) diamond Vickers indenter attached on the load cell of an INSTRON testing machine (with electromechanical actuator, type 1361). The sample containing setup was tightened on an x-y translating table on the top face of the moving piston of the machine. A typical experiment consists of controlling piston displacement at a constant rate of 4.2 ± 0.05 μm per minute both during loading and unloading. A varying loading rate then results on the indenter as was determined experimentally (Figure 2). The loading rate thus increases with applied load initially (minimum detectable load: 10^{-2} N) and becomes roughly independent of it beyond ~30 N. Unloading is carried out similarly with the profile on Figure 2 followed in the reverse sense.

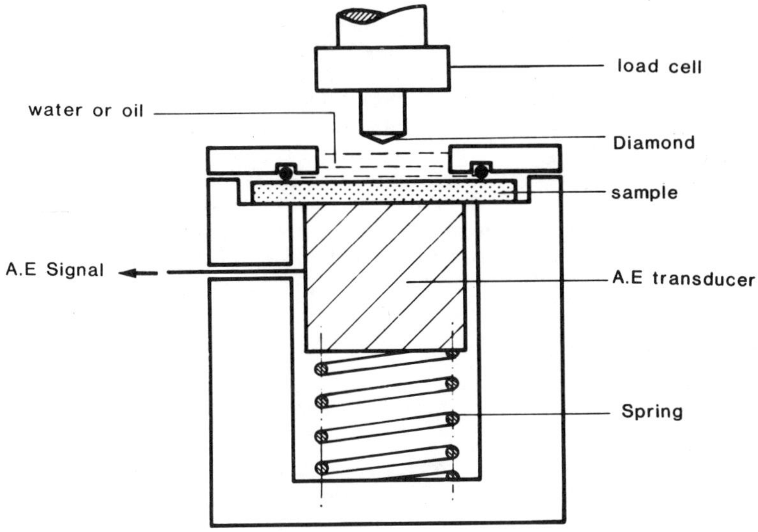

Fig. 1. Experimental device with alumina sample, Vickers indenter, acoustic transducer and environment "pool".

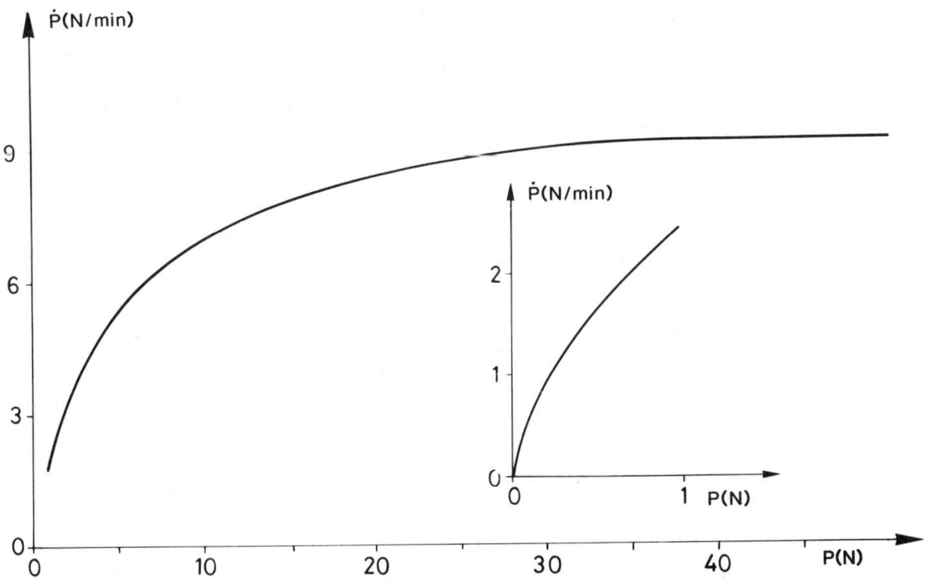

Fig. 2. Loading rate versus load for a piston displacement rate of 4 μm/min.

Acoustic emissions during testing were continuously collected by a Dunegan type D9203A transducer with a resonating frequency of ~150 kHz. The signals delivered by the transducer were processed with a classical Dunegan-Endevco (model 3000) equipment. Signal characteristics of special interest here will include: cumulative ring down counts, numbers of events and event distribution curves versus either count numbers or amplitudes (modules 921 and 920 of the processing equipment respectively). For every set of experimental conditions, i.e. sample, grain size, environment type and maximum applied load, at least eleven different indentations were made and observed in a scanning electron microscope within ten hours after unloading. The quantitative results to be reported therefore will be averages (and standard deviations) of each 11+ set of tests.

III. RESULTS AND DISCUSSION

1. Microscope observations

Experiments were conducted to varying maximum loads up to 50 N maintained for one minute before unloading. Typical aspects of the indentation sites are shown on Figures 3 to 5 on which the following peculiarities may be noted.

In the fine grain size material:

i) At small loads the testing environment plays a significant role on the morphology of damage, and no well-defined radial crack is to be seen

ii) For higher loads the classical radial crack pattern is formed along the diagonals of the prints. Supplementary cracks seemingly originating from the corners and running more or less radially have developed occasionally in the H_2O environment but were never observed in oil.

iii) Surface traces of lateral cracks presumably formed during the unloading stage are found at the highest (50 N) applied load in water.

In the coarse grain size material:

i) Substantial damage mostly in the form of intergranular microcracking leading to grain pull-out is produced at all levels of loading in H_2O but is far less pronounced in oil. No radial crack was formed under small (1 N) loads.

ii) Again supplementary cracks have developed in the H_2O environment only, but more systematically than before.

iii) Lateral cracking has already occurred in water after loading to 25 N.

On the other hand, the crack lengths c of the principal radial cracks were found 10% larger in the large grain alumina for both environments; the half print diagonal a only exhibits a slight increase with grain size. However, as shown on Figure 6 the ratio of crack length c to a does vary depending upon operating conditions. At fine grain sizes c/a increases from ~2 to ~3.8 between 5 and 50 N whereas it reaches ~3.5 at 15 N and remains constant thereafter in the coarser grained material. Finally, it must be pointed out that the traces of all radial cracks on the outside surfaces are about 90% and 80% intergranular for the 1.1 μm and 6,2 μm grain sizes respectively as could be evaluated after thermal etching to reveal the grain boundary network.

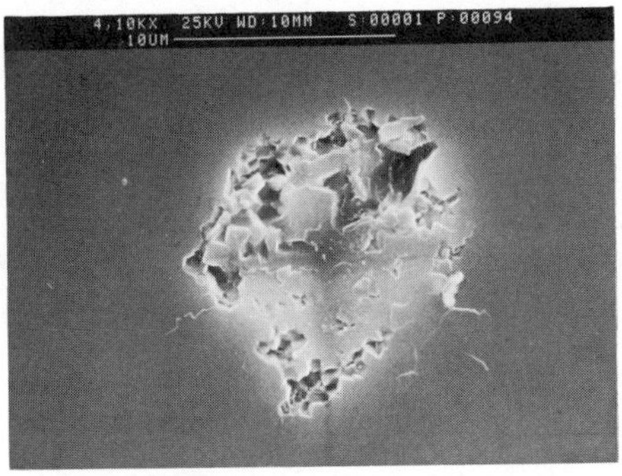

a) d = 1.1 μm oil

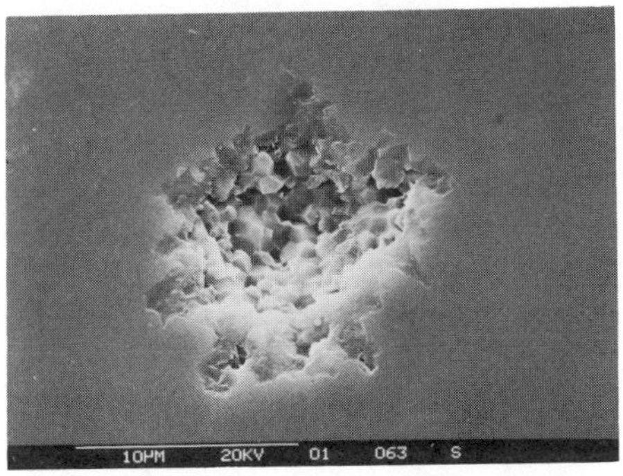

b) d = 1.1 μm water

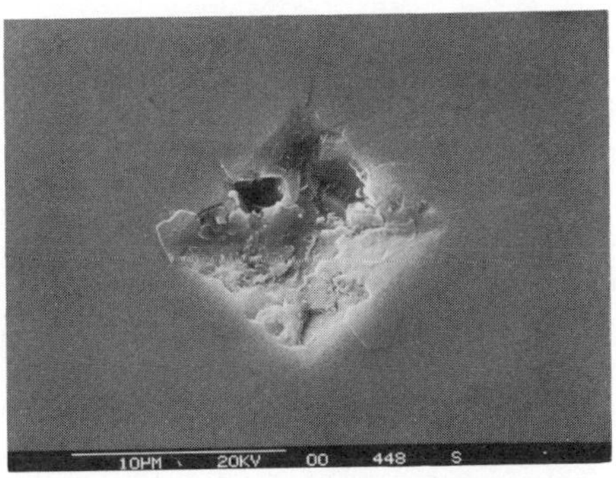

c) d = 6.2 μm oil

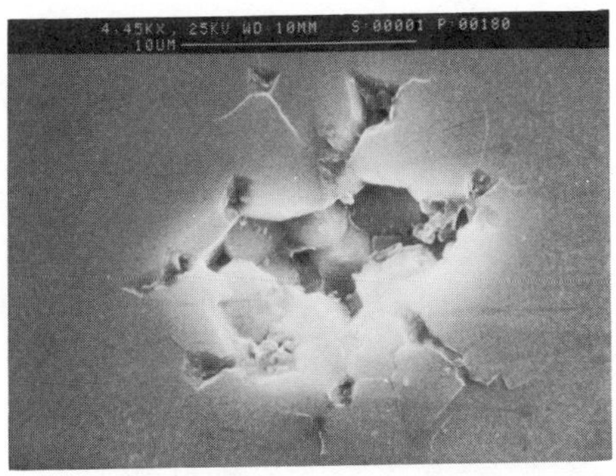

d) d = 6.2 μm water

Fig. 3. Indentation damage as a function of grain size and environment for a 1 N maximum load.

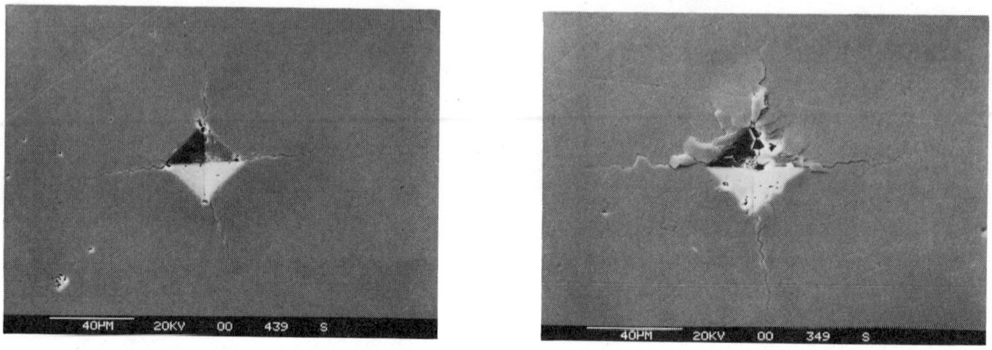

a) oil b) water

Fig. 4. Indentation damage on 6.2 μm grain size Al_2O_3 for a 15 N maximum load, as a function of environment.

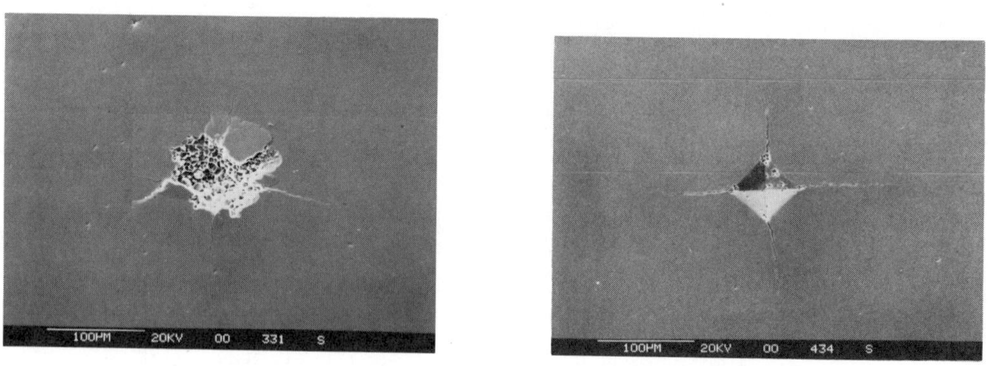

a) oil b) water

Fig. 5. Indentation damage on 6.2 μm grain size Al_2O_3 for a 50 N maximum load, as a function of environment.

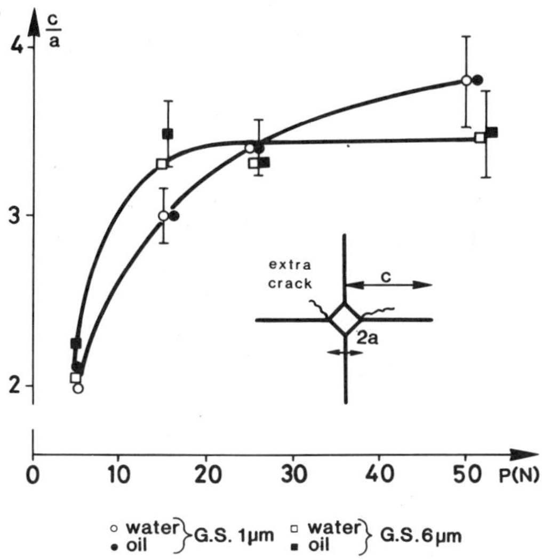

Fig. 6. $^c/a$ ratio versus indentation load and environment at two different grain sizes.

2. Acoustic activity upon loading

Plots of the cumulative numbers of acoustic events versus applied load recorded during the whole loading-hold-unloading cycles are shown on Figures 7 and 8 for the 1.1 μm and 6.2 μm grain sizes respectively.

In the small grain size material (Figure 7), three loading and two unloading stages can be identified in either environments. During stage I the rate of event production with applied load gradually decreases as the latter increases. Then it remains constant (stage II) and increases again to another constant level (stage III) beyond what appears to be a load threshold. It is important also to note that no initial threshold can be identified (given our 10^{-2} N minimum detectable load) below which there would be no acoustic activity. Finally, everything else being equal, testing under water gives roughly twice as many detected acoustic events as under oil.

In the coarse grain size material (Figure 8), the acoustic pattern appears somewhat different. Stage I is barely identifiable and stage II is practically non existent. Here again testing under water gives rise to greater numbers of acoustic events than under oil. But a higher (about one order of magnitude in the event counts) activity is detected here as compared with that in the 1.1 μm Al_2O_3. Other acoustic parameters (Figure 9, Table I) however appear not to be markedly different from those observed at small grain size. And a similar stage III as before can be identified with a characteristic bimodal cumulative amplitude distribution (Figure 9). It is noticeable at lower applied loads: ~7-10 N instead of ~25 N than is the case for the finer grain size alumina.

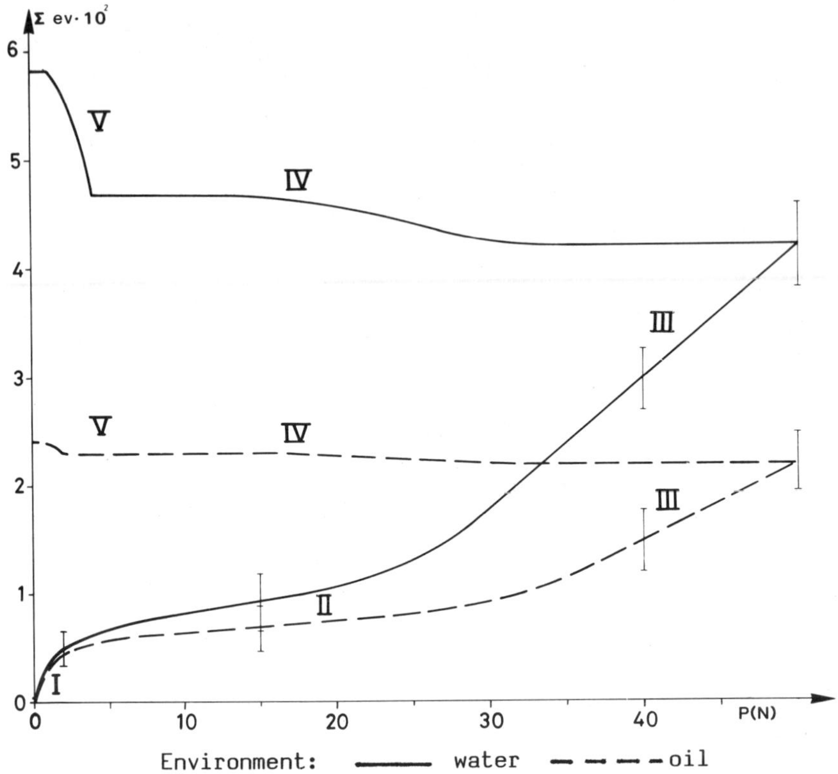

Fig. 7. Accumulated acoustic events during a loading-hold-unloading indentation cycle up to 50 N on 1.1 μm alumina.

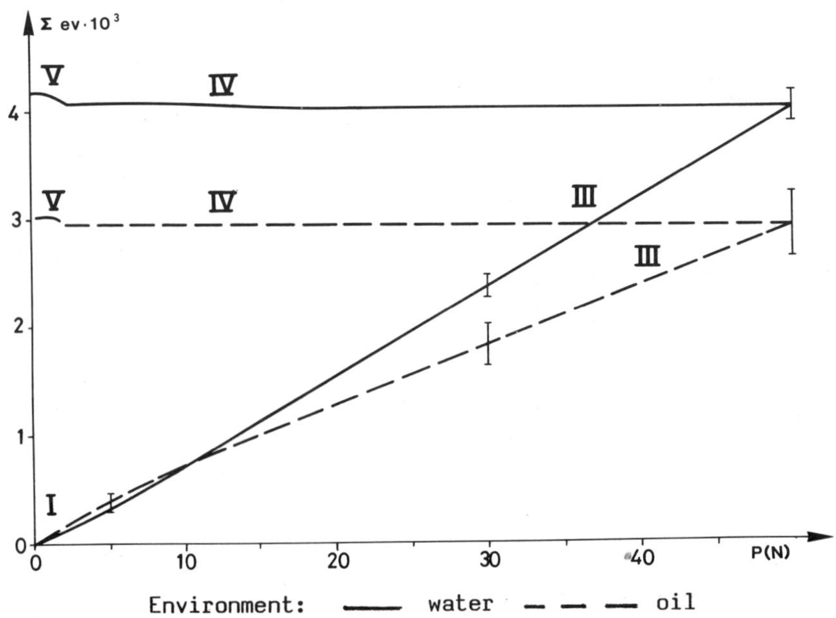

Fig. 8. Accumulated acoustic events during a loading-hold-unloading indentation cycle up to 50 N on 6.2 μm alumina.

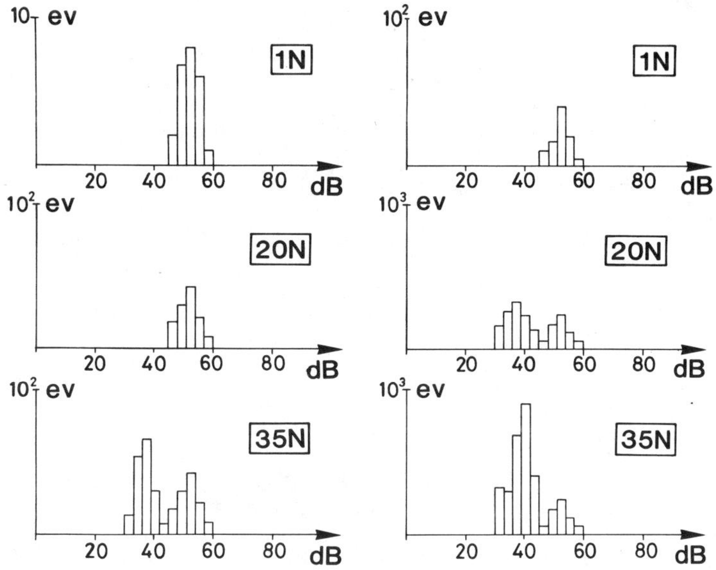

a) 1.1 μm alumina b) 6.2 μm alumina

Fig. 9. Cumulative acoustic amplitude distributions at characteristic loads.

Table I. Average numbers of counts per acoustic event during various stages of acoustic activity (N.D. means not determined).

	d = 1.1 μm		d = 6.2 μm	
	water	oil	water	oil
stage I	9	9	7	3
stage II	4	3	-	-
stage III	17	23	18	N.D.

3. Initial damage generation

Three points seem to be of particular importance when attempting to rationalize the observed surface damage at low applied loads. First of all the shape of the indenter which has been used in the present work was not perfectly pyramidal. SEM observation revealed the top of the diamond piece to be blunt rather than sharp with an approximate spherical tip radius of 2 μm (*). As a consequence the stress initially applied to all samples must have been close to the theoretical hertzian field (20) which is axially symetrical and predicts the following relation for the radius a of the contact circle (**):

$$a^3 = \frac{4}{3}\ \frac{KPr}{E} \qquad [1]$$

with P the applied load

r the sphere radius (2 μm)

$K = \frac{9}{16}\left[(1-\nu^2) + (1-\nu'^2)\right]\frac{E}{E'}$

E, E', ν, ν' the elastic modulus and the Poisson's ratio for the sample and the indenter respectively.

And for the maximum tensile stress component on the sample surface:

$$\sigma_r\ (r=a) = \frac{1-2\nu}{2} \cdot \frac{P}{\pi a^2} \qquad [2]$$

For our minimum detectable load of 10^{-2} N and $E_{Al_2O_3} = 370$ GPa, $\nu_{Al_2O_3} = 0.32$, $E_{diamond} = 723$ GPA, $\nu_{diamond} = 0.25$ it follows: $\sigma_r\ (r=a) \approx 4.8$ GPA. Such a theoretical stress level is several times higher than the macroscopic fracture strength of 700 ± 50 MPa experimentally determined on the 1.1 μm grain size Al_2O_3 in 3-point bending with a 4 cm gauge length. Assuming a fracture surface energy $\gamma = 1\ Jm^{-2}$ (22) the corresponding critical defect length in the Griffith equation is ~0.5 μm. It thus appears that in the fine grain size material and in agreement with experiment, the load threshold is substantially smaller than 10^{-2} N for the first microcrack formation and that the latter is a priori oriented at random with respect to the macroscopic indenter geometry. Given the maximum stress level that appears accessible in this pseudo-hertzian initial loading stage the formation of a whole network of "submicrocracks" is to be anticipated regardless of any other effects due to the testing environment, before the more classical microcrack pattern begins to develop. Observation suggests the same conclusion to be qualitatively valid for our coarser grain size material as well.

---

* On another as-manufactured diamond, we measured a tip radius of about 0.5 μm, which implies a difference of a factor 2 in the stressing conditions. After a few indentations, this radius becomes constant at 2 μm, as it has been observed on several indenters. These differences in tip radius do not have incidence on fracture pattern (Fig. 3) or on the A.E. recording (Fig. 7).

** As recently pointed out by Yoffe (21) this simplified expression is only valid for $a/r \leqslant 2$, i.e. P < 0.1 N in our case. It is nonetheless acceptable here since for P ⩾ 0.1 N the influence of the diamond edges should not be neglected any longer in regard of the tip radius, and a more Boussinesq like stress field (3) then should set in.

Secondly, another effect may be invoked to contribute to the development of the roughly circular damaged zones upon initial loading up to ~1 N (Figure 3). In the present testing procedure the load rates hence stress rates imposed to the samples were rapidly varying in the low load regimes: $\sim 10^{-2}$ to 1 N (Figure 2). Assuming that the local fracture stress σ_f is related to the applied stress rate $\dot{\sigma}$ through the phenomenological equation (23):

$$\sigma_f = c\,(\dot{\sigma})^{1/N+1} \qquad [3]$$

where c is a constant, is equivalent to expecting a greater number of structural defects to develop into microcracks for a given stress when the stressing rate is low than when it becomes large. Such would qualitatively be the case, regardless of whether or not N enters the above expression with its value as determined from macroscopic crack propagation experiments. In other words the apparent local strength of the material is likely to have been lower in the early stages of loading because of a reduced loading rate compared to that at later stages. Under such conditions it is also to be expected that stress corrosion may operate. The observed differences between damage morphologies in water and oil testing environments (Figure 3) are thought to bring confirmation of such effects.

Finally, it must be admitted that the load threshold for cracking upon Vickers indentation of polycrystalline Al_2O_3 is a function of microstructure, notably grain size. We have conducted similar experiments as described here on c axis sapphire single crystals and have recorded no acoustic activity up to ~0.1 N applied load. Such a result is consistent with theoretical predictions (24-25) about median-radial cracking and also with experimental observations on Palmqvist crack initiation in a 25 μm grain size alumina (26), otherwise quite similar to our finer grained materials. Since cracking of any type in the latter has always been observed to be mostly intergranular it follows that the local stress levels that can be sustained by grain boundaries are smaller than those acceptable by single crystals or equivalently by polycrstals with grain size sufficiently larger to appear locally single crystalline in relation with the extent of the applied stress field. According to this argument then a transition should be observed at some intermediate grain size between the mechanical response observed in the present work and a more "single-crystal-like" behaviour (26). In fully dense MgO-doped equiaxed alumina this transitional grain size may be roughly estimated at ~7-10 μm. In other dense single phase ceramic polycrystals it may be different but it is believed that qualitatively similar effects could be observed.

4. Development of indentation cracks

At > 1 N applied loads a set of indentation cracks was observed to develop along the pyramid diagonal planes (Figures 4, 5) corresponding to the acoustic emission regimes II and III (*). Two different sequences of events have been considered in the literature as possibly describing the formation and development of Vickers cracks. They are:

---

* The acoustic behaviours during unloading (regimes IV and V on Fig. 7) have not been studied in detail yet, but are likely to be somehow related to lateral cracking.

a) The initiation of a subsurface penny-like crack developing into a larger 1/2 penny-like crack up to the free surface upon increasing loads (1).

b) the formation of an initial set of Palmqvist cracks subsequently linked into 1/2 penny-like cracks upon increasing loads (27).

Figure 10 schematically shows these two a priori possibilities and the corresponding load levels for the present experimental conditions. But generally speaking it does not seem that a consensus has developed yet as to which if any of either scenarios predominates in ceramics.

Considering our 1.1 μm grain size alumina first, scenario a) on Figure 10 appears not to have taken place because:

i) given the formation of the initial surface damage at the beginning of the loading cycle radial cracking appears more likely to develop from this damaged surface than from below the indenter.
ii) in the load range corresponding to the acoustic stage II (Fig. 7) the measured $^c/a$ ratios (Fig. 6) take up values which usually are considered characteristic of Palmqvist cracks ($^c/a < 3$) (28-29).
iii) when loading is interrupted during stage II and immediately followed by unloading, no acoustic activity is recorded during this unloading (Fig. 11), contrary to what would be expected if a subsurface penny shaped crack were developing into a radial half circular one as per the first scenario.
iv) upon re-loading after unloading during stage II the recorded acoustic activity is resumed with stage II characteristics up to some higher applied load when the acoustic stage III sets in (Fig. 11). Such stage III cannot have another relationship but with the propagation of radial half penny shaped cracks which is the terminal step in both cases.

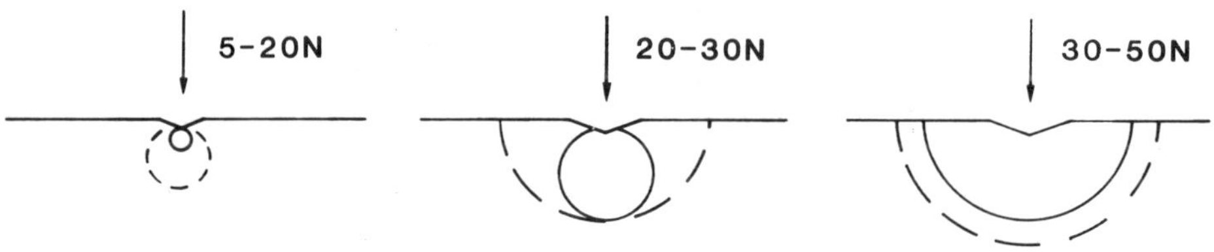

a) subsurface initiation.

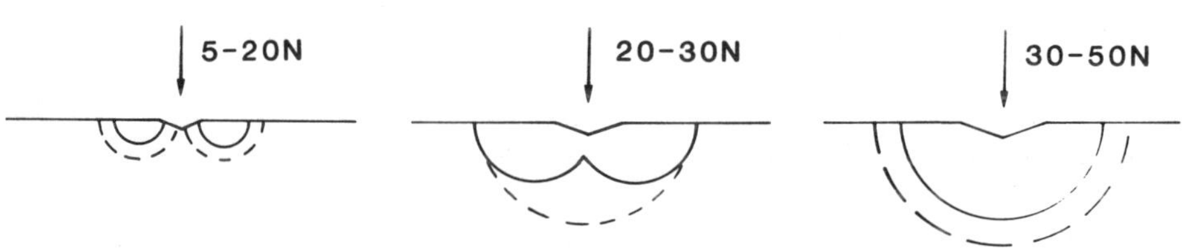

b) initiation via Palmqvist cracks.

Fig. 10. Possible scenarios for Vickers crack formation.

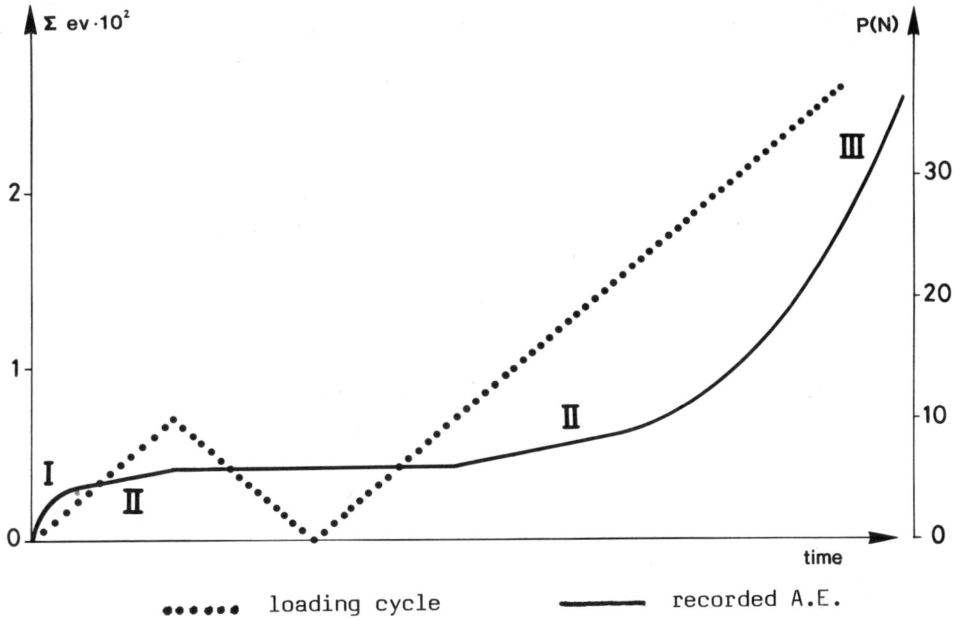

Fig. 11. Acoustic emission during loading, unloading during stage II and reloading experiments on 1.1 μm alumina.

It is proposed therefore that after the initial surface damage formation, Palmqvist type cracks are developed corresponding to the acoustic stage II until they reach a large enough size to link into radial half penny shapes (transition between acoustic stages II and III on Fig. 7).

The sequence of events in the coarser 6.2 μm grain size alumina is not thought to have been fundamentally different from that just outlined. Two acoustic stages again can be identified (Fig. 9, Table I) although stage II now is less well marked (Fig. 8) and is taken over by stage III at smaller applied loads. Similarly the $^c/a$ changes versus P (Fig. 6) are qualitatively similar in both materials. It must be admitted however that the Palmqvist crack propagation step is markedly reduced at coarser grain sizes and may overlap with the initial surface damage formation. Such a difference may be understood when considering the sizes of Palmqvist cracks relative to the scales of microstructure in either case as shown schematically on Figure 12. Since cracking is mostly intergranular, considerably fewer grain facets need be fractured in the 6.2 μm material than in the 1.1 μm one before the maximum Palmqvist size is reached. Now if it is admitted that the minimum cracking step that can be acoustically recorded is the fracture of a grain facet larger than a critical size, then it follows that stage II should be less easily identified as the material grain size is increased up to the level where the initial cracking becomes preferably transgranular and the material again appears structurally homogeneous to the applied stress field.

The reason why higher overall numbers of acoustic events are recorded in the 6.2 μm compared to the 1.1 μm materials respectively (Fig. 7 and 8)

a) 1.1 μm alumina b) 6.2 μm alumina

Fig. 12. Indentation and Palmqvist crack sizes relative to microstructures for a 5 N load.

is not understood at present. It may have a number of origins: e.g. the minimum cracking step acoustically detectable being tested to a microstructural dimension, the formation of extra cracks (Fig. 6), differences in the characteristics of wave propagations from source to transducer in either material, differences in grain boundary compositions hence properties, etc.

In relation to this last item, the indepence of the testing environment which has been evidenced qualitatively in the present work, and which has been observed to be more pronounced in the 6.2 μm material (perhaps because of a presumably 30 x higher impurity concentration at grain boundaries) would be worth studying further.

IV. CONCLUSION

The present study on indentation fracture of fully dense fine grain size alumina is thought to show that:

- up to ~1 N applied loads no well developed individual cracks are to be noted. A damage pattern is formed instead consisting of a whole statistically isotropic network of microcracks the extent and morphology of which are markedly dependent upon grain size testing environment and possibly loading rates. No load threshold for initial microcracking has been observed. The smallest ($\sim 10^{-2}$ N) load that could be measured in the present work already corresponds to a stress level high enough for intergranular fracture to be initiated.

- conventional indentation cracking proceeds in two steps in the indenter diagonal plane. Palmqvist type cracks are formed first presumably as extensions of the initial surface damage. Then as the applied load is increased they link together beneath the surface and may propagate further as a semi-circular like defect. The influence of the testing environment is felt mostly in the early stages of loading and/or in the coarser grain size material.

The present results also may be considered to illustrate some of the limitations in the use of indentation techniques for toughness measurements (13; 18). It is evident that in alumina the phenomenology and micromechanisms of indentation fracture are strongly dependent upon microstructure, notably grain size, and environment. It is the present writers' belief that such could be the case in a whole range of crystalline ceramics as well. It is thus suggested that controlled indentation methods should be used also as means for studying the micromechanical behaviour of materials in detail, and not just as "quick and easy" toughness measuring techniques.

REFERENCES

1. B.R. Lawn, E.R. Fuller, J. Mater. Science 10 (1975) 2016.
2. B.R. Lawn, M.V. Swain, J. Mater. Science 10 (1975) 113.
3. B.R. Lawn, R. Wilshaw, J. Mater. Science 10 (1975) 1049.
4. D.B. Marshall, B.R. Lawn, J. Mater. Science 14 (1979) 2001.
5. B.R. Lawn, A.G. Evans, D.B. Marshall, J. Amer. Ceram. Soc. 63 (1980) 574.
6. D.B. Marshall, B.R. Lawn, A.G. Evans, J. Amer. Ceram. Soc. 65 (1982) 561.
7. B.R Lawn, T.P. Dabbs, C.J. Fairbanks, J. Mater. Science 18 (1983) 2785.
8. B.R. Lawn, K. Jakus, A.C. Gonzalez, J. Amer. Ceram. Soc. 68 (1985) 25.
9. P.K. Gupta, N. J. Jubb, Comm. of Amer. Ceram. Soc. (1981) C 112.
10. D.B. Marshall, J. Amer. Ceram. Soc. 67 (1984) 110.
11. R.K. Govila, J. Amer. Ceram. Soc. 63 (1980) 319.
12. P. Chantikul, G.R.Anstis, B.R. Lawn, D.B. Marshall, J. Amer. Ceram. Soc. 64 (1981) 539.
13. S.S. Smith, B.J. Pletka, in: "Fracture Mechanics of Ceramics, Vol. 6", R.C. Bradt, A.G. Evans, D.P.H. Hasselman and F.F. Lange, ed., Plenum Press, New York (1983) 189.
14. M. Pardavi-Horvath, J. Mater. Science 19 (1984) 1159.
15. R. Warren, H. Matzke, in: "Science of Hard Materials", R.K.Viswanadham, D.J. Rowcliffe, J. Gurland, ed., Plenum Press, New York (1983) 155.
16. K. Niihara, J. Mater. Science Letters 2 (1983) 221.
17. D.K. Shetty, I.G. Wright, P.N. Mincer, A.H. Clauer, J. Mater. Science 20 (1985) 1873.
18. A.G. Evans, in: "Fracture Mechanics Applied to Brittle Materials", ASTM STP 678, S.W. Freiman, ed., ASTM (1979).
19. J.D. Fridez, C. Carry, A. Mocellin, in: Advances in Ceramics, Vol. 10, W.D. Kingery, ed., Amer. Ceram. Soc. Inc. (1984) 720.
20. H. Hertz, "Hertz's Miscellaneous Papers", Mac-Millan, ed., London (1896), chap. 5-6.
21. E.H. Yoffe, Philosophical Mag. A 50 (1984) 813.
22. A. Krell, J. Woltersdorf, E. Pippel, D. Schulze, Phil. Mag. A51 (1985) 765.
23. A.G. Evans, Int. J. of Fracture 10 (1974) 251.
24. B.R. Lawn, A.G. Evans, J. Mater. Science 12 (1977) 2195.
25. J.T. Hagan, J. Mater. Science 14 (1979) 2975.
26. J. Lankford, D.L. Davidson, J. Mater. Science 14 (1979) 1662.
27. J. Lankford, D.L. Davidson, J. Mater. Science 14 (1979) 1669.
28. M.T. Laugier, J. Mater. Science Letters 4 (1985) 207.
29. K. Niihara, R. Morena, D.P.H. Hasselman, J. Mater. Science Letters 1 (1982) 13.

MICROFRACTURE PROCESS DURING FRACTURE TOUGHNESS TESTING IN Al_2O_3 CERAMICS EVALUATED BY AE SOURCE CHARACTERIZATION

Teruo Kishi, Shuichi Wakayama and Shiro Kohara

Faculty of Engineering
The University of Tokyo
Meguro-ku, Tokyo

ABSTRACT

To understand the fracture mechanism in ceramics, quantitative size and mean nucleation rate of microcracks generated in the process zone during stable crack growth were evaluated by the acoustic emission source characterization, using notched 1 inch compact tension specimen of fine and coarse grained alumina. In linear response treatment of elastic wave, source function due to the nucleation of microcracking, which gives the quantitative size of microcracking, could be evaluated by the deconvolution integral of detected AE signals using both dynamic Green's function of a medium and transfer function of transducer. As results, 15-20 μm diameter microcracks were detected at the crack tip during slow crack growth and in addition to these microcracks some larger cracks of about 100 μm were also evaluated at the latter stage of crack extension.

In the case of fine grained alumina, the smaller microcracks results from intergranular crackings due to pores and the larger microcracks from the coalescence of these microcracks. On the contrary, in the case of coarse grained alumina, the former from transgranular cracks and the latter from the transgranular crackings due to extra coarse grains. From these results it was concluded that smaller microcracks have a role to enhance the fracture resistance, while larger microcracks have a possibility to become the origin for stable or unstable fracture. Because of the coalescence process of microcracks, the fracture surface in fine grained alumina takes zigzag configuration, which is related to higher fracture resistance K_{IR} in comparison with that of coarse grained alumina.

INTRODUCTION

Fracture resistance of ceramic materials is increased by the nucleation of microcracking and phase transformation [1-4]. For example, Claussen et al. proposed that an increase of fracture toughness of Al_2O_3 can be attributed to microcrackings formed by t-ZrO_2 to m-ZrO_2 transformation. Those microcrackings are considered to nucleate in the area around a main crack tip [5-7], which is so called as microcracked process zone, damage zone or energy dissipation zone. In this zone, relaxation of stress concentration occurs due to microcrack-macrocrack interaction.

On the other hand, various types of specimens and methods for obtaining plane strain fracture toughness K_{IC} have been carried out, however, at the present time there exists no standardized method for K_{IC} determination of brittle materials such as ceramics [8]. This difficulty depends on the lack of the understanding of microfracture process. Although in some cases final unstable fracture might occur directly from preexistent microcrack, in most cases it occurs from the defect which is formed by growth or coalescence of microcracks [9-10], depending on the shape and size of specimens.

So to evaluate the characteristics of these microcrackings in process zone such as quantitative size and mean nucleation velocity is a most fundamental subject both to understand microfracture mechanism and to know of fracture resistance enhancement, however so far these is no appropriate method to detect such a small size of microcrackings as the order of 10-100 μm.

In this paper acoustic emission source characterization method, which we developed in recent years [11], is applied to evaluate the quantitative size and mean nucleation rate of microcracks and the relation between microcracking and fracture toughness is discussed.

AE SOURCE CHARACTERIZATION

The formalization of AE wave propagation in an elastic medium, such as a fracture toughness testing, is shown in Fig. 1. The stress wave emitted from a microcrack propagates to the surface in a medium with reflection and mode-transformation, and is monitored by a transducer mounted on the surface of the medium. The relationship between the detected wave, V(t), and the source function, D(t), can be expressed as follows [11-13] in the linear response treatment.

$$V(t) = S(t) * G(t) * D(t) = R(t) * D(t) \tag{1}$$

Where G(t) is the dynamic Green's function of a medium, S(t) is the response function of a transducer and R(t) (= S(t) * G(t)) is the combined response function. The symbol, *, indicates the convolution operation, i.e.

$$A(t) * B(t) = \int_0^t A(t-\tau)\cdot B(\tau)d\tau \quad . \tag{2}$$

In the case of mode I cracking (see Fig. 1), D(t) is expressed as the product of the elastic constants of the material and the volume of the microcrack, i.e.

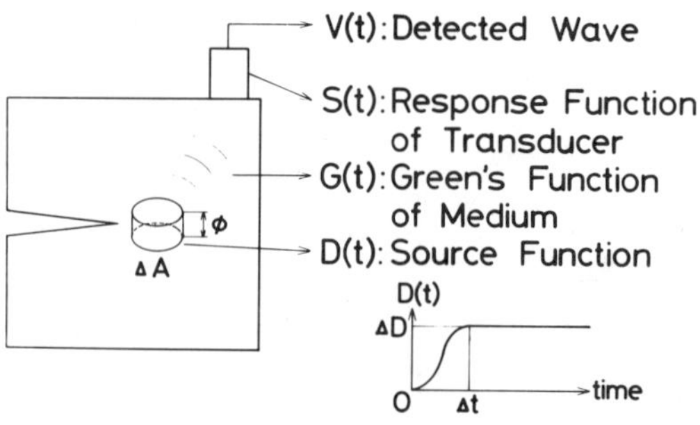

Fig. 1. Formulization of AE wave propagation in an elastic medium. Right lower figure shows a typical shape of source function, of which strength is ΔD and rise time is Δt.

$$D(t) = (\lambda+2\mu)\cdot\Delta A(t)\cdot\phi(t) \quad . \tag{3}$$

Where λ and μ are Lame's constants. $\Delta A(t)$ and $\phi(t)$ are the area and the opening displacement of the microcrack, respectively. A typical shape of a source function is shown in Fig. 1. This curve gives the nucleation of a microcrack, the size of which is proportional to ΔD during the time interval, Δt.

We obtained R(t) experimentally and estimated D(t) as the inverse problem of eq. (1) to evaluate the size and mean nucleation velocity of a microcrack. Furthermore, we can determine the mode of microcracking (see Appendix). The theoretical background of AE source characterization is summarized in Appendix.

EXPERIMENTAL PROCEDURE

Materials

Material properties and chemical composition of materials used are tabulated in Table 1 (a) and (b), respectively. Two kinds of materials are used. One (ADS-80) has a mean grain size of 5 μm and porosity ratio of 9 % and the other (ADS-10) has 20 μm and 5 %, respectively. MgO are used as a binder for both materials. Materials are made by sintering of alumina powder under the atmospheric pressure at 1873 K for ADS-80 alumina at 2023 K for ADS-10 alumina.

Table 1. Specimen (a) material properties and (b) chemical compositions (mass %).

(a) Material properties

	ADS-80	ADS-10
Material	α-Al_2O_3	α-Al_2O_3
Mean grain size / μm	5	20
Density / $kg{\cdot}m^{-3}$	3.57×10^3	3.85×10^3
Porosity / %	9	5
Young's modulus / MPa	2.22×10^5	3.77×10^5
Poisson's ratio	0.235	0.231

(b) Chemical compositions (mass %)

	Al_2O_3	MgO	CaO	SiO_2	Na_2O	Fe_2O_3
ADS-80	92	2.5	1.5	1	0.2	0.02
ADS-10	99.5	0.25	—	0.03	0.01	0.02

Photographs 1 a) and b) show light micrographs of ADS-80 and ADS-10 alumina. In ADS-80, grain size distribute from 2 μm to 40 μm many pores of which sizes are up to 30-40 μm are observed. On the contrary, in ADS-10 porosity is smaller than ADS-80. ADS-10 has a wide grain size distribution, i.e. from 5 μm to 80-90 μm. A remarkable difference of porosity between two materials, as will be shown in following section, expected to result in different cracking processes.

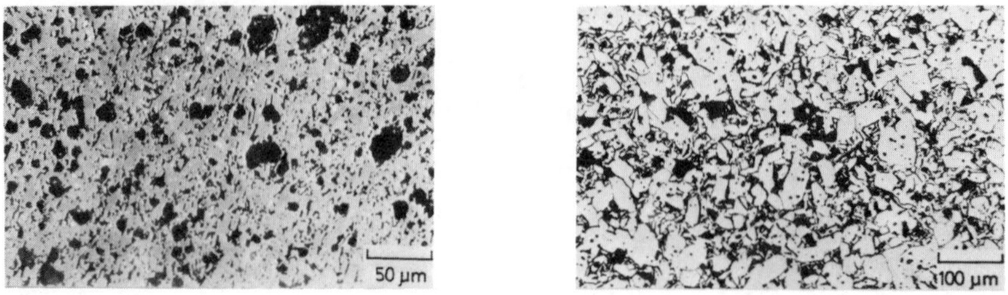

(a) ADS-80 (b) ADS-10

Photo. 1. Light micrographs of (a) fine (ADS-80) and (b) coarse (ADS-10) grained alumina.

Specimens

Fracture toughness testings were carried out using 25 mm (1 inch) thickness compact tension (CT) specimens with chevron (triangular) notches under constant crosshead speed of 1.7×10^{-7} m/s. Figure 2 a) describes a two-dimensional geometry of a CT specimen and Fig. 2 b) shows a chevron notch schematically. The specimens with chevron notches have some unique advantages in fracture toughness testing and are discussed by Munz [8] in detail.

The relation between the resistance of crack extension, K_{IR}, load, P, relative crack length, $\alpha(=a/w)$, and the geometry of specimen and notch can be obtained from equating the necessary and available energy for crack extension, as follows,

$$K_{IR} = \frac{P}{B\sqrt{w}} Y^* \tag{4}$$

where

$$Y^* = \left[\frac{1}{2} \cdot \frac{dC'_{tr}}{d\alpha} \cdot \frac{\alpha_1 - \alpha_0}{\alpha - \alpha_0} \right]^{1/2} \tag{5}$$

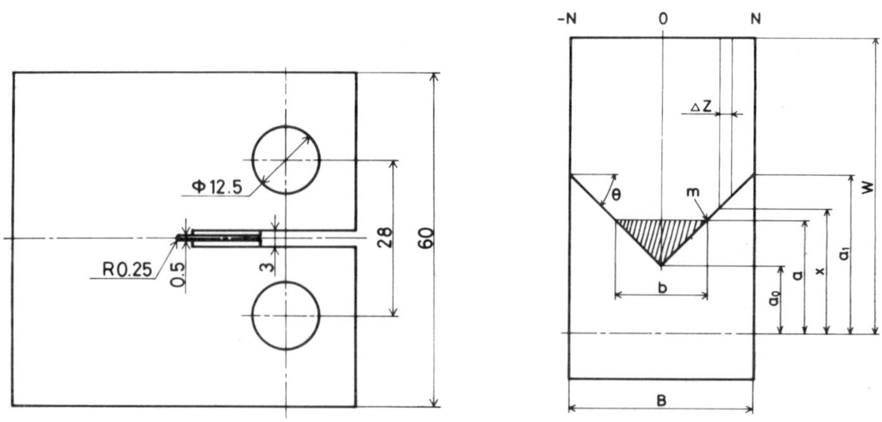

(a) (b)

Fig. 2. (a) Schematical shape of a 1 inch (25 mm) thickness compact tension specimen with (b) a chevron notch.

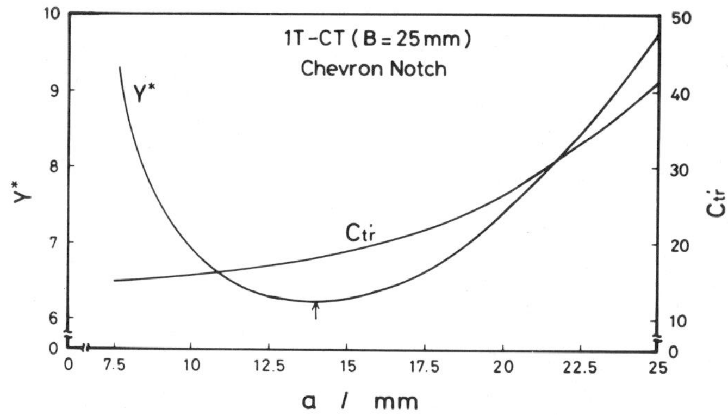

Fig. 3. Relationship between Y^*, C'_{tr} and a.

$\alpha_1=a_1/w$, $\alpha_0=a_0/w$, B and w are notch parameter indicated in Fig. 2 b). C'_{tr} = EBC_{tr} is the dimensionless compliance of the specimen with a trapezoidal crack front and E is Young's modulus. For the present CT specimen (a_0= 7.5 mm, a_1 =20 mm, w=49.5 mm, B=25 mm), Y^* ,C'_{tr} -a relation have been calculated as shown in Fig. 3.

As can be seen from Fig. 3, Y^* -a curve exhibits a minimum at a =14 mm. The greatest advantage of these specimens in the present study is that a sharp crack is created during crack growth and that rather stable crack extension before final unstable fracture, which facilitate the analysis of AE source function, can be given easily.

Furthermore, three points bending tests using specimens of 5 x 5.5 x 30 mm (lower span is 20 mm) have been carried out for estimation of the microcrack nucleation stress, σ_m.

AE Measurement

The Conventional AE Signal Analysis Two 200 kHz resonant type transducers (NAIS:F217M) were employed to measure AE event counts and AE energy. A detected signal was amplified 40 dB by pre-amplifier and analyzed by AE analyzer (PAC:model 3400). Total gain and threshold level were chosen to be 66 dB and 50 μV at the pre-amplifier input terminal. Applied load and crack opening displacement (COD) was also recorded by AE analyzer.

AE Source Characterization Figure 4 describes a schematical diagram of the measurement system for AE source characterization. AE signals were detected by three wide band transducers, which have an effective component till 15 MHz (AET:FC 500). Detected signals were amplified 8 dB by pre-amplifiers(~20 MHz), which were especially established for the present study and have noise level of 80 μV at the input terminal, i.e. low enough for small amplitude signal materials such as ceramics. Signals were amplified 40 dB by main amplifiers (~70 MHz) and A/D converted and collected by 3 ch wave memories (sampling period:50 ns, resolution:8 bits, data length:4 k word/ch. NF:WM-852). One of three channel signals was collected by high frequency (25 ns) high resolution (10 bits) and large (1 M words) wave memory (NF:AE9620), which can record up to 256 signals during a testing. The recorded signals have been analyzed for AE source functions using a desktop computer (HP 9845B). Double shield lines were used for all signal lines to prevent the electromagnetic noise and electronic power was supplied to each equipment through a noise-filter-transformer.

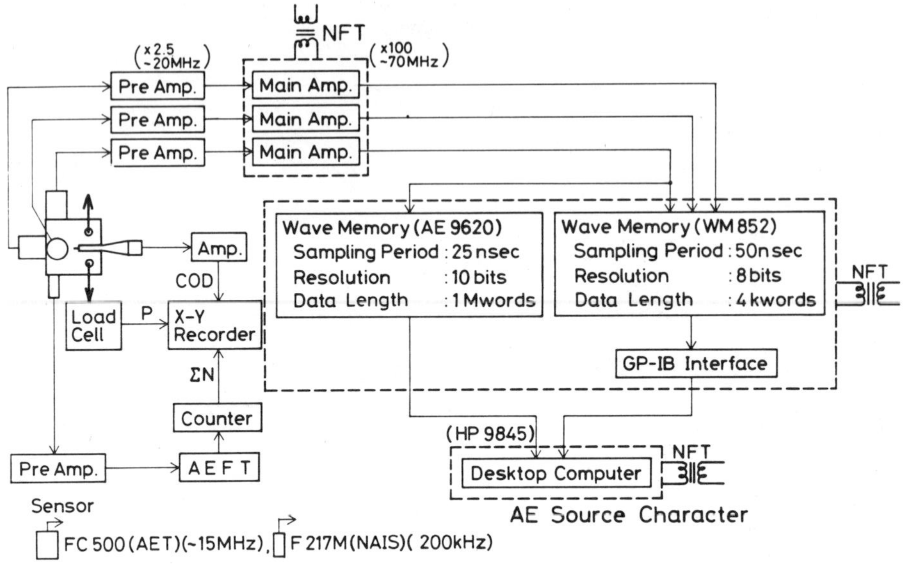

Fig. 4. Measurement system for AE source characterization.

RESULTS

Detected AE Signal Analysis

Figure 5 shows load (P) - COD curves obtained during fracture toughness testings using CT specimens with chevron notches, which were discussed in section 2, of fine grained (ADS-80) alumina (a) and coarse grained (ADS-10) alumina (b), respectively. For fine grained alumina, compliance changed

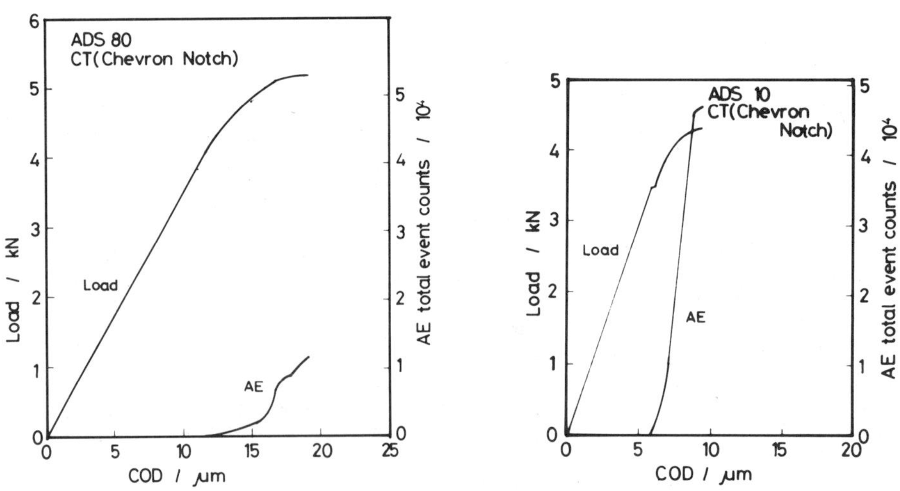

(a) ADS-80

(b) ADS-10

Fig. 5. Load-COD curves obtained by fracture toughness testing of (a) fine (ADS-80) and (b)coarse (ADS-10) grained alumina.

at the load level of 4.1 kN and catastrophic fracture occurred at 5.2 kN. On the contrary, for coarse grained alumina, those load levels were 3.4 kN and 4.3 kN, respectively.

Using eqs.(4) and (5) and Y^*, C'_{tr}- a curves in Fig. 3, apparent K_{IR} value at each load levels can be calculated from compliance calibration under the assumption that unloading would give a straight line to the origin. According to the calculation, for fine grained alumina, a catastrophic fracture occurred at K_{IR} of 5.9 MPa $m^{1/2}$ after the crack extension, Δa, of 8.2mm. On the contrary, for coarse grained alumina, that occurred at K_{IR} of 4.9 MPa $m^{1/2}$ after Δa of 8.5 mm.

AE events were detected during the region of compliance changes. The crack extension was observed by dye penetrant inspection. In this test CT specimen was loaded by two steps. The notch tip was immersed with colored dye only during first loading so that only fracture surface formed by first loading was colored as shown in Fig. 6, which corresponded to the compliance change fairly well.

According to Hoagland and Embury [3], the size of process zone, r_p, can be expressed as

$$r_p = \frac{1}{2}\left[\frac{K_I}{\sigma_c} \cdot \cos\frac{\theta}{2}\left(1+\sin\frac{\theta}{2}\right) \right]^2 \qquad (6)$$

where K_I is the applied stress intensity, σ_c is the critical value of the stress required for microcrack nucleation and θ is the angle from a main crack plane. So the volume of process zone per unit thickness, V, increases with the crack extension in proportion to K_I^4. Now if the density of microcracks, ρ_{mc}, in process zone is uniform and the nucleation of microcracks in process zone is the origin of AE signals, AE total event counts, N, should be related to K_I^4 as follow.

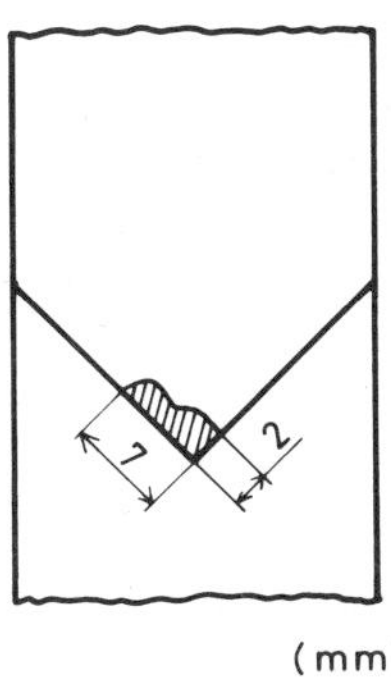

Fig. 6. Dye penetration area of a chevron notch during the first loading in the fracture toughness testing of fine grained (ADS-80) alumina.

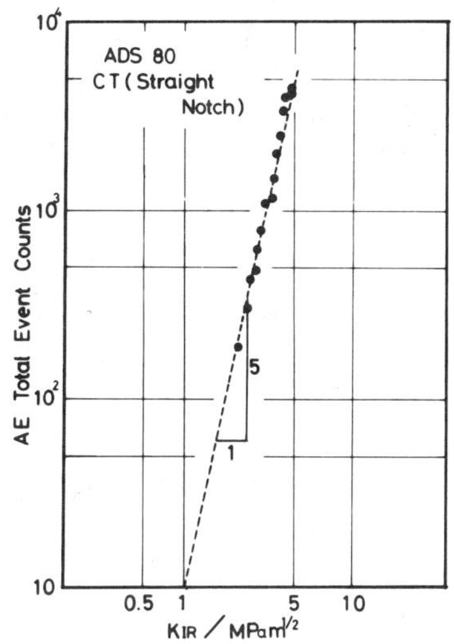

Fig. 7. Relationship between AE total event counts and applied stress intensity factor.

$$N \propto K_I^4 \tag{7}$$

The relation between N and K_I, which is obtained within a CT specimen with straight through notch, is shown in Fig. 7. The slope is about 5 ($\simeq 4$). Therefore it can be concluded that AE events were emitted by the nucleation of microcracks in the process zone.

AE Source Characterization

Figure 8 shows examples of AE waveforms which were detected by three wide band transducers. As seen in the figure, the first motions of three waveforms have same phase signs. Therefore it can be concluded that they were emitted from a mode I opening microcrack, using the principle of AE radiation pattern in Fig. 17.

Figure 9 shows an example of AE source function which was computed by eq. (7). According to the basis of fracture mechanics, the diameter, 2a, and mean nucleation rate, V_c of each microcrack are estimated. In this treatment, the micro crack is considered to be under an applied uniform stress, σ_m, in process zone as shown in Fig. 10.

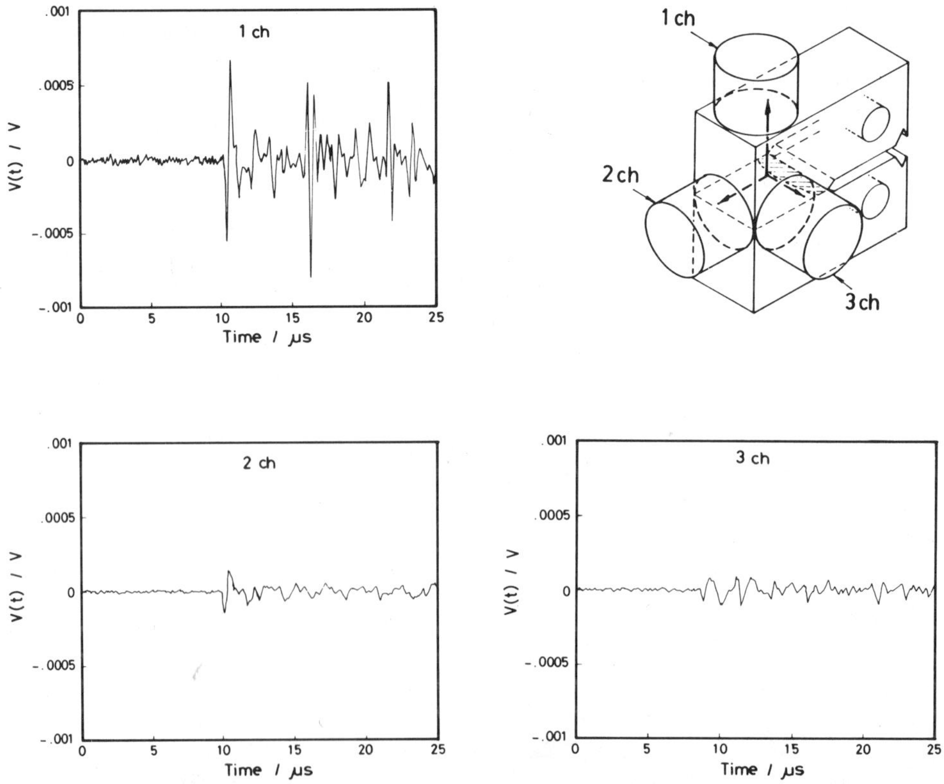

Fig. 8. Waveforms detected by three wide band transducers which were mounted on the positions indicated in the upper figure.

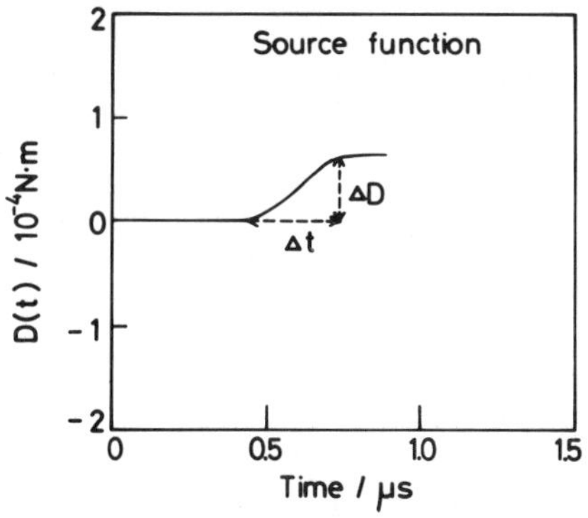

Fig. 9. An example of obtained source function.

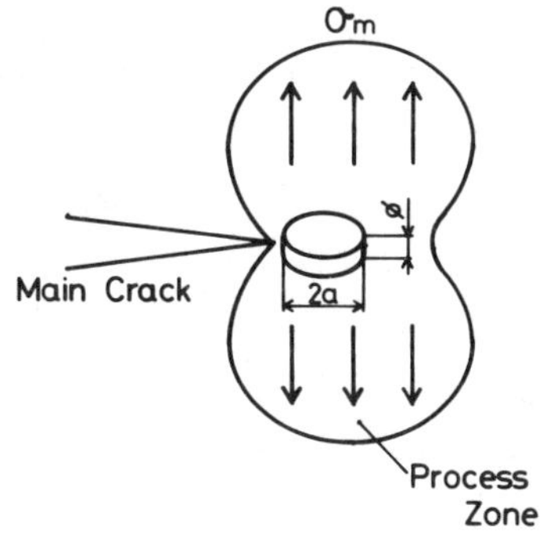

Fig. 10. Microcrack under uniform stress, σ_m, in process zone.

In the case of mode I cracking, the diameter, 2a, and mean nucleation velocity, V_c of a microcrack can be estimated as follows [11].

$$a = \left(\frac{1}{\pi^2 \sigma_m} \cdot \Delta D \right)^{1/3} \quad (8)$$

$$V_c = 2a/\Delta t \quad (9)$$

where Δt is the rising time of source function in Fig. 9.

To evaluate σ_m, three points bending tests were carried out. Figure 11 shows the result of coarse grained alumina. The bending stress at which the AE total event counts and energy increased rapidly (= 210 MPa) can be equal to the microcracking nucleation stress, σ_m. The value of σ_m of fine grained alumina was equal to coarse grained alumina.

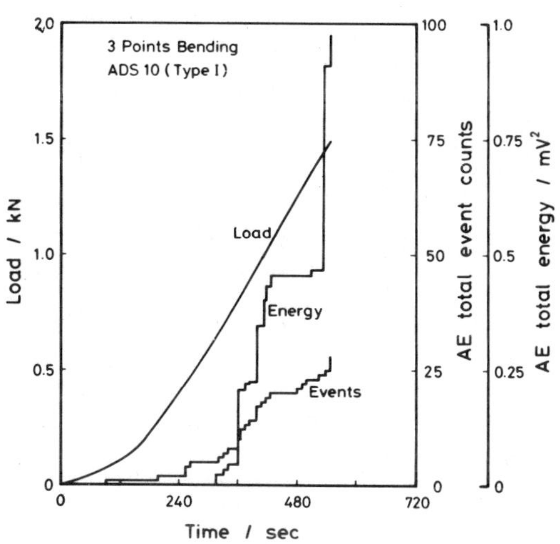

Fig. 11. AE behavior during a three points bending test in coarse grained (ADS-10) alumina.

The Size And Mean Velocity Of Microcracks

The results are summarized in Figs. 12 and 13. In the distribution of the microcrack diameter of fine grained (ADS-80) alumina (Fig. 12 (a)), the peak distribution is observed at the diameter of 36-39 μm. A larger cracking diameter more than 100 μm is also observed in the latest stage of the fracture toughness testing, which will become the origin of final unstable fracture.

On the other hand, in the distribution of fine grained alumina (Fig. 12 (b)), the highest peak is also observed at diameter of 39-42 μm. Remarkable characteristics is the existence of larger diameter crackings of 70-100 μm.

In both cases, it can be concluded that a microcrack with a size of 15-20 μm, which is less than the required value for nondestructive testings suggested by Okada and Sines [9], can be detected quantitatively by AE source characterization technique.

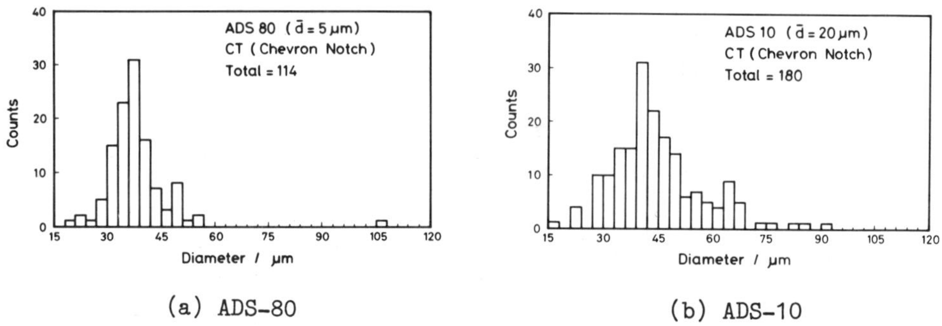

Fig. 12. Distributions of diameter estimated by AE source characterization in (a) fine (ADS-80) and (b) coarse (ADS-10) grained alumina.

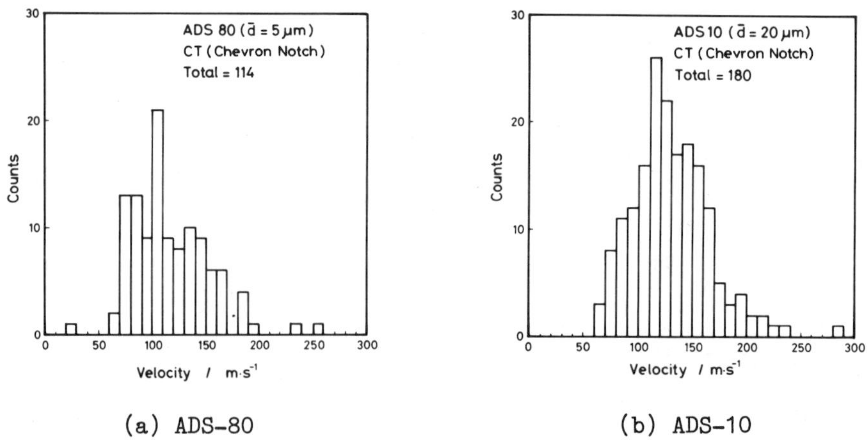

Fig. 13. Distributions of microcrack nucleation velocity estimated by AE source characterization in (a) fine (ADS-80) and (b) coarse (ADS-10) grained alumina.

The distributions of microcrack nucleation velocity are shown in Fig. 13. They show similar tendency except that the values of the velocity of coarse grained alumina are a little larger than that of fine grained alumina. This parameter is expected to be available to estimate the degree of the local embrittlement of materials.

The importance is that the distribution of microcrack diameters changed with the crack propagation, that is, the larger cracks nucleate at the latter part of crack propagation.

Microscopic Fractography

Fracture surfaces coated with Au-Pd deposit were observed by scanning electron microscope (SEM). Remarkable cracking units were shown in Photos 2 and 3, respectively.

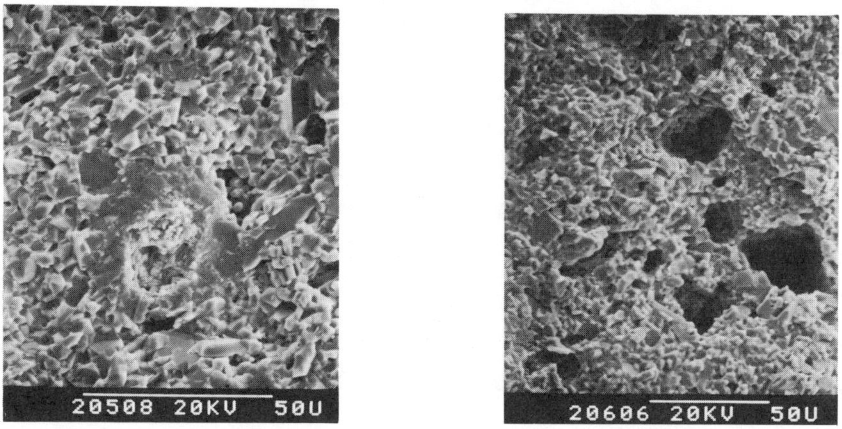

(a) (b)

Photo. 2. SEM fractographs of fine grained (ADS-80) alumina.

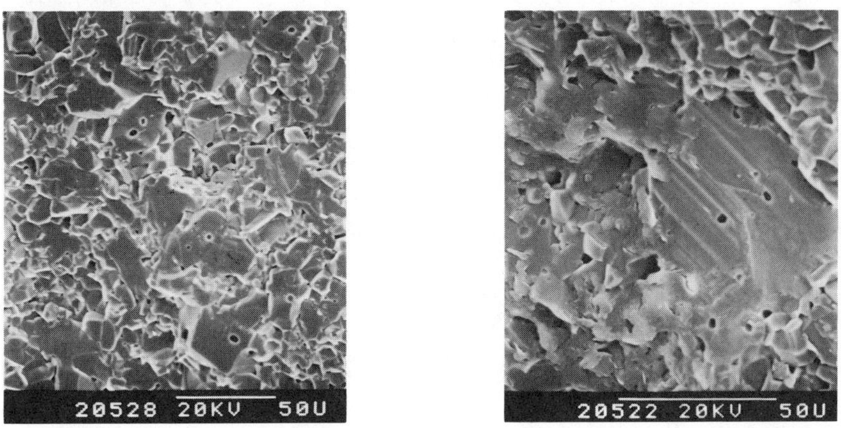

(a) (b)

Photo. 3. SEM fractographs of coarse grained (ADS-10) alumina.

In fine grained alumina, there are two distinguished features. One is a transgranular cracking of a single coarse grain with the size of 30-50 μm (Photo. 2 a)), and the other is the coalescences feature, of which sizes are also 30-50 μm (Photo. 2 b)). The later is usually observed among a few pores which are the microstructural feature of the present material.

On the other hand, in coarse grained alumina, the most of cracking units are transgranular. In Photo 3 a), several transgranular crackings with the sizes of 30-50 μm are observed, which has been observed on the fracture surface corresponding to the initial loading. On the contrary, the cracks, of which size is 70-100 μm, were observed on the fracture surface at the final loading (Photo. 3 b)). These results correspond to the cracking diameter distribution in Fig. 12.

DISCUSSIONS

Crack propagation resistance, K_{IR}, was calculated by Y^*, C'_{tr} - a relationship shown in Fig. 3. Figure 14 shows obtained K_{IR} as a function of crack extension, which is so-called R-curve. K_{IR} values are increasing as a crack length is increasing, except for initial decrease which is considered to be caused by the shape configuration of an artificial notch. It is usually concluded that rising R-curve results from microcrackings around the crack tip in the case of brittle ceramics [3]. Comparing R-curves of fine (ADS-80) and coarse (ADS-10) grained alumina, R-curve of fine grained alumina is higher than that of coarse grained alumina that can be regarded as a result of the difference of microscopic fracture mechanisms in two materials.

As shown in Fig. 15 schematically, in the case of fine gained alumina, peak distribution of microcracking diameter of 30-50 μm is caused by intergranular crackings due to the grain boundary pores, and a hatched large diameter results from the coalescence of these microcracks.

On the contrary, in the case of coarse grained alumina, the peak is related to transgranular cracking dominantly and hatched larger diameters of several cracks correspond to transgranular cracking due to extra coarse grain. In spite of the same size of microcracking in hatched region in two materials, the mean nucleation rate more than 150 m/sec of cracks in coarse grain is much higher than that of about 60 m/sec in fine grain. This result can be explained by the speculation that the rising time of a transgranular cracking is shorter than that of the coalescence process of microcrackings.

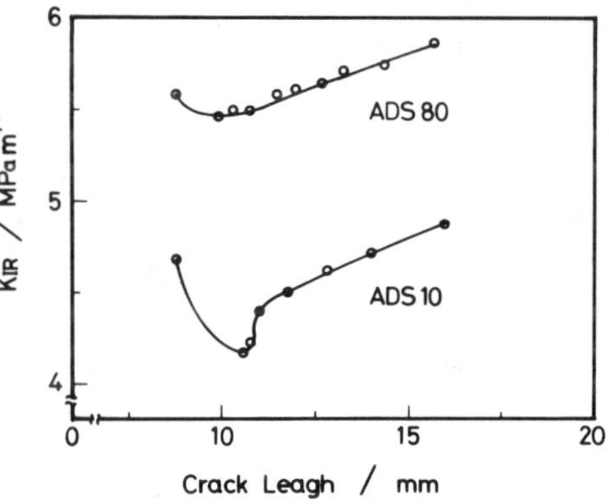

Fig. 14. R-curves of fine (ADS-80) and coarse (ADS-10) grained alumina.

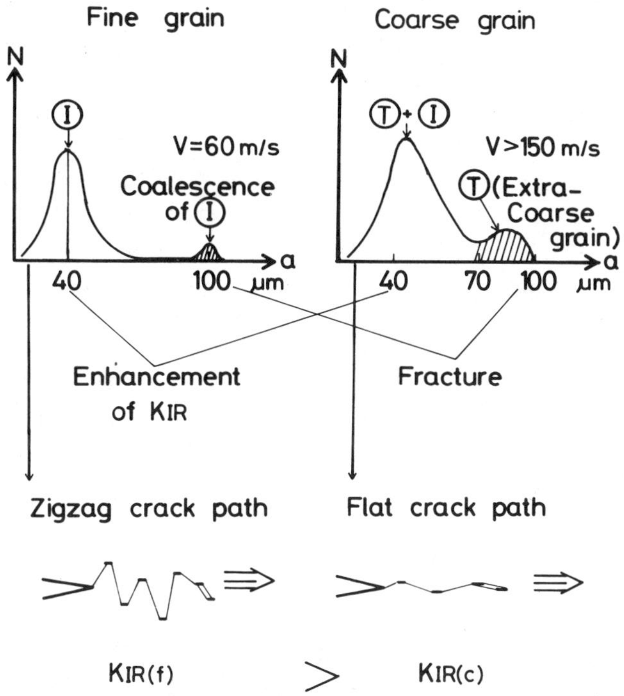

Fig. 15. Microscopic fracture process in alumina. I and T indicate intergranular and transgranular cracking, respectively. Two upper figures show size distributions of microcracks estimated by AE source characterization. $K_{IR}(f)$ and $K_{IR}(c)$ indicate the fracture resistance of fine and coarse grained alumina.

In any case, from the above discussion it can be concluded that small microcrackings, of which diameters are above 40 μm, contribute to the enhancement of fracture resistance and while the hatched larger crackings will become the origin for stable (or unstable) crack extension.

In fine grain, the coalescence of microcracks will control the crack propagation pass, that is, the fracture surface in fine grain alumina become much more zigzag and take extra surface energy. According to the macroscopic fracture surface observation the fracture surface of fine grained specimen is always much more zigzag in compared with that of coarse grained, which explain that K_{IR} of fine grained is higher than that of coarse grained as shown in Fig. 14. Now it's impossible to correlate the microcrackings detected with the quantitative enhancement of fracture resistance, but as a first step, it may be recognized that the smaller microcracks evaluated by AE source characterization have two different roles, that is, the enhancement of K_{IR} and the origin of stable and/or unstable crack extension.

CONCLUSIONS

Quantitative size and mean nucleation rate of microcracks generated in the process zone were evaluated by the acoustic emission source characterization, using notched 1 inch compact tension specimen of fine and coarse grained alumina ceramics. With the result of fractography, following results are obtained:

1. 15-20 μm diameter microcracks were quantitatively detected at the crack tip during slow crack growth by AE source characterization. In addition to these microcracks, same larger cracks of about 100 μm were also evaluated at the latter stage of crack extension.

2. In fine grained alumina, the smaller microcracks results from intergranular cracking due to pores and the larger microcracks from the coalescence of these microcracks. On the contrary, in coarse grained alumina, the former from transgranular crackings, and the latter from the transgranular crackings due to extra coarse grains.

3. It may be concluded that smaller microcracks have a role to enhance fracture resistance, while larger microcracks have a possibility to become the origin for stable or unstable fracture. Because of the coalescence process of microcracks, the fracture surface in fine grained alumina takes zigzag configuration, which is related to higher fracture resistance K_{IR} in comparison with that of coarse grained alumina.

APPENDIX ----- AE SOURCE CHARACTERIZATION

Source Function $D_{jk}(t)$

In a homogeneous, isotropic and elastic solid, a source function due to microcracking, D_{jk}, is expressed by [14]

$$\begin{aligned} D_{jk}(x',t) &= C_{jklm}\cdot\phi_l(t)\cdot\Delta A_m(t) \\ &= [\ \lambda\delta_{jk}\delta_{lm} + \mu(\delta_{jl}\delta_{km} + \delta_{jm}\delta_{kl})]\cdot\phi_l(t)\cdot\Delta A_m(t) \end{aligned} \tag{10}$$

where C_{jklm} is an elastic modulus, λ and μ are Lame's constants and δ_{ij} is Kronecker delta function. $\phi_l(t)$ is the discontinuity of displacement across the surface ΔA_m imbeded in elastic solid (Fig. 1).

In the case of tensile cracking shown in Fig. 1, the only non-zero terms in eq. (1) are

$$\begin{aligned} D_{11}(x',t) &= D_{22}(x',t) = \lambda\phi_3(t)\cdot\Delta A_3(t) \\ D_{33}(x',t) &= (\lambda + 2\mu)\phi_3(t)\cdot\Delta A_3(t) \end{aligned} \tag{11}$$

Linear Response Theory

The displacement due to elastic waves radiated by incremental extension of microcracking can be formulated by the models of seismology [15], dynamic fracture mechanics and dislocation theory in the elastic body.

The general solution $U_i(x,t)$, the displacement in the i-direction on the surface of medium, is expressed using the dynamic Green's function, $G_{ij}(x;x',t)$.

$$U_i(x,t) = G_{ij,k'}(x;x',t) * D_{jk}(x',t) \tag{12}$$

where $G_{ij,k'} = \partial G_{ij}/\partial x_k$.

In the case where the linearity of measuring system, including transducer, is valid, output signal $V(t)$ is related to $U_i(x,t)$ using response function of measuring system including transducer $S_i(x,t)$.

$$V(t) = S_i(x,t) * U_i(x,t) \qquad (13)$$

Dynamic Green's Function G(t) And Response Function Of Transducers S(t)

Dynamic Green's function G_{ij} can be evaluated by three methods, i.e. analytical [16], experimental [17] and numerical methods. In the present paper, G_{ij} is obtained by experimental method combined with a response function of transducer S(t), using breaking pencil lead as a simulated reference signal [11,18].

Pencil lead breaking signal is monopole force, so Green's function for dipole source, $G_{33,3'}$ is given by [11,18]

$$G_{33,3'}(t) = \frac{1}{V_L}\frac{\partial}{\partial t} G_{33}(t) \qquad (14)$$

for the first motion of longitudinal wave with the velocity of V_L. Figure 16 is an experimentally obtained result of the combined response function R(t) of CT specimen.

Fracture Mode Evaluated By Radiation Pattern Of AE Signals

We use the infinite space analytical Green's function of far field longitudinal waves. Figure 17 represents the angular dependence of displacement U_r for mode I and mode II fracture projected on X-Z Plane. For mode III, the same angular dependence as for mode II can be observed provided that U_r is projected on Y-Z plane. This result shows that we can decide the fracture mode of microcracking by measuring radiation pattern of emitted signals [11].

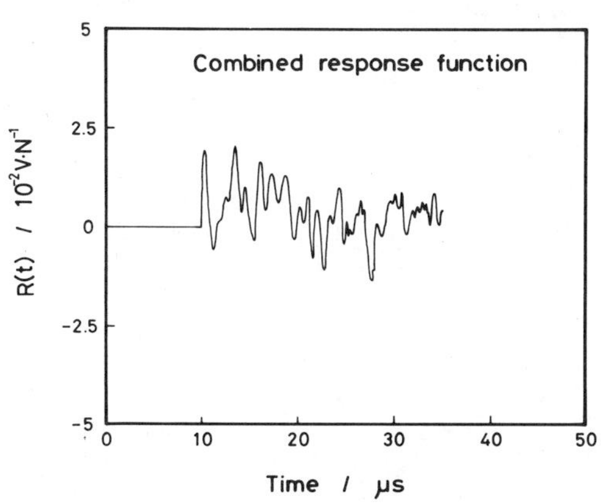

Fig. 16. Combined response function of an elastic medium and transducer, R(t).

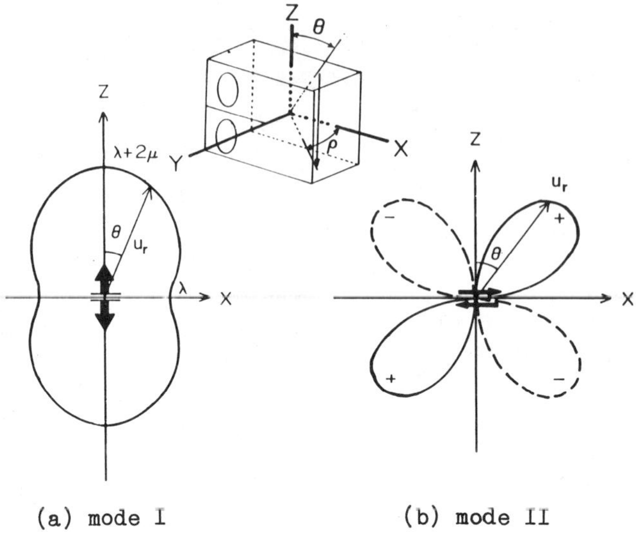

Fig. 17. Radiation pattern of emitted signals due to (a) mode I and (b) mode II cracking .

REFERENCES

1. N. Claussen, J. Steeb and R. F. Pabst, Am. Ceram. Soc. Bul. 56:559 (1977)
2. F. F. Lange, <u>in</u>: "Fracture Mechanics of Ceramics", Vol. 4, R. C. Bradt, D. P. H. Hasselman and F. F. Lange, ed., Plenum Press, New York, 799 (1979).
3. R. G. Hoagland and J. D. Embury, J. Am. Ceram. Soc. 63:404 (1983).
4. A. Krell and W. Kreher, J. Mat. Sct. 18 :2311 (1983).
5. A. J. Gesing and R. C. Bradt, <u>in</u>: "Fracture Mechanics of Ceramics", Vol. 5, R. C. Bradt, A. G. Evans, D. P. H. Hasselman and F. F. Lange, ed., Plenum Press, New York, 569 (1983).
6. A. G. Evans and L. J. Graham, Acta Metall. 23:1303 (1975).
7. F. E. Bresch, K. Frye and Th. Mueller, <u>in</u>: "Fracture Mechanics of Ceramics", Vol. 5, R. C. Bradt, A. G. Evans, D. P. H. Hasselman and F. F. Lange, ed., Plenum Press, New York, 1 (1983).
8. D. Munz, <u>in</u>: "Fracture Mechanics of Ceramics", Vol. 6, R. C. Bradt, A. G. Evans, D. P. H. Hasselman and F. F. Lange, ed., Plenum Press, New York, 1 (1983).
9. T, Okada and G. Sines, J. Am. Ceram. Soc. 66:719 (1983)
10. D. P. H. Hasselman, J. Am. Ceram. Soc. 52:600 (1969).
11. T. Kishi, <u>in</u>: "Progress in Acoustic Emission II", M. Onoe, K. Yamaguchi and H. Takahashi, The Jap. Soc. Non-Destructive Inspection, Tokyo, 302 (1984).
12. H. N. G. Wadley, C. B. Scruby and G. Shrimpton, Acta. Metall. 29:399 (1981).
13. N. N. Hsu and S. C. Hardy, <u>in</u>: "Elastic Waves and Non-Destructive Testing of Materials", Y. H. Pao, ed., Amer. Soc. Mech. Eng., New York, 85 (1978).
14. R. Burridge and L. Knopoff, Bul. of the Seismological Society of America, 54:1875 (1964).
15. K. Aki and P. G. Richards, "Quantitative Seismology" Vol. 1, W. H. Freeman and Company, San-Francisco, 599 (1980).
16. A. N. Ceranoglu and Y. H. Pao, J. Appl. Mech. 48:125 (1981).
17. T. Kishi, K. Ohno and K. Kuribayashi, Hihakaikensa 30:911 (1981).
18. T. Kishi and T. Ohira, Trans. JIM. 24:255 (1983).

EFFECT OF STATIC LOADING ON THE FRACTURE TOUGHNESS OF INDENTATION-PRECRACKED ALUMINA

Zong-Zhe Jin and Juan-Rong Ma

Strength Laboratory
Research Institute of Building Materials
Guan-Zhuang, Beijing

ABSTRACT

After a static loading or fatigue test, the critical stress intensity factor of indentation-precracked specimens of alumina increases by a factor of 30%. The fatigue may effectively remove the residual stresses around the indentation and make K_{IC} approach the real value. In this work, the causes that enable K_{IC} to raise are investigated. The results indicate that residual stresses are removed and a microcrack group is formed at the crack-tip of indentation-fatigue specimens. Although only one microcrack grows to make the specimen finally break, other microcracks can grow and branch more or less before breaking and absorb some energy.

INTRODUCTION

It is convenient to measure the critical stress intensity factor of ceramics for a semi-elliptical surface crack in bending using the indentation-precrack method. Much work has been done both in theory and application up to now [1-6]. In this present work, the effect of residual stress on K_{IC} and the characteristics of static fatigue and low-cycle fatigue for indentation-precracked specimens of 99-Al_2O_3 ceramics are examined. Since the change of the residual stresses, after undergoing a certain fatigue, the bending strength and the critical stress intensity factor are somewhat larger than without fatigue.

The use of the induced indentation crack reduces the scatter of the bending strenths normally found among unindented specimens, caused by the statistical distribution of natural cracks [3]. Because there is a quite large residual contact stress around an indentation crack, the discrete parameter of the fracture stress is still larger [4]. Pertrovic [5] observed that the residual stress may be removed by annealing. Nishikawa [6] observed that the residual stress may be removed by grinding the surface of the indentation specimens. Since it is not very clear that annealing have some effect on the properties of the material, and the grinding may cause new distribution of cracks, we have researched how to remove the residual stress round indentation crack with a static loading and fatigue method.

EXPERIMENTAL PROCEDURE

Sintered alumina of 99% purity was cut into test bars 48 mm long with a cross section of 3 mm depth by 4 mm width. These bars were ground and polished on the tension side. The physical properties of tested material are as follows: the volume density-3.85 g/cm^3, elastic modulus-200 GNm^{-2}, the bending strength-250 MNm^{-2}, the average grain size $\leq$ 30 μm. Some bars were polished with 1 μm grain diamond grit, and others ground with 10 μm silicon carbide grit. As a result of contrast, polishing and grinding indentation specimens, both critical stress intensity factors are approached. Polishing is able to decrease the manual defects on the surface of the specimens and enable us to observe the indentation cracks. It is satisfying that the specimens require to grind finely but not to plish, as the size of the indentation crack is far larger than the grain size and observing the cracks is not necessary. When the cracks have been dyed, we can still measure its size.

The indentation was carried out with a Vickers pyramid indenter at the same loading velocity. The load of the indenter was 100 N. Under this load the lateral cracks are 400-500 μm long, and the outline of the rhombus is clear as radiative cracks appear along the diagonal lines. The expansion of the crack length was measured with an optical microscope and before the fracture test, the indentation cracks were dyed. After the specimens were destroyed, the length and depth of the indentation crack on the fracture surface were measured with a reflex projector. Some fracture surfaces were observed with an SEM.

Three point bending test for all of the test bars was done at room temperature in air with a bending span of 36 mm. The crosshead speeds for loading and unloading are 0.5 mm/min and 2.5 mm/min respectively for low-cycle fatigue, and the cycle loads are 0-0.6 P_f (P_f is fracture load), 0-0.7 P_f, 0-0.8 P_f respectively. The number of low-cycle fatigue cycles was 2,000. Static fatigue was for 48 hours under a load of 0.5 P_f or 0.7 P_f. Fracture tests were done at the crosshead speed 0.05 mm/min.

ANALYSIS OF INDENTED STRESSES

The problem of indentation stresses is one of three dimensional elastic-plastic mechanics. The results of the following were taken by theoretical calculations and microscopic photoeleastic tests [7,8].

When the Vickers indenter is loading on a specimen, $P=2d^2\cdot\sigma_o$, the three principal stresses along Z axis are $\sigma_1=\sigma_\theta=\sigma_r$, $\sigma_3=\sigma_z$. The compressive yield stress is represented by σ_o. The stress distribution and its calculating formulas are indicated in Fig. 1(a). This work makes use of the cylinderical and Descartes coordinates. The X-axis and Y-axis are axes of symmetry for stresses. On the surface of indetation, that is the X-Y plane, there are $\sigma_1=\sigma_\theta$, $\sigma_2=o$, $\sigma_3=\sigma_r$ (see Fig. 1b). The maximum surface stress is $\sigma_1/\sigma_o=-\sigma_3/\sigma_o=0.21$. The hypothetical yield ratio is $\tau/\sigma_o=(\sigma_1-\sigma_3)/2\sigma_o \geq 0.5$.

After unloading, because the deformation cannot entirely recover, residual stresses appear near the indentation (see Fig. 1c). The maximum residual stresses along the Z-axis are $\sigma_{1R}/\sigma_o=\sigma_{\theta R}/\sigma_o=\sigma_{2R}/\sigma_o=\sigma_{rR}/\sigma_o=0.5$. $\sigma_{zR}/\sigma_o=0.4$ is also a tensile stress. It is this tensile stress that makes the elastic/plastic boundary, $Z=\sqrt{2}a$, and produces the semisphere lateral crack. The maximum surface stress is $\sigma_{1R}/\sigma_o=-\sigma_{2R}/\sigma_o=0.1$ (see Fig. 1d).

After static fatigue, because the propagation of the median crack and radial crack the original residual stresses are removed, but a new small residual stress o_{1f} may develop at the crack tip.

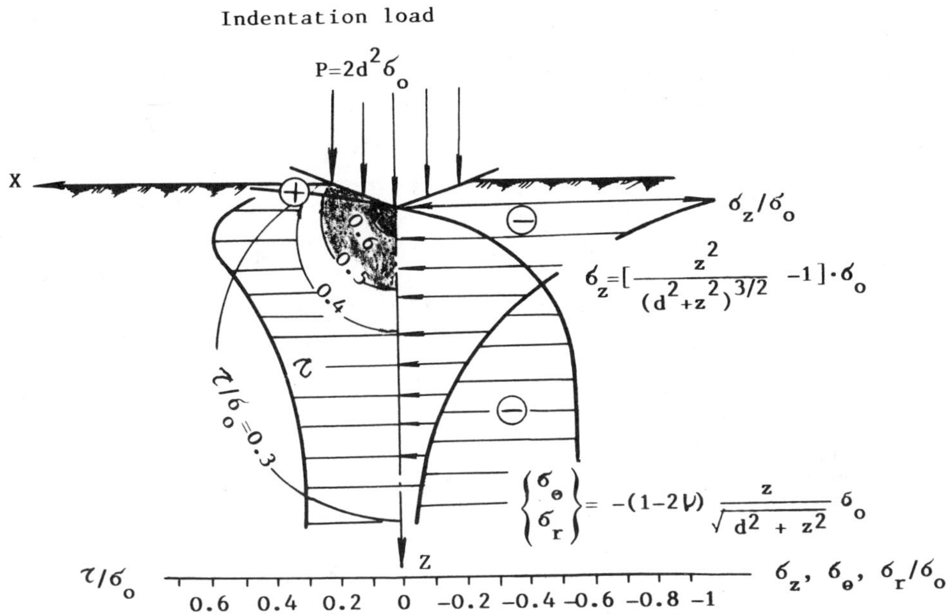

$$\sigma_z = \left[\frac{z^2}{(d^2+z^2)^{3/2}} - 1 \right] \cdot \sigma_o$$

$$\begin{Bmatrix} \sigma_\theta \\ \sigma_r \end{Bmatrix} = -(1-2\nu) \frac{z}{\sqrt{d^2 + z^2}} \sigma_o$$

(a) Principal stresses along the z-axis.

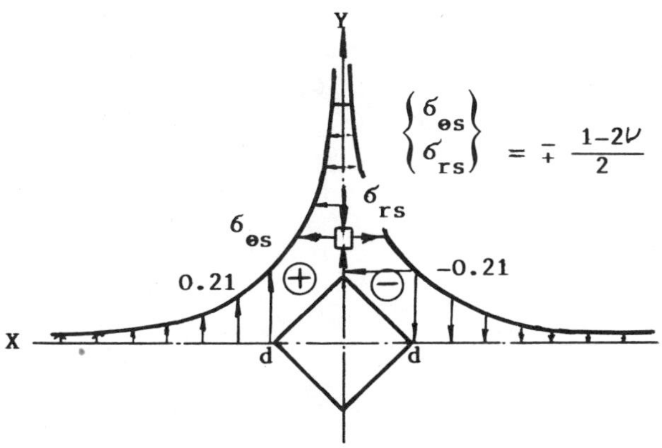

$$\begin{Bmatrix} \sigma_{\theta s} \\ \sigma_{rs} \end{Bmatrix} = \mp \frac{1-2\nu}{2} \cdot \ln \frac{y^2}{d^2+y^2} \sigma_o$$

(b) Surface stress along x- and y-axes.

Figure 1. Stress distribution near the indentation under the indentation load (a), (b).

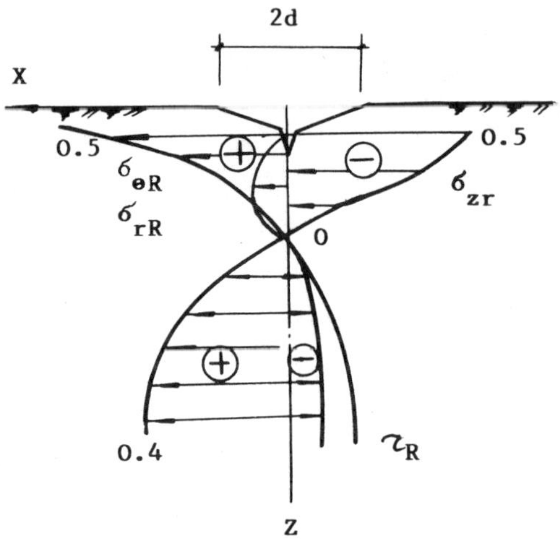

(c) Residual stresses along the z-axis

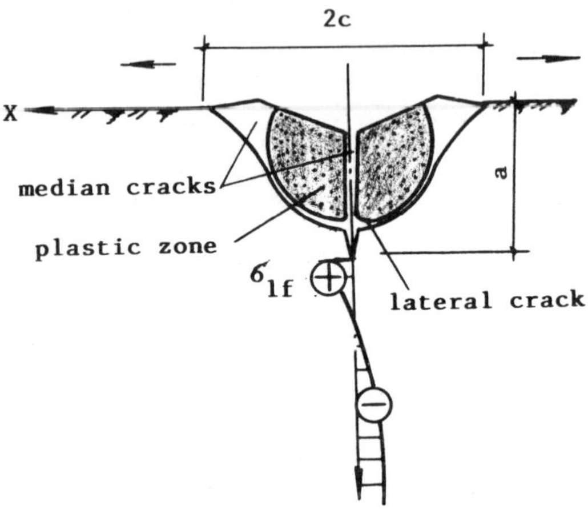

(e) Evaluation of the plastic zone and crack pattern

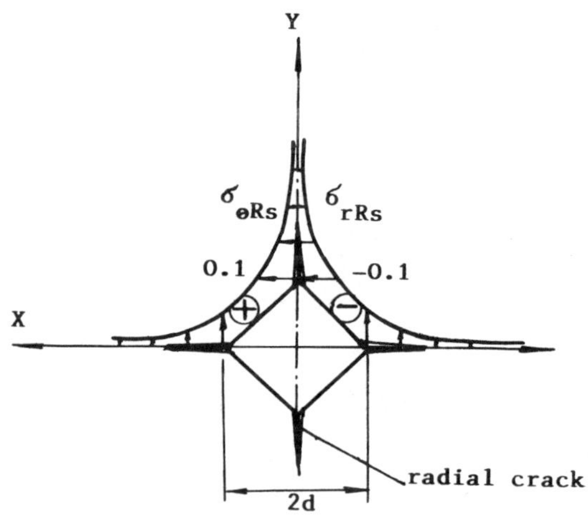

(d) Residual surface stresses along the x- and y-axes

Figure 1. Stress distribution near the indentation under the indentation load (a), (b); after unloading (c), (d) and after static loading (e).

CALCULATION OF K_{IC}

The peripheral stress distribution of a semi-elliptical surface crack on the tension side of the bend specimen requires three-dimensional elastic mechanics. In this work the Shah-Kobayashi [9] approximate solution is adopted. It is rather accurate and simple. According to fracture mechanics theory, the stress intensity factor at the end of the short axis for a semi-elliptical surface crack in bending is given by:

$$K_I = M_e \sigma \sqrt{\pi a}/\sqrt{Q} \tag{1}$$

where M_e is a numerical factor related to the crack and beam geometry, $M_e = M_f \cdot M_b$, M_f and M_b is front and back surface correction factor respectively, the crack depth is far less than the depth of the bar, so $M_b=1$ is taken, Shah puts forward $M_f=1+0.12(1-a/2c)$ which basically summarized various corrections of predecessors. σ is the maximum outer fiber tensile stress and a is the depth of the crack. Q is the shape factor of the crack and is given by:

$$Q = \phi^2 - 0.212(\sigma/\sigma_{ys})^2 \tag{2}$$

where σ_{ys} is the tensile yield stress and $0.212(\sigma/\sigma_{ys})^2$ is a plastic zone correction factor. For the brittle material in this work, this plastic correction factor may be negligible. The value ϕ is the second elliptical integral, which is tabulated in standard mathematical tables for the depth a and the length 2c of cracks.

When the bending stress reaches the fracture stress σ_f, the calculation formula is given by:

$$K_{IC} = [1+0.12(1-a/2c)^2]\ \sigma\ \sqrt{\pi a}/\phi \tag{3}$$

Since the contradiction of residual stress is present in the cases of both the indentation method and indentation-fatigue method, when residual stresses exist, the formula (3) requires a correction term. It is given by:

$$K_{IC} = [1+0.12(1-a/2c)^2]\ \frac{\sqrt{\pi a}}{\phi}\ (\sigma_f + \sigma_R) \tag{4}$$

where $\sigma_R=(0.1\text{-}0.3)\sigma_o$.

RESULTS AND DISCUSSION

(1) Effect of Static Loading on K_{IC}

For the brittle solid indented with the indentation-precracking method, after the load of indenter is removed, a residual stress field has been noted about the indentation [10,11]. Swain [9] considered three principal stresses being in the elastic stress field round an indentation under the action of the indenter load. It is the tensile stress of the three stresses, along the tangential direction of a circle which takes the point where the indenter is in contact with the specimen as the circle center, that has a principal effect to produce the median crack. After unloading, the elastic deformations near the plastic zone can not be entirely recovered, so that it gives rise to a residual stress [10-13]. There is a residual stress perpendicular to the median crack. It is the superposition of both the bending stress and the residual stress, which makes the measured toughness value less than the real value. In this sense K_{IC} is on the low side, which is measured with the usual method for indentation.

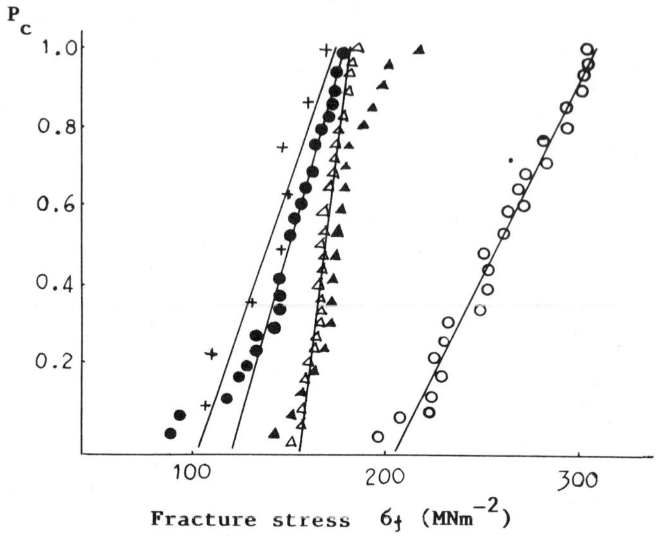

Fig. 2. The distribution of the fracture stresses for 99-Al_2O_3 bending specimens with various test methods. P_c is the cumulative fracture probability.

o - no indented
● - 10 kg Vickers indentation
Δ - 10 kg Vickers indentation and 48 hours static fatigue
▲ - 10 kg Vickers indentation and 2,000 times low-cycle fatigue under 0.8 P_f
+ - 10 kg Vickers indentation and 3 months aging treatment in laboratory

The distribution of the fracture stress values in Fig. 2, shows that the fracture stress of the specimens after indenting is smaller by 30-40% than with no indenting, and that the degree of scatter is clearly reduced. From the test results after fatigue, the fracture stress after the fatigue test is increased by 10-30%, and the magnitude of K_{IC} is increased by about 30%. Besides the increase of the fracture stress, the shape of the indentation crack is nearing a semi-ellipse and the ratio a/c reduces bring about the increase in K_{IC} of the fatigued bending specimens. The scatter range of fracture stresses for brittle materials can be explained by the Weibull statistical analysis [14,15]. Refer to Table 1, the Weibull modulus for the test results in this work may be seen. An increase of the Weibull modulus of K_{IC} for fatigued specimens explains the indentation-precracking specimens which underwent a short time of static fatigue not only removes or reduces the residual stresses at the indentation, but also reduces the scatter of the values. Since the survival rate in low-cycle fatigue under a larger load is very small, and the effects of low-cycle fatigue under smaller load are not clear, this result is not completely clear.

After indentation precracking the specimens underwent a process of static fatigue. The causes for the increase of K_{IC} is that with the crack growing the residual stress is removed, as the residual energy near the indentation is dissipated in the static fatigue test. Another is that the crack-tip produces the branch and the accretion in the static fatigue test,

TABLE 1: K_{IC} AND WEIBULL MODULUS BY INDENTATION-PRECRACKING METHOD AND THE INDENTATION-FATIGUE METHOD

	K_{IC} ($MNm^{-3/2}$)	Standard deviation ($MNm^{-3/2}$)	Number of specimens	Weibull modulus
Indentation-precracking	3.45	0.576	26	7.83
Indentation-static fatigue	4.71	0.414	28	14.4
Indentation-low-cycle fatigue	5.10	1.520	18	3.72

so that a group of microcracks near the crack-tip is formed. These microcracks are able to expand in the fracture test and can absorb more energy than a single crack, thereby they increase K_{IC}.

It is worthy of note that the median crack is a mixed crack of Mode I and II under the action of the indenter load, but the median crack-tip is Mode I under the fatigue load. In a fracture test, the crack-tip of the indented bending specimen is Mode I too; however, a very short loading procedure leads to the microcrack group not fully growing. Therefore, the change of the crack structures in the fatigue test is different from the indentation and fracture tests. For the crack of Mode I, the direction of the maximum principal tensile stress makes an angle of 60° with the directioı of the median crack, and the direction of the principal stress is at 90° (Fig. 3a) [16,17,18]. The direction of crack expansion is usually at right angles to the direction of the maximum principal tensile stress. For polycrystalline ceramics, this direction is also correlated with the microstructure. That is why the branched microcrack group (see Fig. 3b) is formed near the crack-tip. Only when the loading speed is quite slow as in fatigue does the branched microcrack group form.

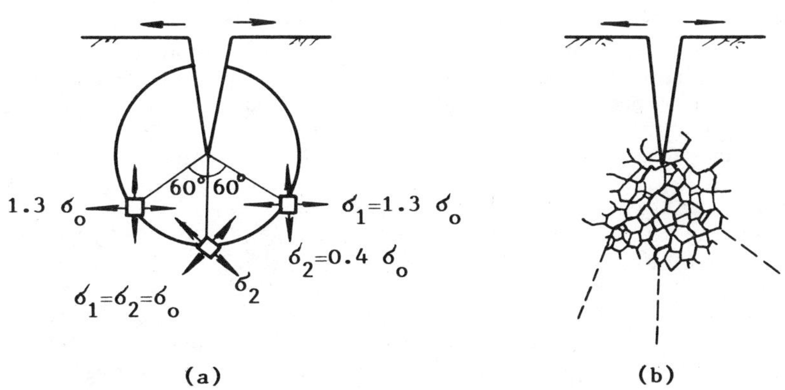

Fig. 3. The position and the direction of the maximum principal stresses at the crack-tip (a). The schematic drawing about crack branching at the crack-tip (b).

(2) Ageing of Residual Stresses

Holland and Turner [19] have pointed out that the residual stresses can be diminished with time by stress corrosion. Swain's test showed that the effect of the residual stress reduces with time, that is an ageing effect. It may be clearly observed after about 3 hours [10]. In this connection, some indented specimens were laid aside and after 3 months the fracture test was done to understand if ceramic materials have a similar ageing effect. The two sets of values, ageing and no ageing, are nearly the same. There is no difference between them by the t-distribution examination. This shows that the residual stress of an indented beam is not reduced with time.

Glass is an isotropic material. Its crack-tip may further develop in the residual stress field and the corrosion environment. Ceramics are generally polycrystalline materials. Complex crystal boundaries are the cracks and can increase the resistance of the crack twisting and branching. Since the energy of the residual stress field near the crack-tip is not large enough to overcome this resistance, the residual stress may not naturally dissapate unless the specimen is subjected to an additional load.

(3) Static and Low-cycle Fatigue

Indentation-static fatigue does not compare with the indentation-low-cycle fatigue in fracture strength (see Fig. 2), however the latter becomes even more scattered. When the load P of low-cycle fatigue is close to the fracture strength P_f, it is easy to observe the expansion of surface cracks. The survival rate for 2,000 cycles is only 20% at $P=0.8P_f$. For changes of surface crack length, the effects of low-cycle fatigue are more marked than for static fatigue. Differences of fracture strength are not large for static loads of $0.5P_f$ and $0.7P_f$.

Because the expansion of the indentation crack is related to the resistance at the crack front, only when the elastic energy at the crack-tip surpasses the work which resists the crack-tip expansion, can the crack expand. But when the elastic energy is too large, the crack runs instantaneously through the cross section. Moreover, the capability of resisting the crack expansion is related to the microstructure of the material. The different changes in the process of low-cycle fatigue are as follow: no expanding, suddenly breaking after some cycles, or random expanding with the increase of fatigue cycles. The phenomenon with which the crack of each specimen expands with increasing fatigue cycles is almost different for each sample. The resistance at the crack front to the crack growth depends upon the loading rate. For high rates of loading, such as in a fracture test, the crack can become critical before serious obstacles resisting crack growth are encountered. For low rates of loading, such as in a fatigue test, more crack growth occurs before failure, increasing the chance of interaction with a serious obstacle. If the crack meets a serious obstacle, then a greater applied load will be necessary for critical crack growth. Therefore, there are various phenomena of crack expansion with increasing fatigue cycles. Pletka [18] has discussed this problem for glass-ceramics.

Fig. 4 shows crack growth for indentations under the fatigue load. These data are too random to make a quantitative analysis, however, the essential aspect of the data is that subcritical crack growth does occur.

Air is a mild corrosion agent. When air, together with a fatigue stress, acts on the crack, the destructibility is more severe than when either of them acts alone on the crack [21,22]. On removal of the residual stress, the mechanisms of both static and low-cycle fatigue are different. The effect that low-cycle fatigue makes the crack expand is better than static

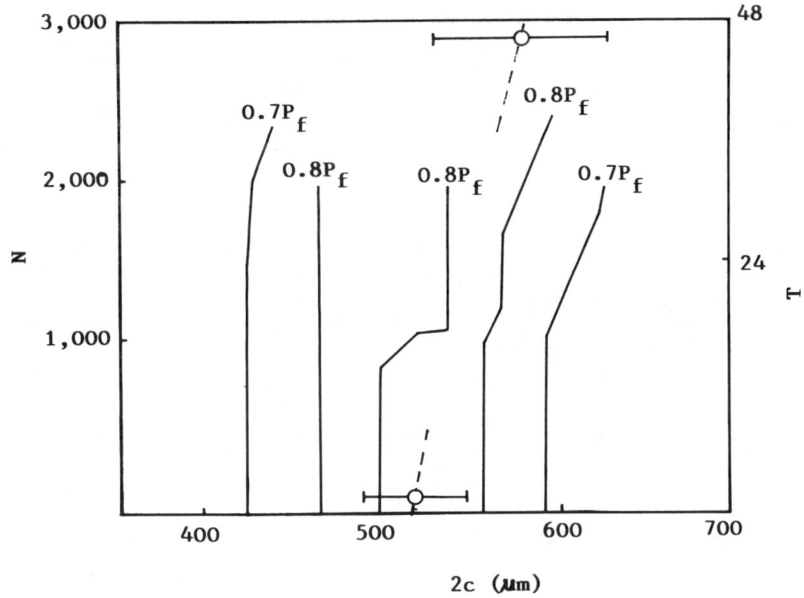

Fig. 4. Typical examples of crack expansion for indented beams in fatigue test; o - mean value of measured crack lengths of static fatigue, those horizontal lines show the standard deviation; 2c - the length of indentation crack; N - the number of low-cycle fatigue; T - the time of static fatigue.

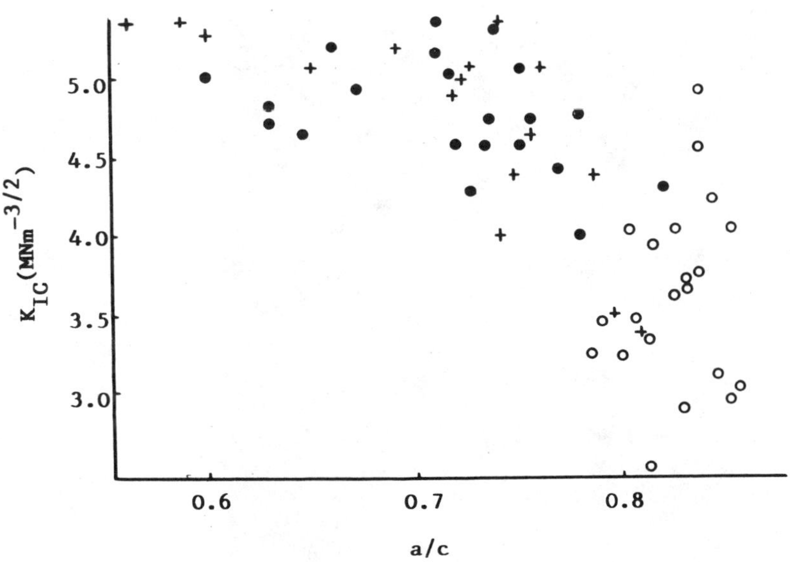

Fig. 5. The relationship between the ratio a/c of the crack size and the critical stress intensity factor, K_{IC}; o - 10 kg Vickers indentation; • - 10 kg Vickers indentation and static fatigue; + - 10 kg Vickers indentation and low-cycle fatigue.

fatigue. The cause lies in that the cyclic loading enables the crack-tip to gather up elastic energy, so that the level of stress concentration is raised and the energy is dissipated by crack growth. But, in the static fatigue test, the process by which the residual stress is removed is the result that air corrosion and the static load act on the crack together.

(4) Fracture Surface Examination

Fracture surface examination was completed by SEM. The indentation cracks are semi-circular in shape, and the crack depth is about 150-200 μm. The ratio of depth to length a/c > 0.8, and the fracture surface shapes are the same as Govila's observation for hot-pressed Si_3N_4. After fatigue testing the shape of the cracks approach a semi-ellipse, the depths of which are about 200 - 300 μm with the ratio a/c nearing 0.7. The increment of the crack length measured before and after the fatigue test is longer than the depth increment, so that the shapes of the cracks approach the hypothesis of the calculation.

(a)
200X

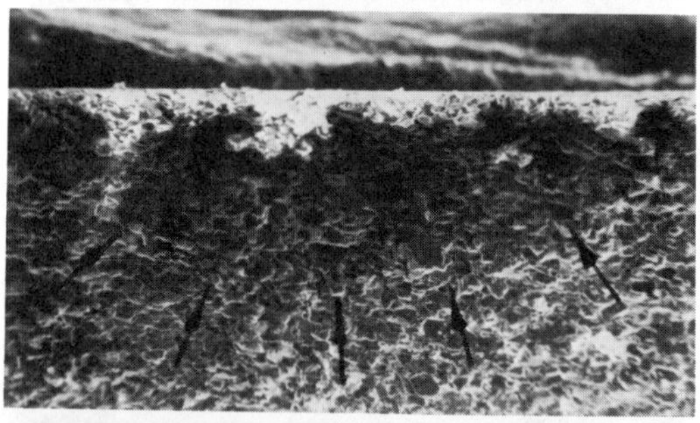

(b)
100X

(c)
100X

Fig. 6. Typical scanning electron micrograph of the fracture surface of a precracked specimen of 99-Al_2O_3 tested at 20°C. Approximate crack depth is 173 μm (a). An indentation-static fatigue specimen, with approximate crack depth of 193 μm (b). An indentation-low-cycle fatigue specimen of approximate crack depth is 204 μm (c). Arrows indicate the boundary of the crack region.

The photographs of Fig. 6 are three typical cases. In some broken specimens, the crack boundary could not be measured with accuracy, but it could be identified after dyeing. This creates about some errors in the calculation of K_{IC}. Especially for low-cycle fatigue, the crack boundary is uneven, which is caused by the differences in resistance to the crack to expand. When calculating K_{IC}, it is recommended to take the deepest point as the crack boundary.

CONCLUSIONS

Because there are residual stresses near the indentation, when the critical stress intensity factor of ceramics for a semi-elliptical surface crack in bending is measured with the help of indentation precracking method, the value of K_{IC} is low. Static fatigue after indenting may effectively reduce or remove the residual stresses, and bring about a microcrack group at the crack-tip so that more energy is absorbed. For 99-Al_2O_3 ceramics, the mean value of K_{IC} is 3.67 $MNm^{-3/2}$ and 4.71 $MNm^{-3/2}$ respectively by means of the indentation-precracking method and the indentation-fatigue method. The indentation-fatigue method increases the measured value of K_{IC} by 28%, this value is close to the true value. In addition, the discreteness of the indentation-fatigue method is smaller than the indentation-precracking method.

The static fatigue method is simple and reliable, its applicability to raise K_{IC} is the best. The effectiveness of crack expansion for low-cycle fatigue is more transparent, but the survival rate is rather low. In the indentation-fatigue test, it is proven that subcritical crack growth does occur for indentation-precracked specimens.

The critical flaws of indentation-precracked specimens is generally a semi-circle in shape. The critical flaws shape of the indentation-fatigue specimens approaches a semi-ellipse, and this agrees even more with the hypothesis for the calcuation of K_{IC}.

The residual stress near indentations does not automatically reduce with time for ceramics. From the point of view of the differences between the microstructure of glass and ceramics, the ageing effects of the two materials are not the same at all.

REFERENCES

1. A. G. Evans and E. A. Charles, J. Am. Ceram. Soc., 59, 371 (1976).
2. K. Niihara, R. Moren and D. P. H. Hasselman, J. Mater. Sci. Lett., 2, 221 (1983).
3. J. J. Petrovic, et al., J. Am. Ceram. Soc., 58 (3-4) 113 (1975).
4. B. R. Lawn, in Fracture Mechanics of Ceramics. Vol. 5, Plenum Press (1982).
5. J. J. Petrovic, et al., J. Am. Ceram. Soc., 59 (3-4) 177 (1976).
6. T. Nishikawa, et al., Yogyo-Kyokai-Shi, 91 (4) 149 (1983).
7. Zong-zhe, Jin, et al., Evaluation on Strength of Microfracture (to be published).
8. Juan-rong, Ma, et al., Microscopic Photoelastic Stress Analysis of Indentation, (to be published).
9. R. C. Shah, A. S. Kobayashi, ASME, On the surface crack. 79 (1972).
10. M. V. Swain, in Fracture Mechanics of Ceramics, Vol. 3, Plenum Press (1978).
11. B. R. Lawn and E. B. Marshall, J. Am. Ceram. Soc., 62 (7-8) 347 (1979).
12. B. R. Lawn and D. B. Marshall, in Vol. 3 Fracture Mechanics of Ceramics, Plenum Press (1978).

13. B. R. Lawn, et al., J. Am. Ceram. Soc., 63 (9-10) 574 (1980).
14. Ayal De S. Jayatilaka, Fracture of Engineering Brittle Material, London, 208 (1979).
15. R. W. Davidge, Mechanical Behavior of Ceramics, Cambridge, 133 (1979).
16. Zong-zhe, Jin, Doctor Thesis, Tohoku University, Japan, 1615 (1982).
17. Zong-zhe, Jin, T. Yokobori, I. Maekawa and A. T. Yokobori, Jr., Proc. Int. Sym. on weak rock, 279 (1981).
18. Zong-zhe, Jin, T. Yokobori, J. Maekawa and A. T. Yokobori, Jr., Proc. Int. Sym. Fracture Mech., 944 (1983).
19. A. J. Holland and W. E. S. Turner, J. Soc. Glass Tech., 21, 383 (1937).
20. B. J. Pletka and S. M. Wiederborn, in Fracture Mechanics of Ceramics, Vol. 4 (Plenum Press (1978)).
21. S. R. Swanson, Handbook of Fatigue Testing (Translated), Shanghai, 151 (1982).
22. S. M. Wiederhorn, in Vol. 2. Fracture Mechanics of Ceramics, Plenum Press (1974).
23. R. K. Govila, J. Am. Ceram. Soc., 63 (5-6) 319 (1980).
24. Zhen-Duo, Guan, Xin-Min, Du, Guisuanyan-Xuebao, 10 (3) 262 (1982).
25. R. F. Pabst, in Fracture Mechanics of Ceramics, Vol. 2, Plenum Press (1974).

NON-LUBRICATED SLIDING WEAR OF Al_2O_3, PSZ, and SiC

J. Derby, S.G. Seshadri and M. Srinivasan

Sohio Engineered Materials Company
Niagara Falls Technology Center
Niagara Falls, New York 14302

ABSTRACT

The non-lubricated wear behavior of Al_2O_3, PSZ, and two grades of sintered alpha SiC has been studied under unidirectional sliding conditions at room temperature. Using a ring on ring geometry, the wear rates were established as a function of normal load and sliding time. Mechanistic understanding is explored by extensive microscopy and profilometry of the wear surface at specific operating regimes. Various wear mechanisms such as asperity shearing, ploughing, and brittle fracture operate at different loads depending on material properties.

INTRODUCTION

Ceramic materials possess excellent wear and corrosion resistant properties and are being assessed by the automotive, seal and chemical processing industries for use in engines, chemical pumps, and wear resistant sleeves. It is important to know the wear behavior of these materials when selecting appropriate applications, therefore tribological studies must be conducted by using suitable test configurations and operating conditions. The proper selection of test conditions allows for the correlation of test results to offer guidance for potential engineering applications.

Wear resistant materials are commonly used in applications where poor lubricating conditions may exist. In seal testing under lubricated conditions, it has been found that the "start up" and "break in" period produce higher frictional forces than "steady state" operating conditions. The increased friction is due to the interaction of asperities, similar to that occurring under dry sliding conditions[1]. Because these periods of high friction produce surface damage, the mechanisms of wear under surface to surface contact are of interest.

In this paper the wear behavior of Al_2O_3, PSZ and two SiC materials with differing densities are evaluated under non-lubricated unidirectional sliding conditions. Mechanistic understanding is explored by using extensive analysis of the wear surface at specific operating regimes. Test results are evaluated in relation to the material properties and their implications in pertinent engineering applications are discussed.

BACKGROUND

I. Mechanisms

An accurate assessment of tribological behavior is difficult due to the complexity of the interactions of a large number of variables. Wear is not an intrinsic material property, but rather a response which is governed by the following factors: (a) the nature of material surfaces in contact including the mechanical response, surface conditions, and contact configuration, (b) lubricant properties and (c) the environment which can influence various surface events such as chemisorption, compound formation, segregation, and reconstruction of the atomic layers.

Buckley[2] has demonstrated that tribological properties are influenced by the presence of surface contaminents which act to reduce surface adhesion in metals. Ceramic materials have lower adhesive forces than metals but may also form surface layers which alter wear and friction properties.

The wear behavior of a material is best understood in terms of a particular wear theory, or the combination of several. Three existing theories invoke the concepts of adhesion, abrasion and brittle fracture. The theory of adhesion was developed by Tabor[3] and has been expanded by Rabinowicz[4]. According to this theory, the deformation of asperities and removal of surface films occur during the initial stages of sliding contact. Junctions are then formed causing the two contacting materials to adhere. During subsequent sliding these junctions are sheared and material is transferred from one surface to another. If the two surfaces are of dissimilar chemical nature, then particles will be transferred from the cohesively weaker to the cohesively stronger material and ultimately the formation of a wear particle occurs. The following schematic describes the adhesive wear model, (figure 1).

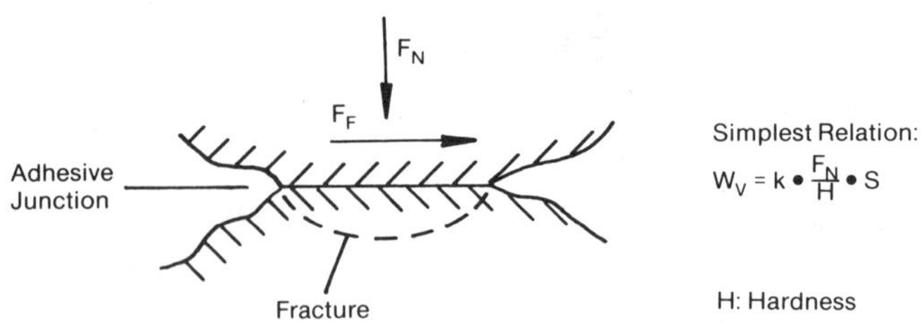

Figure 1: Characteristics of adhesive wear models.

The theory of wear by abrasion considers the effect of a relatively hard asperity sliding against a softer opposing surface. Under the influence of a tangential force the asperity removes the counterface material by the combined effects of "micro-ploughing", "micro-cutting", and "micro-cracking". Most analyses consider the asperity to be sharp as shown in the following diagram, (Figure 2), whereas in reality this is rarely the case.

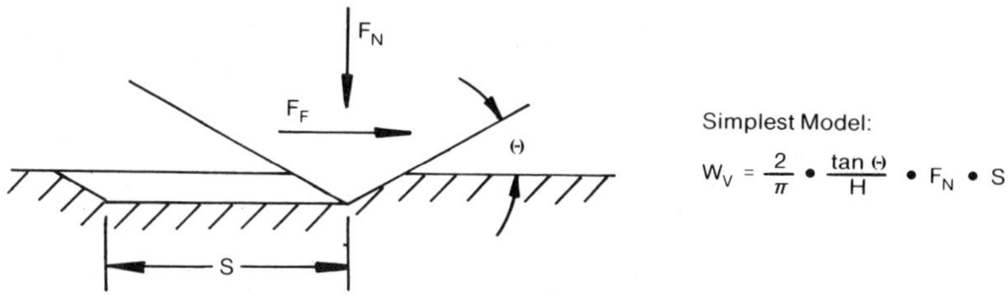

Figure 2: Characteristics of the abrasive wear model.

Ceramic materials exhibit low forces of adhesion and less abrasion than metals. Numerous studies have shown that the wear of ceramics occurs primarily through a brittle fracture process[5-7]. The high brittleness of the ceramics, arising from their low fracture toughness and relatively high hardness, contributes to the formation and propagation of subsurface cracks under dynamic loading conditions. Recurrent loading causes the intersection of cracks and the subsequent formation of wear particles (Figure 3). When pulled from the surface, the particles are trapped at the interface. The specific friction and wear behavior of the test materials is strongly influenced by the wear particles and the changes in their size and shape. At some time, depending on the experimental variables and test configuration, the particles are removed from the system as wear debris.

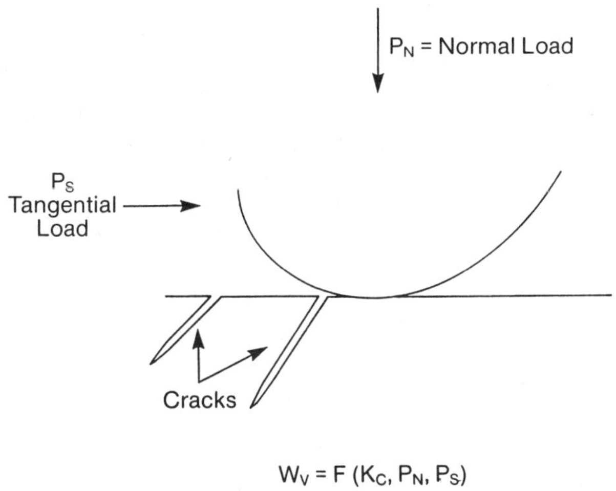

Figure 3. Schematic of the brittle fracture under sliding contract.

II. Recent Work in Ceramics

West[8] has reported on the wear and friction behavior of thirty ceramic materials by using a pin on disc test geometry. The normal load (8.1 N) and sliding velocity (0.3 m/sec) were kept constant during testing. Wear rates were estimated by using indirect microscopic techniques for the pin, and profilometry traces for the disc. The friction coefficients for SiC materials ranged from 0.4 to 0.9 at room temperature and from 0.6 to 1.0 at 800°C. Similar behavior was observed for other materials. The initial surface roughness was not controlled and the reported values for various samples ranged from 3 to 55 CLA (center line average). No correlations were observed between the friction and wear coefficients of any of the materials.

Breznak et al.[9] has studied the wear and friction behavior of SiC and PSZ materials by using flat rectangular specimens under reciprocal sliding conditions. The oscilation frequency used was 10 Hz with a 75 cm stroke, corresponding to an average sliding velocity of 1.5 m/sec, under a normal pressure of 35 kPa. The coefficients of friction generally ranged from 0.3 to 0.5. It was presumed that friction coefficients would be influenced by the initial surface conditions of the samples, yet no significant correlations were observed. Post test inspection of the wear track areas, however, indicated that sample alignment was a problem, which may have adversely affected the observed results.

Lankford [10] employed pins-on-disk type system to study the friction and wear behavior of some candidate diesel engine materials. He suggested that adhesive wear mechanisms causing surface delamination result in disk wear (heavy pin wear) in many combinations involving PSZ, SiC, Si_3N_4, etc.

EXPERIMENTAL METHODS

For the present work, a Falex wear tester was utilized in which a stationary sample seat was pressed against a rotating face by applying various loads to a lever arm (Figure 4). The sliding velocity was held constant at 0.8 m/sec (500 rpm) and all tests were conducted in air at room temperature. Normal loads ranged from 10 to 55 N and sliding distances up to 54 kms were used. All samples were one inch diameter rings which were prepared flat (two Helium light bands) prior to testing. Following each test the samples were carefully weighed and the wear surfaces were examined by using profilometry and scanning electron microscopy.

A procedure was developed to ensure proper mating of the contact surfaces. The sample alignment was measured by comparing profilometry traces of the wear track depths obtained at the four quadrants of the specimen, (Figure 5). If a comparison of the traces indicated misalignment of more than 5.0 μm, then the test was discarded.

Sliding distances were calculated from rpm and test duration. Wear, determined by weight loss measurement, was used to calculate wear rates using the contact area and test duration.

MATERIALS

The relationship of the fundamental properties of a material to its wear behavior is essential for establishing microstructure-property understanding. The test materials were screened for optimal properties and were selected accordingly to enhance the assessment for modelling purposes. The sintered α-SiC (SASC), (SA-80) is a single phase polycrystalline material with an equiaxed fine grained structure (5-8 μm), containing no free silicon. The SASC exhibits excellent corrosion and oxidation resistance and is made from readily available low cost

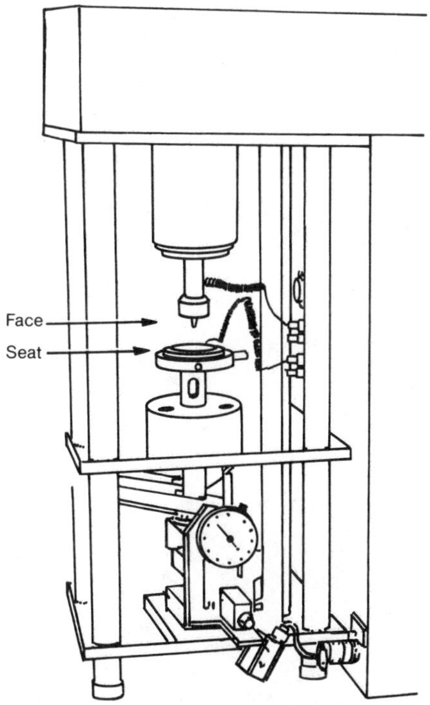

Figure 4. Schematic of sample orientation and test system.

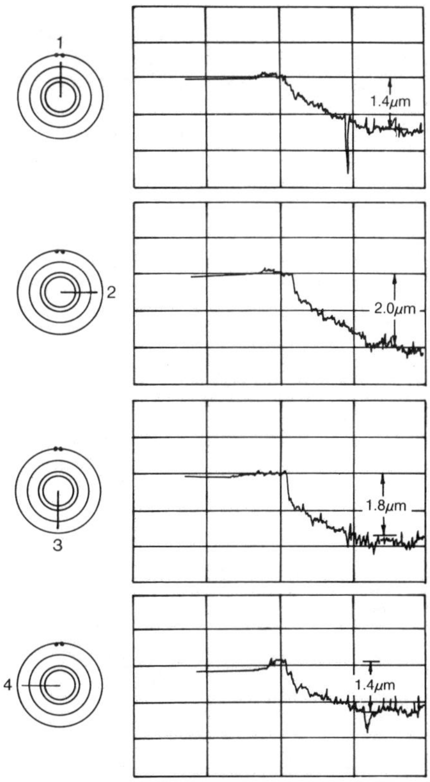

Figure 5. Profilometry traces of wear track (the trace path is shown in the ring).

materials. This material has relatively lower density, has high thermal conductivity and high hardness but lower fracture toughness, depending upon density.

The controlled porosity SiC (SA-80U), is a lower density form of SASC having a uniform distribution of small pores. The increased porosity reduces the hardness but increases the toughness by impeding crack propagation, via crack blunting. The Al_2O_3 is a low porosity ceramic having a density of 3.90 gm/cc. Compared to SASC, the Al_2O_3 has lower hardness but slightly higher toughness. The PSZ is a zirconium oxide stabilized with three mole percent yitrium oxide addition. The average grain size of this material is 0.5 μm and the principal phase is tetragonal with minor monoclinic phases present [11]. Pertinent properties of these materials are listed in Table 1 and their microstructures are shown below (Figure 6).

RESULTS AND DISCUSSION

An accurate assessment of wear behavior requires the establishment of a statistical baseline against which subsequent wear results can be judged. Studies were conducted on SASC to investigate the inherent

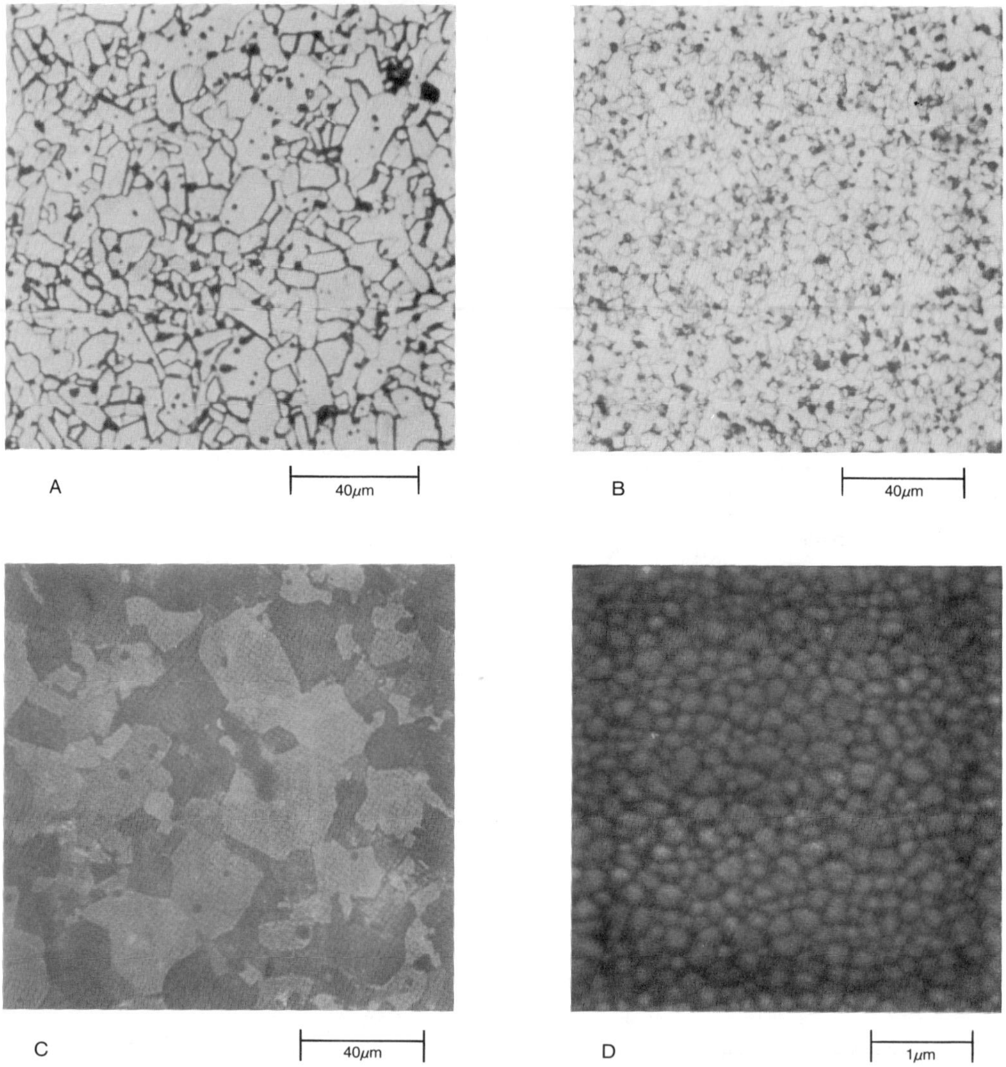

Figure 6. Microstructures of high density SiC (A), controlled porosity SiC (B), Al_2O_3 (C), and PSZ (D).

TABLE 1 TEST MATERIAL PROPERTIES

MATERIAL	DENSITY (g/cm^3)	HARDNESS (GPa)	YOUNG'S MODULUS (GPa)	FRACTURE TOUGHNESS+ (MPa $\sqrt{m}$)
Sintered αSiC (SA 80)	3.16	24.5	400	3.8 - 4.5
Controlled Porosity SiC (SA 80U)	3.00	20.0	360	5.0 - 5.8
Al_2O_3	3.90	18.3	410	5.0 - 6.0
PSZ	6.00	12.7	200	5.5 - 6.5

+ Vickers indentation technique

variation in wear data by using our test methodology. The load and sliding velocity were held constant but the test intervals were allowed to vary. The weight loss data were collected and plotted as a function of sliding time (Figure 7). The data scatter, which was ±9%, reflects the variations in specimen surface preparation and specimen loading. Weight losses for the sample seat were similar to those observed for the matching face. Profilometry measurements indicated that rapid increases in surface roughness had occured during the initial sliding periods, followed by trends of decreasing roughness toward a steady state value.

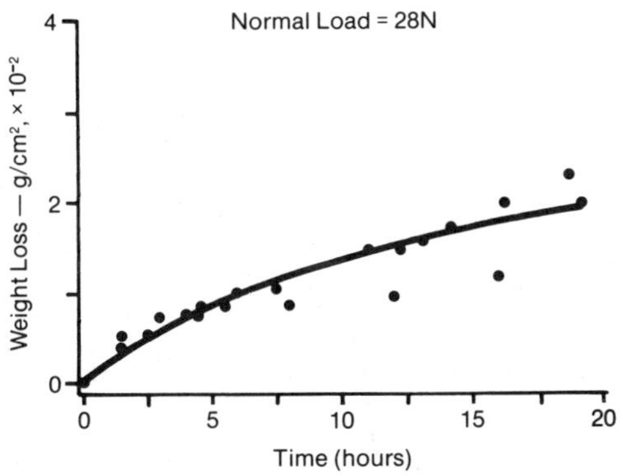

Figure 7. The wear of SA-80 at a normal load of 28N.

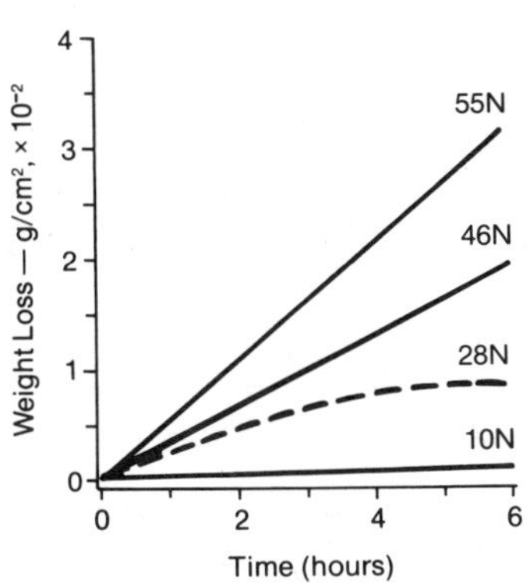

Figure 8. The wear of SA-8 as a function of load.

After establishing a baseline at a set load,additional experiments were conducted on SASC at various loads and the wear results were plotted, (Figure 8). Weight losses for most of the loads exhibited a linear dependence on sliding distance. The intial weight losses, compared to the final, were more pronounced at the 28 N load than for other loads tested. Profilometry measurements indicated that the worn surfaces of samples used for high load tests were much rougher than the low load samples. High loads were also found to produce substantially higher wear in the samples. Some interesting implications of these observations will be addressed later.

Tests conducted at high loads were normally noisy and excessive noise levels were generally found to be accompanied by sample chipping. Whenever chipping behavior was observed, the experiment was terminated and weight loss data during that test interval were discarded.

Wear tests were expanded to include like sliding combinations of Al_2O_3, PSZ and SA-80U. A high load was chosen for test purposes to enhance comparison of the test results. Uniform test procedures were maintained for samples of each material and the wear data were collected and plotted, (Figure 9). The wear behavior of Al_2O_3 at this load was identical to SASC. This was an interesting result since the material properties, especially hardness, of Al_2O_3 and SASC are dissimilar. The SA-80U, however, which is chemically identical to SASC yielded significantly higher wear by comparison. The PSZ, reported to exhibit low wear using pin-on-disk geometries [12], showed the worst performance in this application. The present findings are consistent with the of Breznak et al.[9] obtained on flat plates under reciprocal sliding conditions.

Further investigations were conducted on the wear characteristics of these materials over a range of loads. The data were collected by employing the standard procedures and the wear rates were determined by using linear regression. The wear rates were normalized for contact area and were plotted as functions of normal load, (Figure 10). A comparison of these plots shows that the relative wear behavior of different materials changes for different loads and that various wear processes are most likely operating at different stages.

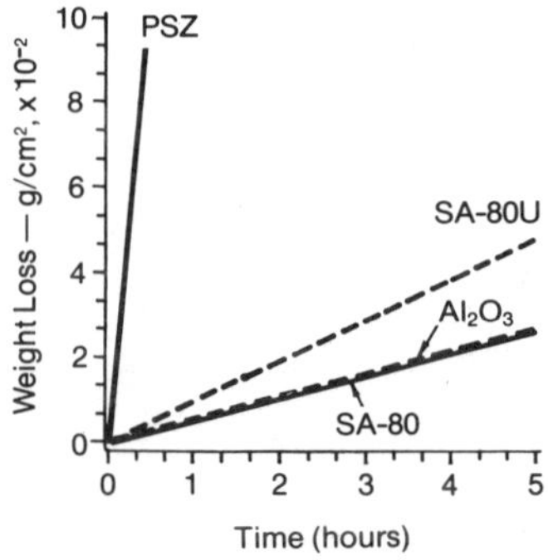

Figure 9. The wear of different ceramic pairs at a normal load of 55 N.

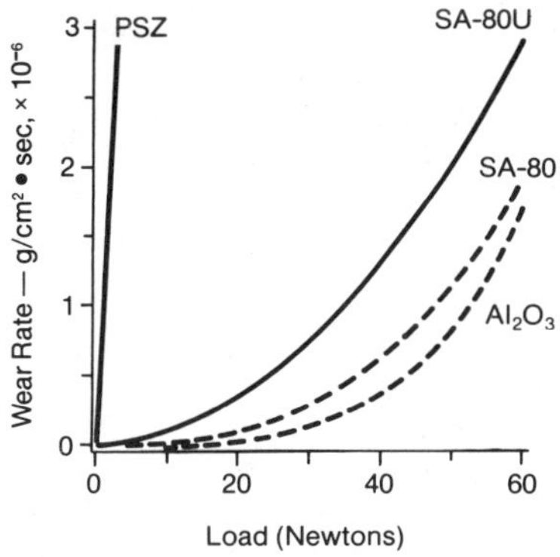

Figure 10. Relative wear rates of different ceramic pairs as a function of normal load.

The SASC specimens tested at the 10 N and 19 N loads were found to have smooth contact surfaces, (0.03 μm R_a). The Scanning Electron Microscopy was used to identify distinguishing features which were indicative of specific wear processes[13]. The contact area created at the 19 N load exhibited micro-ploughing and numerous shallow groves could be observed in the direction of sliding (Figure 11). Some surface porosity and areas of grain pullout were present and minor cracking was also observed. Samples tested at 28 N contained mixed areas of roughness due to transgranular fracture. Contact surfaces at this load tended to become somewhat smoother over extended sliding distances. The early onset of surface roughening explains the high initial weight loss which was observed in the samples at this load (Figure 8). The use of higher normal loads produced proportionally more transgranular fracture and increased surface roughness.

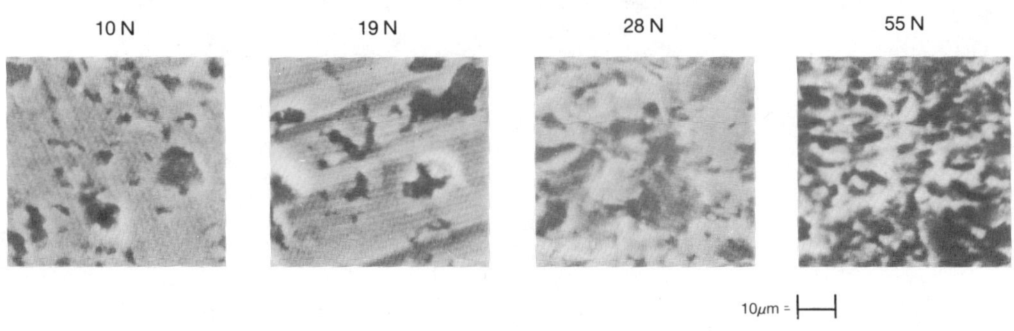

Figure 11. The wear surfaces of SA-80 as a function of increasing normal load. The wear process involves a combination of ploughing, grain cracking and grain pull out.

The wear behavior of SASC is consistent with the theory that threshhold stresses, as predicted by Evans[5], are required for crack propagation. An important indicator of crack initiation and propagation is the brittleness index determined by toughness and hardness. It has been shown that the toughness of a material can be increased through the uniform dispersion of small pores [14], and that the hardness decreases with increasing porosity.

The controlled porosity SiC has exhibited superior wear behavior in lubricated conditions in the past [15-16], owing to the entrapment of lubricating fluid in the surface pores and reducing the fluid evaporation during initial periods of use. In the non-lubricated conditions, this material, however, shows slightly increased wear than the SASC grade material (Figure 10). It is supposed that the wear is caused by the formation of microcracks and their interaction with each other and with the surface porosity. The elongation of pores in the sliding direction [17] in the wear surface generally supports this view (Figure 12) but further study is required to clarify these observations.

It is interesting to note that the wear surfaces produced at low loads of both SiC materials are similar. Polishing and ploughing mechanisms are observed and the microstructure has an etched appearance where the grain boundaries are clearly visible. The use of extended sliding distances and increased load may produce microcracking at the grain boundary, thereby enhancing subsequent grain pullout.

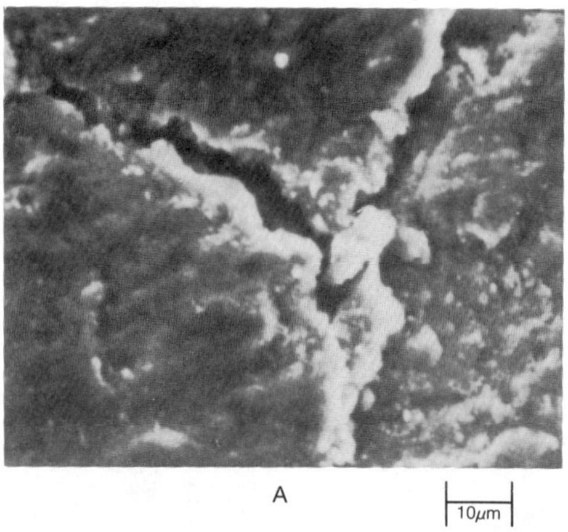

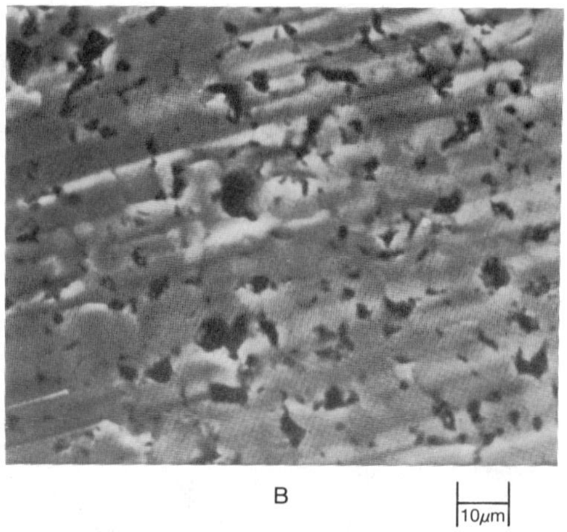

Figure 12. The wear surfaces of SA-80U at high load (A) and low load (B).

The wear rate of Al_2O_3 was also observed to increase exponentially with increasing load. At low loads, the wear rate is similar to SASC. An examination of the wear surface of Al_2O_3 indicates the formation of smaller wear particles (Figure 13), than those observed for the sintered SiC with controlled porosity (SA-80U). As shown in Figure 13, limited scale delamination wear is also observed for Al_2O_3.

Figure 13. The wear surface of Al_2O_3 at a load of 55 N.

Profilometry traces of the wear surfaces for the various materials show some interesting trends. The contact surface roughness of Al_2O_3 and SASC range from 0.3μm R_a at low loads to 0.30 μm R_a at high load. The SA-80U also exhibits smooth surfaces at low loads, (0.03μm), but very rough surfaces are created at high load, (0.63μm R_a). All profilometry traces were obtained by using a 0.8 mm cut off length and no attempt was made to characterize the average waviness of the sample. It is recognized, however, that waviness is an important parameter and that the use of improper cut off lengths for roughness determination can result in significant errors.

The PSZ has unique wear properties in that it exhibits slightly higher toughness to the other materials tested yet it has the lowest wear resistance. The toughness of PSZ occurs due to phase transformations produced from stresses generated at the crack tip. Tetragonal to monoclinic phase transformations act to introduce compressive stresses at the crack tip and to impede crack propagation. The SEM observations of the wear area show no indication of brittle fracture behavior but large wear particles are present, (Figure 14). It is suggested that stresses produced during sliding cause phase transformations to occur in the subsurface layers. In the process, the volumetric changes cause large sheets of wear debris to become separated from the rest of the sample, resulting in high wear rates. Similar high wear results were observed for PSZ by Breznak et al.[9] and Lankford[10]. Breval et al.[18], in subsequent work, conducted x-ray and electron diffraction experiments on wear debris and could not demonstrate conclusively whether tetragonal PSZ undergoes martensitic transformation to the monoclinic polymorph. Lankford[10] has attributed the high PSZ wear due to delamination which is also observed in this work (Figure 14 b).

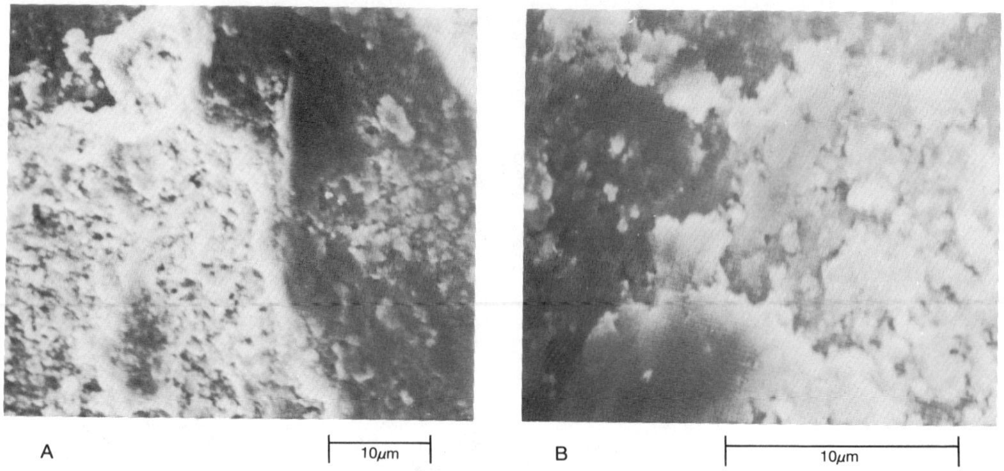

Figure 14. The wear surface of PSZ at a load of 10 N.

The results presented here have some clear implications for engineering application. In situations where ceramics are sought to achieve low wear, it is necessary to choose the proper operating conditions and the correct material combinations to minimize brittle fracture. Although it is the subject of a later publication, it should be pointed out that various tests have been conducted using dissimilar sliding material combinations. Under these conditions, brittle fracture wear was not observed and wear rates were greatly reduced. Studies are in progress to determine the nature of the differences in material properties which are required to avoid brittle fracture wear.

The chemical nature of the contacting surfaces may also play an important role in the wear mechanisms[19]. The formation of oxide layer in the silicon carbide ceramics during wear has been observed in the past[17, 20]. Cranmer[19] has carried out wear tests at higher loads in silicon-based ceramics and observed extensive SiO_2 layer formation. Such surface modifcations alter the friction behavior considerably.

Additional variables such as the vibrations generated at the sliding interface will also influence the wear behavior. The wear rates of SASC were found to increase as the square of sliding velocity in the range 0.08 to 1.60 m/sec. It is generally known that the frequencies of these vibrations depend upon the sliding velocity. Although some limited work has been done in this area[21-22] further investigation of the frequency effects on wear is needed to complete the understanding of the mechanisms.

CONCLUSIONS

Wear testing under non-lubricated conditions using like sliding combinations of silicon carbide, aluminum oxide and PSZ has yielded the following observations:

1. Wear rates are strongly influenced by normal loads.

2. Various wear mechanisms become operable at different load regimes depending on the material properties of the sample.

3. The wear in SiC and Al_2O_3 at high loads is postulated to be due to brittle fracture mechanisms; both in Al_2O_3 and PSZ, delamination wear is also observed.

4. The brittleness index of a material may not be indicative of wear behavior depending on the operable wear mechanism.

ACKNOWLEDGEMENT

The experimental assistance of James Lipa and critical review of the manuscript by T. Sweeting, R. Storm, R. Lashway and D. Ronyak are gratefully acknowledged.

REFERENCES

1. J. Derby and S. G. Seshadri, unpublished work, 1984.
2. D. H. Buckley, "Surface Effects in Adhesions, Friction, Wear and Lubrication", 1981, Elsevier.
3. D. Tabor, "The Hardness of Metals", 1951, Oxford University Press, Oxford.
4. E. Rabinowicz, "Friction and Wear of Materials", 1965, John Wiley and Sons, New York.
5. A. G. Evans, "Abrasive Wear in Ceramics: An Assessment", 1979, NBS Special Publication 562.
6. B. R. Lawn, S. M. Wiederhorn, D. E. Roberts, J. Matr. Sci, 19, 2561-2569 (1984).
7. N. Suh, "Tribophysics", (to be published), personal communication (1984).
8. A. West, F. Carignan, W. Syniuta, R. Cannon, "Evaluation of Improved Stationary Diesel Engines Operating on Residual and Coal Based Fuels", 1983, Report No. DE-AC19-80BC10362.
9. J. Breznak, E. Breval, N. McMillan, submitted to J. Matri. Sci. (1985).
10. J. Lankford, "Sliding Seal Materials for Adiabatic Engines," Phase I Interim Report No. NASA CR-174893 U.S.D.O.E. April (1985).
11. S. M. Kunz, W. D. G. Boecker, and C. H. McMurtry, Extended Abstracts, 87th Ann. Mtg. of Amer. Ceram. Soc., p. 309, (1985).
12. C. S. Yust, "Damage Processes in Sliding Partially Stablized Zirconia Pairs", 1984, Proceedings of the International Conference of Ceramics, Japan.
13. J. Derby, J. MacBeth and S. Seshadri, Proc. I. Mech. E. (London) p. 133-137 (1985).
14. M. Srinivasan, S. G. Seshadri, Abstract in the Ceramic Bulletin, [63], 8, 996, (1984).
15. J. Derby, S. G. Seshadri and M. Srinivasan, 37th Pac. Coast. Reg. Meeting of Amer. Ceram. Soc., San Francisco, Oct. 28-31 (1984).
16. J. Derby, S. G. Seshadri, M. Srinivasan, 38th Annual Mtg. of ASLE, Chicago (1984).
17. J. Derby, S. G. Seshadri, M. Srinivasan, unpublished work, (1985).
18. E. Breval, J. Breznak and N. H. McMillan, submitted to J. Matr. Sci. (1985).
19. K. Miyoshi, D. H. Buckely, M. Srinivasan, Am. Cer. Soc. Bull., 62 4, 494-500, (1983).
20. D. C. Cranmer, J. Matl. Sci., in press, (1985).
21. Ye. A. Kuznetsov, G. W. Gorokhovsky, Wear 73I 41-59 (1981).
22. A. Soom, C. Kim, 1981, ASME Reprint No. 82-Lub-6.

MIXED-MODE FRACTURE OF CERAMICS

J. J. Petrovic

Materials Science and Technology Division
Los Alamos National Laboratory
Los Alamos, New Mexico 87545

ABSTRACT

The mixed-mode fracture behavior of ceramic materials is of importance for monolithic ceramics in order to predict the onset of fracture under generalized loading conditions and for ceramic composites to describe crack deflection toughening mechanisms. Experimental data on surface flaw mixed-mode fracture in various ceramics indicate that the flaw-plane normal stress at fracture decreases with increasing in-flaw-plane shear stress, although present data exhibit a fairly wide range in details of this $\sigma - \tau$ relationship. Fracture from large cracks suggests that Mode II has a greater effect on Mode I fracture than Mode III. A comparison of surface flaw and large crack mixed-mode I-II fracture responses indicates that surface flaw behavior is influenced by shear resistance effects.

INTRODUCTION

The mixed-mode fracture behavior of fracture-initiating flaws in ceramic materials is of importance for two reasons. First, this is the necessary basis of realistic, physically-based multiaxial loading fracture criteria for monolithic ceramics with a statistical flaw distribution (1,2). Secondly, such behavior is intimately related to crack deflection toughening mechanisms in ceramic composites (3). The purpose of the present paper is to review experimental mixed-mode fracture results for ceramics and other brittle materials, and to compare these results to the predictions of present theoretical mixed-mode fracture criteria in order to determine their applicability.

MIXED-MODE FRACTURE THEORIES

The three Modes of fracture are shown in Figure 1. Mode I, the opening mode, exhibits crack surface displacements perpendicular to the crack plane. Mode II, the sliding mode, exhibits crack surface displacements in the plane of the crack and perpendicular to the crack front. Mode III, the tearing mode, exhibits crack surface displacements also in the plane of the crack but parallel to the crack front. Generalized loading conditions will involve combined Modes I, II, and III. The different fracture Modes produce different stress distributions in the material ahead of the crack tip, the intensity of which are governed by the stress intensity factors K_I, K_{II}, and K_{III}.

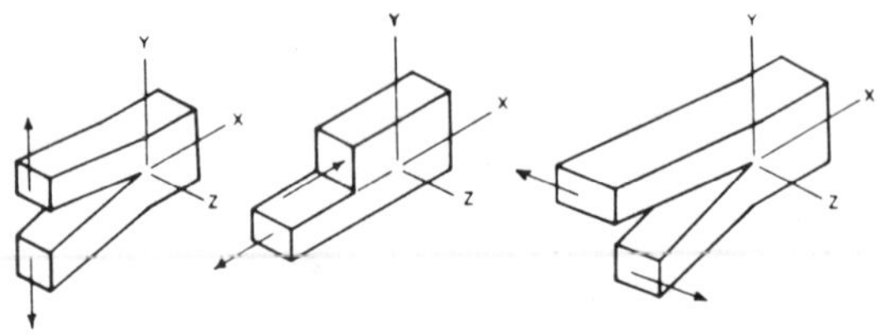

MODE I   MODE II   MODE III

$$\text{Mode I:} \quad \begin{matrix}\sigma_{xx}\\ \sigma_{yy}\\ \sigma_{xy}\end{matrix} = \frac{K_I}{(2\pi R)^{1/2}} f(\theta)$$

$$\text{Mode II:} \quad \begin{matrix}\sigma_{xx}\\ \sigma_{yy}\\ \sigma_{xy}\end{matrix} = \frac{K_{II}}{(2\pi R)^{1/2}} f(\theta)$$

$$\text{Mode III:} \quad \begin{matrix}\sigma_{xz}\\ \sigma_{yz}\end{matrix} = \frac{K_{III}}{(2\pi R)^{1/2}} f(\theta)$$

$$K_I \propto (\sigma_{yy})_R \cdot C^{1/2}$$

$$K_{II} \propto (\sigma_{xy})_R \cdot C^{1/2}$$

$$K_{III} \propto (\sigma_{zy})_R \cdot C^{1/2}$$

C = Crack length

$(\sigma_{ij})_R$ are remote stresses

Figure 1: The Three Modes of Fracture.

The simplest mixed-mode fracture criterion is the coplanar criterion (4). This criterion assumes that mixed-mode fracture will initiate, with crack extension in the plane of the initial crack, when the strain energy release rate in this plane reaches a critical value (this critical value is taken to be a material parameter equal to the pure Mode I critical strain energy release rate). However, coplanar mixed-mode fracture is virtually never observed and so non-coplanar strain energy release rate criteria have been put forward (5-10). In non-coplanar strain energy release rate criteria, the mixed-mode crack is presumed to extend in the non-coplanar direction of maximum strain energy release rate, when this quantity reaches a critical value. Two other criteria of non-coplanar mixed-mode fracture are the maximum tangential tensile stress theory (11) and the strain energy density theory (12). The maximum tangential tensile stress theory assumes that crack extension starts in the plane perpendicular to the direction of greatest crack tip tangential tension σ_θ, with catastrophic fracture occurring when the quantity $(2R)^{\frac{1}{2}}\sigma_\theta$ reaches a critical value. It is interesting to note that the strain energy release rate theory of Nuismer (6) yields the same predictions as the maximum tangential tensile stress theory. The strain energy density theory assumes that crack initiation will start in a radial direction along which the strain energy density is minimum, with fracture occurring when the strain energy density factor reaches a critical value.

The predictions of a number of these criteria for mixed-mode I-II fracture are shown in Figure 2. These theories predict that K_I decreases with increasing K_{II}. It should be noted that these criteria also predict values of the ratio K_{IIC}/K_{IC} for the case of pure Mode II fracture. For the various mixed-mode fracture theories, the range of predictions is $0.63 \leq K_{IIC}/K_{IC} \leq 1.02$.

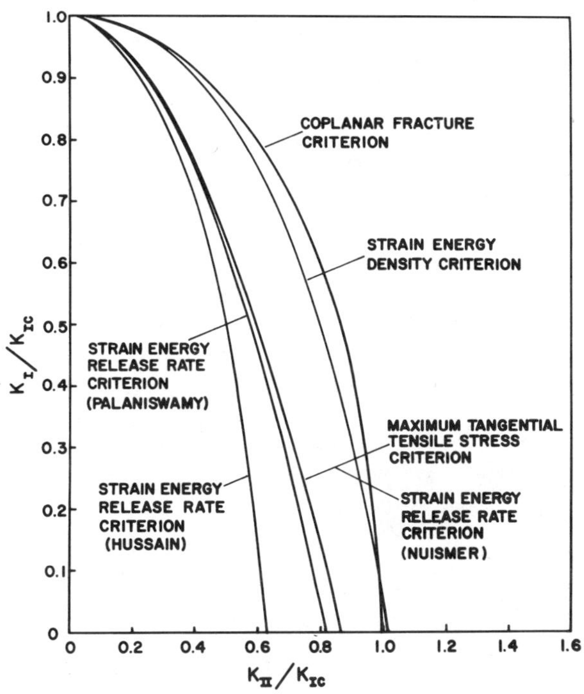

Figure 2: Predictions of Theoretical Mixed-Mode Fracture Criteria.

EXPERIMENTAL MIXED-MODE FRACTURE RESULTS

Indentation Surface Flaws

The mixed-mode fracture of ceramics has been examined largely using indentation surface flaws under various loading configurations. Experimental results of all studies to date are summarized in Figure 3. A key for the experimental data is given in Table 1. In Figure 3, σ_n/σ_o is plotted versus τ/σ_o, where σ_n is the stress normal to the flaw plane, τ is the shear stress in the plane of the flaw, and σ_o is the fracture stress under pure Mode I conditions.

Except for one data point, the data in Figure 3 indicate that σ_n/σ_o at fracture decreases with increasing τ/σ_o. This means that surface flaw fracture is influenced by the presence of an in-flaw-plane shear stress. If this were not the case, σ_n/σ_o would equal one, independent of the

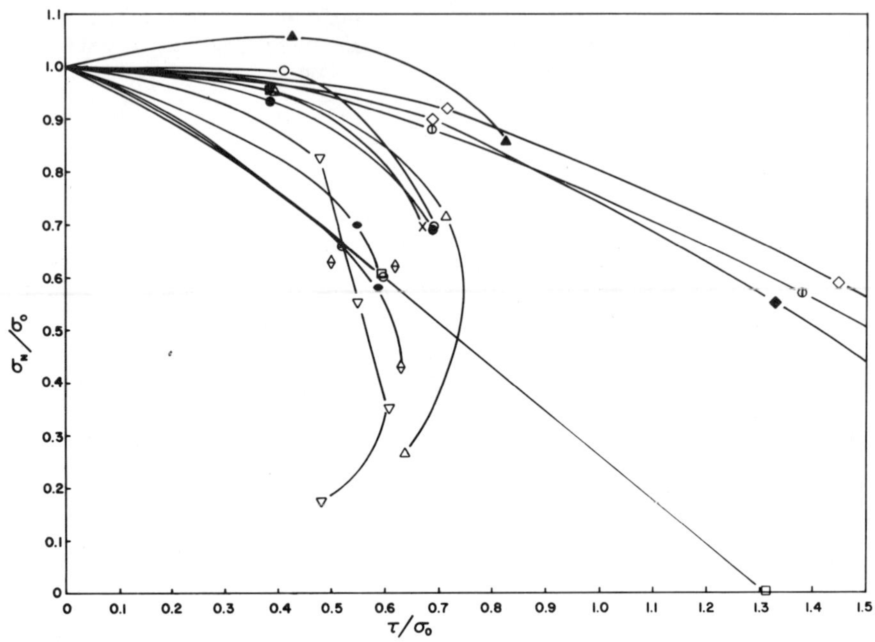

Figure 3: Mixed-Mode Fracture Results for Indentation Surface Flaws.

value of τ/σ_o. The data in Figure 3 clearly demonstrate that a constant normal stress criterion for fracture will be seriously in error, and that the fracture criterion must include mixed-mode loading conditions.

The data in Figure 3 for various ceramic materials and loading configurations exhibit a fairly wide range, rather than lying on a single curve. Most of the data comes from the fracture of inclined surface flaws in bending (13, 14, 15, 17, 18, 19, 20). Even within this single loading configuration, the data spread is significant. At very steep angles of inclination to the bending tensile stress, both σ_n and τ exhibit decreases at flaw fracture, which usually occurs only from the surface tip of the Knoop indentation.

The filled data points in Figure 3 indicate surface flaws which were annealed to remove indentation residual stresses prior to mixed-mode fracture. Marshall (17) has shown that the presence of these residual stresses leads to stable non-coplanar crack growth prior to catastrophic fracture under mixed-mode loading conditions. This change in crack shape might be expected to complicate the local mixed-mode fracture situation. However, the data in Figure 3 do not provide a clear-cut description of residual stress effects on mixed-mode fracture, with some data showing significant effects on σ_n and τ at fracture and other data showing little effect.

Mixed-mode fracture studies of surface flaws have been performed in combined tension/torsion (16,19) and diametral compression (18). For

TABLE 1. Key for Indentation Surface Flaw Mixed-Mode Fracture Results in Figure 3

Sym.	Material	Test Type	Indentation Conditions	Indentation Annealing	Ref.
●	Hot-Pressed Si_3N_4*	Bending	59N Knoop	2h/1200°C	20
○	Hot-Pressed Si_3N_4*	Bending	59N Knoop	None	19
■	Hot-Pressed Si_3N_4*	Tension/ torsion	59N Knoop	1h/1300°C	19
□	Hot-Pressed Si_3N_4*	Tension/ torsion	59N Knoop	None	16
△	Hot-Pressed Si_3N_4**	Bending	25N Knoop	None	13, 14
▽	Soda-Lime Glass	Bending	20N Knoop	None	15
▲	Hot-Pressed Si_3N_4**	Bending	50N Knoop	4h/1200°C	17
	Hot-Pressed Si_3N_4**	Bending	50N Knoop	None	17
◇	Pyroceram 9606	Diametral Compression	50N Knoop	None	18
◆	Pyroceram 9606	Diametral Compression	50N Knoop	4h/820°C	18
○	AVCO Al_2O_3	Diametral Compression	50N Knoop	None	18
○	Pyroceram 9606	Bending	50N Knoop	None	18
⬬	Pyroceram 9606	Bending	50N Knoop	4h/820°C	18
⟠	AVCO Al_2O_3	Bending	50N Knoop	None	18

*Ceradyne 147A Si_3N_4
**Norton NC-132 Si_3N_4

these loading configurations, fracture under pure shear conditions (i.e., zero normal stress on the flaw plane) is possible. For as-indented flaws in hot-pressed Si_3N_4, the author obtained a value of $\tau/\sigma_o = 1.31$ for this condition of surface flaw fracture, although this value may have been influenced by non-coplanar stable crack extension prior to catastrophic failure in pure torsion. Shetty et al. (18) report τ/σ_o values in pure shear of 2.0-2.2 for Pyroceram 9606 and Al_2O_3. They have also indicated a significant difference in the dependence of σ_n/σ_o on τ/σ_o between diametral compression and bending which they ascribe to a stress-state effect.

Large Cracks

The only large crack mixed-mode fracture data presently available for a ceramic material are shown in Figure 4. In this investigation (20), the Mode I-II fracture of hot-pressed Si_3N_4 was examined using circumferentially slotted tubes in combined tension/torsion, while Mode

I-III fracture was studied using circumferentially notched rods in combined tension/torsion.

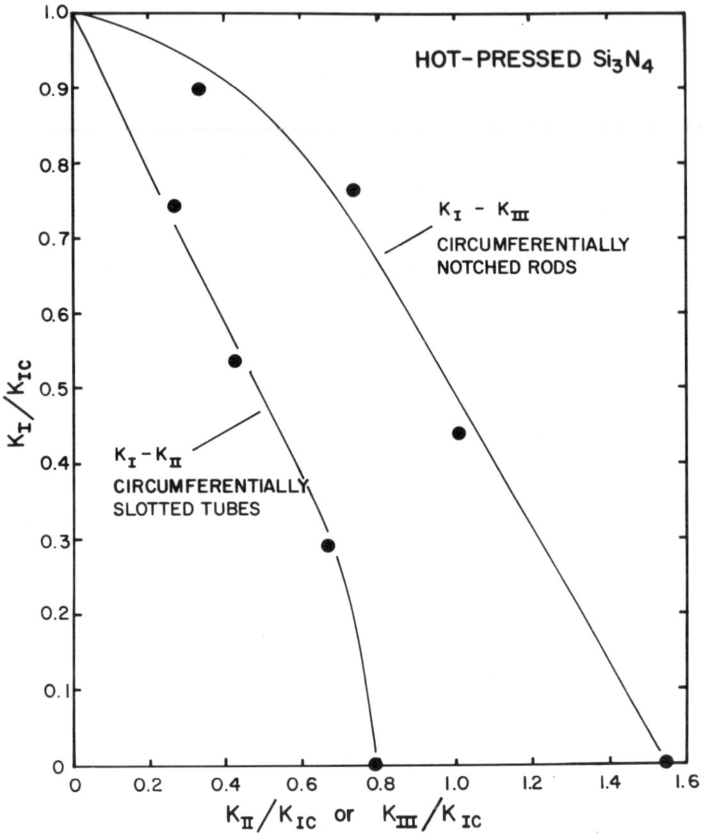

Figure 4: Mixed-Mode Fracture of Large Cracks in Hot-Pressed Si_3N_4.

As may be seen in Figure 4, both Mode II and Mode III influence Mode I fracture, with Mode II conditions having a greater influence. Ueda et al. (21) have observed similar trends for the mixed-mode fracture of brittle polymethylmethacrylate (PMMA). Observed stress intensity factor ratios in Si_3N_4 for pure Mode II and pure Mode III loading were $K_{IIC}/K_{IC} = 0.79$ and $K_{IIIC}/K_{IC} = 1.55$. For the mixed-mode fracture of these large cracks, non-planar fracture was universally observed, as has also been the case for surface flaws. The theoretical non-coplanar fracture theories almost universally predict K_{IIC}/K_{IC} and K_{IIIC}/K_{IC} ratios less than one (20).

With regard to large crack studies on other brittle materials, Awaji and Sato (22) have reported K_{IIC}/K_{IC} ratios in the range of 1.09-1.16 using inclined large cracks in diametral compression for graphite, plaster, and marble materials. For brittle PMMA (21), K_{IIC}/K_{IC} ratios in the range of 0.75-0.9 and K_{IIIC}/K_{IC} ratios of 0.98 have been observed. In metals, various studies (23-26) have indicated K_{IIC}/K_{IC} ratios in the range of 1.0-1.2, with K_{IIIC}/K_{IC} ratios in the neighborhood of 1.2 (23,26).

Comparison of Surface Flaws and Large Cracks

Figure 5 compares experimental values of K_I/K_{IC} versus K_{II}/K_{IC} for circumferentially slotted tubes and annealed Knoop surface flaws in hot-pressed Si_3N_4 (20). Also shown for comparison are a number of

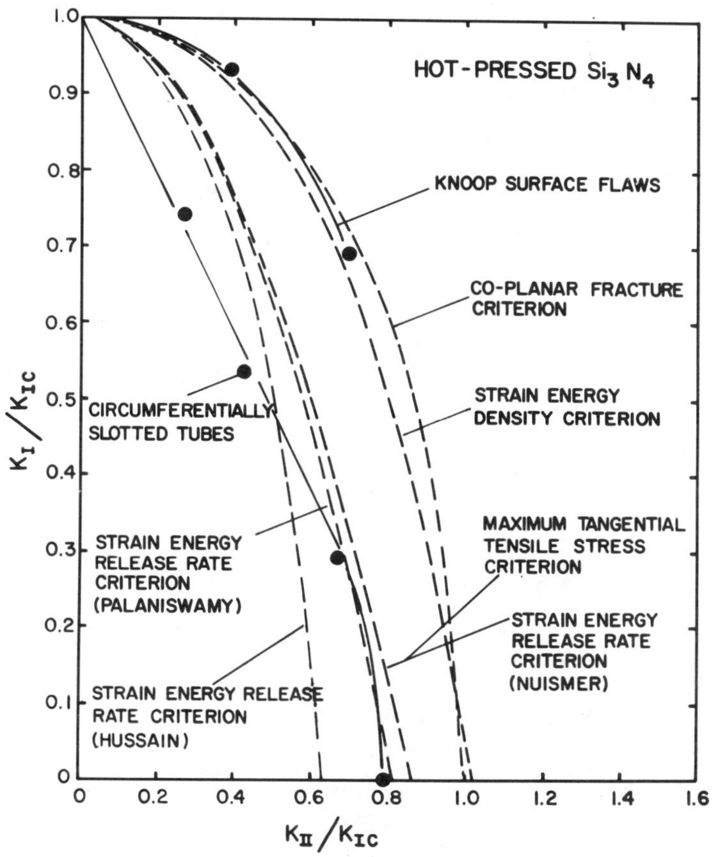

Figure 5: Comparison of the Mixed-Mode Fracture of Circumferentially Slotted Tubes and Annealed Surface Flaws in Hot-Pressed Si_3N_4.

mixed-mode fracture criteria. As may be seen, the mixed-mode fracture data for circumferentially slotted tubes and Knoop surface flaws do not coincide. In general, the surface flaw data exhibits less of an effect of K_{II} on K_I fracture than does the slotted tube data. While the slotted tube data are best described by predictions of the non-coplanar strain energy release rate (5,6,8) and maximum tangential tensile stress (11) criteria, the surface flaw results are best described by the coplanar fracture (4) and strain energy density (12) criteria. Previous investigations (14,17) have also indicated that the mixed-mode fracture of surface flaws is well described by the coplanar fracture criterion, this despite the fact that the fracture response is clearly non-coplanar.

DISCUSSION

The difference in the mixed-mode responses of large cracks and surface flaws shown in Figure 5 is of some insight. Marshall (17) has suggested that surface flaws which are free of indentation residual stresses (i.e., annealed flaws) have crack surfaces that are in contact at asperities. This resistance to shear at the contact points would reduce the effectiveness of the applied shear loading in producing stress intensification at the crack tip, thus effectively reducing the actual K_{II} level for surface flaws under mixed K_I/K_{II} loading below that calculated under the assumption that the surfaces were not in contact. Since the large crack surfaces were not in contact, this would account for the difference in Figure 5.

In a recent study (20), this effect has been modeled by assuming that the in-flaw-plane shear stress for surface flaws is reduced by a multiplicative factor related to the surface roughness and crack opening displacement of the flaw, termed the shear resistance factor S_R. This factor ranges between zero (complete crack resistance to sliding displacements) and one (no crack resistance effects). A likely functional form for the shear resistance factor is the ratio of the surface flaw crack opening displacement (COD) to the average asperity height d. Using the expression of Sneddon (27) for the elastic crack opening displacement of a penny-shaped flaw in an infinite medium under tensile loading, the Mode II stress intensity factor for surface flaws may be expressed as

$$K_{II} = \left(\frac{4}{\pi^{\frac{1}{2}}(2-\nu)}\right)\left(\frac{8(1-\nu^2)\ \sigma_n\ a}{\pi\ Ed}\right)\tau\ a^{\frac{1}{2}}$$

where the second term represents the shear resistance factor. Here, σ_n is the stress normal to the flaw plane, a is the flaw radius, τ is the applied shear stress, E is the elastic modulus, d is the asperity height, and ν is Poisson's ratio.

The above expression indicates that K_{II} for surface flaws depends on both the shear stress τ and the normal stress σ_n. In addition, the flaw size dependence of K_{II} becomes $a^{3/2}$ rather than $a^{1/2}$. Finally, the material variables of elastic modulus E and asperity height d (expected to be some fraction of the material grain size) are introduced. These would suggest that ceramics with high elastic moduli and large grain sizes may be the most susceptible to shear resistance effects. Application of the above expression to mixed-mode surface flaw fracture in hot-pressed Si_3N_4 (20) brings surface flaw responses into better correspondence with the predictions of non-coplanar strain energy release rate fracture theories.

SUMMARY

Experimental mixed-mode fracture responses for ceramics and other brittle materials have been reviewed and compared to predictions of theoretical mixed-mode fracture criteria. Surface flaw results demonstrate a significant effect of in-flaw-plane shear stresses on the flaw-plane normal stress at fracture, indicating that the surface flaw fracture criterion must include mixed-mode loading conditions. Mixed-mode fracture studies of large cracks suggest that Mode II has a greater effect on Mode I fracture than does Mode III. A comparison of surface flaw and large crack mixed-mode I-II fracture behavior indicates that shear resistance effects influence the surface flaw mixed-mode response.

ACKNOWLEDGEMENTS

This work was supported by the Division of Materials Sciences, Office of Basic Energy Sciences, U. S. Department of Energy, under Contract No. W-7405-ENG-36. The author is grateful to M. L. Lovato for assistance in the performance of mixed-mode fracture testing.

REFERENCES

1. S. B. Batdorf and J. G. Crose, J. Appl. Mech., 41:459 (1974).
2. S. B. Batdorf and H. L. Heinisch, J. Amer. Ceram. Soc., 61:355 (1978).
3. K. T. Faber and A. G. Evans, Acta Met., 31:565 (1983).
4. P. C. Paris and G. C. Sih, in "Stress Analysis of Cracks", ASTM STP 381, American Society for Testing and Materials (1965).
5. M. A. Hussain, S. L. Pu, and J. Underwood, in "Fracture Analysis", ASTM STP 560, American Society for Testing and Materials (1974).
6. R. J. Nuismer, Int. J. Fract., 11:245 (1975).
7. W. T. Chiang, in "Fracture 1977", 4:135, University of Waterloo (1977).
8. K. Palaniswamy and W. G. Knauss, in "Mechanics Today", 4:87, S. Nemat-Nasser, ed., Pergamon Press, New York (1978).
9. M. Ichikawa and S. Tanaka, Int. J. Fract., 18:19 (1982).
10. W. Shen and J. D. Lee, Eng. Fract. Mech., 16:783 (1982).
11. F. Erdogan and G. C. Sih, J. Basic Eng., Trans. ASME, Series D, 85:519 (1963).
12. G. C. Sih, Int. J. Fract., 10:305 (1974).
13. J. J. Petrovic and M. G. Mendiratta, J. Amer. Ceram. Soc., 59:163 (1976).
14. J. J. Petrovic and M. G. Mendiratta, J. Amer. Ceram. Soc., 60:463 (1977).
15. S. W. Freiman, A. C. Gonzalez, and J. J. Mecholsky, J. Amer. Ceram. Soc., 62:206 (1979).
16. J. J. Petrovic, in "Fracture Mechanics of Ceramics", 6:63, R. C. Bradt, A. G. Evans, D. P. H. Hasselman, and F. F. Lange, eds., Plenum Publishing Corp., New York (1983).
17. D. B. Marshall, J. Amer. Ceram. Soc., 67:110 (1984).
18. D. K. Shetty, A. R. Rosenfield, and W. H. Duckworth, "Biaxial Stress-State Effects on Mixed-Mode Fracture from Surface Flaws in Ceramics", submitted to the J. Amer. Ceram. Soc., (1985).
19. J. J. Petrovic, unpublished research.
20. J. J. Petrovic, "Mixed-Mode Fracture of Hot-Pressed Si_3N_4", J. Amer. Ceram. Soc., June (1985).
21. Y. Ueda, K. Ikeda, T. Yao, and M. Aoki, Eng. Fract. Mech., 18:1131 (1983).
22. H. Awaji and S. Sato, J. Eng. Mat. & Tech., 100:175 (1978).
23. R. C. Shah, in "Fracture Analysis", ASTM STP 560, American Society for Testing and Materials (1974).
24. A. F. C. Liu, AIAA Journal, 12:180 (1974).
25. D. L. Jones and D. B. Chisholm, Eng. Fract. Mech., 7:261 (1975).
26. L. P. Pook, Eng. Fract. Mech., 3:205 (1971).
27. I. N. Sneddon, Proc. Roy. Soc. London, Series A, 187:229 (1946).

SOME FRACTURE PROPERTIES OF ALUMINA-CONTAINING ELECTRICAL PORCELAINS

Elis Carlström, Roger Carlsson, Anna-Karin Tjernlund and Bengt Johannesson *

Swedish Institute for Silicate Research
Box 5403, S-402 29 Göteborg, Sweden

* Chalmers University of Technology, Dept. of Eng. Metals
S-412 96 Göteborg, Sweden

INTRODUCTION

The mechanical properties of electrical porcelains for high voltage insulators are important as the products are expected to serve for several decades at high tensile loads. The principles and test methods of linear elastic fracture mechanics are well established to explain and measure the long-term mechanical properties of glass as well as many high performance ceramics. These ideas could also be applied to traditional poly-crystalline ceramics but so far they have been used only to a minor extent in this field (1-8).

In the present investigation some alumina-containing electrical porcelains were tested for bending strength, Weibull modulus, slow crack growth exponent and fracture toughness. The work will proceed with more extensive measurements on slow crack growth and life time predictions.

EXPERIMENTAL

Four electrical porcelains used for high voltage insulators and containing three different levels of alumina were studied. The alumina had been added as bauxite and alpha alumina as a partial substitute for quartz. Rods about 150 mm long were supplied by the manufacturer Ifö Electric AB, Sweden and were tested as received. The approximate alumina content and other characteristics of the test rods are given in table 1.

The bending strength was measured by three point bending at five different cross-head speeds in a Zwick 1464 testing machine according to the test parameters given in table 2. The times to fracture have been from 1 s to 4 h approximately.

The Weibull modulus m has been calculated by linear regression according to the least squares method. i/n+1 has been used as the estimator of the fracture probability, where i = rank number at increasing fracture strength and n = total number of samples tested.

Table 1. Materials

	Porcelain A	Porcelain B	Porcelain C	Porcelain D
Alumina content (w/o) within the range *	35-40	45-50	55-60	55-60
Forming method	extrusion	extrusion	isostatic pressing	isostatic pressing and green machining
Youngs modulus (GPa) *	101+-3	112+-3	120	120
Rod diameter (mm)	15.40+-0.02	15.53+-0.02	15.91+-0.42	15.80+-0.07
No. of rods tested **	20	10; 99	10	5

All margins of error refer to the standard deviation.

* Values given by the manufacturer.
** At each cross-head speed.

Determination of the slow crack growth exponent n has been made by linear regression of log (fracture strength) vs log (cross-head speed) (9) according to the least squares method.

Finally the fracture toughness of porcelain A has been measured by three different techniques. With the controlled surface flaw technique (10) the indentation has been performed with a Knoop diamond using a load of 30 N during 30 s on specially prepared test bars 45 x 3.5 x 4.5 mm^3. Afterwards the surface was polished 15-20 µm and 12 bars were tested at four point bending with span 40 mm and cross-head speed 0.2 mm/min.

The direct Vickers indentation technique (10) has been used with a load of 30 N during 30 s on polished surfaces. The crack lengths have been measured in the scanning electron microscope and K_{Ic} values have been calculated for Palmqvist cracks (11). Some test bars were first heat-treated in air at 800°C and slowly cooled in order to reduce any residual compressive stresses that may possibly occur at the surface due to the polishing procedure. The third K_{Ic} measuring method was the bridge indentation technique (12).

Table 2. Parameters for three point bending tests

Span (mm)	100
Edge radius (mm)	5
Cross-head speed (mm/min)	0.002, 0.02, 0.2, 2, 20
Temperature (°C)	22.0+-1.5
Relative Humidity (%)	40-55

RESULTS AND DISCUSSION

Table 3 gives the three point bending strengths together with the resulting Weibull moduli. Figure 1-2 show the Weibull plots for porcelain A and C. For porcelain B a more accurate measurement of the Weibull modulus was also made at a different occasion using 99 test rods (figure 3).

As can be seen from table 3 the Weibull moduli for porcelain A and B are extremely high. The reason for this is the presence of rather large quartz grains in these materials (13). The grains have approximately the same size and originate from the coarse fraction of sand used as a raw material in porcelain A and B. In porcelain B more of the sand has been substituted by bauxite and the strength of the matrix is therefore higher but still the large quartz grains seem to determine the ultimate strength producing a low scatter in strength values and thus a high Weibull modulus.

In porcelain C and D all the sand is substituted by bauxite and alpha alumina. The resulting matrix is consequently stronger but in the absence of the large quartz grains other defects control the strength and the scatter in strength values is much larger and the Weibull moduli have dropped to a level more common to traditional ceramics. The different forming methods may of course also be of importance.

Table 3. Three point bending strength, Weibull modulus and slow crack growth exponent

	Porcelain A	Porcelain B	Porcelain C	Porcelain D
Fracture strength (MPa)				
at cross-head speed				
0.002 mm/min	104+-5	133+-5	159+-20	177+-8
0.02 mm/min	108+-6	131+-5	162+-19	176+-15
0.2 mm/min	112+-4	138+-5	173+-14	206+-5
2 mm/min	121+-5	146+-6	187+-22	211+-15
,99 rods		158+-6		
20 mm/min	135+-3	159+-5	196+-20	231+-17
2-parameter Weibull modulus; m				
at cross-head speed				
0.002 mm/min	20	22	7.2	-
0.02 mm/min	21	26	8.2	-
0.2 mm/min	27	26	12.1	-
2 mm/min	24	24	8.3	-
,99 rods		34		
20 mm/min	41	31	9.1	-
Slow crack growth exponent; n				
at cross-head speeds				
0.02 to 20 mm/min	29	35	34	26
0.002 to 20 mm/min	35	48	40	31

All margins of error refer to the standard deviation.

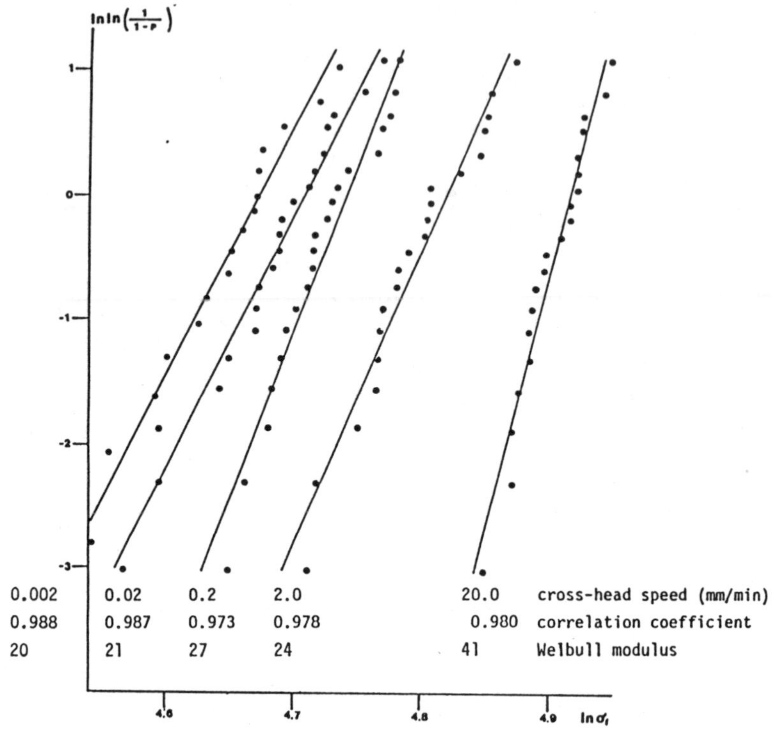

Figure 1. Two-parameter Weibull plot for porcelain A at five cross-head speeds (P = fracture probability).

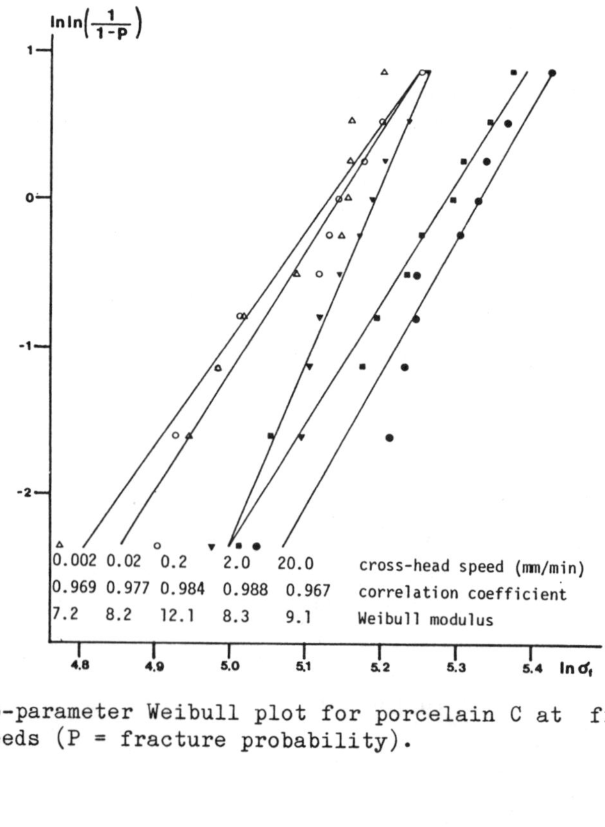

Figure 2. Two-parameter Weibull plot for porcelain C at five cross-head speeds (P = fracture probability).

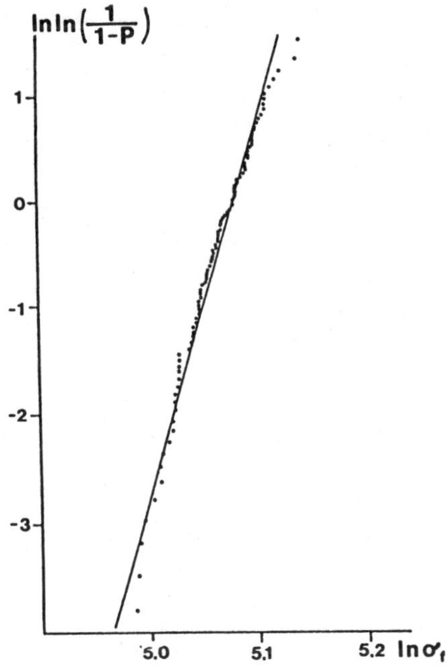

Figure 3. Two-parameter Weibull plot for porcelain B at 2.0 mm/min cross-head speed (P = fracture probability).

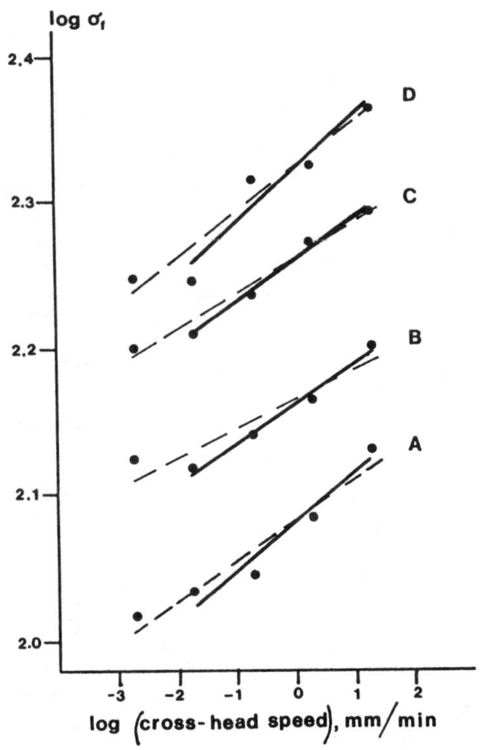

Figure 4. Log (fracture stress) vs log (cross-head speed) for porcelain A, B, C and D. Each point represents 20 samples for A, 10 samples for B and C, and 5 samples for D.

No Weibull modulus have been calculated for porcelain D depending on the low number of test rods.

The log (fracture stress) is plotted against log (cross-head speed) in figure 4. The fracture stresses all fall on a straight line exécept for the 0.002 mm/min value. As no satisfactory explanation has been found for this, the slow crack growth exponent has been calculated with, as well as without, this value (table 3). The slow crack growth exponent increases with the increasing alumina content from porcelain A to B and keeps constant from B to C. The value for porcelain D is uncertain due to the small number of samples tested.

The fracture toughness for porcelain A was measured to 1.1+-0.1, 2.1+-0.2 and 1.5+-0.1 MPa $m^{1/2}$ by the surface flaw technique, the direct Vickers indentation technique and the bridge indentation technique respectively. The high value for the direct Vickers indentation technique was only reduced to 1.9+-0.2 MPa $m^{1/2}$ after heat-treatment so residual compressive stresses at the surface from polishing could not be the main reason for the high K_{Ic} value. Another cause may be the twisting of the crack through the heterogeneous microstructure of traditional ceramics which gives a shorter measured crack length than the real length of the crack.

ACKNOWLEDGEMENT

The authors wish to thank Rolf Ahlgren, Ifö Electric AB, Sweden for supplying the test bars.

REFERENCES

1. M. Matsui, T. Soma and I. Oda, in: "Fracture Mechanics of Ceramics Vol. 4", R.C. Bradt, D.P.H. Hasselman and F.F. Lange, ed., Plenum Press, New York (1978) p. 711.
2. A.G. Evans and M. Linzer, J. Amer. Ceram. Soc. 56:575 (1973).
3. A.G. Evans, S.M. Wiederhorn, M. Linzer and E.R. Fuller Jr., Ceramic Bulletin 54:576 (1975).
4. T. Soma, M. Matsui, I. Oda and N. Yamamoto, J. Am. Ceram. Soc. 63:166 (1980).
5. M.E. Rose and C.M. Burke, J. Am. Ceram. Soc. 67:C-20 (1984).
6. H. Dannheim and H.J. Oel, cfi/Ber. DKG 61:84 (1984)
7. W.G. Clark, Jr and W.A. Logsdon, in: "Fracture Mechanics of Ceramics Vol 2", R.C. Bradt, D.P.H. Hasselman and F.F. Lange, ed., Plenum Press, New York (1974) p. 843.
8. C. Hahn und R. Rubin, in: "Science of Ceramics Vol 12", P. Vincenzini, ed., Ceramurgia, Faenza (1983) p. 591.
9. A.G. Evans, Int. Fract. Mech. 10:251 (1974).
10. G.R. Anstis, P. Chantikul, B.R. Lawn and D.B. Marshall, J. Amer. Ceram. Soc. 64:533 and 64:539 (1981).
11. K. Niihara, R. Morena and D.P.H. Hasselman, in: "Fracture Mechanics of Ceramics Vol. 5", R.C. Bradt, A.G. Evans, D.P.H. Hasselman and F.F. Lange, ed., Plenum Press, New York (1983) p. 97.
12. T. Sadahiro and S. Takatsu, Modern Dev. in Powder Metallurgy 14:561 (1981).
13. E. Carlström and R. Carlsson, in: "Science of Ceramics Vol 13", (presented in September 1985).

TRANSFORMATION TOUGHENING IN THE Al_2O_3-Cr_2O_3/ZrO_2-HfO_2 SYSTEM: THE EFFECT OF ZrO_2-HfO_2 CONTENT

T.K. Brog, J.S. Wallace and T.Y. Tien

Department of Materials and Metallurgical Engineering
University of Michigan
Ann Arbor, MI 48109

ABSTRACT

The fracture toughness, strength, and modulus of Al_2O_3-Cr_2O_3(matrix) /ZrO_2-HfO_2(dispersed phase) composites were measured. The fracture toughness increased with increasing volume fraction dispersed phase whereas the strength and modulus remained nearly constant up to 10v/o dispersed phase and then decreased for larger volume fractions. Multiple toughening mechanisms are apparently active in this system, with microcrack nucleation/growth dominating at high monoclinic contents.

INTRODUCTION

Zirconia toughened alumina (ZTA) has received considerable interest in recent years due to the enhancement in fracture toughness (1-5). The improvement in fracture toughness and strength of zirconia containing materials, as well as the long-term stability, has made possible the prospect of using some of these materials for high temperature structural applications such as for adiabatic engine components.

A modification of ZTA was investigated here to evaluate this material as a preliminary candidate material for engine applications. The matrix consists of an Al_2O_3-Cr_2O_3 solid solution. Additions of Cr_2O_3 were made to both lower the thermal conductivity and to modify the matrix properties, fracture toughness and modulus. Previous work has shown that Cr_2O_3 additions lowers the thermal conductivity (6) and increases the hardness (7). Slight increases in the bulk modulus with Cr_2O_3 additions (0-15m/o) have also been reported (8).

The dispersed phase consists of a ZrO_2-HfO_2 solid solution. Hafnia additions were made to zirconia to increase the tetragonal-to-monoclinic transformation temperature, thus increasing the temperature at which the transformation can be a viable toughening mechanism (9,10). The critical particle size for the transformation, however, decreases with increasing hafnia content. In order to retain a reasonable fraction of the zirconia-hafnia particles in the tetragonal symmetry using conventional processing methods, the maximum hafnia concentration was small, 10m/o.

The present study was undertaken to try to experimentally distinguish the various toughening mechanisms in this system by measuring the fracture toughness, strength, and modulus as a function of zirconia-hafnia volume fraction.

EXPERIMENTAL PROCEDURE

Al_2O_3-5m/oCr_2O_3 and ZrO_2-10m/oHfO_2 solid solutions were made by mixing appropriate amounts of the single oxides and solution annealing at 1350 C for 24 hrs. These solid solutions were attrition-milled in isopropanol for 4 and 8 h respectively. The solid solutions were mixed in a ball- mill, dried, sieved and isostatically pressed (170 MPa). The samples were then sintered at 1550 C for 2 h in a graphite induction furnace (argon atmosphere) and nearly theoretical density (>98%TD) was achieved.

Phase determination was performed on as-sintered surfaces. Integrated intensities of the (111) monoclinic peak (28.5 , Cu K) and the (111) tetragonal peak (30.5) were measured and the tetragonal phase content was calculated using the method of Porter and Heuer (11). The pellets were then prepared for microstructural examination using standard metallographic techniques. The mean linear intercept length was measured using a digitizing tablet and encircling the ZrO_2-HfO_2 particles on SEM micrographs.

Machined bars (50mm x 3mm x 2mm) were annealed at 1550 C (well above the A_F temperature) to relieve residual stress from the machining process thus eliminating the effect of machining on the fracture toughness measurement. The fracture toughness was measured using the ISB (12) technique using single indents (20, 40kg) and fracture toughness values were calculated using the equation from Chantikul et al (12),

$$K_c = (E/H)^{.125}(P^{.33})^{.75}.$$

Unindented bars (ground, annealed) were broken to measure the strength. All measurements were performed in 4 point bending with spans of 9.5 and 19.0mm.

Modulus measurements were made by measuring the sample thickness, density, and the propagation time for longitudnal and shear waves. The sonic elastic modulus and Poissons´ ratio were calculated using

$$E = 3 [V_L{}^2 - 4V_S{}^2/3](1-2 \ )$$

$$= (1-2x^2)/(2-2x^2),$$

where x is the ratio of the longitudinal wave velocity, V_L, to the shear wave velocity, V_S.

RESULTS AND DISCUSSION

The dispersed phase mean linear intercept length increased with increasing volume fraction dispersed phase, as would be expected, due to particle coarsening (Fig. 1). As would be expected for increasing mean particle size, the tetragonal fraction, X_t, decreases with increasing volume fraction dispersed phase, V_v, (Fig. 2). This, however, has also been attributed to loss-of-matrix constraint (4) as V_v is increased.

Measurements of the fracture toughness (Fig. 3) show a monotonic increase with increasing V_v, similar to that reported by Swain and Claussen

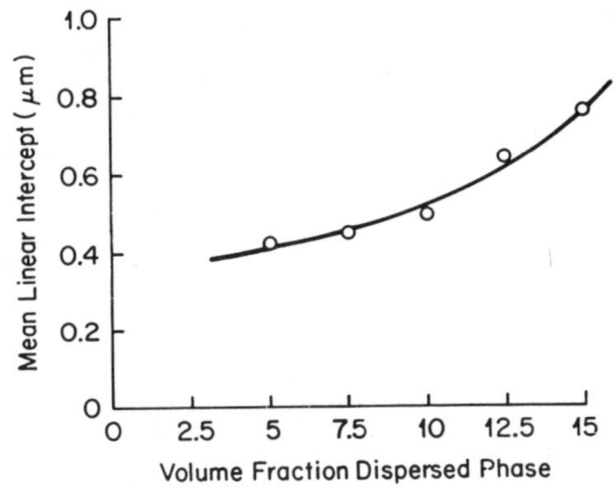

Figure 1. Mean linear intercept length of ZrO_2-10^{m}/oHfO_2 particles as a function of ZrO_2-HfO_2 content.[2]

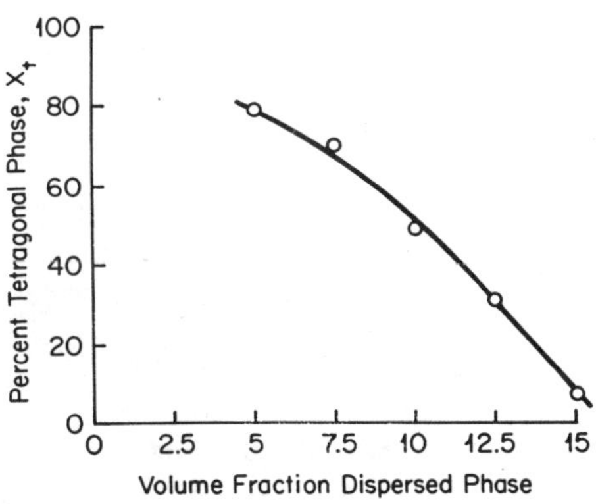

Figure 2. Tetragonal phase fraction of ZrO_2-10^{m}/oHfO_2 particles as a function of ZrO_2-HfO_2 content.

(13) and Hori et al (14) on annealed ZTA samples. While this trend is similar to that expected from crack deflection models (15) there may be contributory mechanisms such as stress induced transformation (SIT) toughening (16,17) and/or microcrack nucleation/growth toughening (3,4,18).

SIT toughening can only occur when tetragonal zirconia particles transform to the monoclinic symmetry in the crack tip stress field. Toughness enhancement has been proposed to occur due to a crack shielding mechanism (19) in which a large compressive zone exists in the region surrounding the crack tip, effectively shielding the crack tip from the applied stress field.

The large volume increase associated with the t-m transformation may cause nucleation and growth of microcracks in the matrix surrounding monoclinic zirconia particles either occurring spontaneously upon cooling from the fabrication temperature or when an external stress field is applied. The nucleation and propagation of these microcracks increases the fracture surface, thus, absorbs energy.

Crack deflection can increase the fracture toughness by deflecting the crack out of the plane normal to the applied stress, thereby reducing the stress intensity applied to the crack tip. Crack deflection appears to be insensitive to stress state and magnitude in a glass-alumina system (20). The theoretical model proposed by Faber et al (15) suggests that the relative toughness enhancement is a function of the volume fraction dispersed phase, the size distribution, and morphology of the dispersed phase.

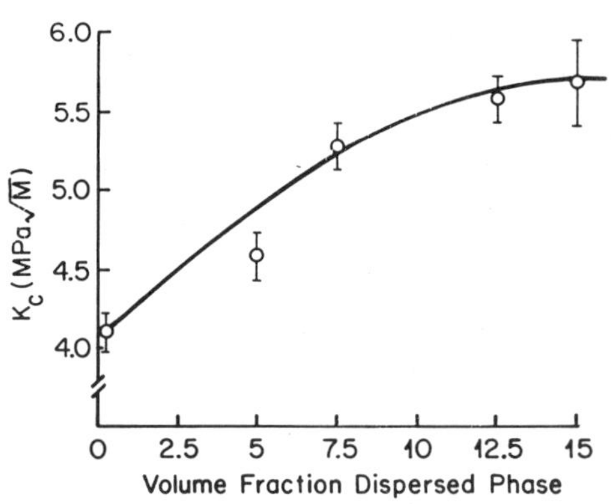

Figure 3. Fracture toughness as a function of ZrO_2-HfO_2 content measured using the ISB (12) method.

There is poor correlation between the fracture toughness and the total tetragonal volume. The fracture toughness does, however, increase with both increasing monoclinic volume and volume fraction dispersed phase. For low volume fractions of ZrO_2 the tetragonal fraction is high and the toughness enhancement is small due to the low tetragonal volume. In this regime SIT toughening and crack deflection are responsible for toughness improvement. For high volume fractions of ZrO_2 the tetragonal fraction is quite low, thus SIT contributions must be small. This suggests that crack deflection and/or microcrack nucleation/ growth are the primary toughening mechanisms. As the volume fraction dispersed phase is increased, the contributions via SIT and microcrack nucleation/growth varies due to the change in tetragonal fraction and volume, and these contributions are superimposed on the enhancement due to crack deflection.

It is difficult to evaluate the effectiveness of each particular toughening mechanism on the basis of fracture toughness measurements alone. Strength and modulus measurements combined with fracture toughness data can be used to identify or detect the presence of SIT and microcrack nucleation/growth toughening.

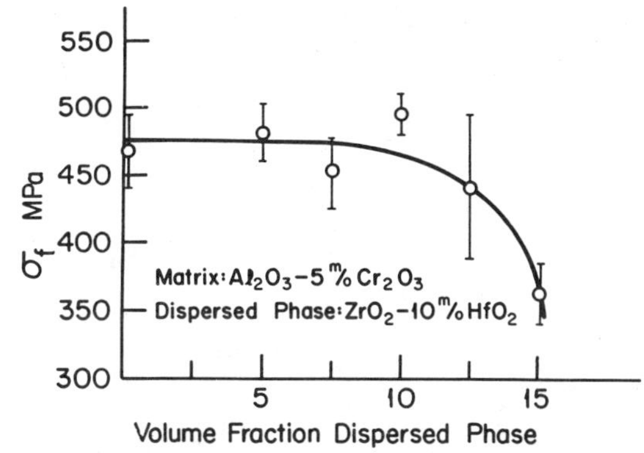

Figure 4. Strength as a function of ZrO_2-HfO_2 content showing the degradation in strength at high ZrO_2-HfO_2 volume fractions.

Measurements of the strength as a function of V_v (Fig. 4) indicate that the strength remains approximately constant with increasing V_v up to 10v/o dispersed phase, above which the strength decreases slightly at 12.5v/o then precipitously at 15v/o. This behavior can be attributed to the existence of microcracks in the samples with high monoclinic contents. These results are similar to those reported by Claussen et al (3) but are not in good agreement with those reported by Hori et al (14) and Becher (5). The strength would be expected to decrease when microcrack nucleation/growth occurs due to the presence of microcracked domains, resulting in a larger critical flaw. Therefore, the strength degradation and the toughness enhancement for large zirconia contents can be attributed to microcrack nucleation/growth. Any toughness enhancement at low zirconia contents can in part be attributed to SIT toughening. SIT toughening may increase the strength by the creation of large compressive zones.

Elastic modulus measurements of composite materials which deviate from the rule of mixtures of the components are evidence for microcracking. Green (4) reported a large deviation from the rule of mixtures for Al_2O_3-ZrO_2 composites which contained high monoclinic fractions and attributed this to microcracking.

Sonic modulus measurements were made on the annealed bars (Fig. 5). The modulus shows a linear decrease with increasing V_v up to 10v/o dispersed phase and a large decrease for the the 15v/o material. One shortcoming of this technique (wave propagation as opposed to dynamic modulus measurements) is that it is impossible to detect microcracking in a sample which occurs when an external stress field is applied.

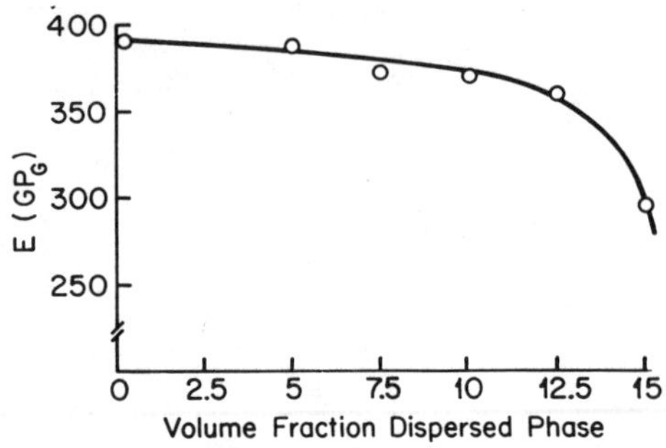

Figure 5. Elastic modulus as a function of ZrO_2-HfO_2 content showing the degradation in the modulus with increasing ZrO_2-HfO_2 content due to microcracking.

The degradation in both the strength and modulus occurs at the same composition, such that a nearly linear relationship exists between the strength and modulus in this material (Fig. 6). Thus, the presence of microcracks in this system contributes to a degradation in both the strength and modulus and an enhancement in the fracture toughness.

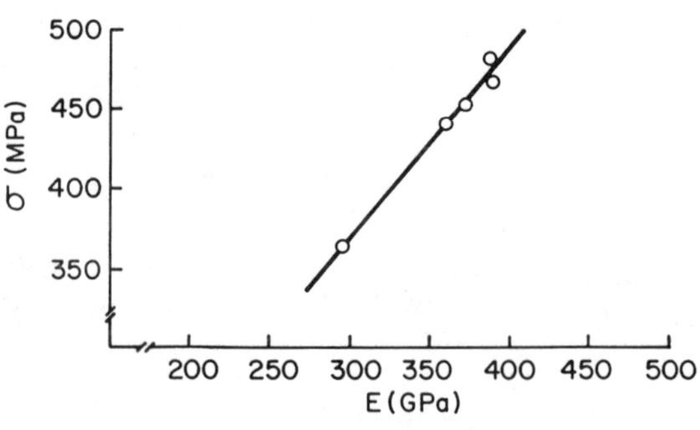

Figure 6. Strength vs. elastic modulus showing the degradation of both the strength and modulus for high ZrO_2-HfO_2 contents due to microcracking.

CONCLUSIONS

1. The dispersed phase particle size increases with increasing volume fraction dispersed phase, resulting in a decrease in the tetragonal phase content.

2. Multiple toughening mechanisms are apparently active in the system.

3. Evidence of microcrack nucleation/growth as a toughening mechanism is substantiated by a decrease in both the strength and modulus at high monoclinic contents.

ACKNOWLEDGEMENT

We would like to thank Dr. Lisa Schioler (AMMRC) for modulus measurements. This work was supported by the Army Materials and Mechanics Research Center and the DOE Office of Vehicle and Engine Research and Development Ceramics Technology for Advanced Heat Engine Program under The Interagency Agreement EC-76-A-1017-002.
The authors thank P. F. Becher for useful discussions.

REFERENCES

1. F. F. Lange, J. Mat. Sci., 17 247-54 (1982).
2. N. Claussen, J. Am. Ceram. Soc., 59 [1-2] 49-51 (1976).
3. N. Claussen, J. Steeb, and W. Pabst, Am. Ceram. Soc. Bull., 56 [6] 559-562 (1977).
4. D. J. Green, J. Am. Ceram. Soc., 65 [12] 610-614 (1982).
5. P. F. Becher, J. Am. Ceram. Soc., 64 [1] 37-39 (1981).
6. D. P. H. Hasselman, R. Syed, and T. Y. Tien, J. Mat. Sci., 20 2549-2556 (1985).
7. R. C. Bradt, J. Am. Ceram. Soc., 50 [1] 54-55 (1967).
8. L. R. Rossi and W. G. Lawrence, J. Am. Ceram. Soc., 53 [11] 604-608 (1970).
9. N. Claussen, F. Sigulinski, and M. Ruhle, Advances in Ceramics, (ed. A. H. Heuer and L. W. Hobbs), Vol. 3, 164-167, The American Ceramic Society, Cols., Ohio, 1981.
10. R. Ruh, H. J. Garrett, R. F. Domagala, and N. M. Tallan, J. Am. Ceram. Soc., 51 [1] 23-27 (1968).
11. D. L. Porter and A. H. Heuer, J. Am. Ceram. Soc., 62 [5-6] 298-305 (1979).
12. P. Chantikul, G. R. Anstis, B. R. Lawn, and D. B. Marshall, J. Am. Ceram. Soc., 64 [9] 539-543 (1981).
13. M. V. Swain and H. Claussen, J. Am. Ceram. Soc., 66 [2] C27-C29 (1983).
14. S. Hori, M. Yoshimura, and S. Somiya, J. Mat. Sci. Letters, 4 413-416 (1985).
15. K. T. Faber, A. G. Evans and M. D. Drory, Fracture Mechahics of Ceramics, (ed. R. C. Bradt, A. G. Evans, D. P. H. Hasselman, and F. F. Lange), Vol. 6, 77-91, Plenum Press, New York, 1983.
16. F. F. Lange, J. Mat. Sci., 17 225-254 (1982).
17. N. Claussen, J. Am. Ceram. Soc., 61 [1-2] 85-86 (1978).
18. N. Claussen, R. L. Cox, and J. S. Wallace, J. Am. Ceram. Soc., 65 [11] C190-C191 (1982).
19. A. G. Evans, in Advances in Ceramics, (ed. N. Claussen, M. Ruhle and A. H. Heuer), Vol. 12, 193-212, The American Ceramic Society, Cols., Ohio, 1984.
20. J. C. Swearengen, E. K. Beauchamp, and J. R. Eagan, Fracture Mechanics of Ceramics, (ed. R. C. Bradt, D. P. H. Hasselman, and F. F. Lange), Vol. 4, p. 973, Plenum Press, New York, 1978.

STRENGTH-TOUGHNESS RELATIONSHIPS FOR TRANSFORMATION TOUGHENED CERAMICS

M.V. Swain

CSIRO, Division of Materials Science
Melbourne, Victoria
Australia 3001

ABSTRACT

Recent studies have shown that the maximum strength of a number of PSZ alloys appears to be limited by the critical stress to initiate the t to m - ZrO_2 transformation. An alternative explanation of these observations is that the peak strength is limited by the substantial R-curve of these tough ceramics. This paper presents evidence to support both approaches. It is concluded that whilst stress-induced transformation provides a simplistic description analogous to the brittle to ductile transition in metals, the inherent R-curve behaviour of cracks is the mechanistically more correct description for the observed maxima in strength-toughness measurements for various PSZ alloys.

INTRODUCTION

Recent studies of the mechanical properties of partially-stabilised-zirconia (PSZ) alloys have shown that maximum strength and toughness do not correlate.[1] Peak strength invariably occurs at values between one half and three quarters of the maximum toughness in, for instance, a sequence of heat treated alloys or alloys of varying composition. This situation has been highlighted recently with the addition of alumina to yttria-tetragonal-zirconia polycrystalline (Y-TZP) materials. The strength of such composite materials increased dramatically whilst the toughness decreased with additions of up to 20 volume % of alumina.[2] Other examples of alloys where this has occured include, magnesia (Mg-) PSZ polycrystals[3], Y-PSZ single crystals[4] and ceria (Ce-) TZP.[5]

The toughness of all the above PSZ alloys is greatly enhanced over those of the fully stabilised zirconia alloys. The toughness increment is due predominantly to the mechanism of transformation toughening,[6,7] although additional mechanisms of crack deflection and microcracking[8] may assist. Within a process zone about the crack tip the stress induced tetragonal to monoclinic zirconia (t to m - ZrO_2) transformation occurs accompanied by both volume dilation and residual shear strain. It is usually considered that this transformation is triggered by a critical stress, tension plus shear, although precise details of the nucleation mechanism and the associated transformation criterion are still poorly understood. The value of this critical stress may be modified by stabiliser content and tetragonal particle size. The magnitude of the transformation stress also dictates the size of the process or transformation zone about the crack tip and the increment of toughness.[7,9] A consequence of this zone development about the crack tip is that an R-curve or rising crack resistance with crack extension occurs until the tip is fully shielded by the process zone.[9] Theoretical estimates of the extent of crack extension prior to attaining steady state toughness are usually some function of the zone size.

Observations of the stress-strain curves for Mg-PSZ and Ce-TZP in flexure show well defined departure from linearity prior to fracture,[1,5] especially for the very tough materials. Associated with this ductile like behaviour, t-ZrO_2 grains or precipitates have been transformed on the tensile surface of the material. That is, the critical stress to initiate the transformation has been reduced to a value comparable to the strength of the material. This behaviour is analogous to the strength of metals being limited by the yield stress. An alternative explanation for the maximum strength limitation with increasing toughness is a consequence of the R-curve behaviour of these materials.[10] The initial flaw size and extent of crack propagation to attain the steady state toughness value determines the maximum strength.

OBSERVATIONS AND DISCUSSION

Strength-toughness measurements

A number of research groups have published mechanical property data for PSZ alloys which appears contradictory in that compositions or heat treatments for maxima in strength and toughness do not agree. Examples of this for three different systems are shown in Figures 1-3.

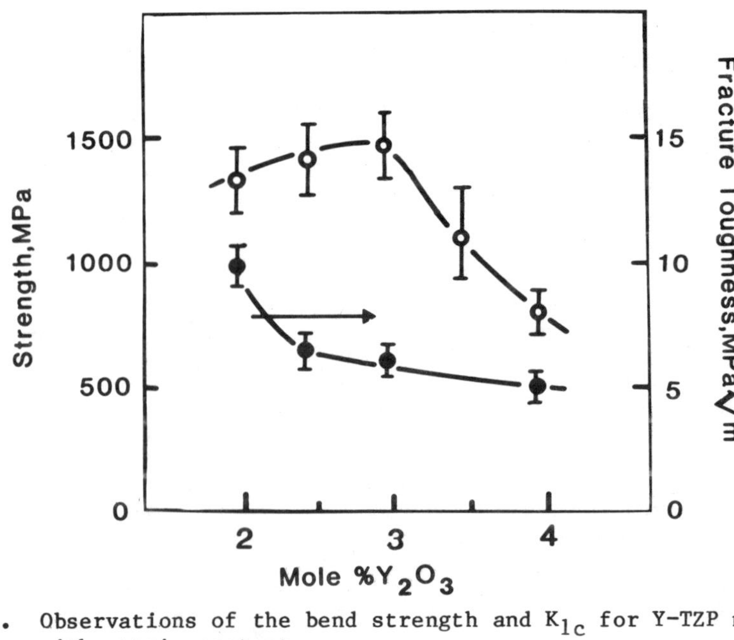

Fig. 1. Observations of the bend strength and K_{1c} for Y-TZP materials with yttria content.

Figure 1 presents data for Y-TZP materials from two different research groups.[11,12] These results indicate that maximum strength occurs at compositions containing between 2.5 and 3 mole % yttria. Whereas the fracture toughness increases monotonically with decreasing yttria content. At maximum strength the K_{1c} value is only 6-7 MPa√m, whilst at peak toughness, 10-11 MPa√m, the strength has decreased to approximately two thirds the peak value. Fracture toughness values were measured from cracks about a heavily loaded Vicker's indentation (500N). A similar discrepancy between strength and toughness may be seen in data reported for Y-PSZ single crystals.[4]

Figure 2 shows results for Ce-TZP alloys.[5] These observations confirm the behaviour of the Y-TZP although the strengths are much lower. Tsukuma and Shimada [5] present further data in their paper indicating that the strength may be increased and toughness decreased by reducing the grain size at constant ceria content. The toughness of some of the low CeO_2 content alloys was reported to be 17 MPa√m whereas the strength of this material was only 200-300 MPa. Peak strength occured at intermediate levels of fracture toughness.

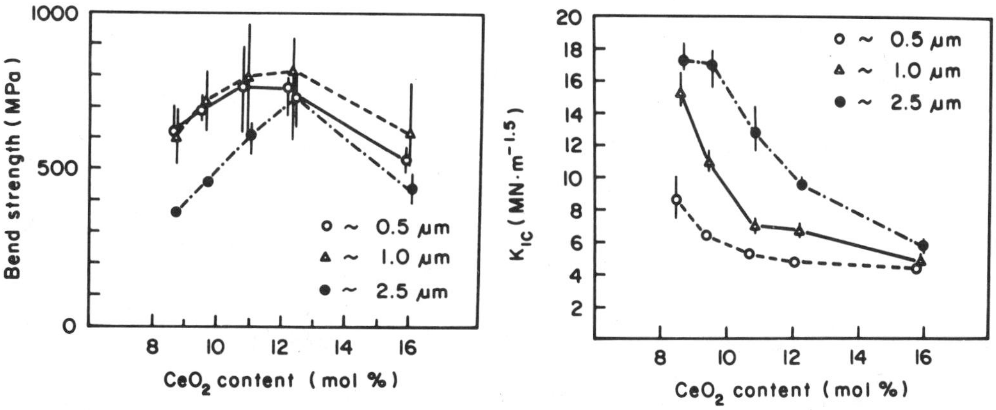

Fig. 2. Observations of stress intensity factor, K_{1c} and bend strength with ceria content. Data taken from Tsukuma and Shimada.[5]

Figure 3 presents data for an Mg-PSZ alloy, containing 9.5 mole % MgO, that has been heat treated at various temperatures below 1000°C.[3] The fracture toughness data were obtained from DCB measurements on large plates. The interesting feature of this figure is that the strength of these alloys is almost independent of toughness for K_{1c} values between 8 and 14 MPa√m. The horizontal axis in Figure 3, the polished surface monoclinic (PSM) content, is a measure of the remaining tetragonal precipitate population. A marked reduction in K_{1c} occurs when the PSM exceeds 25%, and when this parameter reaches 40% there is no remaining tetragonal content and the K_{1c} reduces to values between 4-6 MPa√m.

Stress-strain relationships

Observations of stress-strain behaviour of Mg-PSZ alloys indicates that as the toughness increases these curves become more non linear before failure.[1] Examples of such behaviour for material containing 9.5 mole % MgO that have been sub-eutectoid heat treated at 1100°C[13] are shown in Figure 4. These results were obtained by attaching a small strain gauge (3 x 6 mm gauge section) to the tensile surface of a bar. Additional measurements were made on the tensile and compressive faces of the peak toughness material. Load-unload cycles to higher stresses

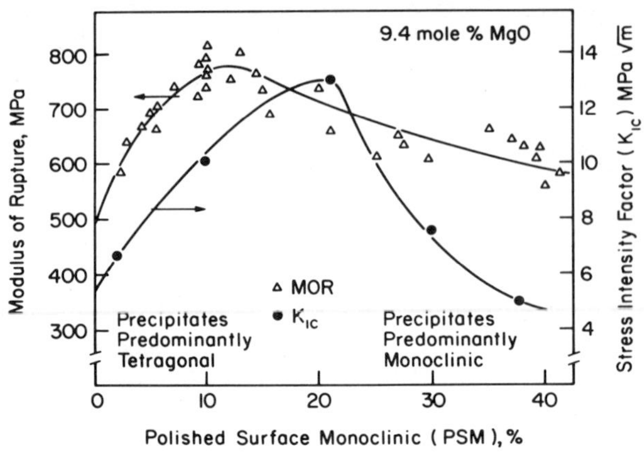

Fig. 3. Relationship between strength and toughness for Mg-PSZ when plotted against polished surface m-ZrO_2 content.

showed that the non-elastic strain was permanent and that the total extent of permanent offset strain was ~ 0.1% for the tensile surface whereas on the compression face only ~ 0.05% was measured. Similar results have been observed by Marshall[14] on material of identical composition and comparable heat treatment. Whilst there is no evidence for transformation on the compressive surface the permanent offset here is a consequence of the transformation on the tensile surface and resultant displacement of the neutral axis.

The extent of permanent deformation prior to fracture is much less than predicted if all the tetragonal precipitates transformed. The total strain, ε_T, due to all the precipitates transforming is equivalent to the dilation on cooling if the temperature drops below the M_s temperature, that is $\varepsilon_T = 1/3\ V_f\ \Delta V$, where V_f is the volume fraction of t-ZrO_2 precipitates (~ 0.4 for the Mg-PSZ alloy) and ΔV is the volume dilation associated with the transformation (0.05), leading to a maximum value of $\varepsilon_T \sim 0.67\%$. Measurements of the increased m-ZrO_2 transformed on the tensile surface after fracture using X-ray diffraction are typically only 4-6% for the peak toughness material. This small increment of m-ZrO_2 accounts for the limited permanent offset observed in Figure 4.

An optical interference contrast micrograph of the tensile surface of the peak toughness material is shown in Figure 5. The isolated clumps or relief in this micrograph are regions of transformed material. Notice that the transformation is far from homogeneous across the surface. Similar localisation of transformed regions is observed on polished sides of the specimen to a depth of 0.5 to 0.7 mm from the tensile surface. The isolated nature of these transformations, which suggests that stress concentrations such as pores, triple points, etc. may assist with initiation of the t → m − ZrO_2 transformation. Notice also in Figure 5 that there are many microcracks on the tensile surface. Some of these microcracks have interacted to form an extended crack of dimension ~ 1.5 mm although the depth of this extended crack is probably comparable to that of the individual microcracks.

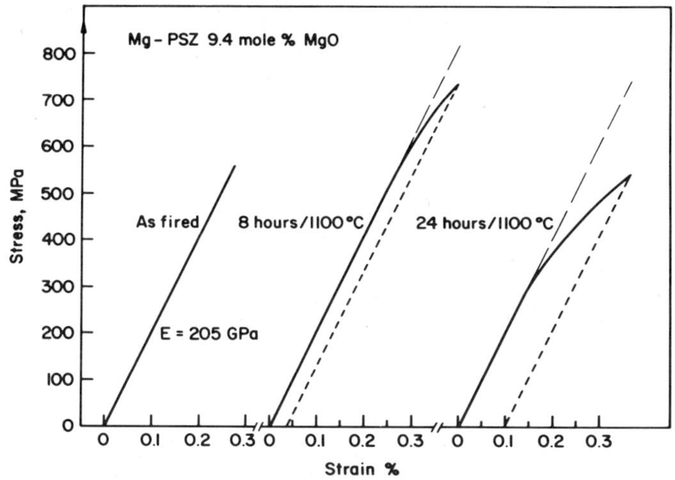

Fig. 4. Stress strain curves of Mg-PSZ materials in the as fired state and heat treated at 1100°C for various times.

Stress-strain curves for Ce-TZP[5] show similar behaviour to the Mg-PSZ. The low toughness materials with high ceria content, display a linear relation to failure whereas the high toughness materials have a marked departure, more abrupt than for Mg-PSZ, from linearity. The extent of permanent offset is small (~ 0.15%) similar to Mg-PSZ implying only a small fraction of this almost completely t-ZrO_2 material has been transformed prior to fracture. If all of the near surface region had transformed, as has been observed in some Y-TZP materials stressed in moist conditions at 200-300°C,[15] then strains in excess of 0.5% should have occured. As yet stress-strain data for other PSZ alloys are not available to confirm whether similar behaviour occurs.

Toughness increment

The toughness increment associated with transformation toughening is given by[16]

$$\Delta K = \eta \, V_f \, \Delta V \, E\sqrt{d}/(1 - \nu) \qquad (1)$$

where d is the transformed zone size, E is Young's Modulus, ν is Poisson's ratio and η a constant. A range of values for η have been reported by Evans and Cannon[16] depending upon the critical condition for initiating the transformation and zone contour about the crack tip. For a critical stress criteria η= 0.22, whereas for shear band determined crack tip contour η = 0.38 and for a critical principal stress criterion $\eta = 0.55\ (1 - \nu)$. In a previous paper[1] the experimental value for this constant for Mg-PSZ was found to be 0.33. An independent accessment using Raman spectroscopy[17] led to a value of or η of 0.4. A plot of data of K_{1c} versus $V_f\sqrt{d}$ for Y-TZP materials with d determined from the X-ray data of reference 11 is shown in Figure 6. An estimate

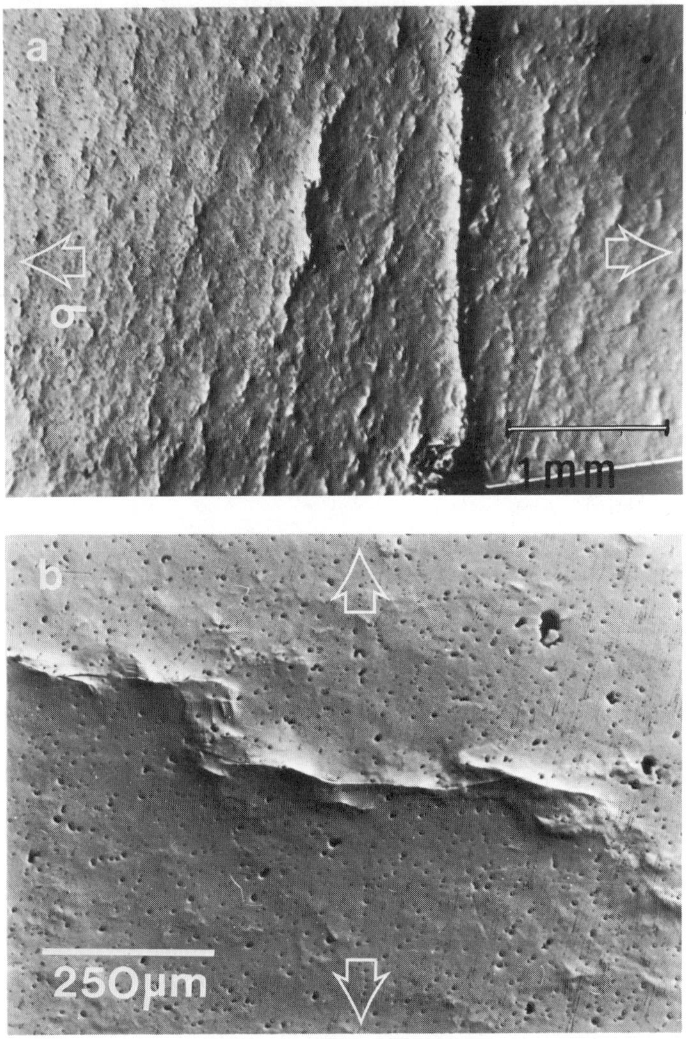

Fig. 5. Optical micrographs using interference contrast of the deformation on the tensile surface of the 24 hour 1100°C heat treated material. a) A low magnification micrograph showing the dispersed nature of the surface transformation and b) higher magnification of the localised linking of microcracks.

of η with $\Delta V = 0.05$, $E = 220$ GPa and $\nu = 0.31$ gives a value of 0.22 in excellent agreement with the critical stress criterion. The difference in values of η for the Y-TZP and Mg-PSZ is probably related to the difference in grain size of these materials. It has been observed that shear bands do occur about the crack tip in peak toughness Mg-PSZ materials.

The implications of the toughness zone size relationship are that the critical stress for the initiation of the transformation may be had from values of the K_{1c} and zone size. The critical stress is given by[16]

$$\sigma_c = \frac{K_{1c}(1 + \nu)}{\sqrt{d}} \quad \left(\frac{\sqrt{3}}{12\pi}\right)^{1/2} . \qquad (2)$$

It is now possible to plot the critical stress versus K_{1c} for a range of PSZ materials if the zone size is known. It is also informative to plot strength data for various PSZ alloys on a similar basis. If the strength is controlled by the flaw size, c_f, then the following simple relationship holds[18]

$$K_{1c} = Y\sigma\sqrt{c_f} \quad (3)$$

where Y is a geometric factor (~ 1.2). Results for various PSZ materials are shown in Figures 7-9. For each of these specific materials the powder processing/sintering conditions were virtually identical and it is assumed that the flaw size is independent of composition, heat treatment, etc. The values of the critical flaw size c_f have been determined for the various PSZ materials from the best fit of the lower toughness materials. These values are in reasonable agreement with SEM observations of fracture origins.

Results for Y-TZP materials and recent data for alumina additions to this material are shown in Figure 7. The low toughness or higher yttria containing material closely fit a critical flaw size of ~ 30 μm except at higher strengths. It may be argued that surface compressive stresses generated during specimen preparation may increase the strength in this region. The effect of the alumina is to significantly increase

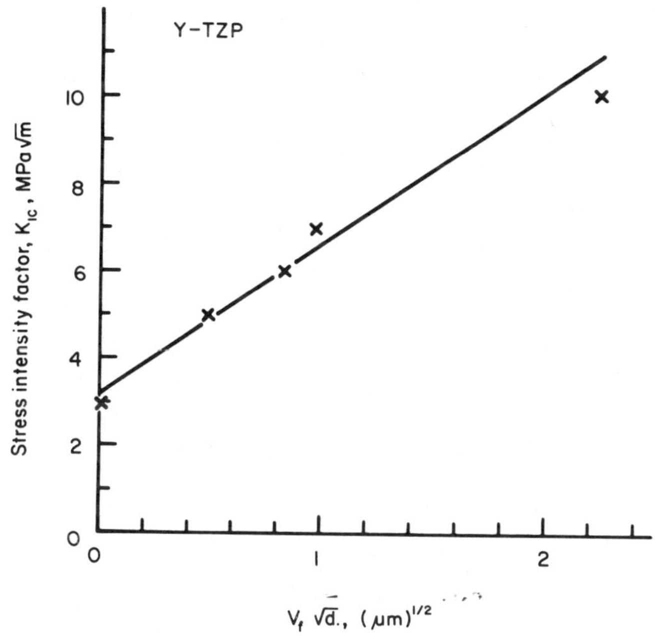

Fig. 6. Relationship between transformed zone depth, d, volume fraction of t-ZrO_2, V_f, stress intensity factor for Y-TZP materials.

the strength by decreasing the critical flaw size, which is supported by recent observations by Tsukuma et al.[19] The broken line through the results in Figure 7 has been calculated from equation (2) using the Y-TZP results and is the critical stress to trigger the $\underline{t} \rightarrow \underline{m} - ZrO_2$ transformation. The maximum strength-toughness results lie very close to this curve and suggest that the strength is limited by the critical transformation stress.

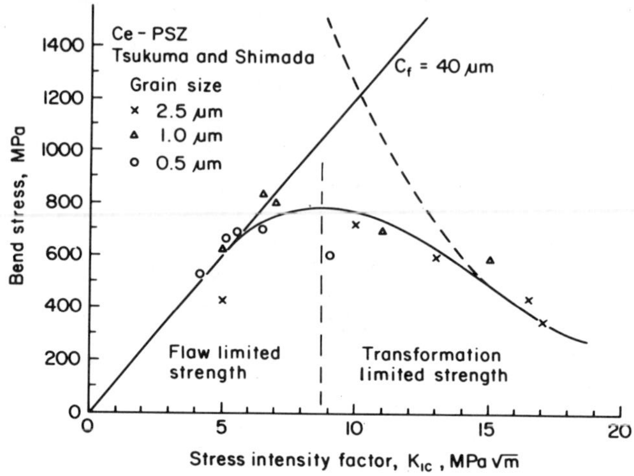

Fig. 7. Bend strength of Y-TZP materials of varying compositions plotted against K_{1c}. The broken line is the predicted critical stress to initiate the $\underline{t} \rightarrow \underline{m}$-$ZrO_2$ transformation. Also included in this figure are recent data for Y-TZP-Alumina alloys.

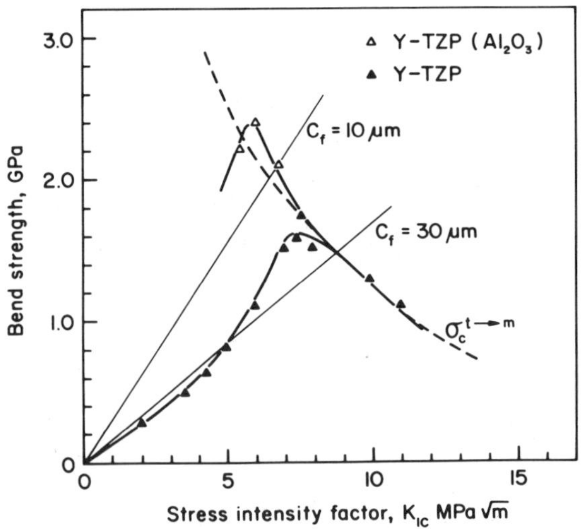

Fig. 8. Bend strength of Ce-TZP materials against K_{1c}. The broken line is extrapolated from the Y-TZP results.

Figure 8 is a plot of the Ce-TZP results replotted from Tsukuma and Shimada.[5] These materials were only sintered and the lower toughness materials imply the strength is controlled by a critical flaw of ~ 40 μm. The strengths of this alloy system are much lower than the Y-TZP materials but the K_{1c} values are much higher. The boundary or critical stress for the $\underline{t} \rightarrow \underline{m}$ -ZrO_2 transformation is not so well defined, nor independently determinable, as for the Y-TZP materials. The boundary plotted in Figure 8 is an extension of the Y-TZP curve.

The Mg-PSZ results shown in Figure 9 are intermediate between the two TZP materials. However as anticipated for much coarser grained materials the critical flaw size is much larger ~ 100 μm. Strengthening due to surface compressive layers is much less apparent than for Y-TZP materials and leads to strength increases from 10 to 15%. The critical locus for the stress to initiate the $\underline{t} \rightarrow \underline{m}$ - ZrO_2 transformation have been calculated from measurements of the zone width and K_{1c} values using a relationship similar to equation (2).

A composite plot of the strength-toughness relationships for the three PSZ alloys is shown in Figure 10. It clearly indicates that the peak strength decreases with increasing toughness and for high toughness materials is virtually independent of flaw size.

R-Curve limited strength

An alternative approach to explain the strength-toughness relationships of the PSZ alloys shown in Figures 7-9 is the known R-curve dependence of the toughness. An example of the theoretically predicted R-curve[7,9] and experimentally measured R-curve for Mg-PSZ[20] is shown in Figure 11. These curves imply that a flaw in Mg-PSZ, or other tough PSZ material, would commence to extend at a relatively low K value but the rising R-curve requires it will not catastrophically extend until the critical stress and flaw length to achieve the steady state K_{1c} value has been reached. Our understanding of R-curves in PSZ

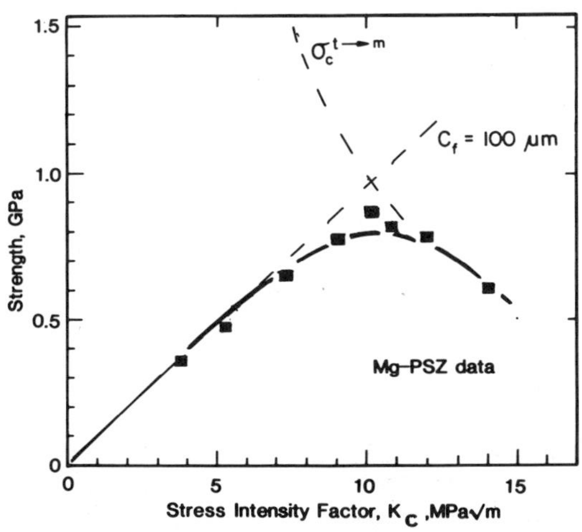

Fig. 9. Bend strength of Mg-PSZ plotted against K_{1c}.

materials is still limited, the results in Figure 11 are for a semi-infinite crack with constant zone size, and experimentally for a crack initiating from a blunt notch. Neither of these situations closely approximate a small flaw with a zone of perhaps comparable size extending to a critical dimension. Figure 5 shows a number of such small flaws with surrounding transformed zones.

Some progress towards a prediction of the strength toughness relationship may be had by considering the R-curve to be bi-linear, that is linear from the matrix toughness value to the steady state toughness value and constant thereafter.[10,19] The extent of crack propogation to reach a steady state K value scales with the zone size whereas the toughness scales with the square root of the zone size. Proceeding with this analysis leads to the prediction of maximum strength occuring when the zone size is given by,[10]

$$d = \left(\frac{\Sigma \ C_f}{\lambda \ K_m} \right)^2 \qquad (4)$$

where Σ is given in equation (1), c_f is the initial flaw size, K_m the matrix toughness and λ the ratio relating crack extension to steady state K to zone size. The interesting feature of equation (4) is that the zone at maximum strength is dependent upon the flaw size. It is readily apparent that the predicted zone sizes, and hence steady state toughness, are least for the Y-TZP (Al_2O_3) and largest for the Mg-PSZ materials. Larger zone sizes than those predicted by equation (4), and hence materials of higher toughness, leads to a reduction in strength which is again in agreement with the observations in Figures 7-10.

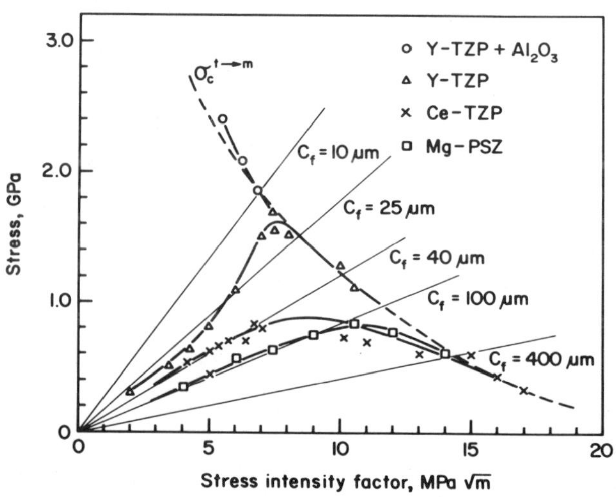

Fig. 10. Compilation plot of the three previous figures showing that the transformation limited results appear to form a locus of the critical stress to initiate the t → m-ZrO_2 transformation. The effect of decreasing flaw size is to decrease the toughness for the stength maxima of a particular alloy.

From the foregoing discussion and results it is apparent there are two approaches for explaining the strength-toughness relationships of transformation toughened PSZ alloys. The stress-strain data combined with calculated critical stresses from toughness-zone size measurements provides compelling evidence for stress-induced transformation to occur at low flexure or tension stresses. However the extent of inelastic deformation resulting from the stress assisted $\underline{t} \rightarrow \underline{m}$ - ZrO_2 transformation is much less than predicted if all the $\underline{t}$-ZrO_2 population were to transform. It appears as though stress concentrations sites be they pores, microcracks, etc may be locations for inhomogenous initiation of the transformation. The predictions based on the R-curve toughness dependence agree well with experimental results and are mechanistically more valid particularly for the materials with very large transformation zone sizes. For materials with very small zone

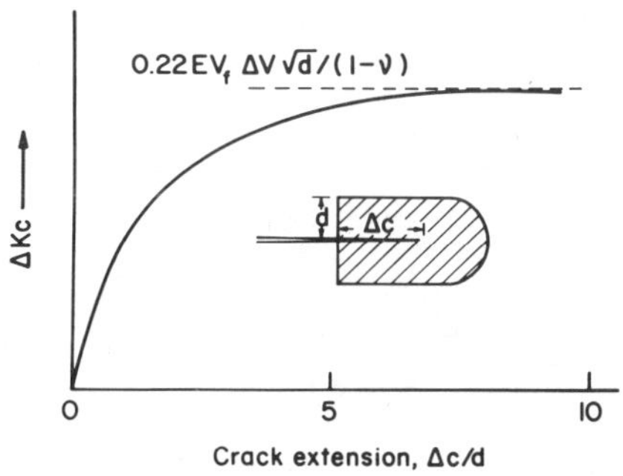

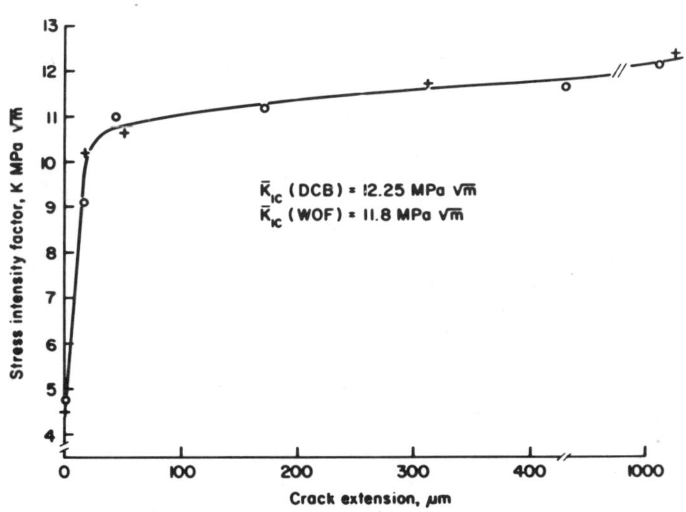

Fig. 11. a) Predicted R-curve behaviour for PSZ materials with η = 0.22.[9] b) Observed R-curve behaviour for a crack initiating from a notch in a notched bend specimen.[20]

sizes, < 1-2 μm, the concept of an R-curve is almost one of semantics or definition. For these materials the two approaches are probably equivalent.

CONCLUSION

Two approaches predict the strength-toughness relationships of transformation toughened zirconia alloys. They are: a transition from flaw limited strength at low toughness to transformation limited strength at high toughness, and the strength of very tough materials is limited by the R-curve behaviour of these materials. Predictions of critical stresses to initiate the $\underline{t} \rightarrow \underline{m}$ - ZrO_2 transformation and measured strengths for a range of PSZ alloys agree remarkedly well supporting the initial proposition. However the limited extent of the ductility, and its inhomogenous nature in very tough materials, questions this fortuitous agreement. On the otherhand, a simple analysis based on the R-curve dependent toughness of these materials does predict the observed strength-toughness relationship. But as suggested the concept of an R-curve for materials with very small zone sizes (< 1-2 μm) is probably meaningless and for these materials the two approaches merge. Finally, despite the simplistic argument of a transition from a flaw size to the transformation-stress strength dependence it does provide a readily appreciated interpretation of results obtained to date and a framework to guide future material development in zirconia toughened ceramic systems.

REFERENCES

1. M.V. Swain, Acta Metallurgica (in press) 1985.
2. K. Tsukuma, K. Ueda, K. Matsushita and M. Shimada, J. Am. Ceram. Soc. 68, (1985).
3. M. Marmach and M.V. Swain, J. Aust. Ceram. Soc. 20, 32 (1984).
4. R. P. Ingel, D, Lewis, B.A. Bender and R.W. Rice, Advances in Ceramics, Vol. 12, 408 (1984).
5. K. Tsukuma and M. Shimada, J. Mater. Sci., 20, 1178 (1985).
6. R.C. Garvie, R.H.J. Hannink and R.T. Pascoe, Nature (Lond.) 258, 703 (1975)
7. R. McMeeking and A. G. Evans, J. Am. Ceram. Soc. 65, 242 (1982).
8. M.V. Swain and L.R.F. Rose, ICF6, Adv. in Fracture Res. Vol. 1 473 (1984) Plenum Press.
9. D.B. Marshall, M. Drory and A.G. Evans. Fracture Mechanics of Ceramics, Vol. 6, 289 (1983)
10. L.R.F. Rose and M.V. Swain to be published.
11. K. Tsukuma, Y. Kubota and T. Tsukidate. Advances in Ceramics, Vol 12, p (1984).
12. K. Kobayashi, H. Kuwajima and T. Misaki, Sol State Ionics 3/4, 489 (1981).
13. R.H.J. Hannink and M.V. Swain, J. Aust. Ceram. Soc. 18, 53 (1982).
14. D.B. Marshall, to be published.
15. M. Mitsui, T. Soma and I. Oda, Advances in Ceramics, Vol 12, 371 (1984).
16. A.G. Evans and R.M. Cannon, Acta Metallurgica (in press) 1985.
17. P.F. Becher et al, unpublished work 1985.
18. B.R. Lawn and T.R. Wilshaw, 'Fracture of Brittle Solids'. Cambridge University Press 1975.
19. K. Tsukuma and M. Shimada. Bull. Am. Ceram. Soc. 64, 310 (1985).
20. M.V. Swain and R.H.J. Hannink, Advances in Ceramics, Vol 12,

TENSILE STRENGTH AND NOTCH SENSITIVITY OF Mg-PSZ

L.W. Stephens [+] and M.V. Swain [++]

[+]CSIRO Division of Manufacturing Technology
[++]CSIRO Division of Materials Science
Melbourne, Australia 3001

ABSTRACT

The tensile strength of peak aged magnesia partially stabilised zirconia (Mg-PSZ) was determined. It was found that unlike flexure strength tests the results were independent of surface finish. The tensile strength was only 60% of the flexure strength. Observations of the fracture surface revealed that often it was not possible to identify the fracture origin. The notch sensitivity of similar Mg-PSZ material was evaluated with notches of varying notch radii. The measured strengths were much higher than predicted on the basis of Neuber's stress concentration factors for notches. Incorporating the Weibull statistics volume under stress relationship into the notch concentration factor led to better agreement with observations, particularly for notches of small radius.

INTRODUCTION

The widespread usage of ceramics in structural engineering applications has been severely restricted to date by the low toughness and notch sensitivity of these materials. In general, stress concentrations at the base of a notch, groove or thread in these brittle materials lowers the strength to virtually unacceptable levels. It is this factor that has prevented the development of such simple means of attachment as nuts and threads between ceramics and other materials. The alternative means of joining ceramics such as epoxy glueing, ceramic-eutectic bonding, brazing, noble metal or reaction bonding, etc. all have serious drawbacks [(1)]. These limitations include, temperature restrictions for operation, finding appropriate eutectics and cumbersome means of joining. Of these methods of joining ceramics the reaction bonding technique appears to have the best strength and is least temperature sensitive [(2)]. However it does require high temperature and pressure to assist in the joining and there are only a limited range of materials that may be so joined.

The development of high performance structural ceramics during the last decade based upon alumina, alumina-zirconia (ATZ), silicon carbide and nitride and more recently partially stabilised zirconia (PSZ) has led to the demand for stronger and simpler means of joining these ceramics to other ceramics and metals. The high strength and toughness of the PSZ and ATZ materials, coupled with their low thermal conductivity and thermal expansion coefficient comparable to cast irons, has encouraged development of experimental programs in adiabatic or insulated engines [3]. Another attractive feature of both PSZ and ATZ materials is that grinding introduced damage increases the strength of these materials [4]. This occurs because the grinding stresses, like the tensile stress field about a crack tip, initiate a martensitic volumetric expanding phase change which leads to a compressive stress at the surface or about the crack tip [5]. Additional tensile stress must be applied to overcome the grinding introduced surface compressive stresses.

The tensile strengths of ceramics are not often measured because of the difficulty of specimen preparation and alignment procedure to ensure there are no additional flexure stresses. For metallic materials minor defects or flaws are usually not as deleterious as for ceramics because plastic flow can occur and reduce the severity of stress concentrations. This last point also accounts for the excellent reproducibility of the strength properties of metallic materials and the typical scatter of ceramic measurements. However recent observations on very tough PSZ materials suggest that these materials also display pseudo-ductility because of stress induced transformation during flexure [6]. This stress induced transformation at stress concentrations might be anticipated to reduce the severity of notches from that predicted by Neuber and subsequent workers [7].

The present work compares the tensile and flexure strength of a peak aged Mg-PSZ alloy material and the influence of notch geometry on the strength. A few additional tests were carried out on a commercially available alumina rod for comparison with the PSZ material.

MATERIALS TESTED

The Mg-PSZ was of commercial origin[+] and contained approximately 9 mole % MgO. The microstructure of this material, shown in Figure 1 consists of relatively large grains (50-60 μm) which are of cubic-zirconia containing approximately 40 volume % of tetragonal zirconia precipitates [8]. The property of this material may be modified by heat treatment and the present material had been aged for the maximum strength condition. Application of stress causes the tetragonal precipitates to transform to monoclinic zirconia.

Upon grinding a layer adjacent to the surface of approx. 10 μm consists of monoclinic zirconia precipitates. The number or volume % decreases almost linearly with depth. The toughness of this material is 9-10 MPa√m at room temperature and decreases to 5 MPa√m at 600°C.

The alumina material of unknown origin consisted of coarse grains bonded with a glassy phase. Typical microstructure of this material is shown in Figure 2. The density of this material was 3.8 x 10^3 Kg.m^{-3} and the toughness was measured as 4.5 MPa√m using the indentation technique.

[+] Nilcra Ceramics Pty. Ltd., Northcote, Victoria.
[++] Diamond Plating Co. Pty. Ltd.

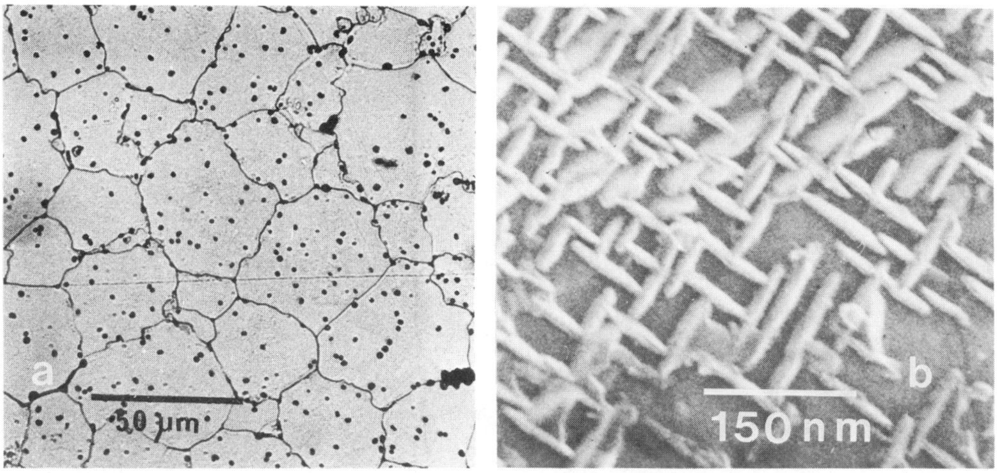

Fig. 1. Microstructure of the Mg-PSZ material showing the large grain size, (a), and the fine tetragonal precipitates within the cubic matrix, (b).

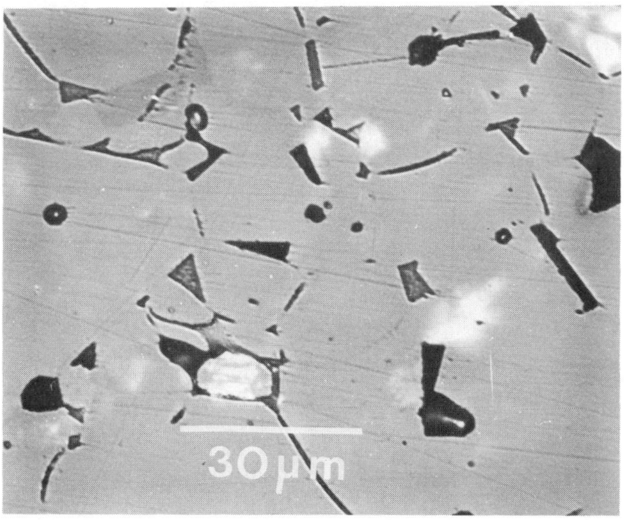

Fig. 2. Polished section of the alumina material showing the presence of glassy material between the grains.

SPECIMEN PREPARATION

Round bars of approximately 10 - 12 mm diameter were available for both materials. Tensile specimens were machined (ground) using 140/170 mesh diamond plated wheels [++] to the geometry shown in Figure 3a. Of the 10 specimens so prepared, 5 were left in the as ground condition ~ 2.5 µm surface roughness, and 5 polished to 0.05 µm surface roughness. At each end of the specimen 10mm was threaded for insertion into copper nuts attached to a differential in a universal testing machine [+]. The remaining specimens were notched with diamond plated wheels of different tip radii. Examples of sectional views of typical notches are shown in Figures 3 b and c. Tests were carried out at

[+] Instron Model TT-C-LM1.6

constant crosshead speed of 20×10^{-6} m.s^{-1} in ambient air at ~ 20°C.

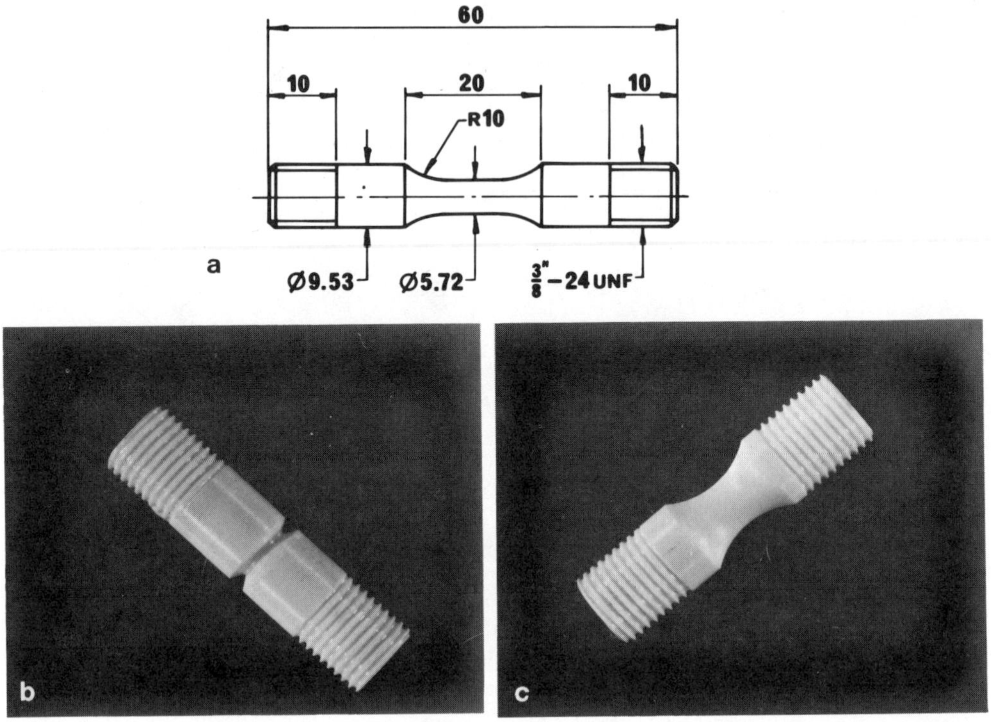

Fig. 3. Cross-sectional view of the tensile test specimens, (a), and photographic views of specimens with notches of 0.3mm and 10mm radii, (b and c respectively).

STRESS CONCENTRATION

The stress concentration, K_t at the base of a notch for a tensile specimen is usually defined as

$$\sigma_{max} = K_t \sigma_{nom} \qquad (1)$$

where σ_{max} is the maximum notch tip stress at the surface and σ_{nom} is the nominal stress for a rod with crossection reduced by the dimension of the notch.

Values for K have been determined by many authors (7,9,10), although the work of Neuber (7) is perhaps most widely known, and are usually available in graphical form as shown in Figure 4. The values are for the maximum principal stress at the base of the notch normal to the plane of the notch. The key parameters are the notch radius, r, remaining section, d, and rod diameter D. The values shown in Figure 4 are for circumferentially notched rods, while values for other geometries are available in handbooks (11) in graphical form.

Another significant feature of these stress concentration factors to be considered is the stress gradient beneath the notch. Neuber (7) pointed out that the fading or decrease in stress from the notch root depended upon the notch geometry. For a bar or plate (2 dimensional) the decrease in stress was inversely related to notch radius $(1/r)$; for a deep circumferential notch an inverse square dependence $(1/r^2)$ was found, and with a shallow notch the initial gradient was proportional to the inverse cube of the notch radius $(1/r^3)$.

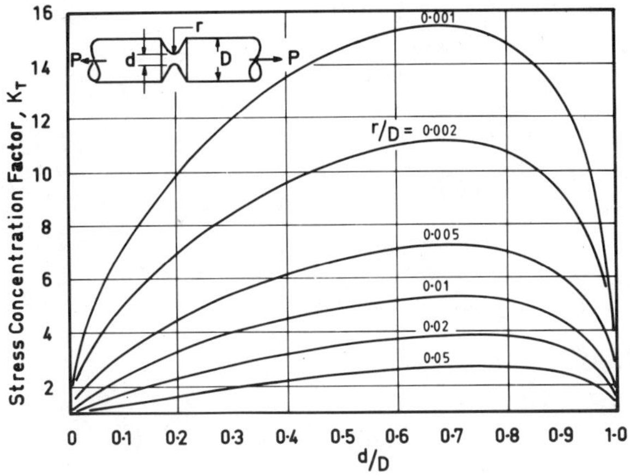

Fig. 4. Stress concentration factors for notched circumferential rods, after Neuber [7].

OBSERVATIONS AND DISCUSSION

Tensile Tests

The tensile strength results for the polished and ground Mg-PSZ specimens are listed in Table 1. They indicate that the strength is independent of the surface condition of the specimens.

Table 1

			Tensile Strength,		MPa
Polished	1	428	Ground	1	469
	2	444		2	463
	3	467		3	463
	4	473		4	463
	5	424		5	408
	X	444 ± 22		X	453 ± 27
	All specimens		X 450 ± 23		

The flexural strength was measured for ground bars 3 mm in cross-section in 4 point bending with inner and outer spans of 7.5 and 19 mm respectively and was found to be 760 ± 44 MPa. The tensile strength is only 59% of the bend strength. Typically a reduction of some 10 to 15% occurs when the flexure specimens are polished because of the removal of the transformed surface compressive layer [4].

Observations of the tensile fracture surfaces revealed that in 3 of the 10 specimens the origins of failure was obviously from some internal defect. Examples of these internal origins are shown in Figures 5a and b. The large pore approximately 250-300 μm in diameter

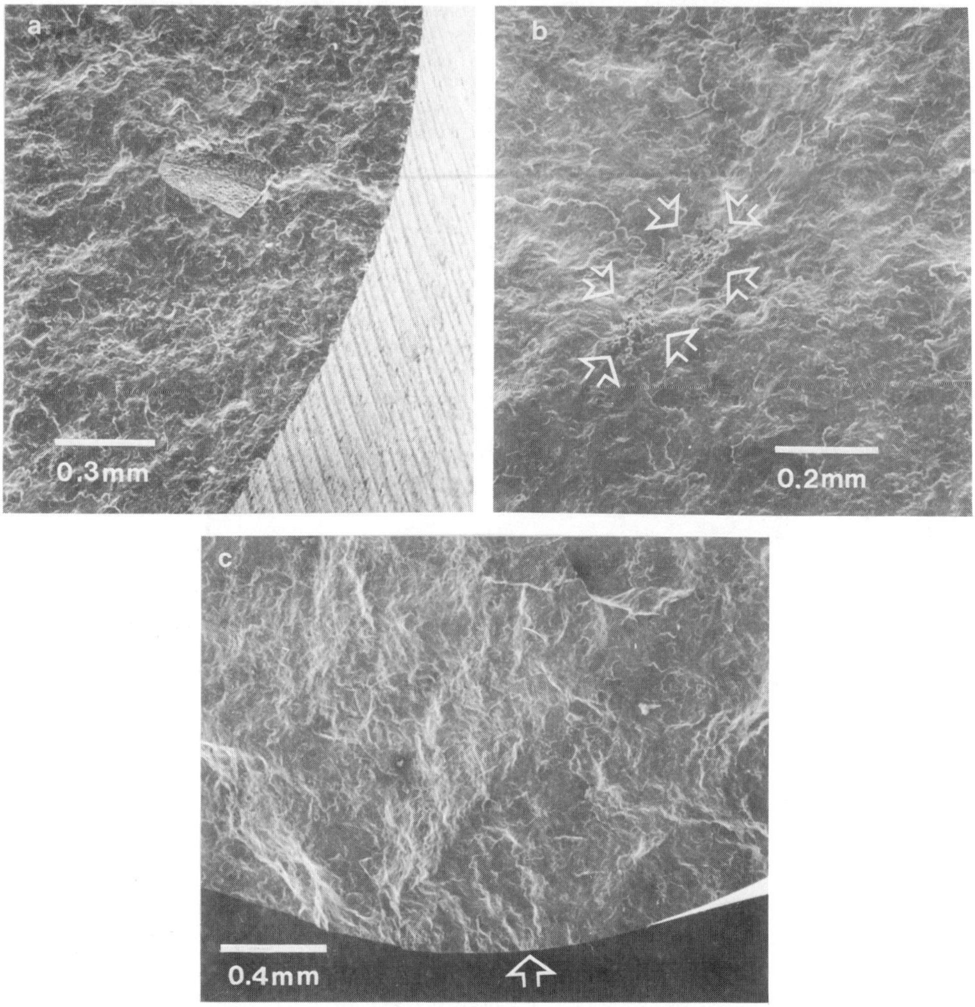

Fig. 5. SEM observations of the fracture origins in Mg-PSZ; (a) and (b) are examples of fracture initiating from internal defects whereas (c) is from a poorly defined surface origin.

consists of a glass phase rich in magnesia and silica. The fracture origin in Figure 5b is a porous region some 200-250 μm consisting of smaller grains, no EDAX analysis was carried out on this area. An example of a fracture not initiating from a well defined area is shown in Figure 5c. This sample was polished after grinding but detailed observations of the fracture did not reveal a specific site for its initiation. The variation in flaw diameter and the fact that both surface and internal fracture sites were observed indicates that the loading system was well aligned and that the material under test was very tolerant to flaws. Estimates of K_{1c} from the flaw size and fracture stress may be made from

$$K_{1c} = Y\sigma \sqrt{c_f} \qquad (2)$$

where Y = 1.1, σ = 450 MPa and c_f = 150 μm leads to a value of only 6 MPa$\sqrt{m}$ which is only half the value anticipated from more traditional estimates of K_{1c} using double cantilever beam techniques. The internal flaws would not be susceptible to stress corrosion cracking which could lower the critical K value. The results do support the concept that crack extension in this material undergoes an R-curve on rising K value with propagation. A more detailed discussion of this point is presented in a companion paper [12]. Additional tests on slightly smaller specimens, 9mm diameter and 25mm long, which were weaker than the above specimens, were also carried out. These materials had a flexure strength of 705 ± 33 MPa and tensile strength of 411 ± 27 MPa. The majority of notched tests were carried out on this material. The tensile strength of these specimens was 58% of the flexural strength, almost identical with the previous material.

Notched Specimens

The strength of the notched specimens, when plotted against notch radii normalised by specimen diameter, are shown in Figure 6. Only specimens in the as ground state were tested because of the difficulty in polishing the notch root region. The results in Figure 6 show that with decreasing notch radius the strength decreases. The limiting notch radius appears to be approximately the specimen diameter for the materials tested. Also shown in Figure 6 are some data points for an alumina rod. The results are too few to be able to draw a conclusion for this material, also the tensile strength was not measured.

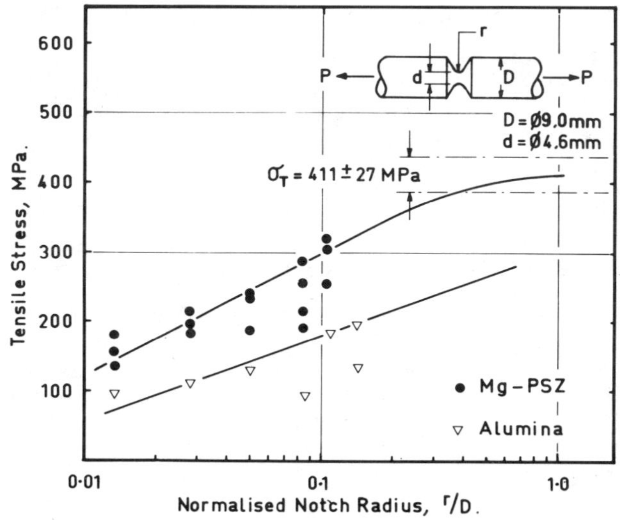

Fig. 6. Tensile strength of Mg-PSZ materials and a few alumina specimens plotted against normalised notch radius.

Additional tests on a slightly stronger Mg-PSZ material have been included in Figure 7 which is a plot of normalised strength versus

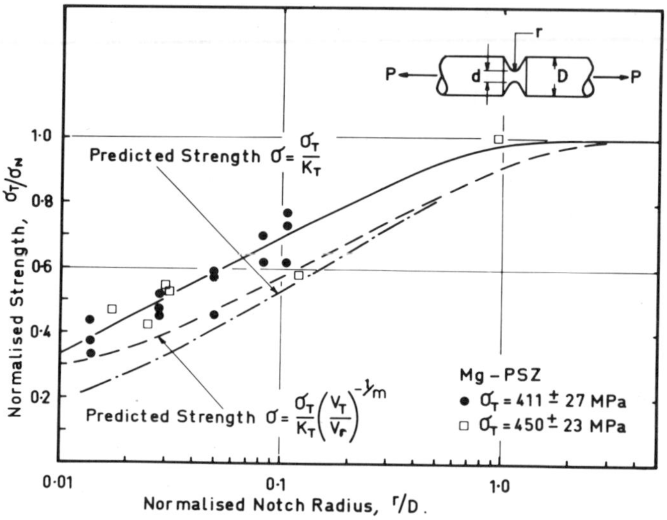

Fig. 7. Normalised strength versus normalised notch radius for all Mg-PSZ tensile specimens. Also plotted is the predicted strength from equation (1) and equation (4).

normalised notch radius. The two sets of data superimpose well, although additional data at sharper and blunter notches than examined would have been useful. Also shown in Figure 7 are the predicted strengths based on the stress concentration factors shown in Figure 4. This curve has been constructed on the basis that the volume under test does not influence the material's strength. However, given the statistical nature of the strength of brittle solids it may be that this approach is inadequate, and Weibull statistics should be employed. The relationship between the strengths of similar materials with different volumes is given by

$$\sigma/\sigma_o = (V/V_o)^{-1/m} \qquad (3)$$

where σ_o and V_o are the reference strengths and volume and m is the Weibull modulus which may be related to coefficient of variation. In the case of notched materials, the volume in question following Weiss (13) may be considered as $\sim r^2\ell$ where r is the notch radius and ℓ the circumference. The reference volume in these materials is the tensile specimen $\sim D^2\ell$. These approximations modify the above expression to,

$$\sigma = \frac{\sigma_T}{K_t}\left(D^2/r^2\right)^{-1/m} \qquad (4)$$

where K_t is the stress concentration factor from Figure 4.

The Weibull modulus for the materials listed in table 1 is ~ 20.

The higher the Weibull modulus the less significant the effect of changing volume. The predicted strength on the basis of equation (4) is also shown in Figure 7. Both predictions provide conservative estimates of the strength.

A more valid fracture mechanics procedure for estimating the severity of a notch is to follow the approach of Miller [14] and Mussler et al [15], who have determined the stress intensity factor for a crack emanating from a notch. The latter authors were able to show that fine grained alumina materials were very notch sensitive whereas coarse grained materials were not. It is necessary to account for the decreasing stress field away from the notch tip, the surface compressive stresses generated during notching, and the R-curve behaviour of these Mg-PSZ materials. The range of the stress concentration factor scales with the notch radius and is typically reduced by a factor of 2 at half the notch radius. The surface compressive stresses fall rapidly with depth and disappear within 10-20 μm of the surface, whereas the R-curve related crack extension to achieve steady state toughness scales with the toughness of the Mg-PSZ. The former two parameters are shown schematically in Figure 8. Whilst it is possible to numerically integrate a Green's function expression [16] to determine the stress intensity factor with such a stress profile along the crack path, the relative uncertainties of the initial flaw size and magnitude of the grinding stresses do not warrant such an approach at this stage.

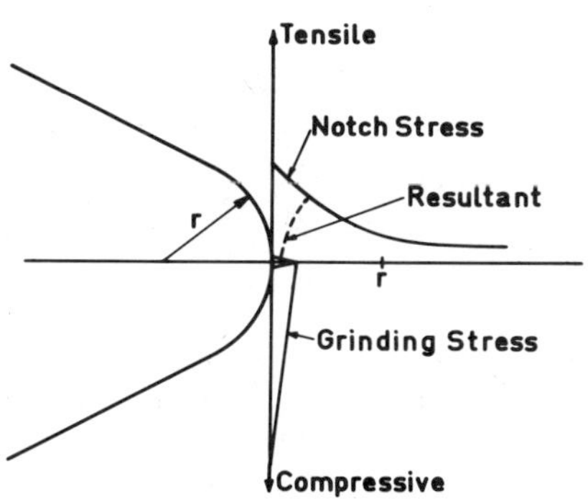

Fig. 8. Schematic diagram of the principal stresses existing about the notch root under load with surface grinding stresses. The maximum occurs just below the surface in the resultant stress field.

The notched fracture specimens were also examined in a scanning electron microscope. Again it was difficult in many cases to readily identify a fracture origin. Examples of the fracture surface, fracture origin and machined surface are shown in Figures 9a, b and c respectively. The latter figure shows that small flaws ~ 10-20μm in length have been formed by diamond grinding. These flaws are parallel to the tensile stress axis and are therefore unlikely to be fracture initiating sites. However, the fracture origin in Figure 9b does appear to be inclined to the machining direction and tensile axis and may have initiated from a larger example of these types of flaws.

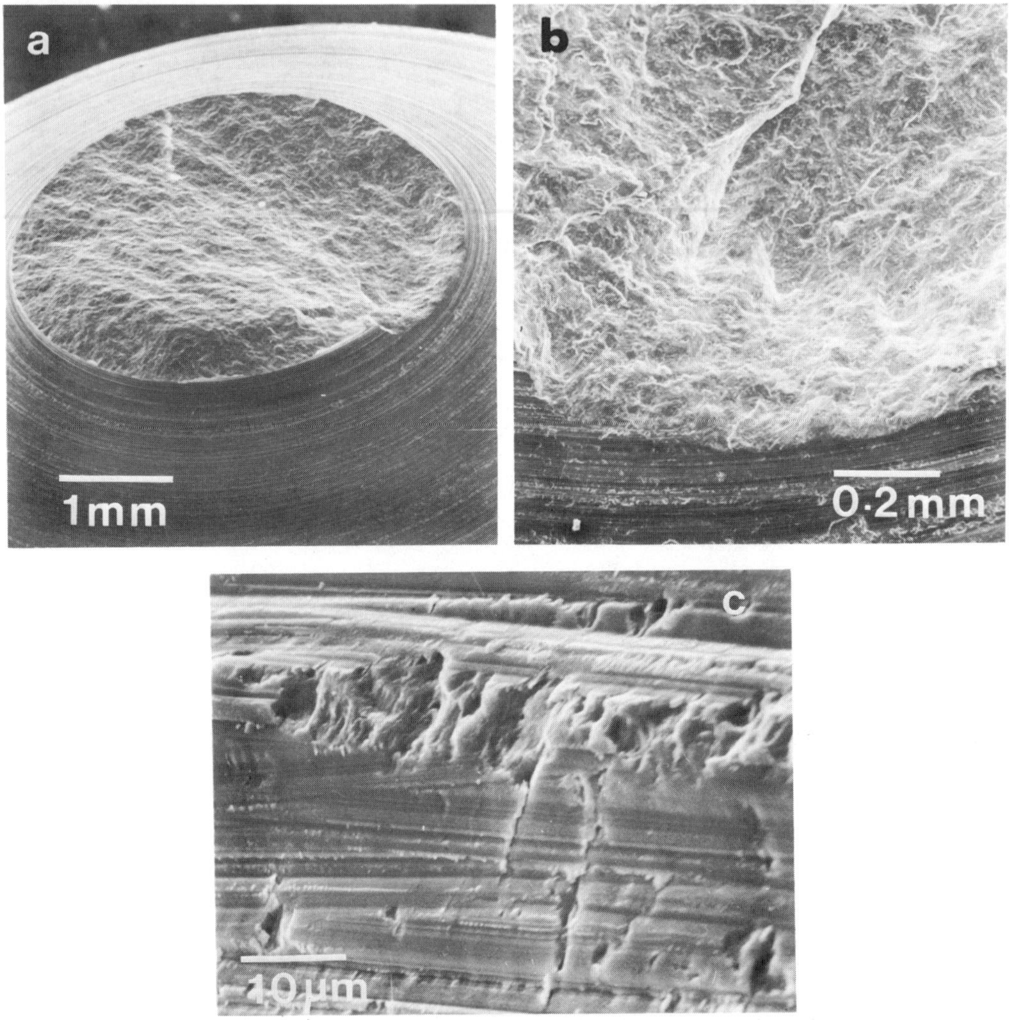

Fig. 9. SEM observations of a fractured notched specimen, (a), and higher magnification of the fracture origin, (b); also shown is the surface cracking generated during grinding, (c).

CONCLUSIONS

The present work has shown that the tensile strength values of Mg-PSZ materials as determined by tensile tests are substantially less than the values obtained using flexure strength measurements. Two sets of results on peak aged Mg-PSZ specimens indicate that the ratio of tensile to flexure is approximately 0.6. The tensile strength appears to be independent of surface condition, unlike flexure strength. The tensile strength also appears to be very insensitive to initial flaw or defect size and often surface fracture origins were difficult to identify. Both sets of tensile strength data exhibited a high Weibull modulus of between 18-20.

Notched specimens of Mg-PSZ decreased in strength with decreasing notch radius but less rapidly than predicted by stress concentration factors. However incorporation of Weibull volume dependence statistics led to better agreement between predicted and observed strengths.

ACKNOWLEDGEMENTS

The authors wish to express their appreciation to Mr. F.B. Polivka of CSIRO for his assistance in specimen preparation and testing. Also to Mr. M. Marmach of Nilcra Ceramics Pty. Ltd. for supply of test materials and for valuable discussions.

REFERENCES

1. M. Erz and H.W. Hennicke. "Ceramics in Advanced Energy Technologies" p 138 - 156. D. Riedel Pub. Co. Dordrecht (1985).
2. H.J. de Bruin, A.F. Moodie and C.E. Warble. J. Mater. Sci., 7 909 - 18 (1972).
3. R. Camo, M. Woods and W. Geary. "Ceramics for Adiabatic Diesel Engines" CIMTEC 4th Energy and Ceramics (1979) p 468.
4. M.V. Swain. J. Mater. Sci., 15, 1577 - 79 (1980).
5. A.G. Evans. "Advances in Ceramics", Vol 12, 193 - 212 (1984). Am. Ceram. Soc., Columbus, Ohio.
6. M.V. Swain, Acta Metallurgica in press (1985).
7. H. Neuber. "Theory of Notch Stresses". Ann Arbor Michigan (1946).
8. R.H.J. Hannink and M.V. Swain. J. Aust. Ceram. Soc., 18, 53 - 62 (1982).
9. C.E. Inglis. Trans. Inst. Naval Archit., 55, 219 (1913).
10. R.E. Peterson. "Stress Concentration Design Factors". (1953) Wiley, N.Y.
11. M.V. Swain and R.H.J. Hannink. "Advances in Ceramics", Vol 12, 225 - 239 (1984) (Am. Ceram. Soc.).
12. M.V. Swain. This volume.
13. V. Weiss. "Notch Analysis of Fracture". Ch III of Fracture III. Edited by H. Liebowitz. Academic Press N.Y. (1971).
14. K.J. Miller. Int. J. Fract. Mechs. 9, 326 (1973).
15. B. Mussler, M.V. Swain and N. Claussen. J. Am. Ceram. Soc., 65, 566 - 72 (1982).
16. B.R. Lawn and T.R. Wilshaw. "Fracture of Brittle Solids". Ch.3 (1975). Cambridge Uni. Press.

FRACTURE MECHANISMS IN LEAD ZIRCONATE TITANATE CERAMICS

S.W. Freiman and L. Chuck
National Bureau of Standards
Gaithersburg, MD 20899

J.J. Mecholsky and D.L. Shelleman
Penn State University
University Park, PA 16802

L.J. Storz
Sandia National Laboratories
Albuquerque, NM 81785

ABSTRACT

Lead Zirconate Titanate (PZT) ceramics can be formed over a wide range of $PbTiO_3/PbZrO_3$ ratios and exist in a number of crystal structures. This study involved the use of various fracture mechanics techniques to determine critical fracture toughness, K_{IC}, as a function of composition, microstructure, temperature, and electrical and thermal history. The results of these experiments indicate that variations in K_{IC} are related to phase transformations in the material as well as to other toughening mechanisms such as twinning and microcracking.

In addition, the strength and fracture toughness of selected PZT ceramics were determined using specimens in which a crack was introduced by a Vicker's hardness indentor. The variation of K_{IC} with composition and microstructure was related to the extent of twin-crack interaction. Comparison of the plot of strength as a function of indentation load with that predicted from indentation fracture models indicates the presence of internal stresses which contribute to failure. The magnitude of these internal stresses has been correlated with electrical properties of the ceramic. Fractographic analysis was used to determine the magnitude of internal stresses in specimens failing from "natural flaws."

INTRODUCTION

Lead-zirconate-titanate (PZT) is a series of solid solutions of $PbZrO_3$ with $PbTiO_3$ as shown in the phase diagram given in Figure 1[1]. The material below the Curie point can exist in a number of crystal structures as designated on the diagram, of which the phases designated F_T (tetragonal), F_R (HT), and F_R (LT) (both rhombohedral) are ferroelectric (FE), and that designated A_o (orthorhombic) is antiferroelectric (AFE). In this paper, we will show that in addition to the fracture mechanisms such as twinning and microcracking which have been identified for another ferroelectric ceramic,

Strength-Fracture Toughness

Most of the critical fracture toughness data reported herein were obtained using the applied-moment-double catilever beam technique[5]. Specimens were nominally 13 x 2 x 40 mm with an ≈ 0.5 mm deep groove cut longitudinally down the specimen to guide the crack. A 14 mm long tapered notch was cut into the top of the specimen as a crack initiator. Prior experience indicates that such a notch yields data equivalent to having an actual crack in the specimen. The data was generally obtained by loading the specimen in air or Fluorinert, FC-77, at a rapid crosshead rate (≈ 0.5 cm/min) on a universal testing machine. Lower temperature tests were performed in Fluorinert which was cooled by the addition of dry ice to the liquid.

Values of fracture toughness were also obtained on the 95/5 compositions using the indentation-strength procedure described by Chantikul et al.[6] Tests were performed on bars nominally 3 mm x 3 mm x 25 mm. A 15 µm diamond final surface finish on the test bars was used in order to eliminate any surface stresses that may have resulted from specimen preparation. After polishing, the specimens were annealed at 250 °C in air. The indentation was placed in the center of the width of the bar using a Vickers indentor at a range of loads, P. A drop of silicone oil was placed on the indentation site to reduce the tendency for slow crack growth. The bars were broken in 3-point flexure (span 19.4 mm) at a crosshead rate of 0.5 mm/min. The specimens were carefully examined after fracture; only those which had clearly failed from the indentation site were analyzed for K_{IC}. K_{IC} was calculated from the portion of the log σ-log P curve which obeyed the relationship predicted from the indentation model. Deviations from this behavior are discussed later in terms of crack-microstructure interactions.

Fractography

Fractographic analysis of the 95/5 PZT specimens failing from "natural flaws", i.e. pores or machining induced defects), was carried out in the scanning electron microscope. Based on the observed flaw sizes and the critical fracture toughness, K_{IC}, determined independently, the level of "internal stress" was calculated and compared to that determined from indentation-strength studies. General fracture morphologies were determined to establish the ratio of intergranular to transgranular failure as a function of PZT composition.

RESULTS AND DISCUSSION

The fracture behavior of PZT will be discussed in two parts. In the first portion, we will discuss effects of general compositional variations in the PZT system on fracture, particularly fracture toughness. The second part of this section will center on a discussion of the fracture mechanisms important to a specific PZT composition; namely, 95% $PbZrO_3$-5% $PbTiO_3$.

General Compositional Effects; K_{IC}

The critical fracture toughness, K_{IC}, of PZT as measured by an applied moment DCB technique is shown in Figure 2 for material having a range of Zr/Ti ratios. The data can be viewed as passing through a series of maxima and minima in K_{IC}. There is a distinct minimum at the morphotropic boundary between the tetragonal and rhombohedral phases. The minimum at the tetragonal-rhombohedral boundary can be observed even though the data came from various materials prepared at different laboratories using three different techniques. Somewhat less well documented minima occur at the other two phase boundaries in the $PbZrO_3$-$PbTiO_3$ system (Figure 1). Previous work[2,3] has

$BaTiO_3$ [2-4], there are additional factors such as the morphotropic phase changes which can contribute directly to fracture. We will show that the concept of "internal stresses" which has been discussed with relation to the fracture of $BaTiO_3$ [2-4], will be broadened to include the possibility of generalized microstructure crack interactions. In addition, a correlation will be established between internal stresses as measured through the electrical characteristics of the material, e.g. polarization loops, and those inferred from the strength measurements.

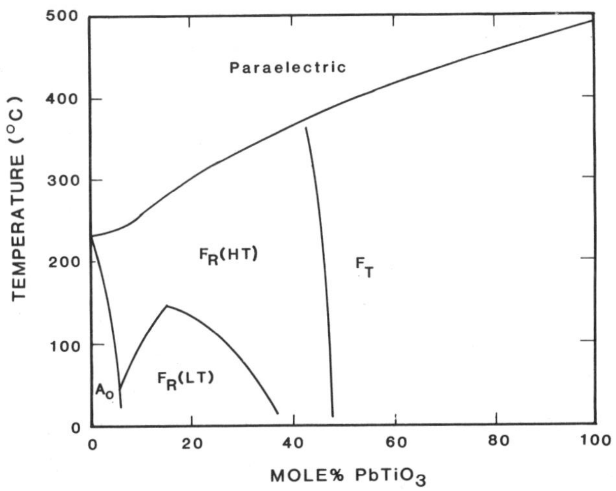

Figure 1. $PbZrO_3$-$PbTiO_3$ phase diagram (after Jaffe, Cook, and Jaffe[1]).

EXPERIMENTAL PROCEDURES

Materials

The PZT materials tested were obtained from a wide variety of sources (Honeywell, Inc.; Sandia National Laboratories; Motorola, Inc.; Nippon Electric; and Plessey, Inc.) The materials were obtained and tested over approximately a ten year period. Although the variability in processing conditions most likely has an effect on the mechanical properties, it is not possible to quantify this relationship. We will show later that controlled composition variations in 95% $PbZrO_3$-5% $PbTiO_3$ (95/5 PZT) can have a significant effect on the fracture process.

The 95/5 PZT materials used in this study were prepared at Sandia National Laboratories from various sources of PbO, ZrO_2, and TiO_2 powders, plus ≈ 0.8 wt.% NB_2O_5. The powders were mixed then calcined at 1000 °C. The ceramic billets were bisque fired at 600 °C for 4 hours then sintered at 1310 °C for 6 hours. The grain size of the fired material was in the range of 5-10 μm for most compositions examined, but some specimens were processed to yield smaller grains, 1-2 μm, as well as larger, ≈ 100 μm. Fracture toughness data was taken on material in the as-fired condition, after electrical poling at 100 °C at a field of 2000 volts/mm, and after the poled specimens had been hydrostatically depoled at room temperature (pressure ≈ 27.6 MPa).

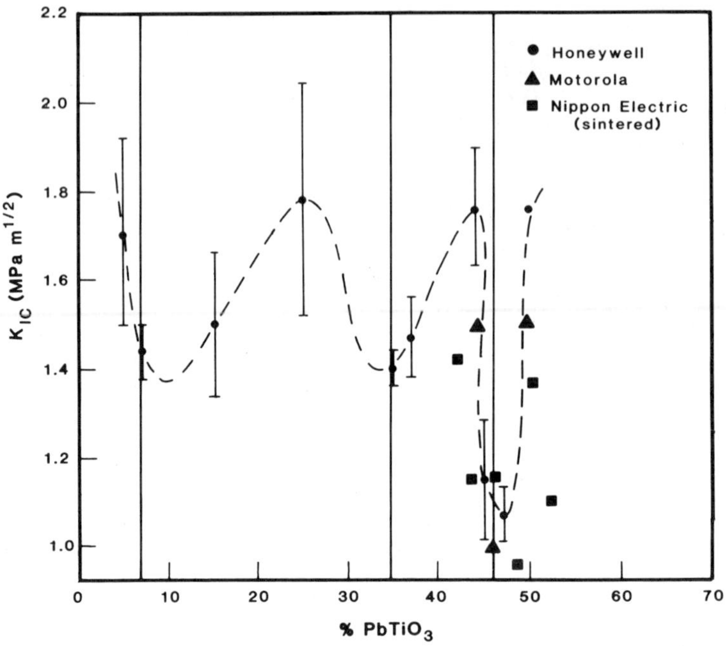

Figure 2. Critical fracture toughness, K_{IC}, as a function of composition in the $PbZrO_3$-$PbTiO_3$ system. Vertical lines indicate the position of the phase boundaries (25 °C) shown in Figure 1.

demonstrated that both crack-twin interactions and microcracking make important contributions to fracture toughness of $BaTiO_3$ in the ferroelectric (tetragonal) state. Neither of these mechanisms are applicable to $BaTiO_3$ in the paraelectric (cubic) state, and therefore a much lower K_{IC} is observed. The fact that the K_{IC} of a commercial PZT* goes from 0.31 MPa m$^{1/2}$ in the cubic state to 0.83 MPa m$^{1/2}$ in the ferroelectric state implies that twinning is important to the toughening process in this material. Further evidence of twinning effects on K_{IC} will be shown later in the discussion on 95/5 PZT.

The data in Figure 2 is in basic agreement with the work of Igarashi and Okazaki[7] who showed that both strength and fracture toughness, as determined by an indentation procedure, pass through a minimum at a Pb/Zr ratio of ~ 46.5/53.5, i.e., the morphotropic boundary. Igarashi and Okazaki also observed that as the Ti concentration was extended beyond the morphotropic boundary it became increasingly difficult to measure the size of the indentation cracks because of a damaged (microcracked) region surrounding the indentation site. They interpreted this damage (microcracking) as being due to internal stresses produced by the increasing distortions of the tetragonal lattice as composition shifts closer to the $PbTiO_3$ end of the phase diagram. These distortions, leading to microcracking, can enhance K_{IC} in direct analogy to effects in $BaTiO_3$. One can calculate the effects of microcracking on trends of K_{IC} with composition in the tetragonal phase regime, following the models developed for Al_2O_3 et al.[8] and $BaTiO_3$ [3]. Rice and Freiman[8] showed that when microcracking is a major source of toughening, the fracture energy γ (= $K_{IC}^2/2E$) can be calculated from the following expression:

$$\gamma = \gamma_p \left(1 - \frac{G}{G_s}\right) + M\overline{\Delta\varepsilon}[(9E\gamma_B G)^{1/2} - \overline{\Delta\varepsilon}\ EG] \qquad (1)$$

---

*TLZ-H, Plessey Corporation, Australia

where γ_p is the fracture energy due to all contributions other than microcracking, i.e. intrinsic plus twinning effects, G is the grain size, G_s is the grain size at which spontaneous failure occurs, $\overline{\Delta\varepsilon}$ is the average strain in the lattice, γ_B is the grain boundary (or single crystal) fracture energy, E is Young's modulus, and M is an empirical constant related to the number of microcracks formed at the crack tip. The PZT composition at which the maximum contribution of microcracking to toughness will occur can be determined. By differentiating Equation (1) with respect to $\overline{\Delta\varepsilon}$, and setting the result equal to zero:

$$\overline{\Delta\varepsilon}_m = \frac{1.5M\gamma_B^{1/2}}{(EG)^{1/2}} \qquad (2)$$

Taking M = 2.5 [8], E = 81 GPa, γ_B = 1 J/m^2 [9], and G = 5 µm yields a value for $\overline{\Delta\varepsilon}_m$ = 0.003. This value of $\overline{\Delta\varepsilon}_m$ is smaller than the distortions which occur in the tetragonal lattice, because a portion of the strain is relieved by twinning[2]. Nevertheless, these calculations imply that the toughness of PZT will be maximized at some intermediate composition in the tetragonal phase-field in agreement with the results of Igarashi and Okazaki[7].

Based upon the above discussion, it can be concluded that the maxima and minima in toughness observed in PZT are due to a trade-off between contributions due to twinning and microcracking. At a phase boundary, twin motion is facilitated, making a somewhat higher energy contribution to this toughening mechanism. However, because of reduced internal strain, the degree of microcracking and therefore its effect on K_{IC} will be significantly reduced. Conversely, the maxima in the K_{IC}-composition diagram represent materials in which the contribution of microcracking to toughness is a maximum. Because microcracking is a more effective toughening mechanism than twinning (as observed in $BaTiO_3$ [3]) the overall toughness reflects its contribution to K_{IC}.

95/5 PZT

Detailed studies of the direct effect of PZT microstructure on both strength and fracture toughness were conducted on 95/5 PZT. This composition is located just to the ferroelectric side of the ferroelectric/antiferroelectric phase boundary shown in Figure 1. The material can be driven into the antiferroelectric state by the application of hydrostatic pressure, yielding a simultaneous release of current. Upon release of the pressure, most of the material returns to its original ferroelectric state, although there is a hysteresis in the pressure-temperature diagram[10] (Figure 3) indicating that there are vestiges of AFE domains remaining after pressure release. We will present evidence that these remaining AFE domains can transform to the FE phase at the crack tip, raising the K_{IC} values above those due to twin-crack interactions or microcracking effects.

The critical fracture toughness of 95/5 PZT is shown in Figure 4 for material in the as-fired, electrically poled, and pressure depoled conditions. The pressure depoled material shows an approximately 50% increase in K_{IC} for all grain sizes than those specimens in the as-fired and poled states. Also, if one assumes that microcracking plays a major part in the toughness behavior, Equation (1) can be used to calculate the grain size at which K_{IC} will be a maximum[3]. The curve obtained from Equation (1) is shown in Figure 4 indicating an expected K_{IC} maximum at a grain size of approximately 40 µm. However, experiments are needed to verify this prediction.

K_{IC} is shown as a function of test temperature for a particular 95/5 composition (≈ 10 µm grain size) in both the poled and pressure depoled condition (Figure 5). The pressure depoled material has a higher toughness over the entire temperature range down to approximately -65 °C where it has

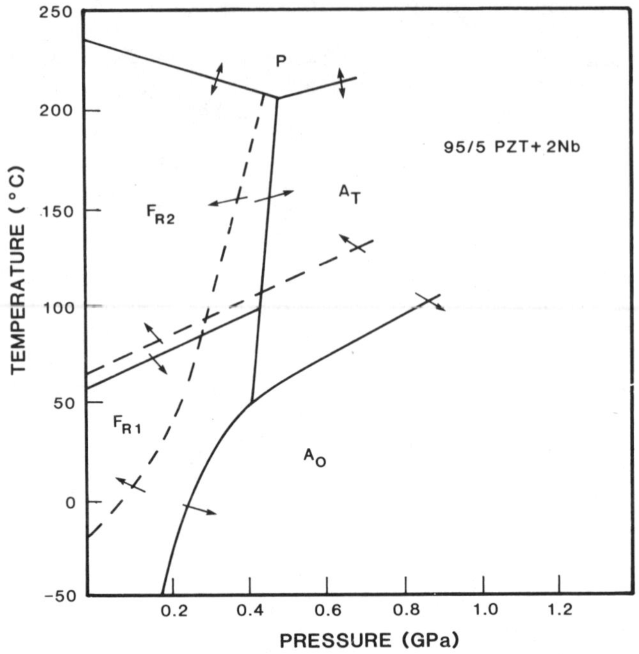

Figure 3. Pressure-temperature phase diagram for 95/5 PZT showing hysteresis in antiferroelectric (A_O) to ferroelectric (F_{R1}) transition (dashed line) as depoling pressure is removed (after Fritz and Keck[10]). The position of the F_{R1} to A_O transition line is very sensitive to composition, which explains the much smaller depoling pressure required for the materials studied herein compared to that in the diagram.

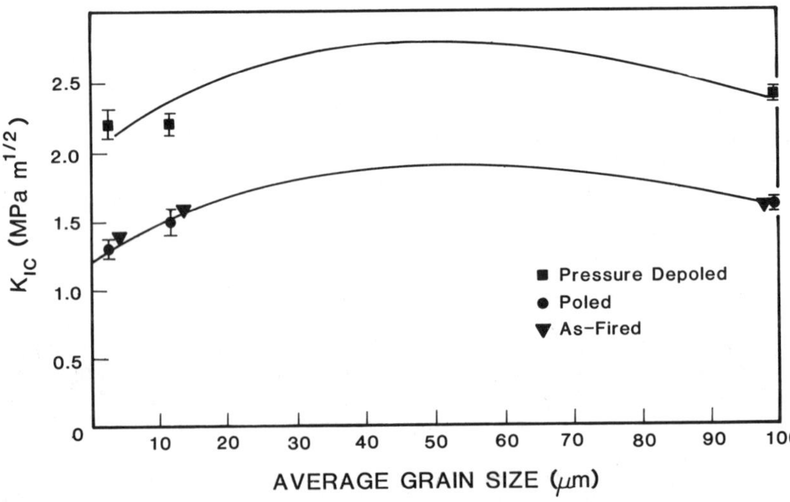

Figure 4. K_{IC} as a function of grain size for a 95/5 PZT after various electrical pretreatments. Curve was drawn to reflect predictions of model for microcracking enhanced toughness[8].

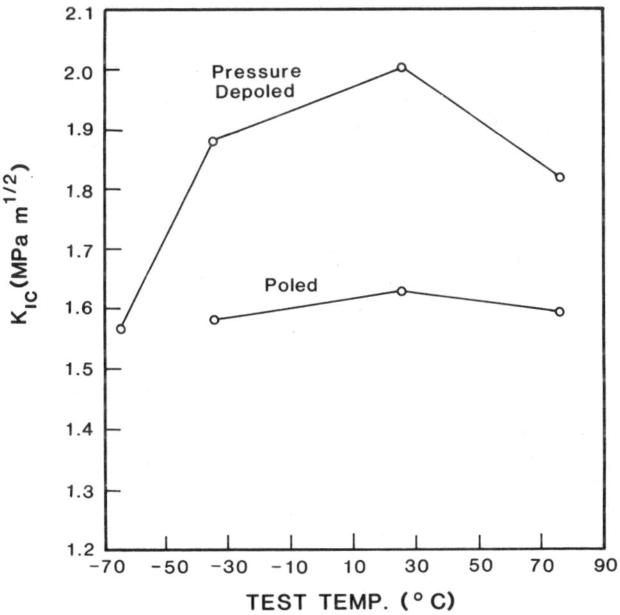

Figure 5. K_{IC} as a function of testing temperature for a 95/5 PZT.

declined sharply. Because twins are present in both the poled and pressure depoled material, a twin-crack mechanism cannot explain this observation. Microcracking probably does contribute somewhat to the toughness, but would not be expected to produce a decrease in K_{IC} at low temperature. We therefore interpret the K_{IC}-temperature behavior in terms of the hysteresis in the antiferroelectric to ferroelectric phase transformation shown in the pressure-temperature-phase diagram for 95/5 PZT (Figure 3). The position of the $A_o \rightarrow F_R$ boundaries for the FE to AFE (solid line) and AFE to FE (dashed line) in Figure 3 show an extensive hysteresis. The AFE → FE transformation after the removal of the hydrostatic depoling pressure becomes more difficult as the temperature is lowered, and requires a negative pressure (tensile stress) below -30 °C. We hypothesize that the transformation of residual AFE domains to FE in the crack tip stress field leads to toughening, in analogy with the mechanism proposed for transformation toughened ZrO_2. Spears[11] has demonstrated remnant AFE phase through electrical measurements, and Kuroda and Heuer[12] have shown the presence of antiferroelectric domains in a pressure depoled material, through the use of transmission electron microscopy. We further hypothesize that the increased barrier to the phase transformation as the temperature is lowered decreases its contribution to toughening. If our hypothesis regarding phase transformation toughening is true, then we should be able to optimize the size and volume fraction of transforming phase. Thus, it is important to continue work to better understand this mechanism of toughening in PZT.

A portion of the study conducted on 95/5 PZT involved a determination of the effect of stoichiometry on fracture toughness and strength. The lead content in the compound $Pb_yNbO_{0.02}(Zr_{0.96}Ti_{0.04})_{0.98}O_3$ was varied from y = 0.96 to 1.02. The material was tested as described in the Experimental Procedure section.

The indentation-strength data obtained from three PZT compositions (y = 0.96, 0.98 (stoichiometric), and 1.01) is shown in Figure 6. The data was analyzed according to Equation (3)[8]:

$$K_c = \eta_v (E/H)^{1/8} (\sigma P^{1/3})^{3/4} \quad (3)$$

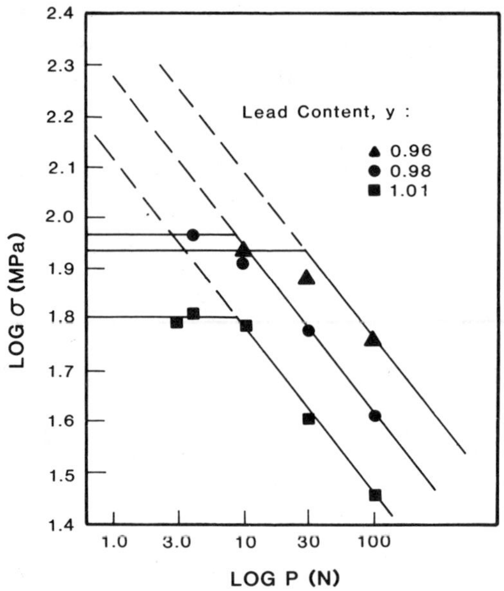

Figure 6. Log fracture stress versus log indentation load, P, for 95/5 compositions. $Pb_yNb_{0.02}(Zr_{0.96}Ti_{0.04})_{0.98}O_3$. Based on indentation model[6], lines of slope -1/3 were drawn through points indented at 100N. Horizontal lines represent strength plateaus observed at small P.

where E is elastic modulus, H is hardness, σ is the fracture strength of bars indented at load, P, and η_v is an empirical constant determined to be equal to 0.59[8]. The dashed lines in Figure 6 were drawn with slope = -1/3 through the data for large load P. From the position of these curves, a value of K_{IC} can be calculated according to Equation (3). Such calculations yield the values of K_{IC} for the three materials shown in Table 1. The toughest material in terms of resistance to macroscopic crack growth is that with the lowest lead content. This is also the composition in which the highest ratio of transgranular to intergranular failure is observed (Figure 7) and therefore the composition in which the most twin-crack interaction will take place. As in $BaTiO_3$[2], this twin-crack interaction is a fundamental mode of toughening of PZT in the ferroelectric state. Comparison of indentation-fracture data at room temperature and 250 °C yielded a ratio of K_{IC}(25 °C)/ K_{IC}(250 °C) = 2.2, for a 95/5 material, similar to that discussed earlier for a commercial PZT composition near the morphotropic boundary.

In addition to having a direct correlation to fracture toughness, the transformation of both $BaTiO_3$ and PZT from the cubic to the ferroelectric state can also produce local microstructural stresses which can combine with applied stresses to cause mechanical failure. The effect of these stresses on fracture have been demonstrated in $BaTiO_3$ through the use of indentation-fracture procedures[4]. As shown in Figure 6, (for clarity, only three of the six curves obtained are included) the deviation of the data from the $P^{-1/3}$ curve at small indentation loads can be ascribed to the direct effect of microstructural stresses on flaw extension. The different compositions were ranked according to their microstructural effect on strength by plotting $\sigma_{plateau}/\sigma_{(P=100N)}$ as a function of stoichiometry (Figure 8), where $\sigma_{(P=100N)}$ is used as normalizing factor to account for shifts in macroscopic fracture toughness. The largest effect of microstructure was found to occur at the greatest deviations from stoichiometry. While it is still believed that it is the internal stresses in the material which add directly to the applied stress, other phenomena such as increasing K_{IC} with increasing flaw size,

Table 1. Fracture Toughness and Internal Stress for 95/5 PZT's

Lead Content	Failure Stress (MPa)	$K_{IC}(MPam^{1/2})^*$	Effective Internal Stress (MPa) **
0.96	75 ± 3	1.7	20±3
0.98	74 ±10	1.3	13.5±0.6
1.01	45.7 ± 0.1	1.1	25 ± 1

* From indentation-strength measurements

** Calculated from $\langle \sigma_I \rangle = \sigma_{failure} - \frac{0.8\ K_{IC}}{C^{1/2}}$ (C = flaw size)

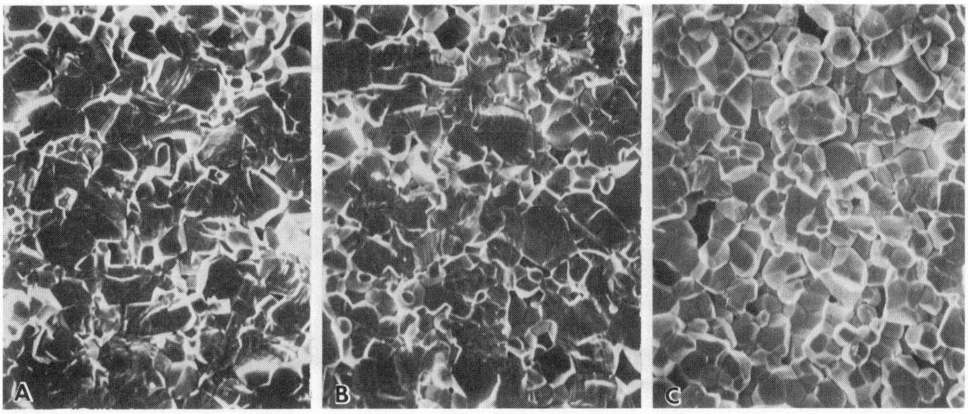

Figure 7. Scanning electron micrographs of fracture surfaces of 95/5 specimens showing variation in ratio of transgranular to intergranular failure. (A) Pb content, y = 0.96; (B) y = 0.98; (C) y = 1.01. Note indications of twin-crack interaction on larger grain faces in (A) and (B).

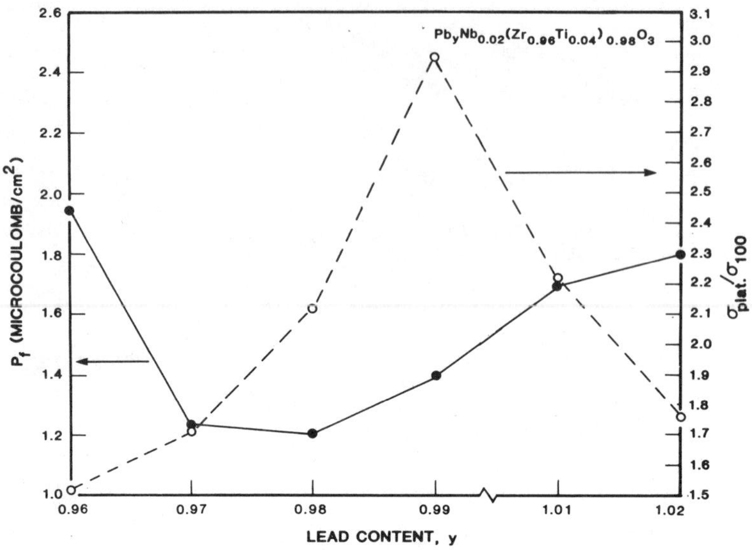

Figure 8. Both level of the strength plateau (Figure 6) and asymmetry of the electrical polarization loops, P_f, for 95/5 PZT as a function of stoichiometry. Strength maximum at approximately the point of smallest loop asymmetry is interpreted in terms of a minimum in the internal stress level.

i.e. "R" curve behavior cannot be ruled out. From a practical viewpoint, these data suggest that the optimum material for use in an actual device would be one near stoichiometry even though the critical fracture toughness is not a maximum here.

Fractographic analysis was performed on failures which occurred from "natural" flaws, i.e., flaws not generated by the Vicker's indentation. Using an expression developed to explain failure in $BaTiO_3$ [2], it was possible to calculate internal stress levels in the three 95/5 PZT materials from K_{IC}, flaw sizes, and strength. These calculations (Table 1) indicate that the levels of internal stress follow the same ranking as the microstructural effect shown by the plateau in the strength.

A correlation has also been obtained between the asymmetry of the electrical polarization curves for these PZT compositions[13] and the level of internal stress. The more asymmetric is the polarization, the greater is the predicted value of stress in the material[13]. The asymmetry of the polarization loops for 95/5 PZT (Figure 8) is seen to be a minimum at essentially the same compositions at which the microstructure has the least effect on strength, i.e. the largest value of $\sigma_{plateau}/\sigma_{(P=100N)}$.

SUMMARY

Fracture in PZT is shown to be a complex process. The resistance of this material to crack growth is dependent on both its chemical composition and microstructure. The competition between toughening due to twin-crack interactions and that due to microcracking is hypothesized to be responsible for the maxima and minima in the K_{IC}-composition plot for PZT. Easier twin motion at the phase boundaries and concomitant reduction in stress levels leading to less microcracking induced toughening is responsible for the minimum in K_{IC} at the morphotropic boundaries. Detailed studies carried out on a

95/5 PZT indicate that in addition to microcracking and twinning the phase transformation from an antiferroelectric to a ferroelectric phase at the crack tip also makes a significant contribution to fracture toughness.

It was shown that indentation-fracture techniques can be used to separate effects of microstructure on toughening and strength. Indentation-fracture data obtained on 95/5 materials demonstrated that K_{IC} is a maximum at the stoichiometric composition and that internal stresses generated by the paraelectric to ferroelectric phase transformation contribute directly to a reduction in the measured strength. These internal stresses could also be correlated with changes in the electrical polarization-voltage curves for the materials. It was shown that the combined effects of toughness and internal stress must be considered in choosing a material for an actual component.

ACKNOWLEDGMENTS

The support of the Office of Naval Research and Sandia National Laboratories during the course of this work is gratefully acknowledged. We would also like to thank R.C. Pohanka and R.H. Dungan who contributed many helpful ideas to these studies. The help of T.L. Baker and D. Heukeroth in collecting the data is also greatly appreciated.

REFERENCES

1. B. Jaffe, W.R. Cook, Jr., and H. Jaffe, "Piezoelectric Ceramics," Academic Press, New York (1971).
2. R.C. Pohanka, S.W. Freiman, and B.A. Bender, J. Am. Ceram. Soc. 61:72 (1978).
3. R.C. Pohanka, S.W. Freiman, K. Okazaki, and S. Tashiro in 5:353 "Fracture Mechanics of Ceramics," R.C. Bradt, A.G. Evans, D.P.H. Hasselman, and F.F. Lange, ed., Plenum, New York (1982).
4. R.F. Cook, S.W. Freiman, B.R. Lawn, and R.C. Pohanka, Ferroelectrics 50:267 (1983).
5. S.W. Freiman, D.R. Mulville, and P.W. Mast, J. Mater. Sci. 8:1527 (1973).
6. P. Chantikul, G.R. Anstis, B.R. Lawn, and D.B. Marshall, J. Am. Ceram. Soc. 64:539 (1981).
7. H. Igarashi and K. Okazaki, Japan-U.S. Seminar on Dielectric and Piezoelectric Ceramics (1982).
8. R.W. Rice and S.W. Freiman, J. Am. Ceram. Soc. 64:350 (1981).
9. R.W. Rice and R.C. Pohanka, J. Am. Ceram. Soc. 62:559 (1979).
10. I.J. Fritz and J.D. Keck, J. Phys. Chem. Solids 39:1163 (1978).
11. R.K. Spears, Ferroelectrics 37:653 (1981).
12. K. Kuroda and A. H. Heuer, Proc. of 41st Annual Meeting of Micros. Soc. of America, San Francisco Press, San Francisco, CA (1983).
13. R.H. Dungan and L.J. Storz, submitted to J. Am. Ceram. Soc.

LOADING-UNLOADING TECHNIQUES FOR DETERMINING FRACTURE PARAMETERS OF BRITTLE MATERIALS UTILIZING FOUR-POINT BEND, CHEVRON-NOTCHED SPECIMENS

Thomas B. Troczynski and Patrick S. Nicholson

Ceramic Engineering Research Group
Department of Metallurgy and Materials Science
McMaster University, Hamilton, Ontario, Canada

ABSTRACT

A technique is described for determining the threshold stress intensity factor K_t at which a stationary crack just begins to move. If the crack velocity is described by $v \sim K^N$, K_t can be measured via repeated cyclic loading stable-fracture unloading (UN) experiments. The resistance to fracture calculated for each cycle approaches the threshold (or initiation) value as the full fracture of four-point bend, chevron-notched specimens is approached. The experimentally measured resistance-to-fracture-initiation values for SiC was ~ 6J/m^2, 30 J/m^2 for yttria partially stabilised zirconia, ~ 10 J/m^2 for yttria fully stabilised zirconia, and ~ 80 J/m^2 for graphite. The R-curve behaviour of a PSZ- β Al_2O_3 composite was analysed by the UN technique. The nonelastic energy dissipation rate for this composite approached 600 J/m^2.

I. INTRODUCTION

The classical Griffith analysis [1] of the energy balance in a cracking body gives the thermodynamic equilibrium equation;

$$dU/dA = 0 \tag{1}$$

where U is the total energy of the system and A is the fracture area. When equilibrium (1) is reached, the crack just starts to grow, with equal thermal activation barriers for healing and advancing [2]. When conditions are favourable, the crack will advance with diminishing activation energy until the critical point is reached whereat the activation barrier disappears [2]. The latter condition results in the critical stress intensity factor K_c (or fracture toughness) and clearly this may be several times larger than the threshold stress intensity factor K_t at the Griffith condition (1).

The purpose of the present work is to establish a procedure to measure K_t for brittle materials. The technique utilises repeated loading/slow- fracture/unloading (UN) cycles of four-point bend, chevron-notched (CN) bars (Fig. 1). CN specimens are employed due to the ease of crack initiation at the chevron tip and the subsequent fracture stability. An analytical treatment of the CN specimen's compliance and stability was first provided by Bluhm [3,4], Sakai and Yamasaki [5,6] and more recently by Troczynski and Nicholson [7]. The application of CN specimens in fracture mechanics was proposed by Barker [8] and developed further by Munz et al. [9-11]. Subcritical crack growth has been neglected in the analytical treatment of CN specimens. The present work focuses on subcritical crack growth in CN specimens. If the critical conditions are reached during the tests the specimen fractures completely thereby making cyclic loading-unloading impossible.

It is assumed, that the crack velocity, v, is described by the power equation;

$$v = v_c (K_I/K_c)^N \tag{2}$$

within the subcritical crack growth range (i.e. for the stress intensity factors, K_I, between K_t and K_c. v_c is the crack velocity at the critical condition and N is a constant. At the moment of fracture initiation (i.e. $K_I = K_t$), the crack starts to move with the threshold velocity v_t. If the materials' parameters K_c, K_t, v_c, v_t, N are known (or can be estimated), equation (2) can be integrated to yield the variation in the stress intensity factor with crack extension. Such an integration has been performed for the four-point bend, CN specimens, for both constant displacement and constant loading rate conditions [12]. The main steps and results of these procedures are presented in this paper. If equation (2) is valid, the resistance to fracture obtained in repeated loading-unloading cycles approaches the threshold (or initiation) value as full fracture is approached. The technique is utilised to determine the resistance to fracture initiation of silicon carbide, graphite, fully and partially stabilised zirconia and a β-Al_2O_3/ partially stabilised zirconia composite.

II. MODEL DEVELOPMENT

In the loading of a CN sample utilising a constant cross-head displacement rate (Fig. 2), the total compliance of the system is composed of that of the machine (C_M) and that of the specimen (C_S) thus the load P is given by

$$P = \frac{D_S + D_M}{C_S + C_M} = \frac{\dot{D} \cdot t}{C_S + C_M} \tag{3}$$

where D_S is the deflection of the specimen, D_M the deflection of the machine, $\dot{D}$ the deflection rate, and t is the time of loading. For an ideally-sharp chevron triangle tip (OPR, Fig. 1) fracture will initiate at any load immediately after loading commences (Point O, Fig. 2). In reality, the chevron tip is blunted by some Δx (Fig. 1) and the sample initially loads in the linear regime (OT, Fig. 2), i.e. there is no crack growth for times up to $t = t_o$. At $t = t_o$ (Point T, Fig. 2), the crack starts to grow with $v = v_t$ and $K_I = K_t$. The total deflection at this time is D_O, the load is P_O and now, a new count of time, τ, is initiated. Equation (3) then becomes;

$$P(\tau) = \frac{\dot{D} \cdot (t_o + \tau)}{C_s + C_M} = \frac{\dot{D}\tau + D_o}{C_S + C_M} \tag{4}$$

where $D_O = \dot{D} \cdot t_o$.

It can be shown [10, 12] that the stress intensity factor, K_I, for the CN specimen is given by;

$$K_I = \frac{P}{B\sqrt{W}} \left[\frac{BE}{2} \frac{dC_S}{dx} \frac{x_1 - x_0}{x - x_0} \right]^{1/2} \tag{5}$$

where the geometric parameters B, W, x, x_1, x_0 are included in Fig. 1 (E is Young's modulus divided by $1-f^2$ and f = Poisson ratio). Substitution of equations (4,5) into (2) results in;

$$v = v_c \left[\frac{\dot{D} \cdot \tau + D_0}{K_C (C_S + C_M)} \right]^N \cdot \left[\frac{E}{2BW} \frac{dC_S}{dx} \frac{x_1 - x_0}{x - x_0} \right]^{N/2} \tag{6}$$

and integration of this equation gives the time, τ, after which the crack will grow to any desired length, x, i.e.;

$$\tau = \frac{1}{\dot{D}} \left\{ \left[\frac{\dot{D}(N+1)}{F} \cdot I(x) + D_0^{N+1} \right]^{\frac{1}{N+1}} - D_0 \right\} \tag{7}$$

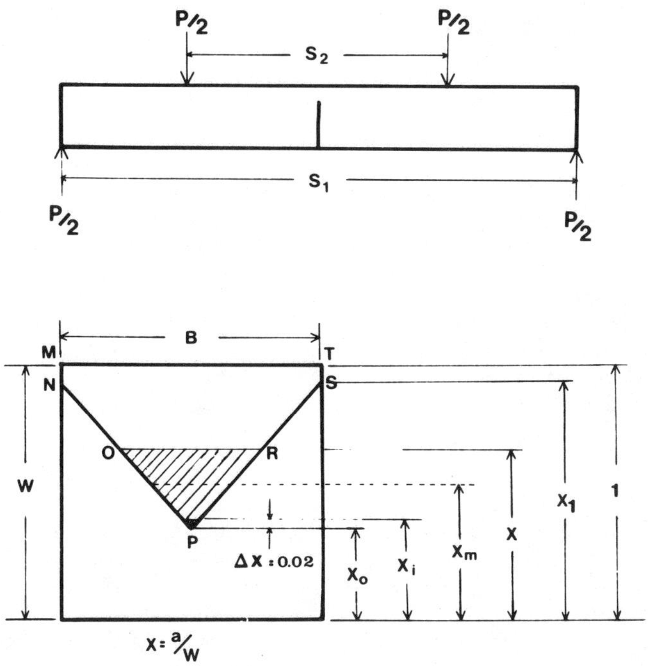

Fig. 1: Geometry of the four-point-bend chevron notched specimen.

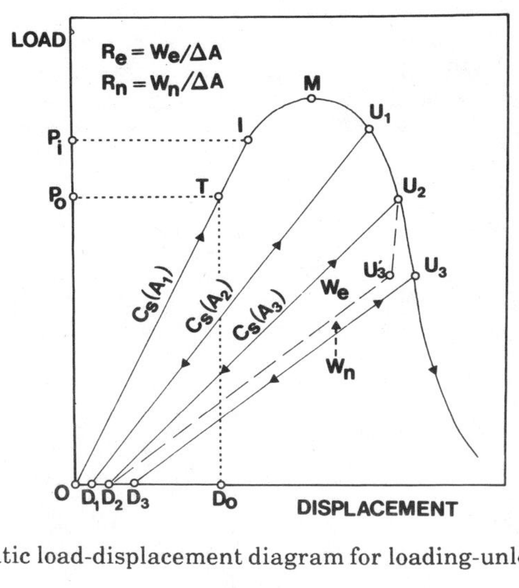

Fig. 2: Schematic load-displacement diagram for loading-unloading experiment.

where F is a constant for a given material and specimen geometry and I(x) is a function resulting from numerical integration, i.e.

$$I(x) = \int_{x_i}^{x} \left[\frac{(x-x_0)(C_S + C_M)^2}{dC_S/dx} \right]^{N/2} dx \qquad (8)$$

and

$$F = \frac{v_c}{W} \left[\frac{E(x_1 - x_0)}{2BW\,K_C^2} \right]^{N/2} \qquad (9)$$

Substitution of the value of τ obtained from equation (7), into equations (4,5) yields the variation of stress intensity factor K_I with crack length or fracture area [12].
Cyclic loading/stable-fracture/unloading experiments will now be considered for two types of materials, i.e. high stiffness, medium toughness (eg. SiC) and low stiffness and toughness (eg. graphite). For both materials it is assumed that $v_c/v_t = 10^5$; and E = 400 GPa, $K_c = 3$ MPa $\sqrt{m}$, N = 30 for SiC and E = 11 GPa, N = 10, $K_C = 1.1$ MPa $\sqrt{m}$ for graphite. Incorporation of these parameters into equation (2) yields $K_t = 2.0$ MPa $\sqrt{m}$ for SiC and $K_t = 0.3$ MPa $\sqrt{m}$ for graphite. Utilising test system parameters of $C_M = 0.1$ µm/N, bending spans $S_1 = 25.4$ mm, $S_2 = 12.7$ mm and B = W = 4 mm, K_I was calculated as a function of the percentage fracture area, A. In Fig. 1, for example, at crack extension x,

$$A = ORP/MNOPRST$$

The predicted variation of K_I (between K_t and K_c) with A for SiC is shown in Figure 3. The consecutive curves (1 to 7), represent the K_I variation in a hypothetical UN test (Fig. 2), for specimens with initial crack lengths, x_i, at 0.42, 0.50, 0.55, 0.60, 0.65, 0.70 and 0.80 respectively ($x_o = 0.40$). The broken line (curve #8) is expected when $x_i = 0.90$. The test proceeds according to the pattern indicated in Fig. 2 (cf traces in Fig. 3). The specimen with an initial fracture area, A_1 (i.e. compliance $C_s(A_1)$) loads along line $OTIMU_1$ until the fracture area reaches A_2. It is then unloaded and the cycle repeated along line $D_1U_1U_2D_2$ to create new fracture area A_3 etc. The respective lines in Fig. 3 (shown solid only for $K_t \le K_I$) are $A_1U_1A_2$, $A_2U_2A_3$, etc. The dotted lines in Fig. 3 plot the K_I values for uninterrupted loading and critical conditions occur at points M, N and S (these lines extend beyond $K_I = K_C$ from equation 2. The $K_I \ge K_C$ range is not considered in the present work). In Fig. 2, for simplicity, it is assumed that the loading and unloading paths coincide. In general some permanent deformation D_1, D_2 etc. may occur. For non-zero permanent deformation, the total work performed in a given cycle (eg. $D_2U_2U_3D_3$ for the third cycle in Fig. 2) is divided into elastic and nonelastic parts (W_e, W_n) as proposed by Eftis et al. [13] and applied by Sakai et al. [14]. Based on the load-displacement graph, the resistance to fracture, R_e, is calculated as W_e per unit area created and the nonelastic energy dissipation rate, R_n, as W_n per unit area created. Since only the W_e input is transformed into new surface area (i.e. the compliance increase), the relation;

$$W_e = \bar{G}\,\Delta A \qquad (10)$$

is proposed, where $\bar{G}$ is the mean value of the strain energy release rate in a given cycle (i.e. the resistance to fracture, R_e, calculated for a given cycle, neglecting the kinetic energy). The correspondence between Fig. 2 and Fig. 3 is now expressed through G. If. Fig. 3 is plotted on an axis system the strain energy release rate G (from $G = K_I^2/E$) versus the fracture area, then $\bar{G}$ is given by;

$$\bar{G} = \int_{A_i}^{A_{i+1}} G\,dA \,/\, (A_{i+1} - A_i) \qquad (11)$$

This integral is simply the area under the G-A curve for the loading-unloading cycle between fractured area A_{i+1} and A_i. As shown in Fig. 3, the value of $\bar{G}$ (or R_e - as calculated from the load-displacement diagram) for SiC initially increases and then decreases towards the threshold (or initiation) value (G_t or K_t) as full fracture A = 100% is approached. Similar plots for graphite (Fig. 4) show that $\bar{G}$ (or R_e) steadily decreases to the threshold (initiation) value as full fracture is approached. For graphite, none of the K_I-A curves cross the critical $K_I = K_C$

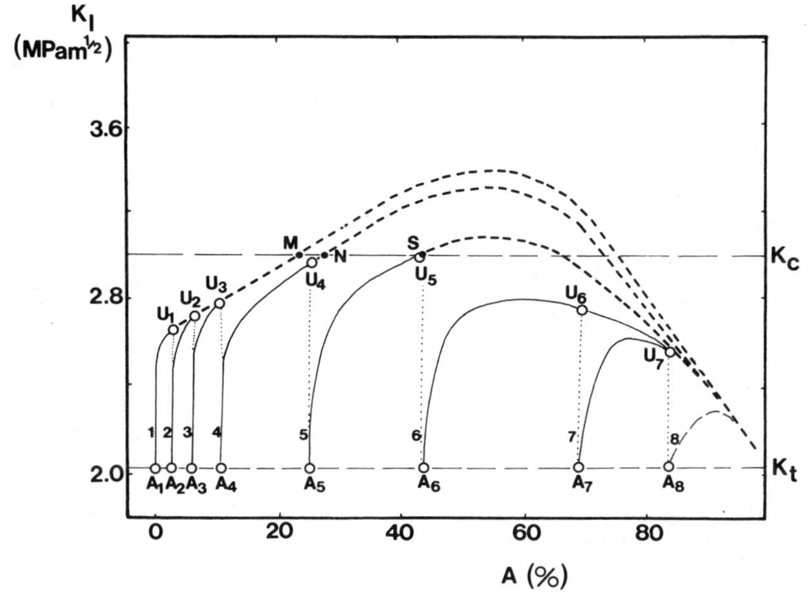

Fig. 3: Stress intensity factor against fracture area for loading-unloading of SiC.

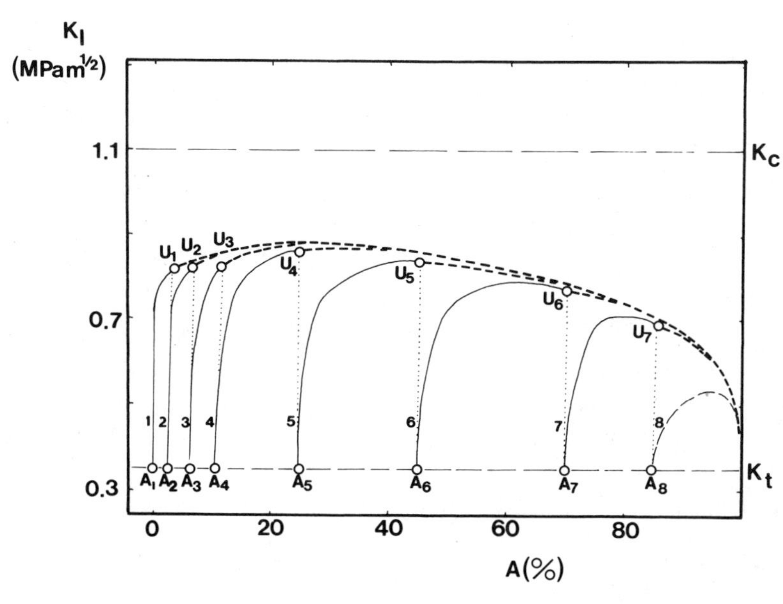

Fig. 4: Stress intensity factor against fracture area for loading-unloading of graphite.

line, even for continuous loading up to full fracture. This reflects the crack growth stability of graphite. The predicted decrease of $\overline{G}$ towards the threshold value explains the observed decreasing work-of-fracture with the crack extension [14,15].

The model was tested via loading-unloading experiments on silicon carbide, graphite and stabilised zirconium oxide ceramics.

III. EXPERIMENTAL PROCEDURE

The experimental arrangement must assure fracture stability. This requirement was fulfilled in the present work by utilisation of CN bend specimens, a hard piezoresistive load cell* and slow (0.1 μm/s) constant cross-head displacement rates+. The compliance of the testing system was 0.06 μm/N. The load was transmitted to the specimen through sapphire rollers over bending spans of 25.4 mm and 12.7 mm. The displacement was measured by an LVDT$ with a practical resolution of 0.1 μm. The specimens were chevron-notched with a 150 μm thick diamond coated blade. The SiC¢ bars (4x4x30 mm) were prepared for testing by the manufacturer and the graphite++ specimens were cut from large blocks and machined to 5x5x30 mm.

The loading faces were finished by 400 grit polishing paper. Stabilised zirconium oxide** samples (8 wt.% Y_2O_3 PSZ and 12 wt.% Y_2O_3 FSZ) were mold pressed, isopressed and sintered for 6 hours at 1450°C. The loading faces were polished and the compressive stress surface layer (resulting from the notch cutting procedure) was etched away by cold concentrated HF ($\frac{1}{2}$ hour). The composite PSZ–β–Al_2O_3 specimens were prepared by mixing 20 vol.% of ground β-alumina (120 μm mean particle size) with 80 vol.% powder PSZ. The shaped bars were sintered for 10 hours at 1450°C. The resulting microstructures are discussed elsewhere [16].

R_e values were calculated directly from the load-displacement curves as schematically shown in Figure 2. The fracture area increment ΔA was obtained via the specimen compliance increase [3].

IV. RESULTS AND DISCUSSION

Representative load-displacement curves are shown for each type of material tested and plots of the resistance to fracture (R_e) versus percentage of the fracture area were compiled for several specimens. The load-displacement curve for SiC is shown in Fig. 5 (only loading paths are shown for clarity). The brittleness of this material is demonstrated by the fast drop of load with consecutive cycles and the negligible permanent deformation (the ~ 1 μm of permanent deformation observed is attributed to the testing system). The resistance to fracture, R_e, compiled for this material (4 specimens) is shown in Fig. 6. High stress concentrations at the chevron tip for small values of A result in an initially large (~ 20 J/m^2) value of R_e, which decreases thereafter. Subsequently, the R_e vs. A curve increases and then decreases again, as predicted by the model (Fig. 3). At A ~ 80%, the crack length is ~ 95% of the sample width and measurements are difficult to perform. It appears that R_e decreases to ~ 6 J/m^2 as full fracture is approached.

Similar plots for graphite are shown in Figures 7 and 8. The permanent deformation (~ 15 μm) was recorded as expected from the elastic-plastic nature of this material [14]. The resistance to fracture vs. A (plotted for three specimens, Fig. 8) decreases continuously to ~ 80 J/m^2 as full fracture is approached. Again the model predictions (Fig. 4) appear correct.

---

* Type TC-2000-500 Kullite, Semicond Prod Ltd.
\+ Compression testing machine WF10052, Wykeham Farrance Inc.
$ Type 100 DC-D, Schaevitz Eng.
¢ Hexalloy, Carborundum
++ AXM, POCO Corp.
** Type ZYP, Zircar Prod Inc.

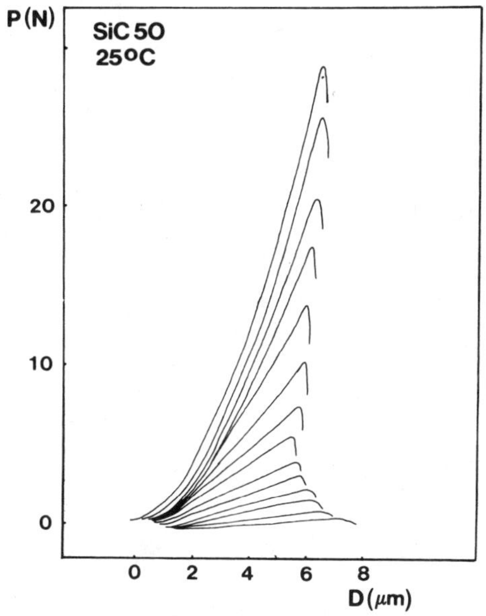

Fig. 5: Load-displacement diagram for SiC.

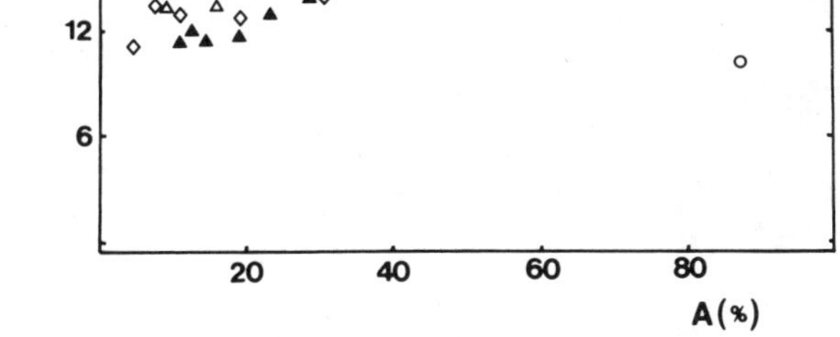

Fig. 6: Resistance to fracture of SiC.

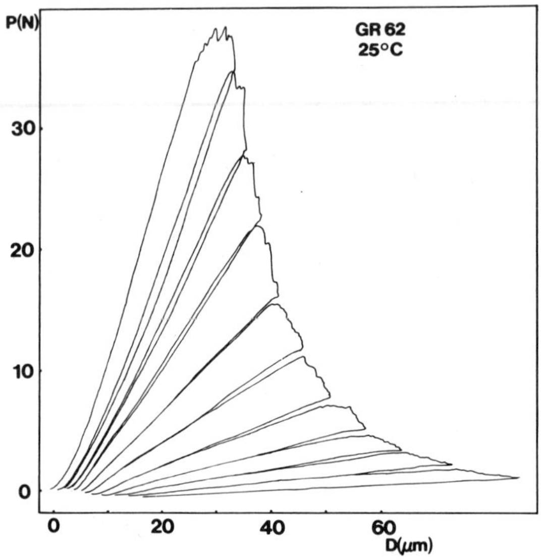

Fig. 7: Load-displacement diagram for graphite.

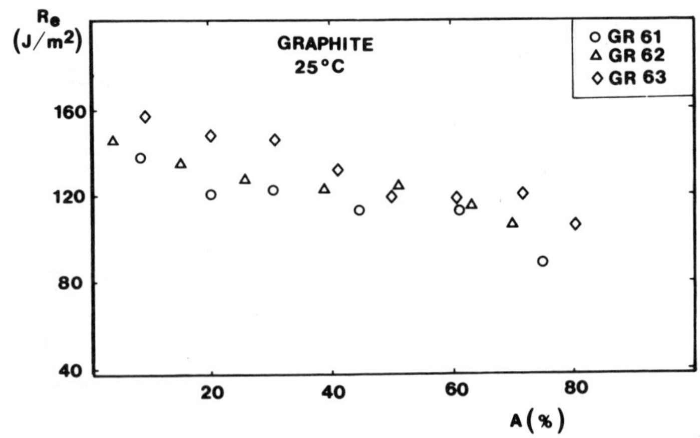

Fig. 8: Resistance to fracture of graphite.

FSZ (fully-stabilised-zirconia) results are shown in Figures 9 and 10. The observed total permanent deformation (~ 4 µm) is attributed to indentation of the sample surface by the sapphire knife edges. The resistance to fracture is approximately constant (~ 10 J/m^2) with crack extension. The Young's modulus of the specimen (which decides the shape of R_e vs. A curve) is intermediate between that of SiC and graphite (i.e. ~ 200 GPa vs. ~ 400 GPa and ~ 11 GPa), which explains why the R_e vs. A curve (Fig. 10) is also intermediate between that of SiC (Fig. 6) and graphite (Fig. 8).

The load-displacement plot for PSZ (partially-stabilised-zirconia) is shown in Fig. 11 and the resistance to fracture R_e for four specimens is plotted in Fig. 12. The decrease of R_e with A (towards ~ 30 J/m^2) is more pronounced, than for FSZ. The closed symbols are estimations of the nonelastic energy dissipation rate, R_n. The value of R_n for PSZ oscillates around 5 J/m^2 and decreases towards zero as full fracture approaches.

The results for the PSZ–β Al_2O_3 composite are shown in Figures 13 and 14. Extensive permanent deformation is evident and this is unusual for a ceramic material. In the final loading cycles almost all of the total energy input is nonelastically dissipated (i.e. a negligible compliance increase was observed). This suggests R-curve behaviour and this is further demonstrated by the results shown in Fig. 19. R_e increases from ~ 50 to ~ 200 J/m^2 while R_n increases from ~ 50 to ~ 600 J/m^2 as full fracture approaches. Clearly equation (1) cannot be applied to this material and the resulting R_e variation cannot be predicted by the model. Microstructural studies of the composite fracture surface morphology are under way and should lead to explanation of the R-curve behaviour.

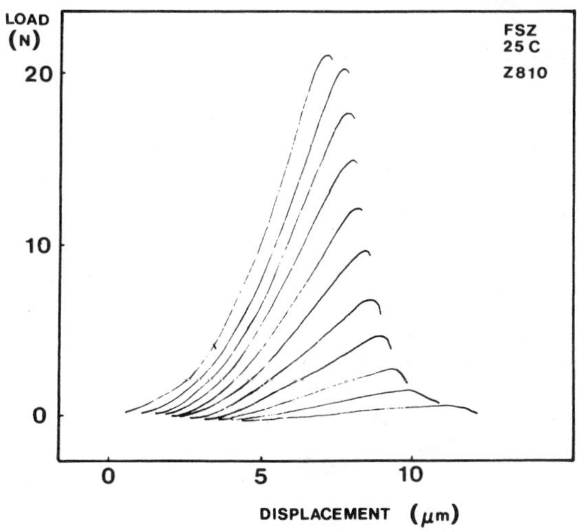

Fig. 9: Load-displacement diagram for FSZ.

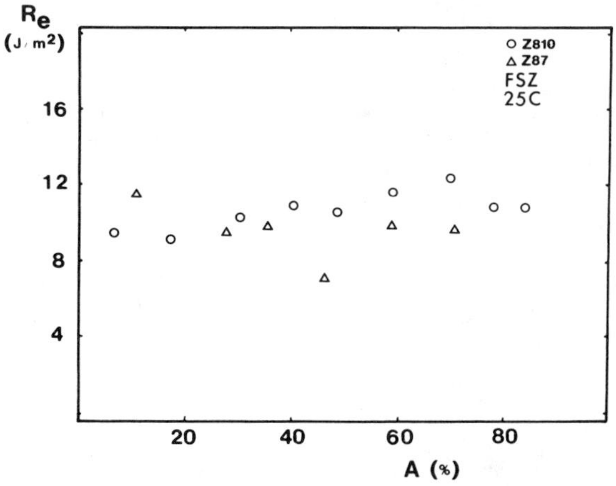

Fig. 10: Resistance to fracture of FSZ.

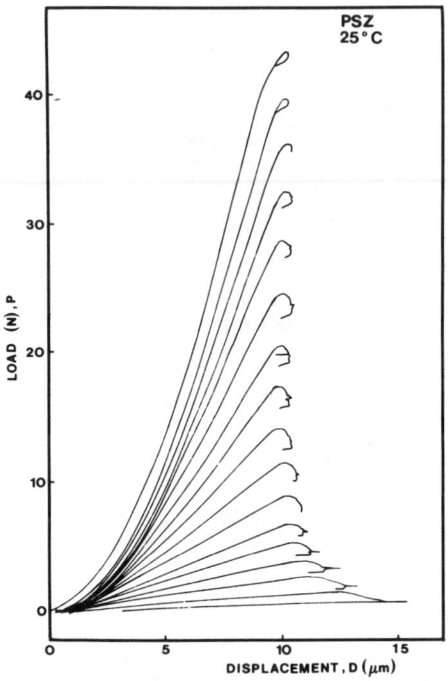

Fig. 11: Load-Displacement Graph for PSZ. UN Technique.

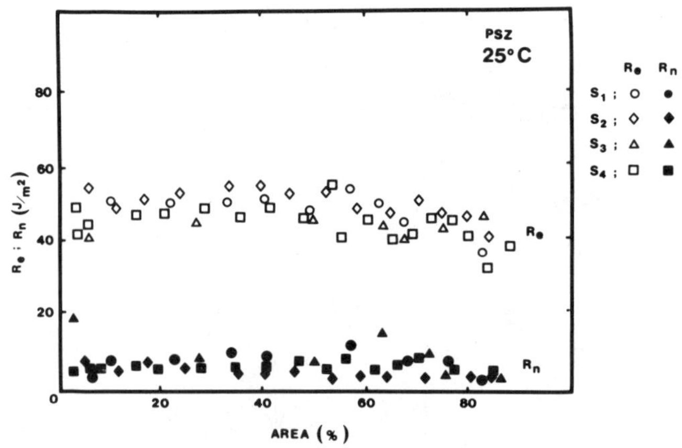

Fig. 12: Resistance to fracture (Re) and nonelastic energy dissipation rate (Rn) for PSZ (4 specimens).

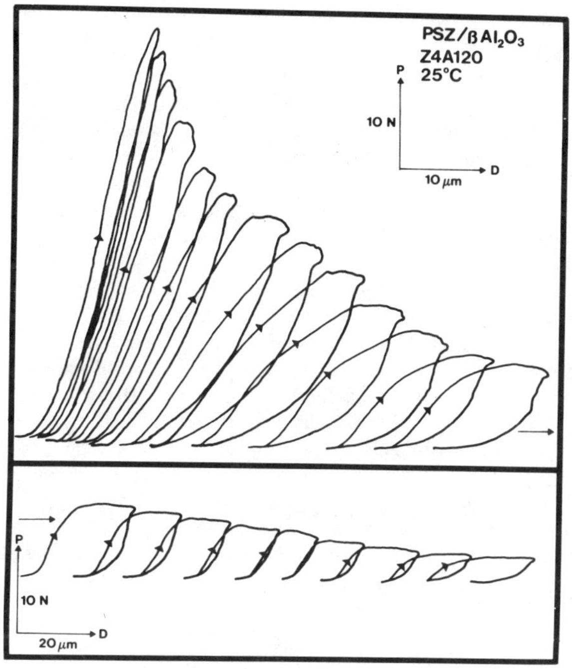

Fig. 13: Load-displacement graph for the composite: PSZ + 20 vol% βAl_2O_3 120 μ m size.

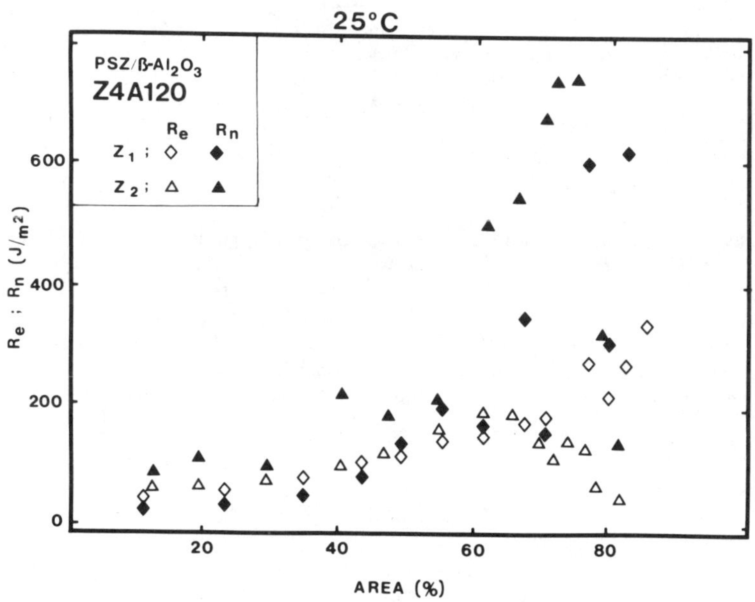

Fig. 14: Re and Rn values for the composite: PSZ + 20 vol% βAl_2O_3 120 μ m size (2 specimens).

V. CONCLUSIONS

A model has been proposed wherein the threshold stress intensity factor K_t, defined as that which initiates crack growth in brittle materials, has been identified and a method for its measurement suggested. If crack growth between the threshold (K_t) and critical (K_C) stress intensity factors is described by a power law, then K_t can be measured by multiple loading/stable-fracture/unloading experiments on one specimen, provided its geometry guarantees stable crack growth.

Subcritical crack growth is modelled in a chevron-notched, four-point bend bar sample. The variation of the stress intensity factor against fracture area in consecutive cycles is related to the work-of-fracture terms, R_e, obtained directly from load-displacement data. The model predicts that the R_e value approaches the resistance-to-fracture for the material as the crack approaches 100% of the sample cross-section and has a value close to the surface energy term envisaged by the Griffith analysis. Experimental results and model predictions coincide well, except when the material exhibits R-curve behaviour (PSZ–β Al_2O_3 composite).

The measured resistance-to-fracture of SiC was ~ 6 J/m^2, of graphite ~ 80 J/m^2, of fully stabilised ZrO_2 ~ 10 J/m^2, of partially stabilised ZrO_2 ~ 30 J/m^2. For the PSZ–β Al_2O_3 composite R_e approaches ~ 200 J/m^2 and the nonelastically dissipated energy R_n approaches ~ 600 J/m^2.

REFERENCES

1. A.A. Griffith, Phil. Trans. Roy. Soc. (London), A221, 163 (1921).
2. E.R. Fuller, Jr. and R.M. Thomson, pp. 507- 548 in Fracture Mechanics of Ceramics, vol. 4, Ed. by R.C. Bradt, D.P.H. Hasselman and F.F. Lange, Plenum Press, New York (1978).
3. J.I. Bluhm, Eng. Fract. Mech., 7, 593-604 (1975).
4. J.I. Bluhm, pp. 409-417 in Fracture 1977, Vol. 3, Edited by D.M.R. Taplin, University of Waterloo Press, Waterloo, Ontario, Canada, 1977.
5. M. Sakai and K. Yamasaki, J. Am. Der. Soc., 66 (5), 371-375 (1983).
6. M. Sakai and K. Yamasaki, J. Am. Cer. Soc., 66 (5), 376-379 (1983).
7. T.B. Troczynski and P.S. Nicholson, to be published, J. Am. Cer. Soc., 1985.
8. L.M. Barker, Eng. Fract. Mech., 9 (2), 361-369 (1977).
9. D. Munz, R.T. Bubsey and J.L. Shannoon, Jr., J. Am. Cer. Soc., 63 (5-6), 300-305 (1980).
10. D. Munz, J.L. Shannon, Jr. and R.T. Bubsey, Int. J. Fr., 16, R137-R141 (1980).
11. D. Munz, R.T. Bubsey and J.E. Srawley, Int. J. Fr., 16 (4), 359-374 (1980).
12. T.B. Troczynski and P.S. Nicholson, to be published, J. Am. Cer. Soc., 1985.
13. J. Eftis, D.L. Jones and H. Liebowitz, Eng. Fr. Mech., 7, 491-503 (1975).
14. M. Sakai, K. Urashima and M. Inagaki, J. Am. Cer. Soc., 66 (12), 868-874 (1983).
15. K.W. Davidge and D. Tappin, J. Mat. Sci., 3, 165-173 (1968).
16. J. Glass and P.S. Nicholson, to be published in Comm. Am. Cer. Soc. (1985).

APPLICATION OF THE POTENTIAL DROP TECHNIQUE TO THE FRACTURE MECHANICS OF CERAMICS

Thomas B. Troczynski and Patrick S. Nicholson

Ceramic Engineering Research Group
Department of Metallurgy and Materials Science
McMaster University, Hamilton, Ontario, Canada

ABSTRACT

An Electrical Potential Drop (PD) technique is used to follow crack growth in chevron-notched, four-point bend fracture toughness specimens of conductive ceramics. The potential drop in the vicinity of the notch was successfully modelled and the resulting routine was verified at room temperature using graphite as a model material and was applied at high temperatures to zirconium oxide ceramics. The resistance to brittle fracture of partially stabilised zirconia doped with 8 wt% Y_2O_3 decreased to ~3 J/m^2 at 1100°C and ~1 J/m^2 at 1300°C. Dispersion of β-Al_2O_3 particles in the PSZ matrix enhanced the overall resistance to fracture of the composite to a maximum of 1800 J/m^2 at 1300°C.

I. INTRODUCTION

The Potential Drop (PD) technique has been used to study fracture and fatigue in electrical conductors. The first calibration results were reported by Anctil et al. [1]. The analytical solution of Laplace's equation, governing the potential distribution in the vicinity of a crack was attempted by Johnson [2] and Gilbey and Pearson [3]. The equipotential linc distribution around the crack was investigated by Richtie et al. [4] who suggested the optimal conditions for the current input and potential probe positions. Aronson and Richtie further optimized this technique by finite element analysis [5]. The non-zero volume crack was considered by Clark and Knott [6] and Knott [7] and the sources of error in the technique were reviewed by Druce and Booth [8]. It was concluded that the method is highly sensitive to crack growth and variations in the potential-probe position close to the crack has relatively little effect on the results.

The cited works assumed a notch with a straight-through (ST) crack front. Druce and Booth [8] considered the errors resulting from crack bowing in ST specimens. The solution of Laplace's equation for polygonal domains was undertaken by Papmichael and Whiteman [9,10]. They suggested that complicated boundary conditions result in serious difficulties when an analytical approach is employed. They proposed the utilisation of finite element methods [10] or purely experimental calibration (also suggested by [7,8].)

Frequently, the crack front is not ST, the most obvious case being the chevron-notched specimen (CN) (Fig. 2A). In a CN fracture test, the crack initiates at the triangular tip of the chevron and proceeds stably for a distance characteristic of the sample geometry and independent of the material [11].

The advantage of the PD technique for conductive ceramics as compared to its use for monitoring crack growth in metals is its sensitivity. The resistivity of PSZ at 1000 - 1300°C, for example, is in the 10 ohm.cm range thus, a potential drop ~ 10 mV is experienced if ~ 1 mA current passes through the sample. The resistivity of metals, on the other hand, is in the microohms.cm range resulting in a potential drop of ~ 1 μV on passage of ~ 1A. Such large current densities can also result in excessive heating of the specimen.

The purpose of the present work is to extend the plane solution for electrical potential drop [12] to the chevron notched shape, verify the analysis utilising model graphite chevron notched specimens and subsequently examine the high temperature fracture behaviour of PSZ and PSZ-β-Al_2O_3 composites via loading/stable-fracture/unloading experiments.

II. MODEL DEVELOPMENT

Modification of the solution of the Laplace's equation for the electrical potential distribution around a crack, [12], gives the dimensionless crack length, x, as;

$$x = \frac{2}{\pi} \cos^{-1}\left[\frac{1 - p_1^2}{p_2 - p_1^2 p_3} \right]^{1/2} \quad (1)$$

where

$$p_1 = \frac{\exp\left[\frac{u(x)}{2k}\right] - 1}{\exp\left[\frac{u(x)}{2k}\right] + 1} \quad (2)$$

$$p_2 = \text{sech}^2\left[\frac{\pi}{2} \cdot \frac{y}{W}\right] \quad (3)$$

and

$$p_3 = \text{sech}^2\left[\frac{\pi}{2} \cdot \frac{d}{W}\right] \quad (4)$$

The parameters of these equations are illustrated via the geometry of the ST specimen (Fig. 1). u(x) is the electrical potential drop between surface points E,F (distance = 2y), for a normalized crack length x. Constant dc current is applied between the surface points A and G (distance apart = 2d). The proportionality constant, k, is found by measurement of $u(x_o)$, for a known initial crack length $x = x_o$. In particular,for $x_o = 0$, and $d > 2W$ (the normal experimental situation for bend bars), it follows from equation (1) that;

$$k = \frac{u(o)}{2y} \cdot \frac{W}{\pi} \quad (5)$$

Equations (2) and (5) suggest that the drop of potential against crack growth data should be stored in the dimensionless form suggested by Anctil et al. [1], i.e.;

$$Q(k) = \frac{u(o)}{u(x)} = \frac{R(o)}{R(x)} \quad (6)$$

where R(o), R(x) are the electrical resistances of the reference (unnotched) and ST notched samples, respectively. R(x) is always measured across the crack, between lines $Y = \pm y$ (current is applied between lines $Y = \pm d$), Fig. 1. R(o) is measured either as R(x), if the initial notch has not been cut, or between any pair of lines at C,D (2y′ distance apart) whilst the current is applied between lines $Y=0$ and $Y=-d$ (Fig. 1).

If the notch cannot be approximated by the straight-through geometry, solution (1) of the Laplace equation is invalid and the potential distribution around the crack must be

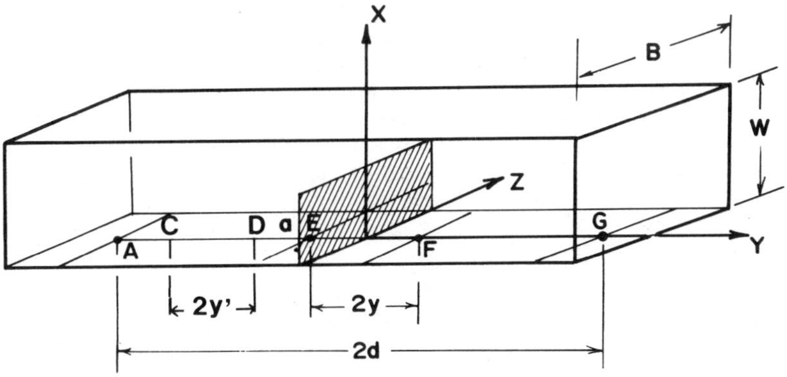

$$x = \frac{a}{W}, \quad c_2 = \frac{2y}{W}, \quad c_3 = \frac{2d}{W}, \quad c_1 = \frac{2y'}{W}$$

Figure 1: Geometry of the straight-through notched (ST) specimen.

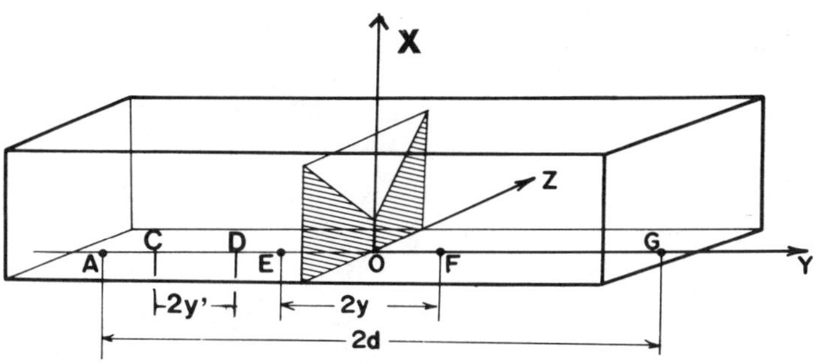

Figure 2(a): Geometry of chevron specimen.

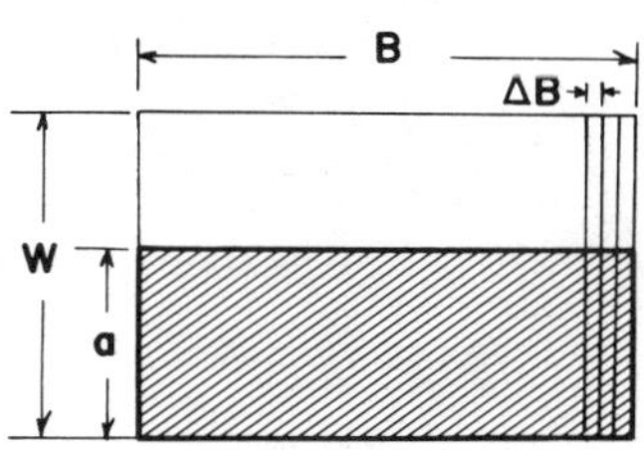

$x = a/W$

$A_o = B.(W-a)$

$A_i = \Delta B.(W-a)$

Figure 2(b): ST specimen cross-section.

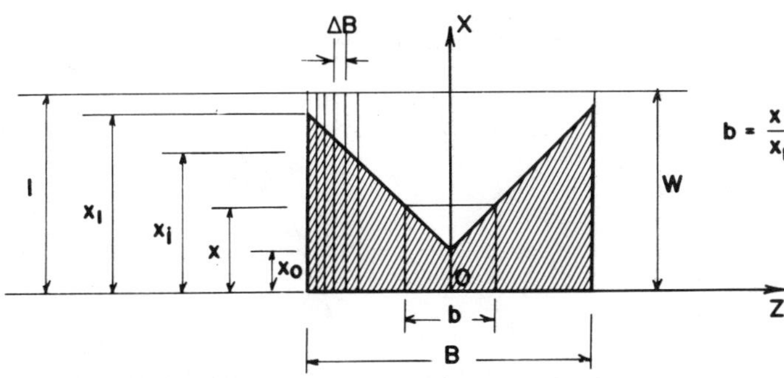

$$b = \frac{x - x_o}{x_1 - x_o} . B$$

Figure 2(c): Chevron specimen cross-section.

described by the three-dimensional equation. To avoid the complicated boundary conditions involved in the three-dimensional solution for the CN specimen, a semiempirical, simplified approach is taken and subsequently verified by experimental calibration.

The ST specimen is sliced in a direction normal to the crack front (Fig. 2b), and the system treated as a set of resistors connected in parallel. The interaction between the "slices" in contact in the real specimen is neglected and the validity of this assumption is tested experimentally. Due to the specimen symmetry, the interlaminar current density in the Z-direction (Fig. 2a,c) should be negligible. The proposed equivalent circuit therefore possesses interlaminar resistances much larger than those along the slice, i.e. the parallel configuration.

If the ST sample is "sliced" into n slices, each having the same notch length x and resistance $R_i(x)$ then, from equation (6);

$$Q(x) = Q_{ST} = \frac{R(o)}{R(x)} = \frac{R(o)}{n \cdot R_i(x)} \tag{7}$$

Since the current density in any plane, parallel to the XY plane, is the same at each point along an arbitrary line (x_o, y_o), the following relation is valid;

$$1/n = A_i/A_o \tag{8}$$

where A_i is the area of the uncracked section of the slice i, and A_o is the area of the uncracked part of the specimen (Fig. 2b). Accordingly;

$$Q_{ST} = \frac{R(o)}{R_i(x)} \cdot \frac{A_i}{A_o} = n \cdot \frac{R_i(o)}{R_i(x)} \cdot \frac{A_i}{A_o} \tag{9}$$

or

$$Q_{ST}(x) = \sum_{i=1}^{n} Q_{iST}(x) \tag{10}$$

where $Q_{iST}(x) = Q_{ST}(x)/n = R(o)/R_i(x)$. For the ST specimen $Q_{ST} = R_i(o)/R_i(x)$ and equations (9,10) are always fulfilled. The slicing procedure for the chevron-notched specimen (Fig. 2c) however, results in a series of approximately ST samples of changing area $A_i(x_i)$ and resistance $R_i(x_i)$. Therefore, equation (9) must be applied to each slice with a particular $Q_{iST}(x_i)$ and $A_i(x_i)$.

The definition of the Q parameter for the chevron-notched specimen is;

$$Q_{CH}(x) = \frac{R(o)}{R(x)} \tag{11}$$

and, by approximating the system as a set of parallel resistances, the following relationship results from Ohm's law;

$$Q_{CH}(x) = \sum_{i=1}^{n} \frac{R(o)}{R_i(x)} = \sum_{i=1}^{n} Q_{iST}(x_i) \tag{12}$$

or, from equations (9) and (10);

$$Q_{CH}(x) = \sum_{i=1}^{n} Q_{ST}(x_i) \cdot \frac{A_i}{A_o} \tag{13}$$

For any crack length $x > x_o$, part of the chevron crack front (of length equal b, Fig. 2c) can be expressed as ST.

So, considering the specimen and the notch geometry [13], equation (13) simplifies to;

$$Q_{CH}(x) = \frac{2\cdot\ (1-x)(x-x_0)\cdot\ Q_{ST}(x) + I(x)/N}{V(x)\cdot\ (x_1 - x_0)} \tag{14}$$

where N is the number of slices necessary to describe the chevron shape (here taken as N = 500) and V(x) is the geometric parameter characteristic of a given specimen and crack length. V(x) and I(x) are defined as follows;

$$V(x) = 2 - x - x_1 + \frac{(x-x_0)(x_1 - x)}{x_1 - x_0} \tag{15}$$

and

$$I(x) = \sum_{i=n_b}^{n_B} [Q_{ST}(x_i)(1 - x_i)] \tag{16}$$

The summation in the equation (16) proceeds along the tapered part of the crack front (Fig. 2c), between $x_i = x$ and $x_i = x_1$ (for slice number n_b and n_B respectively).

Equation (14) was verified for a number of graphite*, chevron-notched samples of different geometries and probe locations. The bars (approximately 10x10 mm and 40 to 100 mm) were polished on 400 grade sandpaper and saw-notched to give an 0.8 mm-wide slot. The reference resistance R(o) was measured between the same points as the variable resistance R(x), before the initial notch was cut. For convenience, the distances 2y', 2y and 2d are normalized to the specimen width W, to give the dimensionless parameters C_1, C_2, C_3 respectively (Fig. 1).

Accurate electrode separation was obtained via the technique proposed in [3], i.e. the parameter C_1 was adjusted such that the initial optically-measured crack length ($x = x_o$) was equal to the value obtained from the solution of equation (14). The adjustment procedure resulted in a C_{1eff} value. Subsequently, the chevron triangle was saw-blunted, to simulate an advancing crack front, and the new notch length was measured by a travelling microscope with a 0.2 mm accuracy. The potential drop across the crack varied between 0.1 to 2.0 mV for a surface applied 200 mA dc current†. Typical experimental results are compared with the numerical calculations in Fig. 3. The agreement is satisfactory, the majority of the scatter in the experimental results being attributed to the difficulties of optical measurement of the notch length.

III. EXPERIMENTAL PROCEDURE

The chevron-notched, four-point bend ceramic specimens (bending spans S_1 = 22.8 mm, S_2 = 10.7 mm) were cycled through loading/stable-fracture/unloading experiments [14], in a constant displacement rate** machine. The constant current was pushed through the specimen (approximately 5x5x30 mm) and the potential drop, u, was measured across the advancing crack. Typical experimental setup and output are shown in Figure 4. On the load against potential drop graph, fracture initiates at point S but, due to the slow initial crack advance, the first increase in the potential drop is recorded at point I. Unloading starts at point T and in the last stage of unloading the potential drop decreases due to partial closure and electrical contact of the fractured surfaces. From the increment of u at the moment of unloading (T), the fracture area and specimen compliance (from Bluhm's slice model [15]) is

---

* type AXM, POCO.

† Potentiostat type 251, HB Thompson.

** Compressive testing machine type WF10052, Wykeham-Farrance Eng Ltd., with 500 lb. load cell, Transducer Inc.

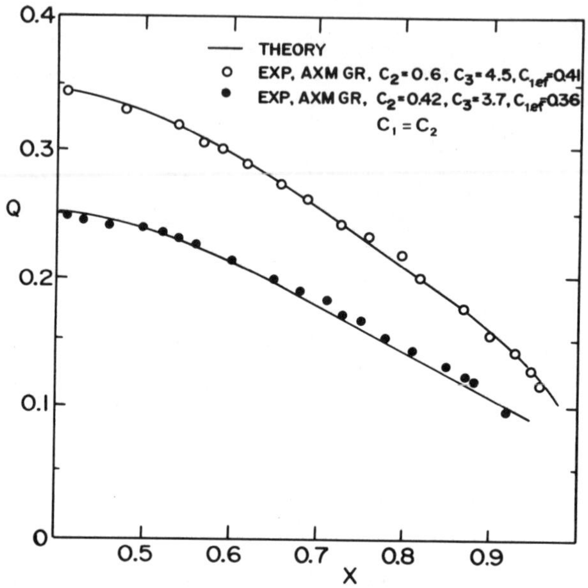

Figure 3: Calibration results of PD method for chevron specimens.

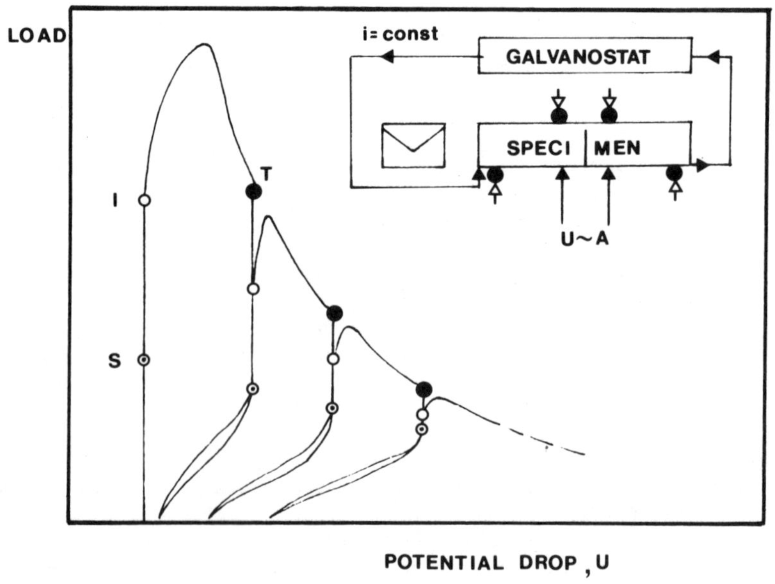

Figure 4: Schematic output of the PD technique.

obtained. The values of Young's modulus (necessary for the compliance calculation) were obtained from four-point elastic bending of unnotched bars. From the load, compliance and fracture area increments, the resistance to fracture, R_e, was calculated [14]. It must be emphasised that the PD technique results in a value of the resistance to fracture responsible for the specimen compliance increase (i.e. as per the definition provided by Eftis and coworkers [19] and applied by Sakai et al. [17]. A nonelastic energy dissipation rate (resulting in permanent deformation of the specimen) can be calculated only if the load-application points displacement is simultaneously recorded [14]. It can be shown that, as full fracture is approached in consecutive cycles (Fig. 4), R_e approaches the resistance to fracture initiation G [14].

Three types of material were investigated. POCO graphite (at 25°C), 8 wt% Y_2O_3 partially-stabilised-zirconia†† (PSZ) and PSZ-β alumina composites (ZA). The PSZ materials were tested between 1000 and 1300°C. Commercial PSZ powder was mold-pressed, isopressed and sintered for 6 hours at 1450°C. The ZA composite was prepared by mixing 20 vol.% of ground β-alumina (mean particle size 120 μm and 50 μm) with 80 vol.% of PSZ and sintering pressed bars for 10 hours at 1450°C. The resulting microstructures are presented elsewhere [16]. All bars were polished on the loading faces and chevron-notched with a 150 μm thick diamond blade. The test temperature was controlled within ± 2°C.

IV. RESULTS AND DISCUSSION

The accuracy of the CN crack length calculation via potential drop was verified for graphite samples (Fig. 3). For this material a load vs PD plot is shown in Figure 5 and a load vs displacement can be found in [14]. Both plots generate the materials' resistance to fracture, R_e, and the results for three graphite specimens are shown in Figure 6 (applied currents were 200 mA (GR 67), 100 mA (GR 65) and 0 mA, i.e. only the load-displacement graph was recorded for GR 62). The open and closed circles for specimen GR 67 result from the simultaneously measured PD and load-displacement for the same sample. Good agreement exists between the two methods. It appears however, that relatively large currents interact with the propagating crack and in the case of the specimen GR 67, (i = 200 mA), the current density at A = 50% reached ~ 5 A/cm^2, effectively heating the sample web. As the applied current decreased to zero, the measured resistance to fracture agreed with the literature values. Due to the heating phenomena, the current was decreased by two orders of magnitude (1 to 3 mA) for the high temperature studies of the partially stabilised zirconias.

The load versus potential drop results for PSZ tested at 1100°C are shown in Fig. 7 and the resulting R_e (for two specimens) in Fig. 8. As full fracture is approached (A = 100%), the resistance to fracture decreases to ~ 3 J/m^2 (i.e. at least an order of magnitude smaller than at room temperature [18]). PSZ tested at 1300°C (Fig. 9) did not exhibit crack closure upon unloading. The constant potential drop lines i.e. the vertical unloading paths, suggest that plastic deformation of the material has occurred. Indeed, partially fractured specimens remained permanently bent upon removal from the furnace. The resistance to fracture of PSZ at 1300°C (plotted for two specimens) is shown in Fig. 10 and is ~ 1 J/m^2 as full fracture is approached.

The load versus potential drop for a PSZ-β-Al_2O_3 composite (20 vol.% of 120 μm large particles) tested at 1000°C is shown in Fig. 11. The decrease in the potential drop upon unloading, which continued even at zero load, indicates a relaxation of the residual elastic stresses between particles and the fracture surface. This phenomenon also causes crack growth during unloading, i.e. the loading and unloading paths do not coincide. The resistance to fracture calculated from Fig. 11 and for another specimen tested at 1300°C, are presented in Fig. 12. At 1300°C the R_e for the composite is similar to that of PSZ (i.e. ~ 1 J/m^2) but at 1000°C it is an order of magnitude larger, i.e. ~ 11 J/m^2 (the data for PSZ at 1000°C are not available, since all specimens tested at this temperature broke catastrophically).

†† Type ZYP, Zircar Prod. Inc.

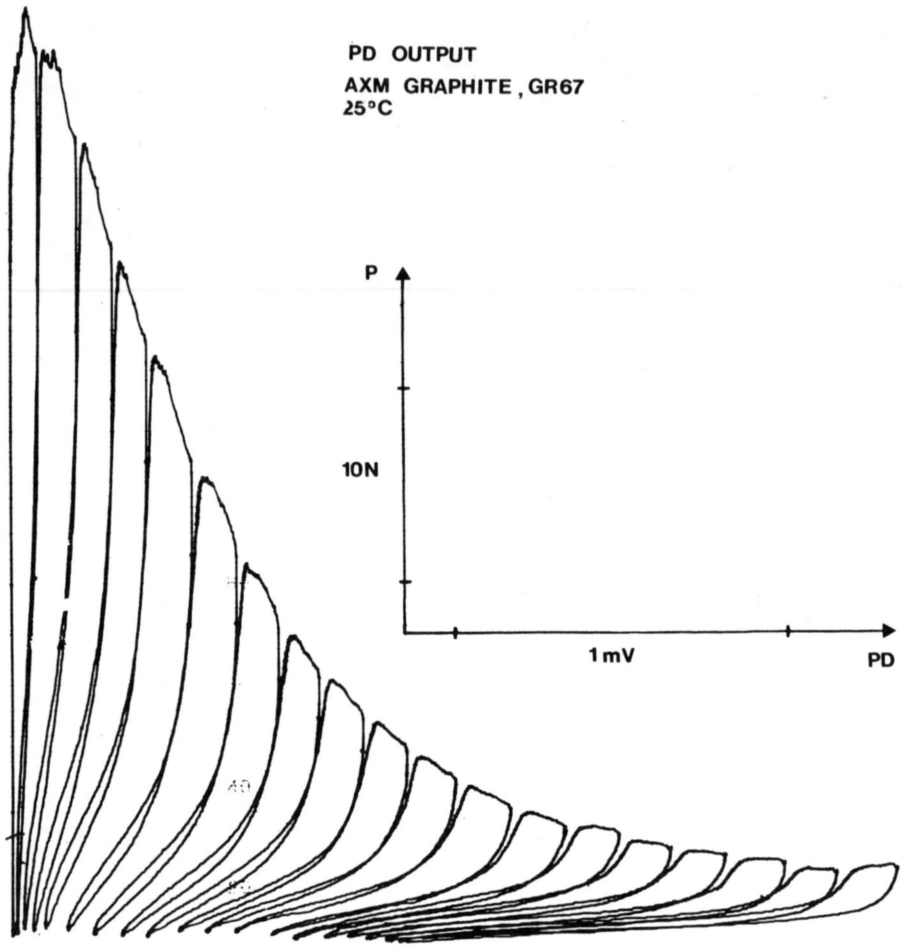

Figure 5: Real output of PD technique for graphite.

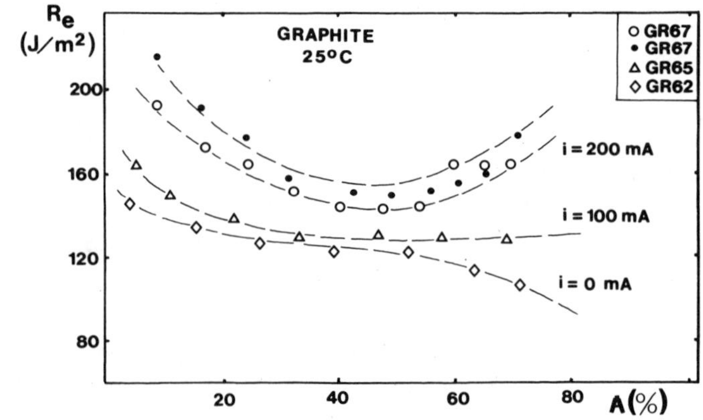

Figure 6: Resistance to fracture of graphite at different current applied.

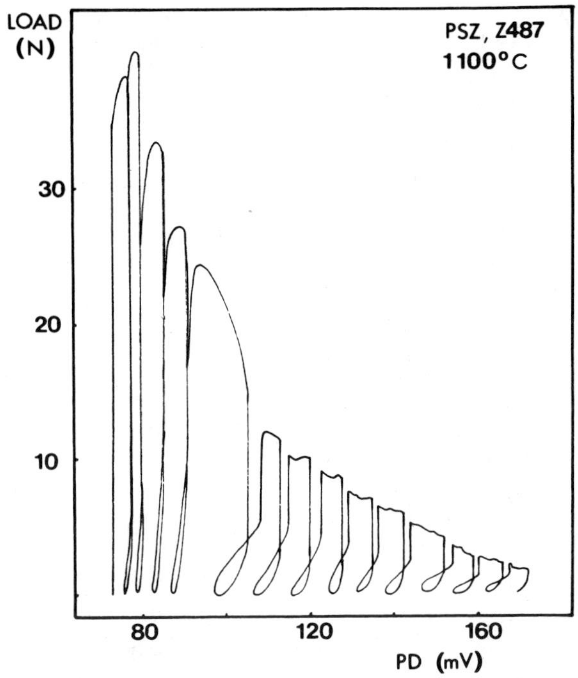

Figure 7: PD output in loading-unloading experiment for PSZ at 1100°C.

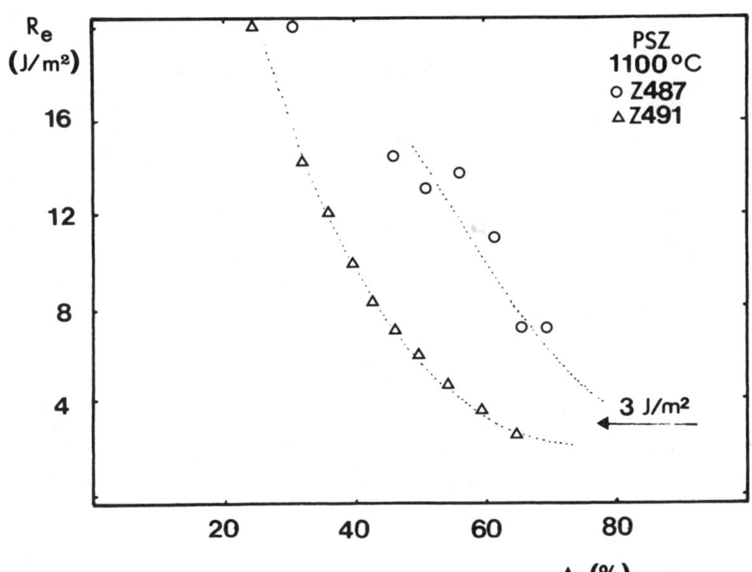

Figure 8: Resistance to fracture of PSZ at 1100°C.

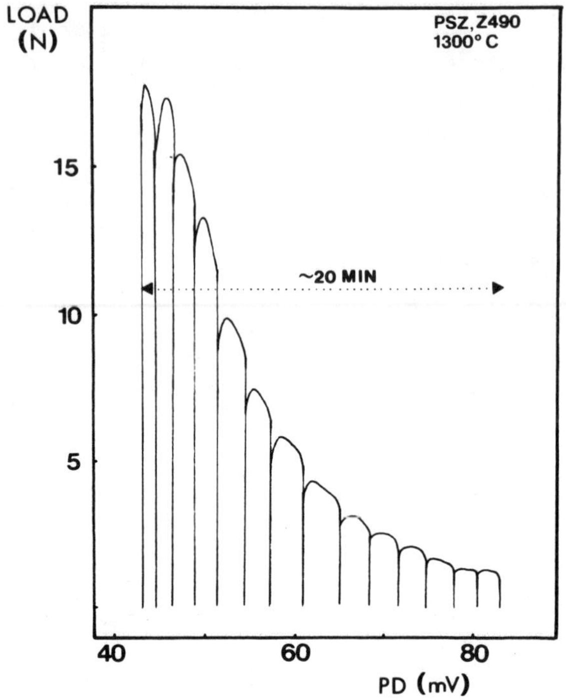

Figure 9: PD Output for PSZ at 1300°C.

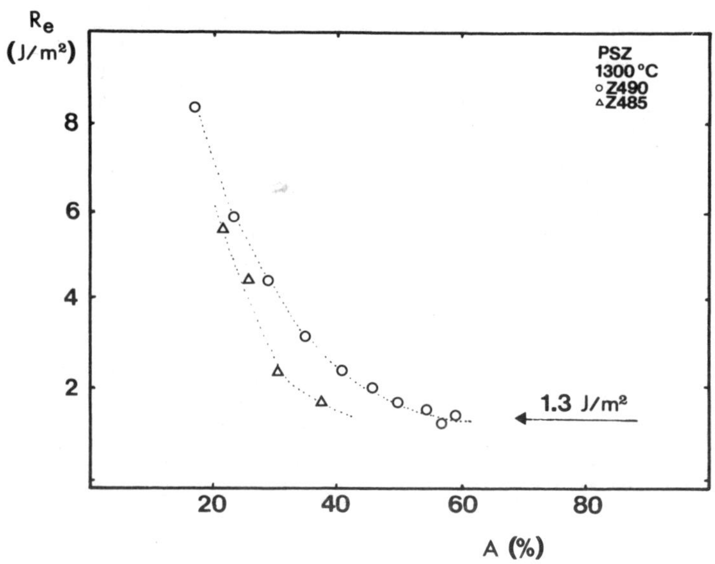

Figure 10: Resistance to fracture of PSZ at 1300°C.

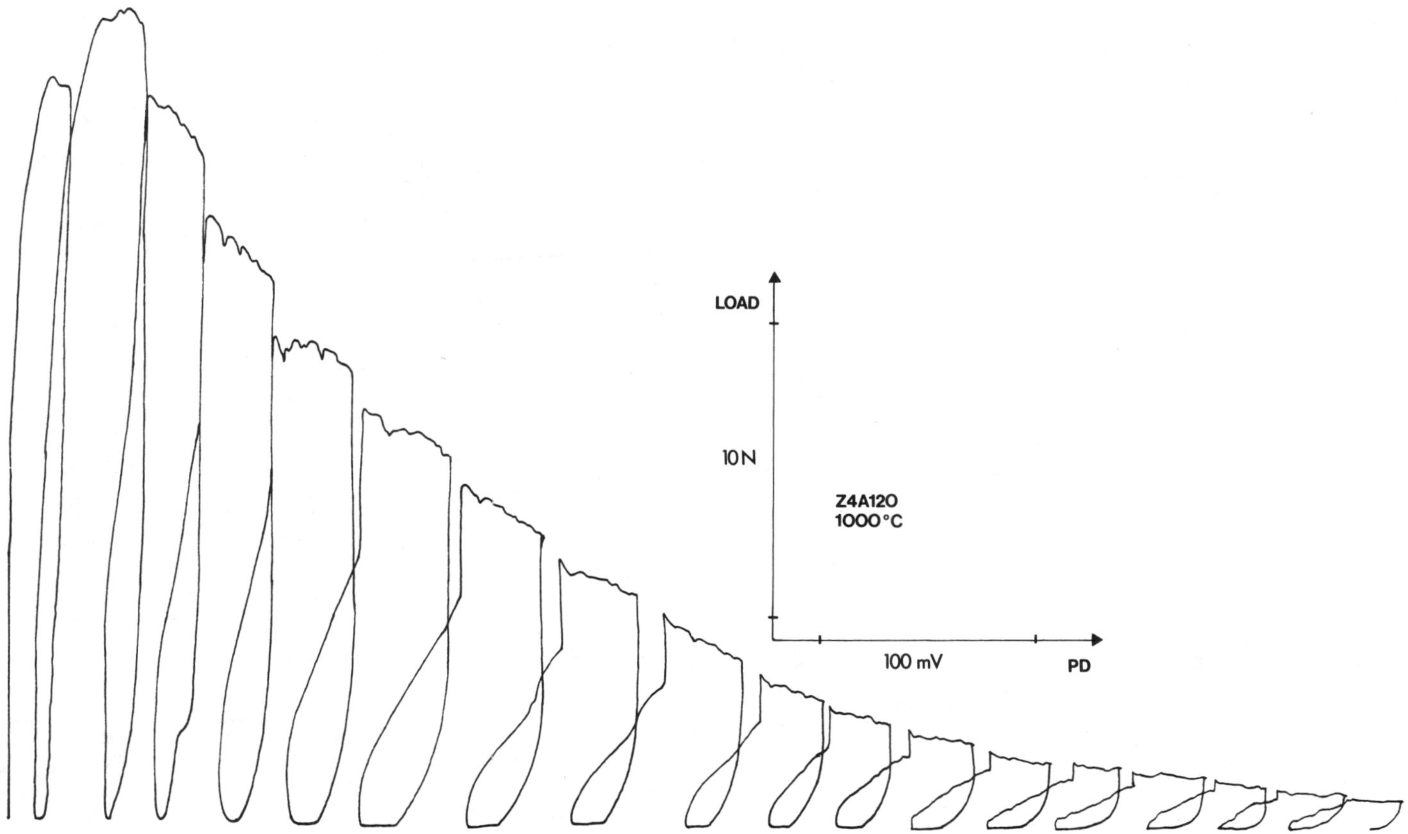

Figure 11: PD output for Z4A120 at 1000°C.

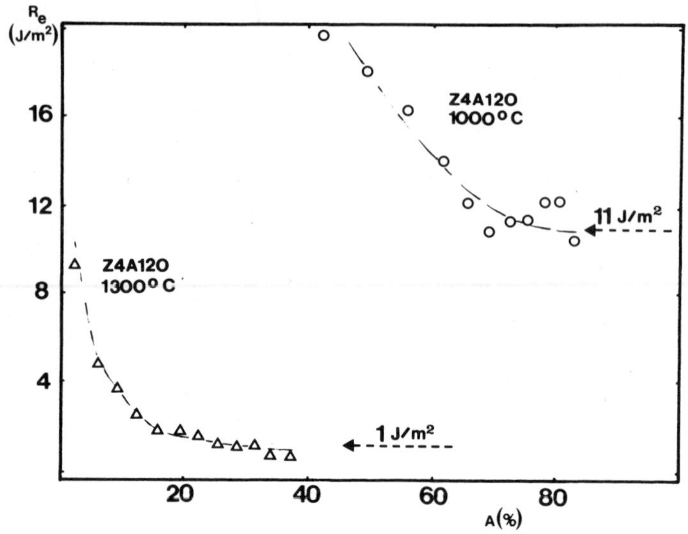

Figure 12: Resistance to fracture of PSZ-20 vol % βAl_2O_3 (120 µm size) composite, at 1000°C and 1300°C.

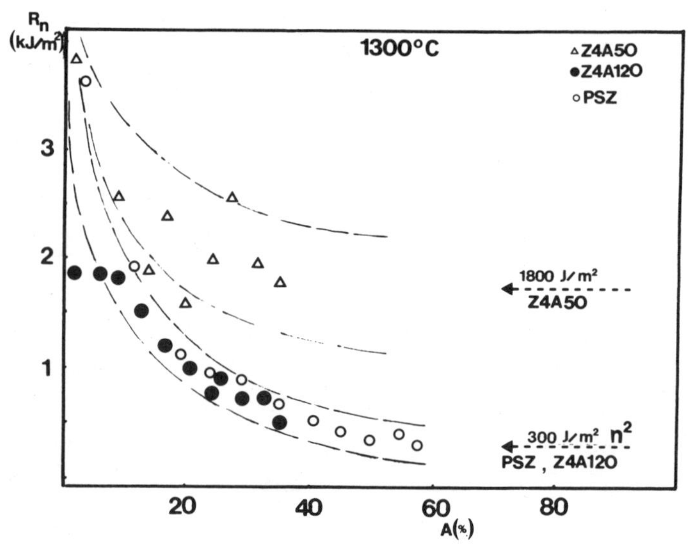

Figure 13: Total input of work per unit area to fracture PSZ and PSZ-βAl_2O_3 composites.

It is interesting to compare the values of R_e (i.e. energy spent to create unit surface area excluding work expended in permanent deformation) to the total input of work needed to break the specimen, R_n. The latter quantity is obtained from the load-potential drop diagrams and the cross-head displacement in each cycle. If it is assumed that the elastic energy stored in the testing system at high temperatures is negligible, the total input of work in each cycle corresponds to the cross-head displacement multiplied by the mean load. The fracture area increment was calculated from the potential drop. The 1300°C results are shown in Fig. 13. R_n values for PSZ and Z4A120 are similar and decrease towards 300 J/m^2 as full fracture approaches. R_n for the Z4A50 composite (PSZ + 20 vol.% β Al_2O_3, 50 μm particle size) is significantly larger (~ 1800 J/m^2) and it is suggested that the smaller β Al_2O_3 particles dispersion-harden the plastically deforming PSZ matrix.

V. CONCLUSIONS

A semi-empirical procedure which leads to a simple relationship between the potential drop and the crack length for a chevron-notched bar specimen has been developed. Calibration of this technique gives good agreement between the theory and the experimental results. The deviation of the simplified model from the real potential distribution in the chevron-notched specimen is corrected for by the C_{1eff} calculation.

The PD technique was used to determine the resistance to fracture of materials tested in cyclic loading/stable-fracture/unloading experiments. Comparison of the PD results for the model material (graphite) with direct load-displacement measurements showed good agreement. The PD technique was then applied to the evaluation of crack growth and fracture resistance in zirconium oxide ceramics at high temperature. The values of resistance to fracture were ~ 3 J/m^2 for PSZ at 1100°C, ~ 1 J/m^2 for PSZ at 1300°C, 11 J/m^2 for PSZ-β Al_2O_3 at 1000°C; ~ 1 J/m^2 for PSZ-β Al_2O_3 at 1300°C. Plastic deformation was found to accompany the fracture of yttria-stabilized PSZ at 1300°C and the incorporation of 20 vol.% 50 μm β Al_2O_3 particles "dispersion hardened" PSZ at 1300°C. As a result, the total input of work per unit area to fracture the latter composite was ~ 1800 J/m^2 as compared with 300 J/m^2 for pure PSZ and the PSZ + 20 vol.% β Al_2O_3 (120 μm particle size).

REFERENCES

1. A.A. Anctil, E.B. Kula and E. DiCesare, Proceedings, Am. Soc. Testing Mats. 63, 799-809 (1963).
2. H.H. Johnson, Mat. Res. and Standards, 5, 422-445 (1965).
3. D.M. Gilbey and S. Pearson, RAE Tech. Rep., No. 66402, Farnborough, U.K. (1966).
4. R.O. Richtie, G.G. Garret and J.F. Knott, Int. J. Fr., 7, 462-467 (1971).
5. G.H. Aronson and R.O. Richtie, J. Test. Eval., 7 (4), 208-215 (1979).
6. G. Clark and J.F. Knott, J. Mech. Phys. Sol., 23, 265-276 (1975).
7. J.F. Knott, "The Measurement of Crack Length and Shape During Fracture and Fatigue", Edited by C.J. Beevers, EMAS, U.K.,113-135 (1980).
8. S.G. Druce and G.S. Booth, "The Measurement of Crack Length and Shape During Fracture and Fatigue", Edited by C.J. Beevers, EMAS, U.K.,136-163 (1980).
9. J.R. Whiteman and N. Papamichael, J. Appl. Math. Phys., 23, 665, 1972.
10. N. Papamichael and J.R. Whiteman, J. Appl. Math. Phys., 24, 305 (1973).
11. L.M. Barker, Eng. Fract. Mech., 9 (2), 361-369 (1977).
12. R.J. Cooke and J.L. Robinson, Internal Report, Dept. of Physical Metallurgy and Science of Materials, University of Birmingham, U.K. (1971).
13. T.B. Troczynski and P.S. Nicholson, to be published, Am. Cer. Soc. Bull., 1985.
14. T.B. Troczynski and P.S. Nicholson, this volume.
15. J.I. Bluhm, Eng. Fract. Mech., 7, 593-604 (1975).
16. J. Glass and P.S. Nicholson, to be published in J. Am. Cer. Soc. (1985).
17. M. Sakai, K. Morashima and M. Inagaki, J. Am. Cer. Soc., 66 (12), 868-874 (1983).
18. T.B. Troczynski and P.S. Nicholson, to be published in Communications of J. Am. Cer. Soc. (1985).
19. J. Eftis, D.L. Jones and H. Liebowitz, Eng. Fract. Mech., 7, 491-503 (1975).

A COMPARISON OF VICKERS AND KNOOP INDENTATIONS IN SODA-LIME GLASS

J. E. Ritter, Jr., F. M. Mahoney, and K. Jakus

Mechanical Engineering Department

University of Massachusetts, Amherst, MA 01003

ABSTRACT

Flaw and strength characteristics of Vickers and Knoop indentations in soda-lime glass were studied over a wide range of indenter loads. Vickers indentation produced ideal, well-developed deformation/crack patterns over the entire load spectrum; whereas, Knoop indentation patterns contained significant amounts of extraneous surface damage and radial cracks were consistently well-developed only at high indenter loads. Despite these complications, both Knoop and Vickers results showed very good fit to the indentation fracture mechanics model. However, unlike results reported for silicon nitride and glass-ceramic materials, Vickers and Knoop radial cracks on soda-lime glass were of comparable length for the same indentation load and equivalent crack lengths yielded equivalent fracture strengths independent of indenter geometry. Strength behavior of annealed flaws suggests that the flaw shape parameter for Knoop indentations is less than for Vickers indentations (0.21 vs. 0.28). Aging test results indicate greater relief of the indentation-generated residual stress in Knoop flaws than in Vickers flaws.

INTRODUCTION

The use of controlled flaws in studying the fracture behavior of brittle materials is now well established. The Vickers and Knoop indenter geometries have been used to produce controlled flaws because of their commercial availability for microhardness testing. The Vickers geometry produces two well-defined, perpendicular radial cracks and has been widely used in the experimental verification of indentation fracture analysis.[1,2] The Knoop geometry generates only a single radial crack and consequently has been used in fracture studies involving oriented flaws.[3,4,5] However, application of Knoop indentation is uncertain because the plastic zone extends over most, or all, of the surface trace of the radial crack and this may invalidate the basic assumption in the failure analysis where the residual stress field around the indentation is approximated as a point force located at the crack center.

Two recent studies have compared the flaw and strength characteristics resulting from Knoop and Vickers indentation. Marshall[6] concluded in his study on silicon nitride that the fracture toughness and strength characteristics for both indenter geometries is well described by indentation analysis as long as the crack length to impression size ratio is greater than 1.3. On the other hand, Petrovic[7] found that for a glass-ceramic material fracture toughness depended on the indentation load, suggesting either a toughness-crack size dependency or that the treatment for residual stress needs further refinement.

The purpose of the present study is to make an in-depth comparison of the flaw and strength characteristics of Knoop and Vickers indentation in soda-lime glass. Soda-lime glass is used in this study because it is free of any microstructural variations that may complicate the results and because crack patterns can be readily seen. Data is analyzed in terms of indentation fracture mechanics theory and compared to similar studies using silicon nitride[6] and glass-ceramic.[7]

EXPERIMENTAL PROCEDURE

Soda-lime glass plates (57x57x3mm) were cleaned in an acetone bath, annealed at 520°C for 24 hours, and then randomized. Controlled surface flaws were introduced with Vickers indentations over a load range of 2.45 to 98.1 N using a standard microhardness tester.[†] For Knoop indentations the load range was 39.2 to 392.4 N with indentation loads greater than 100 N requiring the use of a Universal testing machine.[††] All indentations were done under ambient conditions and the relevant crack dimensions were measured in the optical microscope attached to the microhardness tester.

Following indentation samples were strength tested either as-indented, annealed, or after aging in distilled water for periods ranging from 15 seconds to 24 hours. After annealing and aging the crack dimensions were remeasured. The indent of each sample was covered with mineral oil following each treatment to minimize any crack growth before strength testing.

The strength of all indented samples was determined at a crosshead speed of 1 cm/min using a concentric ring bend fixture in conjunction with the Universal testing machine. The effects of fatigue were minimized by testing the samples with the indent covered by mineral oil. Strength was calculated from the equation derived in Ref. 8.

[†]Tukon, Measurements Systems Div., ACCO Industries, Inc., Bridge port, CT.

[††]Instron Corp., Canton, MA.

RESULTS AND DISCUSSION

As-Indented Crack Dimensions

Well-defined radial cracks were produced over the entire load spectrum for Vickers indentations. Similar to a previous study[1] the load range was set by crack initiation at the lower bound (2.45 N) and extensive chipping at the upper bound (98.1 N). For Knoop indentations 40N represents a crack initiation threshold and above 400 N the samples would often fracture during indentation. Marshall[6] also found that the crack initiation load for Knoop indentations was significantly higher than for Vickers indentations (20-40 N vs. 5-15 N). Between 40 and 400 N the Knoop indenter generated a variety of undeveloped to well-developed crack systems. At one extreme only the plastic indent impression was visible with no cracking evident.* At the other extreme lateral cracks extended around the Knoop indent impression intersecting the surface to form bifurcation cracks leading outward from the indent and radial cracks extended outward from the tips of the indent. Above 150 N well-developed crack systems were consistently reproduced; however, this was accompanied by extensive chipping and crushing around the plastic impression zone.

By modeling the indentation stress field and radial crack system as a center-loaded crack, it can be shown that the radial crack dimension after indentation (c_o) is:[9]

$$c_o = (\chi_r P/K_c)^{2/3} \tag{1}$$

where P is indenter load, K_c is critical stress intensity factor, and χ_r is the residual stress constant that is dependent on indenter geometry and material properties. Based on radial crack size measurements of Vickers indentation data over a wide range of ceramic materials, the constant χ_r was found to be:[1,9]

$$\chi_r = \xi_v^R (E/H)^{1/2} \tag{2}$$

where E is elastic modulus, H is hardness, and ξ_v^R is a material independent constant equal to 0.016.

The surface trace dimensions of radial cracks generated by the Vickers and Knoop indenters are plotted as $P/c_o^{3/2}$ vs. P in Fig. 1. The data for both Vickers and Knoop indentations in soda-lime glass fall along the same

---

*This type of subthreshold indent only occurred in 10% of the samples indented above 40 N and none above 100 N. Since in subsequent strength tests failure did not occur from these indents, they were excluded from the data analysis.

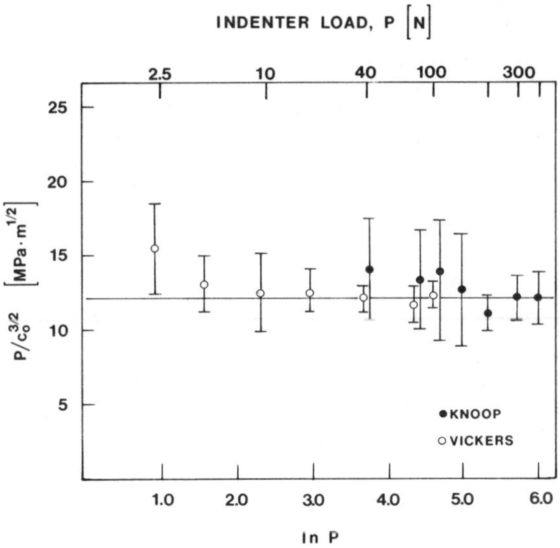

Fig. 1. Indenter load/radial crack size data for Vickers and Knoop Indentations in soda-lime glass. Data points and error bars are mean and standard deviation; horizontal line represents Eq. (1).

horizontal line, indicating that under equivalent indenter load both Vickers and Knoop indenters produce radial cracks of similar size and that an equivalent level of residual stress is present in both indentations. The constant value of 12.00 ± 0.50 $MPa{\cdot}m^{1/2}$ agrees well with the value of 12 $MPa{\cdot}m^{1/2}$ reported previously for Vickers indentations in soda-lime glass.[1] The larger scatter in Knoop values between 40 and 150 N indenter loads is due to the variety of underdeveloped crack systems generated that undoubtedly lead to a greater scatter in the residual stress present, i.e. χ_r values. It is also important to note that no deviation from the horizontal line is evident in surface trace measurements in Fig. 1 as crack sizes exceed 1/10 the specimen thickness (above 75 N). This is consistent with data obtained by Marion[10] who found that crack propagation was not affected by the lower free surface as long as crack depths did not exceed 1/3 the specimen thickness.

The results of the radial crack sizes may also be used to calculate the critical stress intensity factor based on Eqs. (1) and (2), knowing the elastic modulus and hardness. Hardness values as a function of indenter load are show in Fig. 2. Both Vickers and Knoop hardness are calculated as the applied load divided by the projected area of the indentation impression.[9] With both Vickers and Knoop indentations, hardness becomes load-invariant above a certain threshold load that is thought to correspond to the transition from elastic contact to fully-developed plasticity.[9] This threshold appears to be about 5 N for Vickers indentations and 100 N

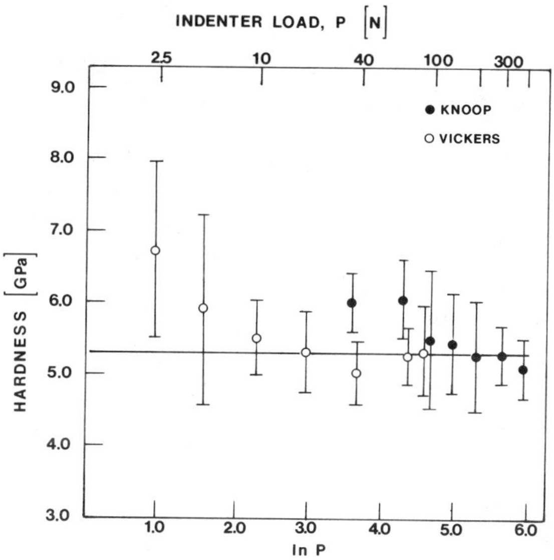

Fig. 2. Vickers and Knoop load-invariant hardness as a function of indentation load. Data points and error bars are mean and standard deviation; horizontal line represents the average value.

for Knoop indentations. Average values of hardness determined from Vickers and Knoop indentations are similar (5.3 GPa) and agree well with literature values for soda-lime glass.[1] Taking E = 70 GPa,[1] χ_r is calculated from Eq. (2) to be 0.058 for the Vickers indentations and based on the results of Fig. 1 it is believed that this value is the same for Knoop indentations. Using this result and taking $P/c_o^{3/2}$ to be 12.00 $MPa{\cdot}m^{1/2}$, Eq. (1) gives $K_c = 0.70 \pm 0.03$ $MPa{\cdot}m^{1/2}$ which is in good agreement with the generally accepted value of 0.75 $MPa{\cdot}m^{1/2}$ for soda-lime glass.[1] It is important to emphasize that indentation crack data generated by both Vickers and Knoop indenters result in similar values for the critical stress intensity factor of soda-lime glass.

The equivalent residual stresses, i.e. χ_r, generated by both indenter geometries in soda-lime glass differs from crack size results on silicon nitride[6] and glass-ceramic.[7] With these two materials radial crack sizes for Knoop indents were considerably larger than those for Vickers indents at similar loads, indicating a significantly greater residual stress in Knoop indents than in Vickers. Marshall[6] related the residual stress difference in silicon nitride to the relative volumes displaced by the two indenters. The equivalence of χ_r for soda-lime glass could be related to the deformation/fracture patterns produced by indentations. For both silicon nitride and glass-ceramic, Vickers and Knoop indenters produced ideal crack patterns with no evidence of chipping. Similarly, ideal crack patterns were produced by Vickers indentations in soda-lime glass; however, Knoop indentations had considerable chipping and crushing around the impression zone. Presumably with soda-lime glass the extraneous surface damage around the Knoop indent causes a residual stress relief such that Knoop and Vickers residual stress levels are similar.

Strength Degradation

The mechanical failure from radial cracks under the combined influences of the residual stress and a normal applied tension (σ_a) has been analyzed in detail.[2] This approach predicts stable, equilibrium crack growth prior to catastrophic failure due to the presence of the residual stress. This equilibrium crack response is described by:[2]

$$\sigma_a = [K_c/(\pi\Omega c)^{1/2}][1-\chi_r P/K_c c^{3/2}] \tag{3}$$

where c is crack size and Ω is a crack geometry parameter. The failure condition is defined by the maximum value of $\sigma_a(c)$ in Eq. (3):[2]

$$\sigma_m = [(27/256)K_c^4/\chi_r(\pi\Omega)^{3/2}]^{1/3}P^{-1/3} \tag{4}$$

Values of the indentation strength parameter ($\sigma_m P^{1/3}$) for both indenters are plotted in Fig. 3 and fall along the horizontal line as predicted by Eq. (4). The result that both Vickers and Knoop values fall on the same constant value line (120.5 ± 5.1 $MPa{\cdot}N^{1/3}$) gives additional support to the contention of equivalent residual stress in both indentations. The large strength variability for Knoop indentations between 40 and 150 N is probably due to variations in flaw types as previously mentioned. The constant value of $\sigma_m P^{1/3}$ determined here for Vickers and Knoop indentations agree very well with values reported by Chantikul et al[2] (120 $MPa{\cdot}N^{1/3}$) for Vickers indentation in soda lime glass.

Strength results may also be used to calculate fracture toughness. Rearranging Eq. (4) gives:

$$K_c = \eta_v^R(E/H)^{1/8}(\sigma_m P^{1/3})^{3/4} \tag{5}$$

where $\eta_v^R = [(256/27)(\pi\Omega)^{3/2}\xi_v^R]^{1/4}$. The constant η_v^R has been calibrated for Vickers indentations over a wide range of ceramic materials and it was found to equal 0.59 ± 0.12[2]. Substituting this value and ($\sigma_m P^{1/3}$) = 120.5 $MPa{\cdot}N^{1/3}$ into Eq. (5) gives K_c = 0.94 $MPa{\cdot}m^{1/2}$, independent of indenter geometry. This value is essentially the same as that determined by Chantikul et al.[2] Note that this K_c determined by the indentation-strength technique is significantly higher than that determined by crack size, or more conventional means, e.g. double cantilever beam. It is believed that this is related to the fact that candidate materials for the calibration

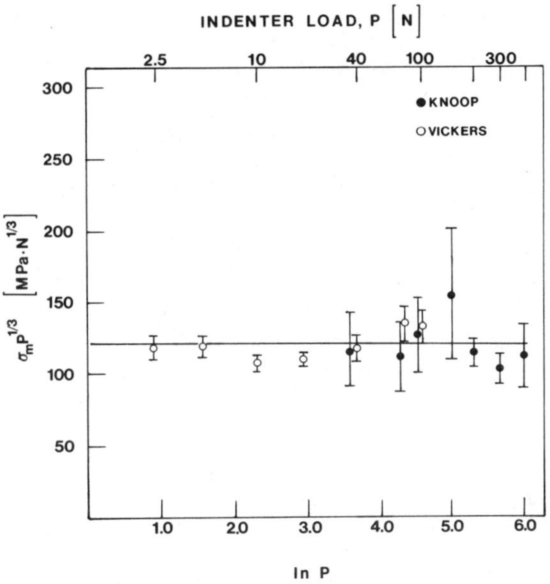

Fig. 3. As-indented strength degradation data for Vickers and Knoop indentations in soda-lime glass. Data points and error bars are mean and standard deviation; horizontal line represents Eq. (5).

were as close as possible to an ideal elastic/plastic material which resulted in an upper bound value for η_V^R.[11]

Converse to results in this study, data on silicon nitride[6] and glass-ceramic[7] exhibited different strength ($\sigma_m P^{1/3}$) and crack size ($P/c_o^{3/2}$) parameters for Knoop and Vickers indentations. For silicon nitride the ratio of the Knoop to Vickers strength parameters was comparable to the ratio of crack size parameters (about 0.7). For glass-ceramic Knoop strength and crack size parameters were less than Vickers at a given indentation load; however, the parameters were not invariant with respect to indentation load, as alluded to in the introduction. In all cases (soda-lime glass, silicon nitride, and glass-ceramic) the relative strength results between the indenter geometries are consistent with their respective crack size results and the corresponding indentation residual stress.

Aging and Annealing

With both Vickers and Knoop indentations strength rapidly increased on aging in water until an upper plateau is reached after about 10^3 s. For the Vickers indentations the strength of this fully-aged condition was about 75% of the annealed strength (see Table 1) and for Knoop indentations about 70% (see Table 2). With Vickers indentations this study, as well as others[2,12,13], observed that aging is accompanied by an increase in both the radial and lateral crack sizes; however, with Knoop indentations no radial crack growth was detected although lateral crack sizes showed a

Table 1. Summary of average radial crack sizes and strengths of Vickers indentations in soda-lime glass. Strengths were measured under inert conditions. Numbers in parentheses represent one standard deviation

INDENT LOAD (N)	As-indented		Fully-aged		Annealed	
	c_o (μm)	σ_m (MPa)	c'_o (μm)	σ'_m (MPa)	c_a (μm)	σ_i (MPa)
2.45	29.3 (3.8)	87.9 (6.4)	37.1 (2.6)	92.8 (1.6)	30.1 (3.1)	120.6 (7.2)
4.90	52.0 (4.9)	70.4 (4.2)	69.1 (4.6)	72.2 (2.6)	59.8 (3.5)	95.7 (3.0)
9.81	85.0 (12.2)	50.8 (2.3)	118.0 (8.2)	59.2 (1.8)	102.8 (4.3)	76.3 (3.0)
19.6	134.2 (10.7)	40.8 (2.0)	192.4 (8.3)	47.3 (2.8)	169.8 (8.9)	64.3 (3.7)
39.2	220.5 (11.3)	34.9 (2.7)	307.2 (13.9)	37.1 (4.8)	272.1 (17.0)	51.8 (2.9)
78.4	354.7 (24.5)	31.6 (2.7)	468.2 (45.5)	33.0 (4.8)	446.5 (36.5)	42.3 (1.9)
98.1	399.3 (19.9)	38.7 (2.4)	526.8 (66.0)	30.2 (4.0)	505.2 (47.3)	38.8 (3.6)

Table 2. Summary of average radial crack sizes and strength of Knoop indentations in soda-lime glass. Strengths were measured under inert conditions. Numbers in parentheses represent one standard deviation

INDENT LOAD (N)	As-indented		Fully-aged		Annealed	
	c_o (μm)	σ_m (MPa)	c'_o (μm)	σ'_m (MPa)	c_a (μm)	σ_i (MPa)
196	677.4 (48.4)	19.8 (1.6)	604.2 (103.3)	25.1 (2.6)	576.7 (91.1)	39.5 (4.7)
294.3	839.0 (71.0)	15.6 (1.4)	826.0 (96.1)	21.1 (3.8)	841.1 (130.4)	35.4 (5.6)
392.4	1015.7 (100.5)	15.3 (3.1)	990.8 (111.0)	20.6 (2.2)	954.3 (116.3)	26.6 (3.5)

significant increase (about 30%). The one difference between the crack systems of Vickers and Knoop indentations is that with Vickers indentations lateral cracks are initially much smaller than radial cracks but with Knoop indentations lateral cracks are initially about the same size. In the fully-aged condition (24 hours in water) lateral cracks in Vickers indentations are about the same size as the radial cracks but with Knoop indentations lateral cracks become considerably larger than the radial cracks. Thus, it may be possible that the extensive lateral cracks associated with Knoop indentations restrict radial crack growth.

The results of this study support the conclusions of Lawn et al[12] who attribute the increase in strength due to annealing and aging to a relief in the indentation residual stress. With aging this relief of residual stress is accomplished through environmentally-assisted growth of lateral cracks since the increase in strength is accompanied by a corresponding increase in lateral crack size. From Eq. (4) it is thought that residual stress relief decreases χ_r so that it should be replaced by χ_r' where $\chi_r' < \chi_r$. However, from the results in Tables 1 and 2 it can be seen that lateral crack growth during aging does not completely relieve the residual stress since the fully-aged strength is less than the annealed strength.

Since annealing completely removes the residual stress ($\chi_r = 0$), annealed, inert strength (σ_i) is given by:

$$\sigma_i = K_c/(\pi\Omega c_a)^{1/2} \tag{6}$$

where c_a is the annealed radial crack length. Rearranging Eq. (6) affords calculation of the fracture toughness-flaw geometry parameter:

$$K_c/(\pi\Omega)^{1/2} = \sigma_i \cdot c_a^{1/2} \tag{7}$$

For Vickers indentations an average value of $K_c/(\pi\Omega)^{1/2} = 0.80 \pm 0.08$ MPa·m$^{1/2}$ is obtained from Table 1. This agrees quite well with the value (0.77 ± 0.05 MPa·m$^{1/2}$) found by Marshall and Lawn[14] for annealed Vickers flaws in soda-lime glass. Assuming $K_c = 0.75$ MPa·m$^{1/2}$, an empirical flaw shape parameter (Ω) of 0.28 is calculated.* Interestingly, if this value is substituted into the definition of η_v^R in Eq. (5) and taking $\eta_r^R = 0.016$ as determined by crack size measurements, the constant η_v^R is calculated to be 0.59, which corresponds exactly to the value calibrated over as-indented strength measurements for a wide range of ceramic materials. This result implies that the flaw shape parameter is the same for indented-annealed samples and as-indented samples.

Data from annealed Knoop flaws (Table 2) gives an average value of $K_c/(\pi\Omega)^{1/2} = 0.93 \pm 0.11$, which corresponds to an empirical flaw shape parameter $\Omega = 0.21$. This lower Ω value for Knoop flaws suggests that these flaws have a lower surface correction factor than do Vickers flaws.

*Note that Ω for a half-penny surface flaw is $1.12\,(4/\pi^2) = 0.45$.[15]

It is interesting to compare the annealed strength to the as-indented strength. Assuming that no residual stress relief occurs after indentation and prior to strength testing for the as-indented samples, combining Eqs. (1), (4), and (5) gives:[2]

$$(\sigma_m \cdot c_o^{1/2})/(\sigma_i \cdot c_a^{1/2}) = 0.47 \tag{8}$$

For Vickers indentation the average ratio of the as-indented strength-crack size parameter to the annealed parameter is 0.65 ± 0.06 based on the data in Table 1 and from Table 2 this average ratio for Knoop indentations is 0.53 ± 0.09. The somewhat higher experimental ratios than the ideal ratio of 0.47 indicates that some residual stress relief occurred, undoubtedly related to the formation of the lateral cracks. The more extensive lateral cracking evident in Knoop indentations would seem to suggest a higher ratio due to greater residual stress relief for Knoop as opposed to Vickers indentations. However, as mentioned earlier, the large amount of lateral cracking appears to restrict radial crack growth in Knoop indentations, which could result in higher annealed strengths and a lower ratio.

As discussed by Lawn et al[12] the increase in strength on aging can be understood with reference to Eqs. (3) and (4). The aging environment extends the radial crack to a value $c_o' > c_o$ and reduces the residual stress intensity to a value $\chi_r' < \chi_r$ by means of lateral crack extension. The reduction in residual stress intensity shifts the point of instability, leading to higher fracture strengths ($\sigma_m' > \sigma_m$) and less precursor crack growth ($c_m' < c_m$). If $c_o' < c_m'$, Eq. (4) predicts an as-indented/aged strength ratio of

$$\sigma_m/\sigma_m' = (\chi_r'/\chi_r)^{1/3} \tag{9}$$

On the other hand, if $c_o' > c_m'$ Eqs. (3) and (4) predict a strength ratio of

$$\frac{\sigma_m}{\sigma_m'} = \frac{[(27/256)K_c^4/\chi_r]^{1/3}P^{-1/3}}{[K_c/(c_o')^{1/2}][1-\chi_r'P/K_c(c_o')^{3/2}]} \tag{10}$$

Since Knoop-indented samples displayed no evidence of radial crack growth (on the surface) during aging, it is believed that their strength ratio should be given by Eq. (9). Taking $\chi_r = 0.058$ and $K_c = 0.75\ \mathrm{MPa{\cdot}m}^{1/2}$ gives an average $\chi_r' = 0.025 \pm 0.003$ for Knoop indentations. Data for Vickers indentations given an average $\chi_r' = 0.047 \pm 0.007$ based on Eq. (9) and 0.037 ± 0.003 based on Eq. (10). It is not possible to determine experimentally if c_o' is greater than or less than c_m', however, both estimates of χ_r' for Vickers flaws are greater than χ_r' for Knoop flaws, which indicates that the aging of Knoop indentations in water relieves a greater proportion of the indentation residual stress than aging of Vickers indentations. This difference could be related to the fact that with aging of Knoop indentations lateral cracks become considerably larger than the radial cracks, making the system more compliant.

SUMMARY

Both as-indented crack size and strength results show that the residual stress associated with Vickers and Knoop indentations is the same in soda-lime glass. Thus, equivalent indentation loads yield similar radial crack lengths and strengths for both indenters. Also, fracture toughness values as measured from Vickers and Knoop indentations are similar; however, toughness values measured by the strength method are significantly greater than that measured from as-indented crack sizes. It is thought that this higher strength-toughness value is related to the calibration procedure used in developing this test. Annealing and aging Vickers and Knoop indentations increases strength through residual stress relief. Whereas annealing removes residual stress through relaxation phenomena, aging relieves residual stress through lateral crack growth, making the system more compliant.

ACKNOWLEDGEMENTS

This research was supported by NSF Grant No. DMR-8311699.

REFERENCES

1. G. R. Anstis, P. Chantikul, B. R. Lawn, D. B. Marshall, J. Am. Ceram. Soc., 64, 533 (1981).
2. P. Chantikul, G. R. Anstis, B. R. Lawn, D. B. Marshall, J. Am. Ceram. Soc., 64, 539 (1981).
3. J. J. Petrovic and M. G. Mendiratta, J. Am. Ceram. Soc., 59, 163 (1976).
4. S. W. Freiman, A. C. Gonzales, and J. J. Mecholsky, J. Am. Ceram. Soc., 62, 206 (1979).
5. D. B. Marshall, J. Am. Ceram. Soc., 67, 110 (1984).
6. D. B. Marshall, J. Am. Ceram. Soc., 66, 127 (1983).
7. J. J. Petrovic, J. Am. Ceram. Soc., 66, 277 (1983).
8. F. F. Vitman and V. P. Pukh, Zavodskaya Laboritoriya, 29, 863 (1963).
9. D. B. Marshall, J. Am. Ceram. Soc., 67, 57 (1984).
10. R. H. Marion, Fracture Mechanics Applied to Brittle Materials, 103 (1979).
11. R. F. Cook and B. R. Lawn, J. Am. Ceram. Soc., 66, C-200 (1983).
12. B. R. Lawn, K. Jakus, and A. C. Gonzales, J. Am. Ceram. Soc., 68, 25 (1985).
13. D. B. Marshall and B. R. Lawn, J. Am. Ceram. Soc., 63, 532 (1980).
14. D. B. Marshall and B. R. Lawn, J. Mater. Sci., 10, 113 (1975).
15. J. C. Newman, Jr., and I. S. Raju, Engng. Fracture Mech., 15, 185 (1981).

CERAMIC-TO-METAL BONDING FROM A FRACTURE MECHANICS PERSPECTIVE

Pierre O. Charreyron, Noel J. Bylina and James G. Hannoosh

Norton Company
Worcester, MA 01606-2698

ABSTRACT

Preliminary experiments on SiC and Mo beams illustrate the typical fracture mode of ceramic/metal attachments. A finite element model of the experiment confirms the presence of a stress singularity at the bond edge. A fracture mechanics approach is proposed to predict joint feasibility and strength. The approach is illustrated on two simple geometries: plane strain rectangular sections bonded along one edge and butt-to-butt axisymmetric cylinders. A parametric study of these examples provides design guidelines on adherend geometry and elastic properties. The results can also be presented under the form of compatibility maps.

INTRODUCTION

The development of reliable ceramic/metal or ceramic/ceramic joints is critical for various applications: structural ceramics in heat engines, biomedical prosthesis and electronic components.

Two aspects need to be simultaneously considered:

The adhesion problem - Structural ceramic materials such as SiC, Si_3N_4 or ZrO_2 show little chemical affinity to each other or to metals. In addition, poor wettability prohibits using braze alloys developed for high temperature metals.

The mismatch problem - In the vicinity of the joint, the elastic properties: Young's modulus, Poisson's ratio and thermal expansion coefficient change very abruptly. Thermal expansion mismatch is most critical for heat engine components as they must withstand very severe temperature cycles (up to 1500°C amplitude). The discontinuity results in very high stresses at some locations of the interface.

Historically, most of the research effort has been devoted to the adhesion aspect. Twentyman [1] and Tallman et al.[2] review the current state of the art. Progress on reducing the mismatch aspect, on the other hand, seem to stop at the recognition of the potential of interlayers and transition pieces.[3, 4, 5] Little attention, however, has been given to other ways of reducing the stresses.

The primary purpose of this work is to model the mechanics of ceramic/metal attachments and to derive generic design guidelines.

PRELIMINARY EXPERIMENTS

Thin beams of silicon carbide and molybdenum were brazed at 1000°C. Unfortunately, none of the samples survived cooldown to room temperature. In all of the samples, fracture originated at one of the beam corners (Figure 1). This crack propagated inside the ceramic following a path parallel to the bond surface. This suggests that the real cause of fracture is not insufficient adhesion of the braze but excessive stresses in the silicon carbide.

The respective thermal expansion coefficients of Mo and SiC are 5.5 and $4.3 \times 10^{-6}\ ^{o}C^{-1}$. The attachment was successfully achieved by adding a beam of Tungsten with intermediate expansion of $4.7 \times 10^{-6}\ ^{o}C^{-1}$.

THE STRESSES IN A CERAMIC/METAL BEAM GEOMETRY

This reproducible failure mode cannot be accounted for in view of previous mechanical models found in the literature [6, 7]. Most of these models are variations of Timoshenko's analysis of bimetallic thermostats.[8] This theory suggests failure should be normal to the beam axis and originate on the ceramic free surface where bending stresses are maximum. But, as Timoshenko himself points out in the original reference, it does not apply to the end regions of the beams, where the stress picture is more complex.

Fig. 1 Ceramic/Metal Beam Attachment
Left: A-Sic/W/Mo B-Sic/Mo (failed)
Right: Detail of Fracture origin.

The only way of computing these stresses is to solve the general equations of elasticity. Unfortunately, analytical solution becomes fairly cumbersome as soon as more than one material is involved, and solutions are restricted to a few simple geometries.[9, 10, 11, 12] Erdogan[11] showed in particular that elastic mismatch can often result in a stress singularity (theoretically infinite stresses) at the bond corner.

Finite element analysis provides an easier way to look at the stress in the end regions, provided mesh resolution is sufficient. The preliminary experiment was modeled within the plane stress assumptions. Figure 2 shows a distorted contour plot of the maximum principal stresses. The darker regions correspond to locations of higher tensile stresses. In the center region, stress contours are parallel to the beams axes and are in perfect agreement with Timoshenko's analytical results. But the maximum tensile stresses are located at the bond corner within the ceramic. Due to the singular effect, the maximum stress computed there depends on the mesh resolution and is of little quantitative significance.

This stress pattern is in excellent agreement with the observed fracture mode in the SiC/Mo samples.

THE FRACTURE MECHANICS APPROACH

Griffith's theory of fracture states the conditions for a crack to propagate in a homogeneous solid. It can be readily generalized to the extension of a debond between adherends. The criteria for adhesive fracture remains:

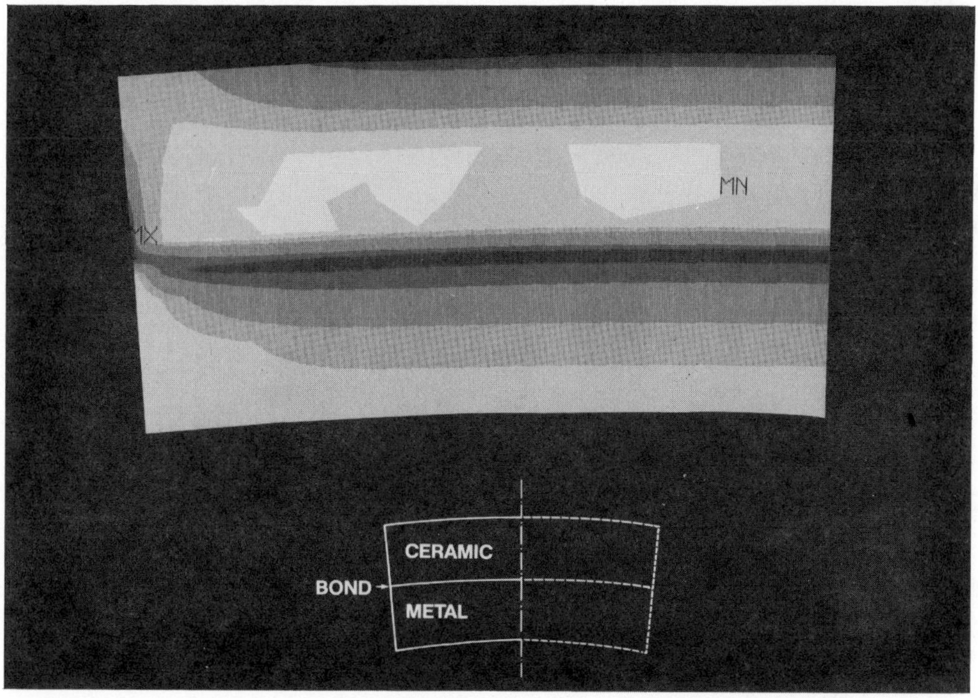

Fig. 2 Maximum Principal Stress Contours in Ceramic/Metal Beam Attachment (distorted plot).

$$G = \frac{\partial(F-U)}{\partial a}$$

with F work performed by external forces

U elastic strain energy

a fracture area

G_c critical energy release rate

For a more comprehensive treatment of adhesive fracture mechanics, the reader is referred to Anderson et al.[13]

The critical energy release rate is the physical property characterizing the adhesion or toughness of a specific bond. It is homogeneous to a surface energy. When a joint involves a brittle adherend, there is a possibility that the bond might be tougher than the adherend itself. In that case, the crack propagates within the ceramic as was observed in our experiments. A practical consequence is that the ceramic toughness is an upper limit of the desirable bond adhesion. There is very little data available on critical energy release rates of ceramic/metal or ceramic/ceramic bonds. With only slight modifications, conventional fracture toughness tests (DCB, short bar) are suitable to bond toughness measurements.[14] But if residual stresses due to differential thermal contraction are significant, the data should not be simply reduced with the standard stiffness calibration curves. A few examples of adhesion values found in the literature are given in Table I.

TABLE I ADHESIVE ENERGY RELEASE RATES OF CERAMIC METAL BONDS

Substrate/Coating	$G_{I,C}$ ($J.m^{-2}$)	Test Method	Reference
PSZ/Ag-37Cu-7Ti	120-170	DCB	(15)
Al_2O_3/Brazing Filler Metals	40-120	DCB	(15)
SiC/Brazing Filler	60	DCB	(15)
Al_2O_3/Pt-Au	2-4	DCB	(16)
Al_2O_3/Nb	9	SENB	(17)
Si_3N_4/Zr	12	SENB	(17)
Ti/Al_2O_3*	2.5-5	SENB	(18)
Stainless Steel/Al_2O_3*	2.5-5	SENB	(18)
Steel/Al_2O_3*	10-15	DCB	(19)

* Plasma sprayed

DCB = double cantilever beam
SENB = single edge notched beam

The energy release rate, G, depends directly on the stress distribution around the attachment. Therefore, it can be computed as a function of joint geometry, adherends mechanical properties and thermal or mechanical loads. Analytical calculation is generally not possible, but the crack closure (or virtual crack) method offers a convenient numerical alternative.[20, 21] The method is basically a numerical implementation of Irwin's[22] approach. The energy released when the crack length increases by a small increment, Δa, is equal to the external work it would take to bring the crack back to its original position. Figure 3 is a schematic of the finite element mesh near the crack tip: the bond is modeled by a very thin row of elements. Incremental debonding is simulated by removing bond elements one after the other. The energy potentially released is calculated for each new incremental crack advance. If it exceeds the critical value G_c, the crack will spontaneously progress. Otherwise it will stop. Pre-existing cracks are inherent to any ceramic/metal joint. They could be microstructural defects in the ceramic phase or small areas of loose adhesion if, for example, the surfaces were not perfectly clean before bonding. It is finally the size of these inherent flaws that determines whether the bond holds or not.

The method is now illustrated on two simple examples:

EXAMPLE 1 - <u>Plain Strain Rectangular Sections Bonded along their edge</u>

Let us consider two plane strain rectangles of zirconia and nodular cast iron bonded along their longest edge. The model dimensions and material properties used for the analysis are summarized in Table II. An ideal bond of zero thickness is assumed. The bonding temperature is 1000°C and the sections are freely supported during cooldown to room temperature.

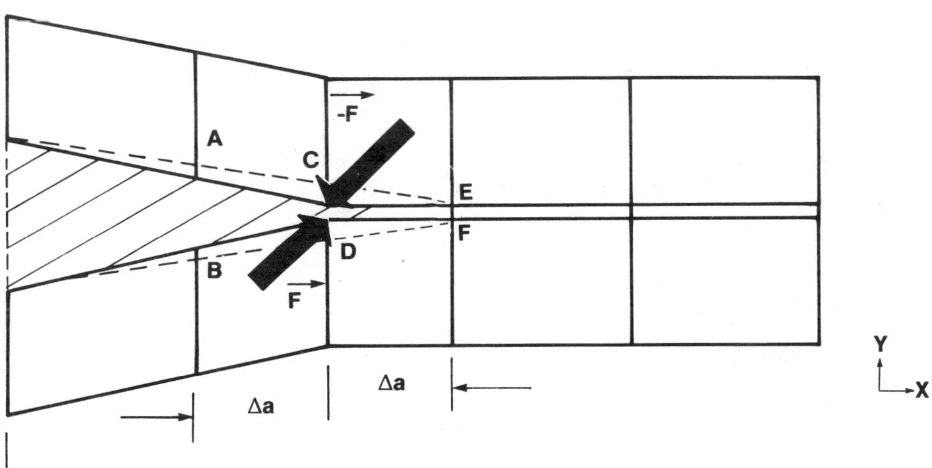

$$G_I = \frac{F_y (U_{y,D} - U_{y,c})}{2\Delta a}$$

$$G_{II} = \frac{F_x (U_{x,D} - U_{x,c})}{2\Delta a}$$

Fig. 3 Calculation of Strain Energy Release Rate by the Crack Closure Method.

TABLE II REFERENCE MODEL DATA

		ZIRCONIA	NODULAR CAST IRON
E	Young's Modulus (GPa)	137.9	168.0
υ	Poisson's Ratio	.26	.31
α	Average Thermal Expansion Coefficient $\times 10^{-6}$ $(^{o}C)^{-1}$ between $20^{o}C$ and $1000^{o}C$	10.8	13.7
H_i	Height (mm)	12.5	12.5
R or L/2	Half-length or radius (mm)	50	50

Bonding temperature : $T = 1000^{o}C$
Operating temperature : $T = 20^{o}C$

Using the crack closure method, we can compute the variations of energy release rate, G, when a debond propagates along the bond line starting from one edge. A fairly coarse mesh with debond increments of approximately 1mm gives the global profile over the total bond length. As shown in Figure 4a, G increases up to a debond length of approximately 5mm and stays constant from there on.

Increased mesh resolution is required for the initial part of the profile, which corresponds to crack lengths comparable to inherent debonds. The curve obtained for a refined increment of 0.125mm is a smooth continuation of the coarse mesh profile (Figure 4b).

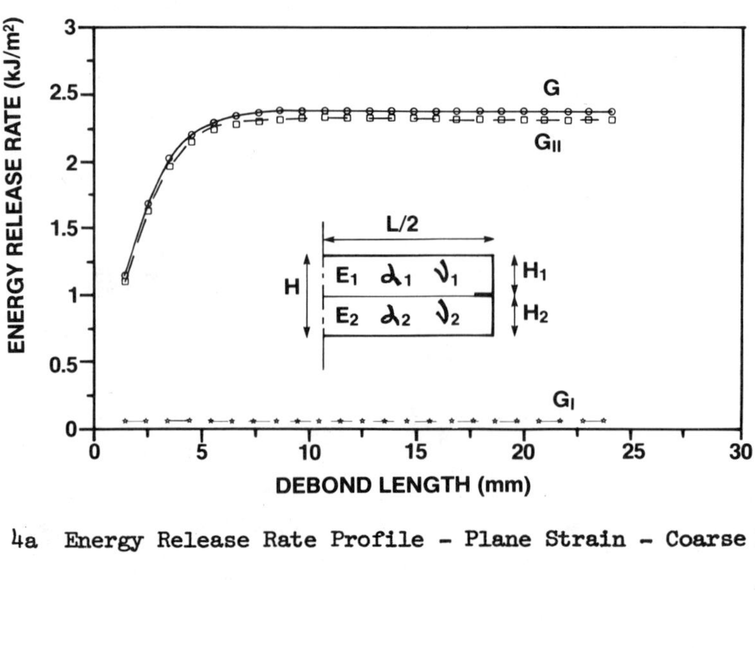

Fig. 4a Energy Release Rate Profile - Plane Strain - Coarse Mesh.

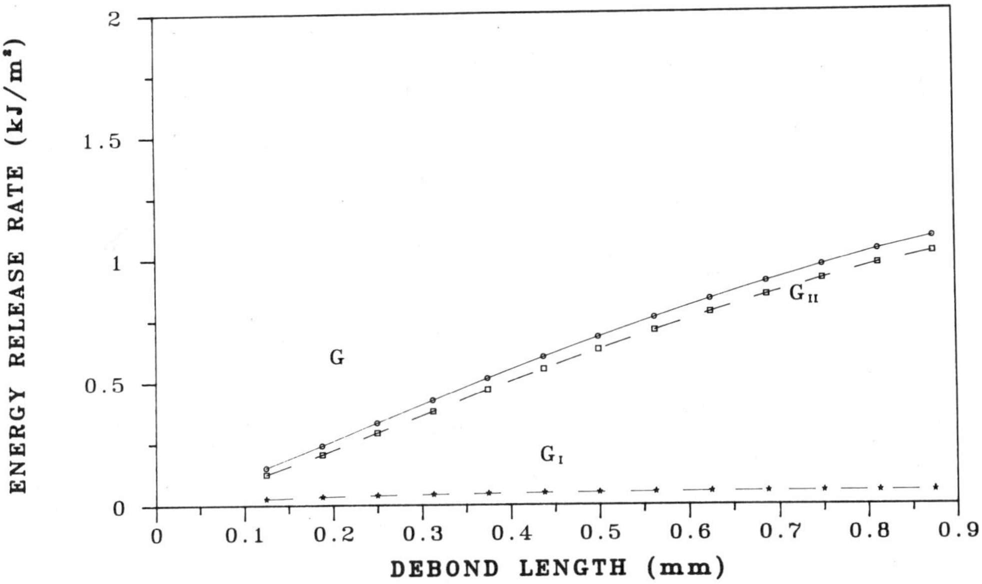

Fig. 4b Energy Release Rate Profile - Plane Strain - Refined Mesh.

Also plotted on Figure 4 are the respective contributions of mode I and II to fracture. As could be expected, shear is the dominant mode when the loads result from thermal expansion mismatch only.

The shape of the energy release rate profile can be interpreted by comparing the energy distribution in the sections at various crack propagation steps. In-plane strain energy density contours corresponding to respective debond lengths of 0 and 20mm are plotted on Figure 5a and 5b. The very dense contours around the debond tip mark the energy concentration due to singular stresses. No absolute value should be attached to the peak values as they certainly depend on the mesh resolution. They, however, indicate a sharpening of the singularity as the cracks moves in. This is to be related to the rising part of the profile on Figure 4a. At some point, the contour pattern becomes fully established and just translates along the bond line. This steady regime corresponds in turn to the plateau observed in Figure 4. Note also that the parallel strain energy contours in the far-field concur with Timoshenko's analytical predictions.

EXAMPLE 2 - <u>Butt-to-Butt Axisymmetric Cylinders</u>

With minor modifications to the finite element input data, the plane strain sections can be converted to axisymmetric butt-to-butt cylinders. The crack is assumed circumferential and originating at the bond edge. This assumption, although physically questionable, allows the retention of the advantage of symmetry, and keeps the cost of computations down.

The shape of the energy release rate profile (Figure 6) is similar to that of the plane strain case, except for the following differences:

- A slowly rising energy release rate replaces the previously constant plateau characterizing the steady propagation regime. This is because the rate is inversely proportionial to the decreasing area increment.

- As the crack tip nears the cylinder centerline, there is virtually no more energy to be released and therefore the rate falls back to zero.

- Overall rate values are slightly higher than in the plane strain case. The difference accounts for the relaxation of hoop thermal stresses.

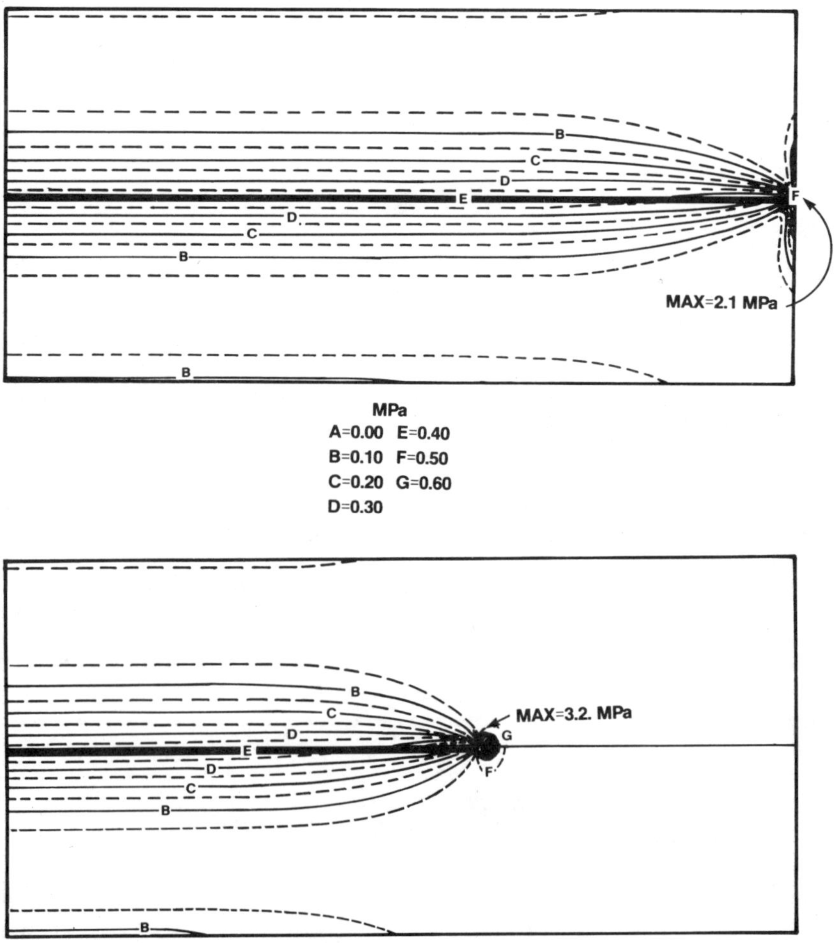

Fig. 5 In-plane strain energy density contours - Plane strain
a) prior to fracture (Top)
b) for a debond length of 20mm (Bottom).

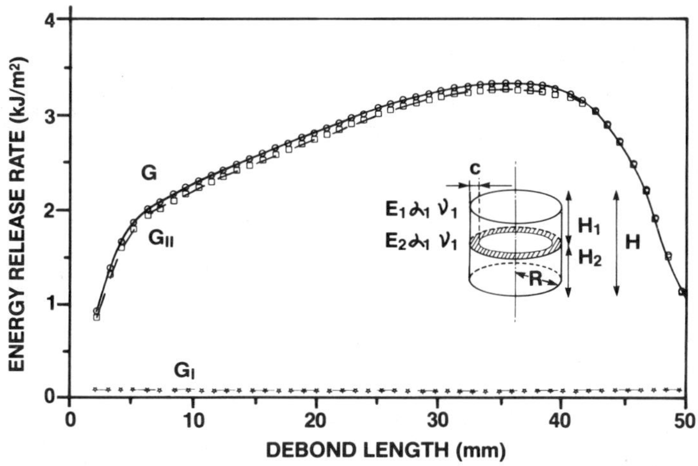

Fig. 6 Energy Release Rate Profile - Axisymmetric.

PARAMETRIC STUDIES

The purpose of this section is to investigate how adherends dimensions and elastic properties affect the energy release rate in our simple examples. This should ultimately provide guidelines for the design of more complicated joints. Throughout this parametric analysis, dimensions or material properties, when not explicitely specified are those of Table II. Physically, the most meaningful energy release rate value is that which corresponds to the worse inherent bond defect at the joint corner. However, the defect size depends strongly on processing parameters such as surface preparation. It is really a characteristic of the bonding process, on the same level as the critical energy release rate, G_c. Therefore, the maximum value of the energy release rate profile, G_{MAX} is preferred as a comparison basis.

Material Properties Effect

G_{MAX} values for a few material pairs of industrial interest are compiled in Table III. These values are quite high compared to cohesive toughness of available ceramic materials.

The most significant factor is thermal expansion mismatch $\Delta\alpha$. Figure 7 shows G_{MAX} varying like $(\Delta\alpha)^2$, independently of the mean thermal expansion α_{avg}, as long as the adherends are free of constraints.

TABLE III EFFECTS OF ADHERENDS MATERIAL SYSTEM RATE ON MAXIMUM ENERGY RELEASE

Material System	α_1/α_2	E_1/E_2	G_{max} (kJ/m^2) Plane Strain	Axisymmetric
W/SiC	1.14	.85	.32	0.41
W/Si_3N_4	1.38	.19	.85	1.79
NCI/ZrO_2	1.27	1.23	1.23	2.24
IN903/SiC	2.08	.36	8.31	13.20
446SS/SiC	2.92	.48	27.83	50.00
446SS/Si_3N_4	3.50	.64	29.40	57.70
IN738/SiC	3.08	.44	32.20	---
IN738/Si_3N_4	3.70	.60	33.78	---
NCI/Si_3N_4	3.80	.56	36.05	---
304SS/Si_3N_4	5.5	.64	98.00	---
5456Al/Si_3N_4	7.2	.23	127.75	---

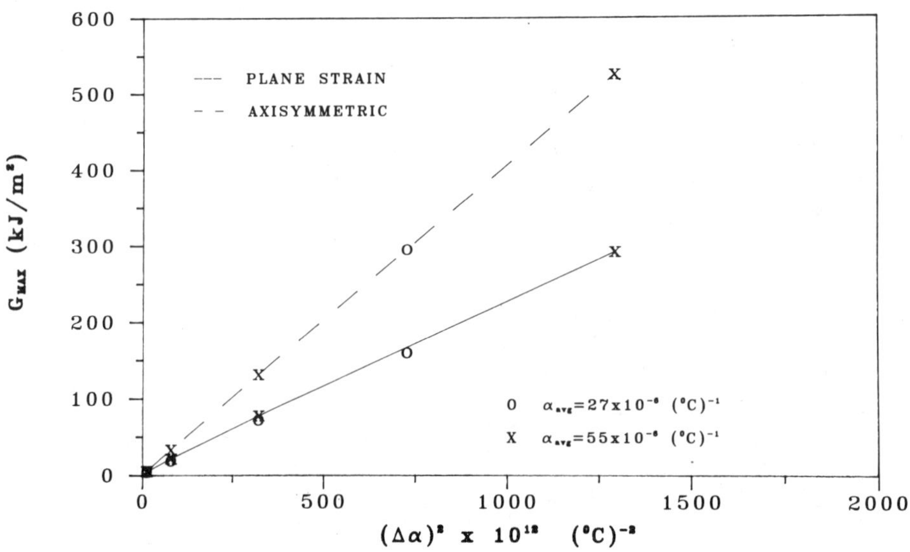

Fig. 7 Thermal Expansion Mismatch Effect.

TABLE IV Poisson is Ratio Effect (Plane Strain)

υ_1/υ_2	G (kJ/m^2)
0.28/0.28	1.75
0.26/0.31	2.24
0.24/0.49	7.35

The stiffness effect is illustrated in Figure 8: G_{MAX} is roughly proportional to the mean Young's modulus, but decreases as the stiffness ratio departs from unity.

Finally, metals and ceramics also differ by their Poisson's ratio. Table IV shows that this mismatch should not be underestimated.

Geometry Effect

The effect of adherends height has several facets. Figure 9 shows the energy release rate profile when the adherend heights are varied, but kept equal. The bulk value of G (i.e., far from the end) is practically proportional to the total height, H. A simple explanation is that the amount of energy released by a crack increment is proportional to the volume of thermally stressed material. But simultaneously, an increase in height also accentuates the end perturbations to the point where the steady regime disappears for aspect ratios (2H/L)larger than 1.

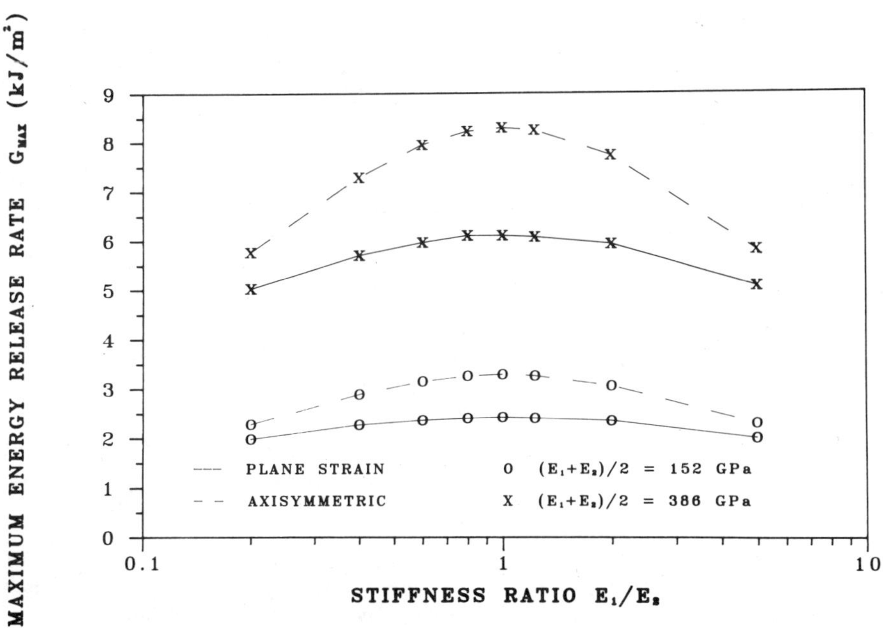

Fig. 8 Stiffness Effect.

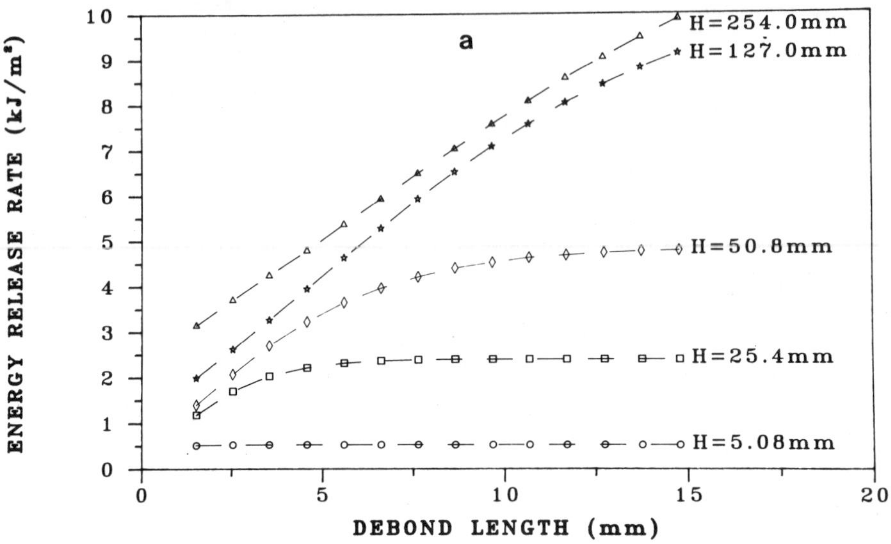

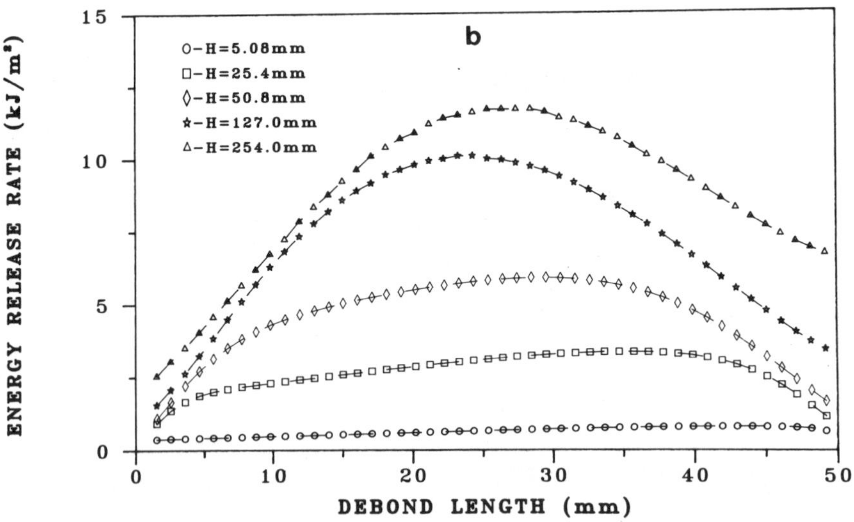

Fig. 9 Total Height Effect
a) Plane Strain (Top) b) Axisymmetric (Bottom).

End effects are not sensitive to variations of the relative height as long as the total height stays the same (Figure 10), but on the other hand, the general energy level is increasingly reduced as the height ratio departs from unity. A practical consequence is that coatings should be less vulnerable to debonding than massive adherends.

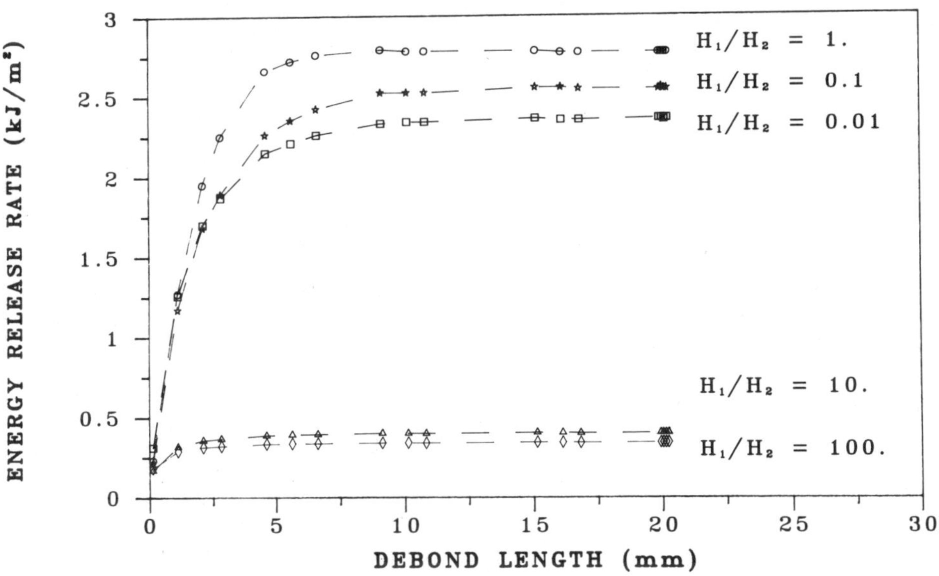

Fig. 10 Relative Height Effect - Plane Strain.

Finally, many of the results of the parametric studies could have been predicted qualitatively by considering the approximate model of two parallel trusses of stiffness E_iH_i attached to each other. This amounts to neglecting bending and end effects. One can then show that the strain energy released per unit length of trusses is:

$$G_{MAX} = \frac{E_1 H_1 + E_2 H_2}{2} \quad \frac{n}{(n+1)^2} (\Delta\alpha \; \Delta T)^2$$

where $n = \dfrac{E_1 H_1}{E_2 H_2}$

ADHEREND COMPATIBILITY ASSESSMENT

A practical application of the fracture criterion is to predict the range of materials that can be joined to any given ceramic material, for instance, Zirconia. In principle, this would require information about the bond critical energy release rate, G_c, and the largest inherent debond.

The special case when the bond is tougher than one of the adherends (e.g., the SiC beam in our experiments) is not uncommon. The fracture criteria should then be expressed in terms of the cohesive energy release rate and inherent flaw size of the most brittle constituent. In practice, the assumption of cohesive fracture provides the absolute compatibility limit of an adherend system, independently of the joining technique efficiency.

The criteria can be presented in the form of a compatibility map where the thermal expansion coefficient and Young's modulus of the candidate adherend are the x and y coordinates. Figure 11 pictures such a chart for Zirconia in the reference geometry of Table II. The hacked region maps the domain of non-compatible materials. It includes several important industrial materials such as 304 Stainless Steel, some Aluminum alloys and Silicon Nitride. The compatibility domain would be even more reduced if the bond itself was not as tough as the ceramic. The dashed line delimits the domain for a bond adhesion G_c of 50 J/m^2 and inherent debond length of 125μm. Note, however, that the boundaries can be displaced by changing the geometry of the adherends or cooling temperature differential, ΔT.

DISCUSSION

For simplicity, the adherends were assumed perfectly elastic. Local yielding at the crack tip is likely to happen but, as in cohesive fracture mechanics, the local plastic dissipation can be lumped into the elastic adhesion value G_c. In addition, some bulk plastic flow is not impossible if the metal is soft and the ceramic tough. But this event is not desirable in structural attachments undergoing thermal cycling.

In this study, a perfect bond with no thickness was assumed. This is of course, an idealization. It was shown in other instances[23] that the bond thickness affects adhesion. A third material layer could be easily included in this approach. Ductility of the bond layer is another aspect to consider.

In the two examples, the only loads were those induced by differential thermal strains. One could as well add any constraints, external actions

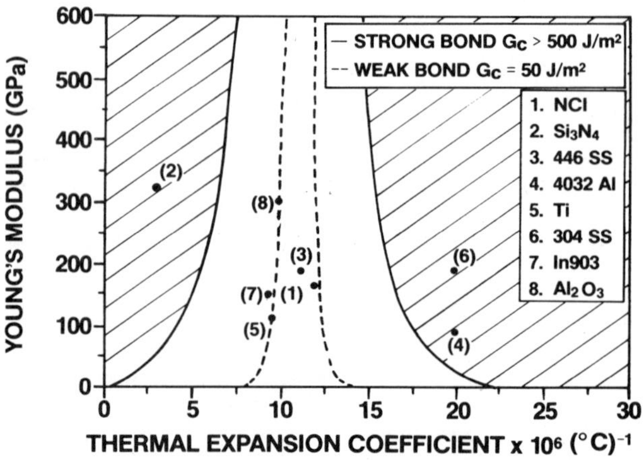

Fig. 11 Compatibility Map for ZrO_2 (bonded at 1000°C.)

OBSERVED CHANGES IN FRACTURE STRENGTH FOLLOWING LASER IRRADIATION AND ION BEAM MIXING OF Ni OVERLAYERS ON SINTERED ALPHA-SiC

K. L. More and R. F. Davis

Department of Materials Engineering, Box 7907
North Carolina State University
Raleigh, NC 27695-7907

ABSTRACT

Pulsed laser annealing and ion beam mixing are currently being investigated as surface modification techniques in an attempt to enhance the mechanical properties of polycrystalline α-SiC. Thin Ni overlayers (20nm - 100nm) are evaporated onto the SiC surface. The specimens are subsequently irradiated with pulses of a ruby or excimer (KrF or XeCl) laser or bombarded with high energy Si^+ *or* Xe^+ *ions. Following laser irradiation, the fracture strength of the SiC is increased by as much as ≈50%, but after ion beam mixing, no strength increase is observed. Weibull statistics are used to characterize the SiC fracture strength data before and after surface modification. Cross-sectional transmission electron microscopy (X-TEM), scanning electron microscopy (SEM), secondary ion mass spectroscopy (SIMS), and Rutherford backscattering (RBS) techniques are used to characterize the modified SiC in order to determine the cause for the increase in fracture strength and the extent of mixing as a result of the surface modification.*

INTRODUCTION

Failure in ceramic materials occurs due to the propagation of a crack or critical flaw existing at or near the tensile face of the specimen. As ceramic materials continue to be substituted for metals in structural uses, an understanding of the strength controlling flaws and microstructure is necessary for the design of ceramic parts. Alpha-SiC is one of the primary candidate materials for high strength and high temperature applications. However, during processing and subsequent machining, surface-related defects are introduced which limit the overall strength of this ceramic.

Surface modification techniques are presently being investigated in an attempt to enhance the mechanical properties of the α-SiC. The strength can be increased in two ways: (1) a change in the size and/or position of the critical flaw or (2) the introduction of a residual compressive stress at and/or near the surface. Since the two surface processing methods, pulsed laser irradiation and ion beam mixing, are both non-equilibrium processing techniques, materials interactions can be induced at the surface which are not possible with conventional thermal techniques. Thus, conditions (1) or (2) noted above can be created as a result of this surface modification and the strength of the SiC can be increased

If a substrate of material A is coated with a thin layer of material B and subsequently irradiated with laser pulses, depending on the properties of B and at sufficient energy densities, the energy of the laser is absorbed by the B electrons and rapidly converted into heat. As this heat is rapidly transferred to the substrate, A and B can mix via liquid diffusion. If the melting threshold is not exceeded, the two may mix as a result of solid state mass transport. Laser irradiation involves extremely rapid heating and cooling rates ($\approx 10^9$K/s). As such, unique surface microstructures can be created and then preserved, thus altering the physical properties of the α-SiC.

In contrast, ion beam mixing involves the acceleration of high energy ions in a potential to a degree whereby they are implanted into the sample. Under the correct conditions, the implantation will cause mixing of a coating and its substrate.

The techniques of ion beam mixing and laser irradiation are relatively new processing methods, but each has been extensively reviewed.[1-4] These techniques have most recently been applied to metal-semiconductor[5-7] and metal-insulator[8,9] systems, but investigations have also been made into the surface modification of β-SiC thin films grown in the authors' laboratory[10-12] and α-SiC.[13,14]

EXPERIMENTAL PROCEDURE

The material under investigation is an ≈98% dense α-SiC[+] produced by pressureless sintering of a mixture of powders of submicron SiC and densification aids containing B (≈0.4 wt%) and C (≈0.5 wt%) at a temperature of ≈2373K. The sintered SiC plates are subsequently cut into standard flexure test bars (0.64 cm X 0.32 cm X 5.08 cm) and the surfaces sequentially ground using 100, 400, and 600 grit diamond wheels. The edges of each bar are bevelled at 45° so that edge cracks and flaws are no longer critical for failure. The SiC surfaces are sputter cleaned in a vacuum of approximately 10^{-7} Torr followed by the evaporation of thin Ni overlayers (20 nm - 100 nm). The maximum specimen temperature reached during cleaning and Ni deposition does not exceed 323K. Each side of a bar is coated separately; the other sides are masked by glass slides to prevent additional deposition.

Ion beam mixing is achieved using either 350 keV Xe^+ ions at a dose of 2.5 X 10^{16} Xe^+/cm^2 or 140 keV Si^+ at a dose of 1 X 10^{17} Si^+/cm^2. The ion beam is ≈2.54 cm in diameter; thus, to ion beam mix the entire specimen, there must be considerable overlap of the beam spots. Laser irradiation is conducted using one of three lasers: a pulsed ruby laser (λ = 693 nm, τ = 25 ns, nominal pulse energy density = 1.2 J/cm^2), a KrF excimer laser (λ = 248 nm, τ = 45 ns, pulse energy density = 1.85 J/cm^2 and 3.0 J/cm^2), and a XeCl excimer laser (λ = 308 nm, τ = 30 ns, pulse energy density = 1.85 J/cm^2 and 3.0 J/cm^2). The surface of each specimen is sequentially irradiated with laser pulses until all four sides have been completely annealed (the size of the laser beam is dictated by the energy density used).

Each modified SiC bar is broken in 4-point flexure to determine the fracture strength and compared to the fracture strength values similarly obtained for the as-received SiC. Inner and outer span lengths of 1.27 cm and 2.54 cm, respectively, and a crosshead speed of 8.5 X 10^{-4} cm/s are used for the fracture research. A slow crosshead speed may be used, since slow crack growth has been found to be negligible at room temperature for

[+] HEXOLOY®, SOHIO Chemicals and Industrial Products Co., Niagara Falls, NY sintered α-SiC.[15] These fracture data are further analyzed using Weibull statistics.[16]

The surfaces of the laser processed and ion beam mixed SiC are examined using X-TEM samples prepared from small test bar sections by mechanical thinning to ≈75 μm, dimpling, and subsequent low energy (5-6 kV), low angle (12°) ion beam thinning using Ar ions. The TEM results are compared to the chemical analyses derived from SIMS (CAMECA IMS-3f ion microprobe) depth profiles and RBS. The SIMS profiles are obtained using an ≈50 μm O^+ primary beam at a beam current of ≈1700 namp and an accelerating voltage of ≈14.8 kV. A 1 MeV He^+ beam is used for the RBS analysis. Critical flaw identification and fracture surface analysis of the as-received, laser-treated, and ion beam mixed SiC is conducted via SEM.

RESULTS AND DISCUSSION

Microstructural Analysis

The microstructure of the as-received α-SiC is shown in the polished section of Figure 1(a) and consists of fairly equiaxed α-SiC grains having an average grain size of ≈8.5 μm, smaller and darker B_4C grains, and particles of free C within the SiC matrix. Fracture in SiC is transgranular, as shown in Figure 1(b).

An ≈100 nm, as-deposited Ni overlayer on the SiC is shown in Figure 2. From the X-TEM results, the Ni coating is found to be fairly uniform in thickness across the SiC surface, but is non-conformal around rough surface areas.

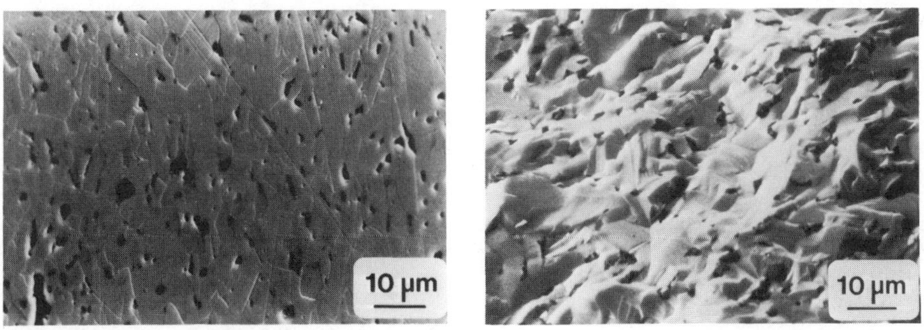

(a) (b)

Figure 1. Microstructure of sintered α-SiC: (a) a polished section and (b) a fracture surface.

For the ion beam mixing experiments, the SiC was coated with 50 nm of Ni. The bars were initially bombarded with high energy Xe^+ ions, but to prevent bubble formation at the interface and for the compatibility of using a self-ion, the bars were later ion beam mixed using Si^+ ions. Cross-sectional TEM was used to examine the surface of the SiC following ion beam mixing. The resulting surface structures are shown in Figure 3. As noted on the micrographs, the surface after ion beam mixing using either Xe^+ ions (Figure 3(a)) or Si^+ ions (Figure 3(b)) consists of three distinct regions: (a) a residual polycrystalline Ni layer, (b) an amorphous mixed region (as determined by electron diffraction), and (c) the SiC substrate. The stoichiometric ratio of Ni/Si in the amorphous mixed regions (this contains no phase information, but is strictly a stoichiometric ratio), as determined from RBS, approach Ni_2Si and Ni_3Si_2 for the Xe^+ and Si^+ bombarded specimens, respectively. Actually, from SIMS depth

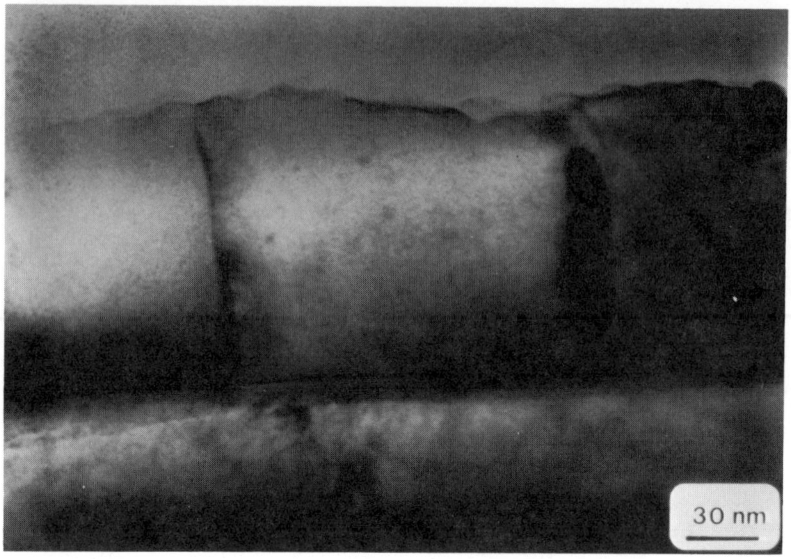

Figure 2. 100 nm as-deposited Ni overlayer on SiC.

profiles, as shown in Figure 4, the Ni concentration does not remain constant throughout the mixed layer, but changes from high at the surface (where the Ni remained polycrystalline) to virtually zero at the mixed layer - SiC interface.

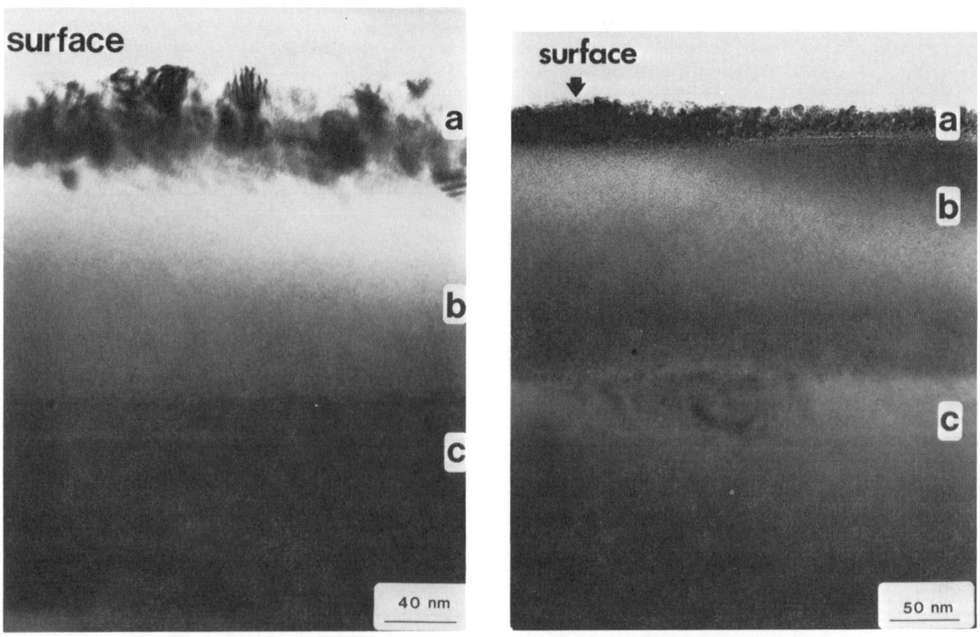

(a) (b)

Figure 3. Cross-sectional TEM micrographs of ion beam mixed Ni-coated SiC using (a) Xe^+ ions and (b) Si^+ ions.

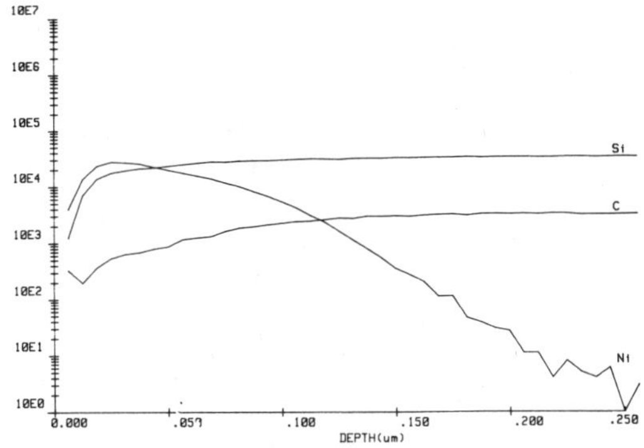

(a)

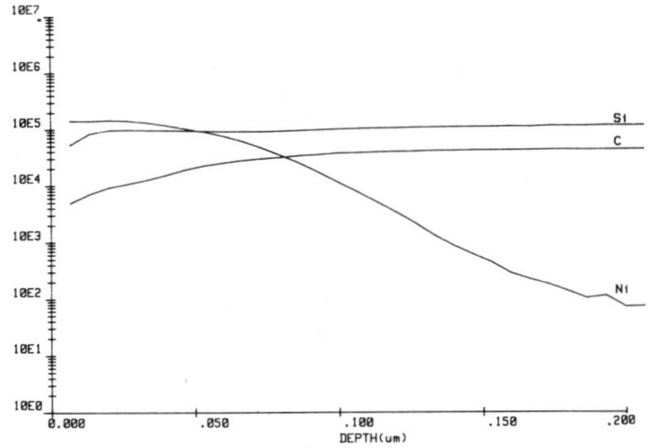

(b)

Figure 4. SIMS depth profiles for ion beam mixed, Ni-coated SiC using (a) Xe^+ ions and (b) Si^+ ions.

Prior to laser annealing the Ni-coated SiC, laser "glazing" experiments were conducted on as-received, uncoated SiC using the KrF excimer laser at a pulse energy density of 3.0 J/cm^2. The resulting microstructure following irradiation is shown in Figure 5. The surface of the SiC becomes extremely faulted and plastically strained, as is shown by the electron diffraction pattern. It appears that the surface of the SiC melts; i.e. the energy density of the laser is sufficient to exceed the melting threshold of the SiC. Thermodynamically, one would expect the SiC to sublime at atmospheric pressure, but due to the extremely rapid heating and cooling rates, it is possible that the SiC melts at the surface. As heat is transferred from the surface to the underlying crystal, microtwins form to compensate for the stresses produced by the high thermal gradients and are "quenched" in during cooling.

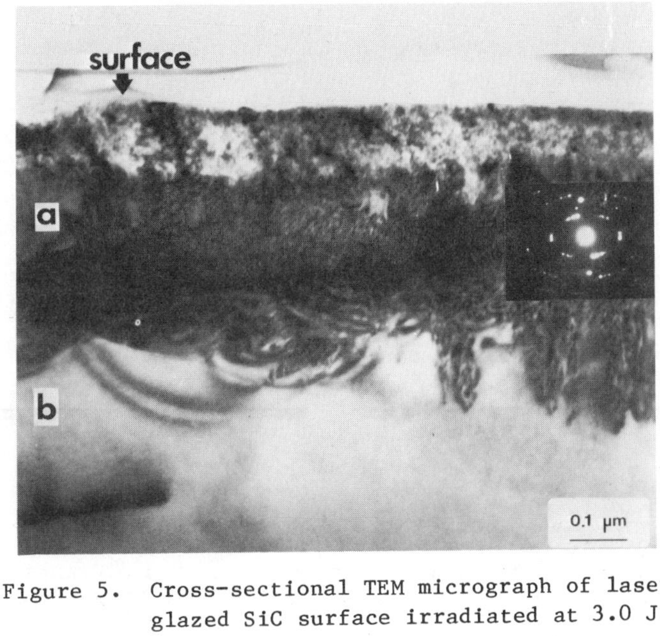

Figure 5. Cross-sectional TEM micrograph of laser glazed SiC surface irradiated at 3.0 J/cm^2.

If the SiC is coated with 100 nm of Ni and subsequently irradiated with either excimer laser at a pulse energy density of 1.85 J/cm^2, the resulting microstructures appear as shown in Figure 6. The surface structures of each laser annealed specimen are very similar; both consist of a Ni-SiC mixed surface region ≈150 nm in thickness and a heavily faulted SiC substrate. The energy absorbed during irradiation was sufficient to cause complete melting of the Ni overlayer. Although the SiC did not melt, the temperatures reached during irradiation were high enough to cause interdiffusion of this material into Ni (the solubility of SiC in molten Ni is moderately high). This mixing effect can occur at as low a temperature as 1073K, as determined from annealing the Ni-coated SiC in Ar and subsequently examining the specimen surface with RBS. The SiC substrate is highly faulted and resembles the structure found in the sub-surface, microtwinned regions of the laser glazed SiC. The SIMS depth profiles of Figure 7 show that the ion yield of Ni in the mixed region is slightly greater than either Si or C; at the interface, the Ni concentration falls to zero, indicating that there is virtually no interdiffusion of Ni into the SiC. Thus, limited interdiffusion does occur to an extent whereby the bonding between the Ni and SiC substrate is enhanced. At higher pulse energy densities, ≈3.0 J/cm^2, the resulting surface microstructures are very similar to those obtained at the lower energy densities except that there is more interdiffusion of SiC into the Ni overlayer.

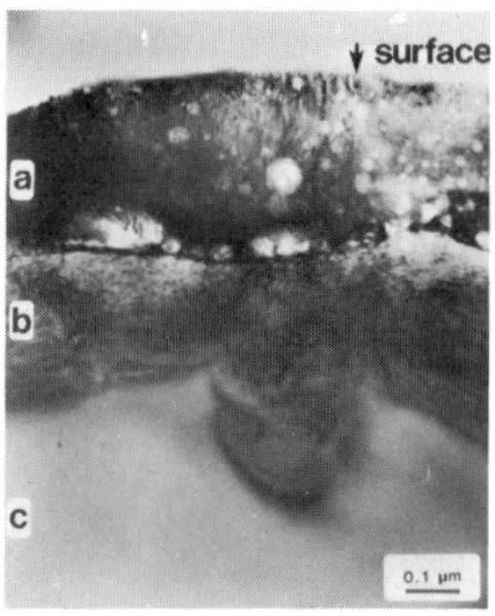

(a)

(b)

Figure 6. Cross-sectional TEM micrographs of 100 nm Ni-coated SiC irradiated at a pulse energy density of 1.85 J/cm^2 using (a) a XeCl laser and (b) a KrF laser.

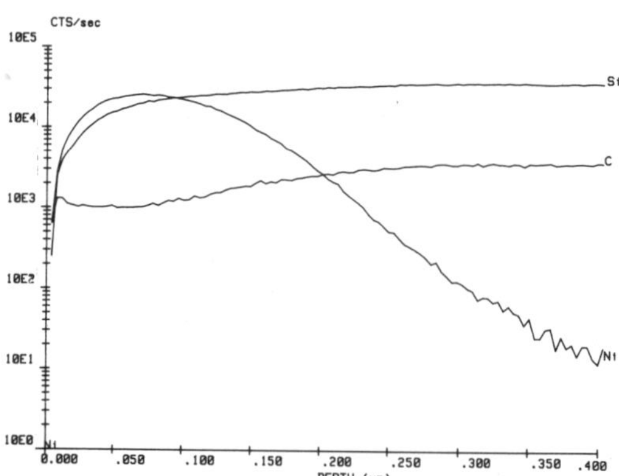

(a)

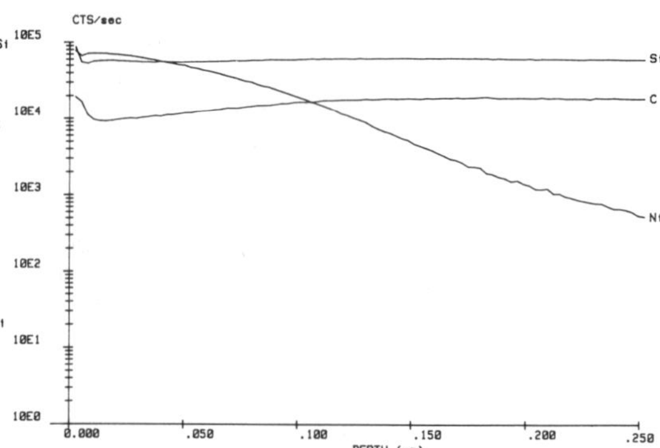

(b)

Figure 7. SIMS depth profiles for 100 nm Ni coated SiC irradiated at a pulse energy density of 1.85 J/cm^2 using (a) a XeCl laser and (b) a KrF laser.

SiC specimens coated with a 100 nm Ni layer were also irradiated with pulses of the longer wavelength ruby laser at a nominal pulse energy density of 1.2 J/cm^2. The resulting microstructures are shown in Figure 8. The Ni and the SiC do not mix at the interface following pulsed ruby laser annealing, as is evident in the SIMS depth profile of Figure 9. All of the energy of the laser beam is absorbed by the Ni overlayer and used to heat it to ever increasing temperatures. The Ni melts, and as shown in the X-TEM micrographs, flows into small surface cracks and voids. The SiC substrate remains virtually unchanged; the heat transferred to the bulk crystal was not sufficiently high to cause extensive faulting, as was the case following irradiation with the excimer lasers operated at a greater energy density. Unlike the surface layer after excimer laser annealing, the Ni thickness is not uniform across the surface following ruby laser irradiation; there are areas where there is no Ni and other areas, such as in the near surface regions of the flaws, where the Ni is much thicker than the original 100 nm. A comparison between the ruby and KrF laser irradiated surfaces is shown in Figure 10; there are no bare SiC areas on the surface after KrF laser annealing, as is the case after ruby laser irradiation.

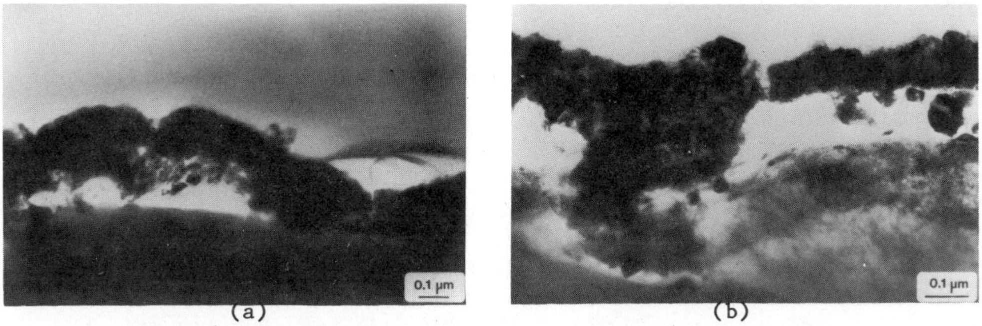

(a) (b)

Figure 8. Cross-sectional TEM micrograph of 100 nm Ni-coated SiC irradiated with a pulsed ruby laser at a nominal pulse energy density of 1.2 J/cm^2: (a) Ni surface layer and (b) Ni infilling a surface void.

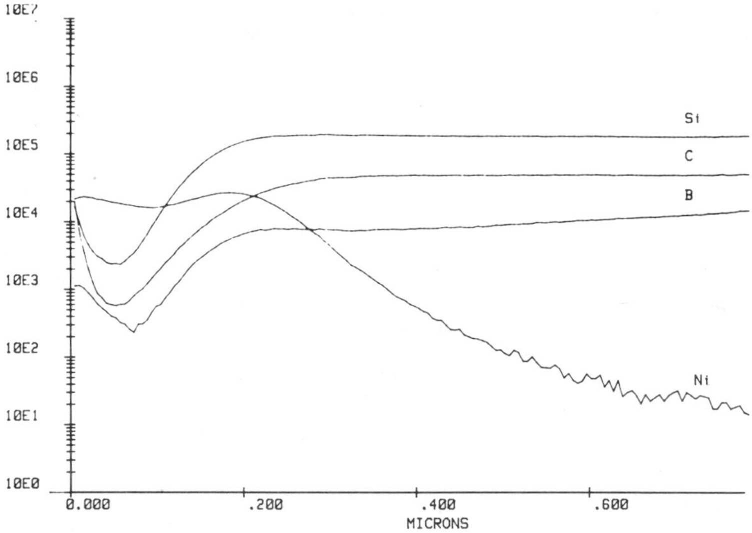

Figure 9. SIMS depth profile of 100 nm Ni-coated SiC irradiated with a pulsed ruby laser.

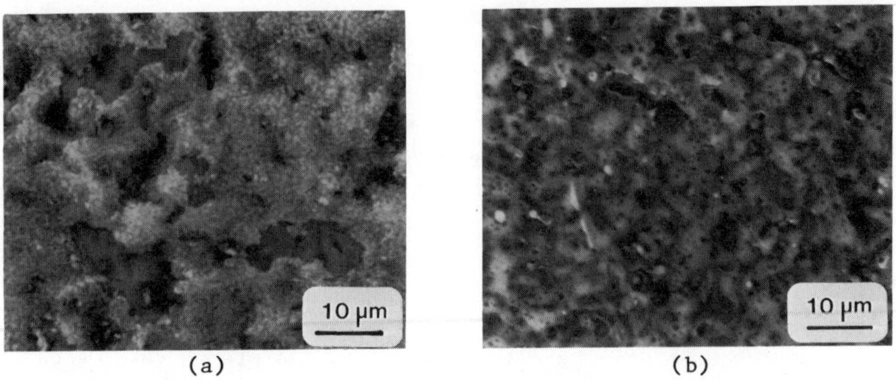

Figure 10. SEM micrographs of laser irradiated surfaces using (a) a pulsed ruby laser and (b) a KrF excimer laser.

Analysis of Strength Data

The most general approach used to analyze the frequency distribution in fracture problems is the Weibull distribution. This distribution is expressed as:

$$F(X) = 1 - \exp[-((X - X_0)/(X_1 - X_0))^m] \qquad (1)$$

where F(X) is the cumulative failure probability, X is the fracture strength test value, X_0 is the minimum test value, X_1 is the characteristic test value, and m is the Weibull modulus. The Weibull modulus is a measure of the flaw size variation and distribution and characterizes the scatter in the strength values. The cumulative failure probability, F(X), is expressed as:

$$F(X) = i/(N + 1) \qquad (2)$$

where N is the number of test values and i is the rank in strength (usually taken in order of the lowest strength to the highest strength such that the highest strength has rank N). By setting X_0 equal to zero, which simplifies the analysis, and taking a double logarithmic transformation, equation (1) can be written as:

$$\ln\ln[1/(1 - F(X))] = m\ln X - m\ln X_1 \qquad (3)$$

Thus, a linear least squares fit between $\ln\ln[1/(1 - F(X))]$ vs. $\ln X$ will yield a slope equal to m.

The fracture stength data for the as-received SiC is plotted in Figure 11 and is compared to that for the as-coated (100 nm), unirradiated SiC. The Weibull modulus, m, for the as-received SiC was found to be ≈10.8 for a coefficient of determination of 0.98 (98% of the variation in the data can be explained by equation (1)). The value of m for the as-coated, untreated SiC was ≈4.2, indicating a much greater scatter in the strength values. However, coating the SiC with Ni does not change the average fracture strength of the SiC.

The number of test specimens used is very important for valid statistical representation. Usually, 25 or more test specimens are necessary.[16] Unfortunately, availability of material and cost generally dictate the number of samples in a test population, as was the case with the present experiments. For these investigations, 10 specimens was the maximum, with

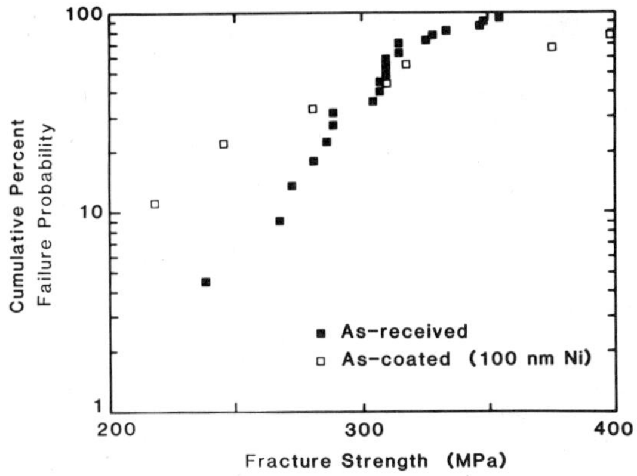

Figure 11. Fracture strength results comparing as-received SiC to as-coated, untreated SiC.

most experiments using as few as three. Thus, the majority of the data cannot be validly represented by Weibull statistics. Instead, the values presented are nominal values supported by data obtained via SEM.

Following ion beam mixing, there was no observed change in fracture strength. Ion beam mixing was thus dismissed as a viable surface modification technique to increase the fracture strength, after processing only four samples.

Laser glazing the SiC resulted in no significant change in fracture strength also. However, if the SiC surface was irradiated consecutively twice within 5s, there was a decrease in strength of ≈20%.

Laser irradiation of Ni overlayers on SiC, on the other hand, considerably enhanced the fracture strength of the SiC. After irradiation with the pulsed KrF excimer laser, an ≈20% increase in overall strength was observed, as illustrated in Figure 12(a). The strength increases observed following KrF laser irradiation were virtually the same for different processing variables; i.e. if the pulse energy density was increased to 3.0 J/cm^2 or if the initial Ni thickness was decreased to 20 nm, the increase in fracture strength remained about the same. Figure 12(b) shows fracture strength results obtained after irradiation with the XeCl laser. In this case, there were observed differences in strength data when the processing parameters were changed. There was no change in strength following irradiation of the 100 nm Ni-coated surface at a pulse energy density of 1.85 J/cm^2; a nominal increase of ≈14% at an energy density of 3.0 J/cm^2 (only three specimens were processed and there was considerable scatter in the strength values); and an increase of ≈20% when the Ni overlayer was 20 nm. The differences in strength values appear to be due to an inhomogeneous XeCl laser beam which caused variations in mixing across the irradiated surface area; i.e. the energy density in the inside of the pulsed beam is considerably greater than the energy density on the outside of the beam. When the Ni thickness is only 20 nm, mixing can still occur across the entire irradiated area within a single pulse, whereas if the Ni thickness is 100 nm, the extent of mixing across the irradiated area may vary. Thus, the effect appears as an increase in strength when the Ni thickness is much less.

The greatest enhancement in strength, however, was obtained as a result of pulsed ruby laser irradiation. The fracture strength results

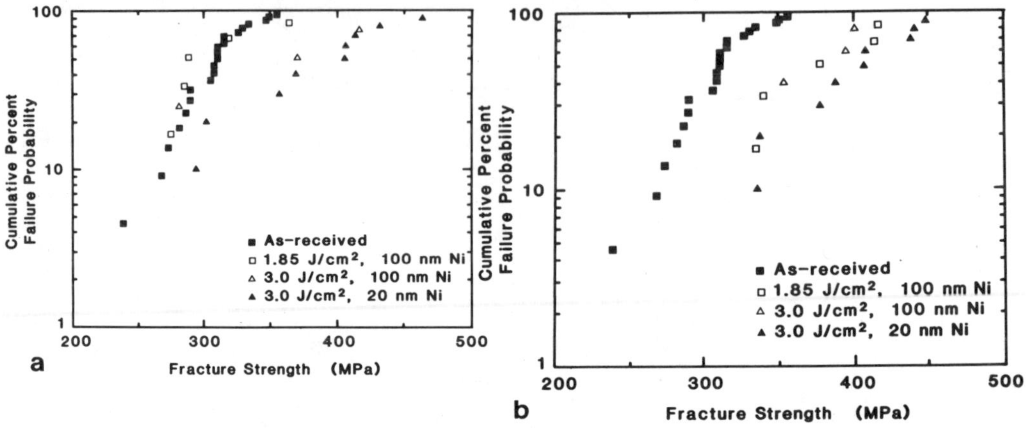

Figure 12. Fracture strength results comparing as-received SiC to Ni-coated SiC irradiated with (a) a KrF laser and (b) a XeCl laser.

are plotted in Figure 13. The ruby laser irradiated SiC showed nearly a 50% increase in strength.

The fracture strength data for the different modification techniques are summarized in Table I.

Fractographic Analysis

Scanning electron microscopy was used to identify the fracture origins in the as-received and surface modified SiC. Failure was found to occur at surface flaws in all the specimens. The fracture origins in SiC are easily identified; standard mirror-mist and macroscopic crack branching fracture features are evident on the fracture surface which can be used to locate the flaw[17-19] since the fracture mirror represents a "scaled up" image of the flaw.[20] The surface-related flaws in the as-received material are a product of processing and machining; the most common processing flaws are large SiC agglomerates or surface porosity, the latter of which is a result of inhomogeneous binder distribution prior to sintering. Since the SiC is ground using 600 grit diamond as the final machining step, surface cracks are also prevalent. Common flaws in the as-received material are shown in Figure 14.

The general expression relating the critical flaw size to the strength is:

$$K_{Ic} = Y\sigma_c c^{\frac{1}{2}} \qquad (4)$$

where σ_c is the theoretical fracture strength, K_{Ic} is the critical stress intensity factor, c is the critical flaw size, and Y is a constant dependant on flaw geometry and position. Thus, a smaller critical flaw size will result in a higher fracture strength. A change in flaw position from surface to sub-surface will also cause a strength increase. A change in flaw size and/or position will result in a shift in the Weibull curve, as illustrated above.

Following ion beam mixing, an examination of the fracture surfaces of the specimens showed that there was no change in the flaw type, as compared to the as-received material. Since the flaw size and position remained unchanged following modification, the strength remained the same.

Table I. Summary of fracture strength data before and after surface modification.

Surface Modification Technique	No. of Specimens	Average Fracture Strength (MPa)	Standard Deviation
As-received	20	310	29
100nm Ni, untreated	8	319	70
50nm Ni, IBM* 1×10^{17} Si/cm^2	2	307	19
50nm Ni, IBM 2.5×10^{16} Xe/cm^2	2	312	35
100nm Ni, RLI** 1.2 J/cm^2	4	462	42
100nm Ni, KLI+ 1.85 J/cm^2	5	376	39
100nm Ni, KLI 3.0 J/cm^2	3	382	19
20nm Ni, KLI 3.0 J/cm^2	9	396	40
100nm Ni, XLI++ 1.85 J/cm^2	5	307	36
100nm Ni, XLI 3.0 J/cm^2	3	356	56
20nm Ni, XLI 3.0 J/cm^2	9	383	54
laser glazed, 3.0 J/cm^2	6	343	37
laser glazed twice, 3.0 J/cm^2	3	246	66

* Ion Beam Mixed
** Ruby Laser Irradiated
\+ KrF Laser Irradiated
++ XeCl Laser Irradiated

Ion beam mixing only serves to implant Ni into the SiC causing the surface to become amorphous; the temperatures reached during the process do not exceed 323K (although extreme temperatures are reached within the individual collision cascades)[4] so there is no thermal contraction of the

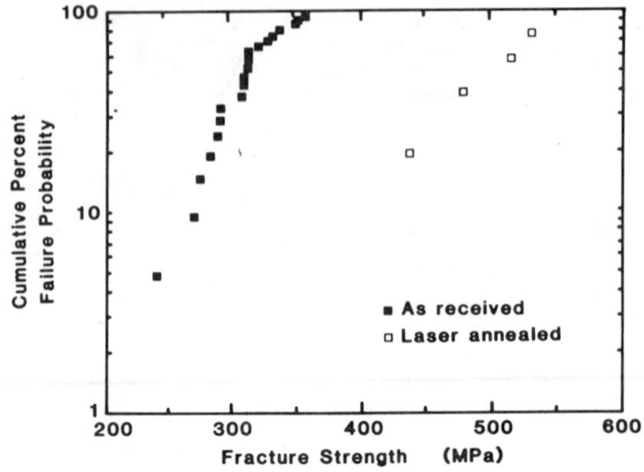

Figure 13. Fracture strength results comparing as-received SiC to ruby laser irradiated SiC.

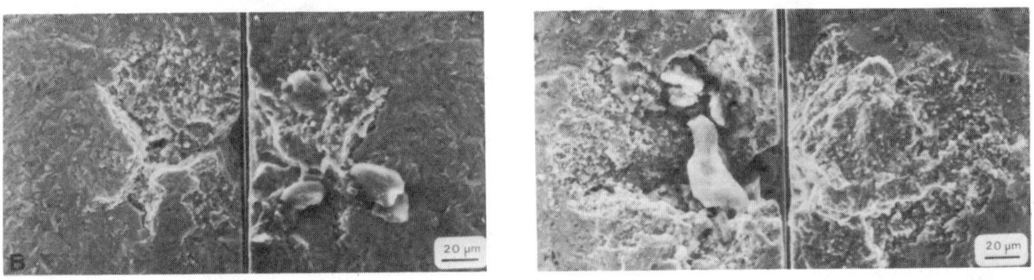

(a) (b)

Figure 14. Common surface flaws found in the as-received SiC: (a) a large SiC agglomerate and (b) a surface crack and open porosity.

metastable surface alloy produced as a result of ion beam mixing; therefore, there is no residual compressive stress induced at the surface to cause a strength increase.

There was also no significant strength increase after laser glazing. In this case, sufficiently high temperatures were apparently reached to cause local surface melting of the SiC. The SiC did not "flow" on the surface, however. Consequently, there was no change in flaw size or position as a result of processing. The ≈20% decrease in strength following double irradiation was possibly due to small cracks, already present at the surface, but made larger and more severe as a result of thermal shock during the second heating. The critical flaws found in these double irradiated specimens were surrounded by extensive secondary fracture at the surface and were considerably larger than the flaws in the starting material.

Irradiation with the KrF excimer laser also had little effect on the flaw size or position; the critical flaws were located at the surface and were relatively large, as shown in Figure 15. There was, however, a significant increase in fracture strength following irradiation as a result of an induced residual compressive stress at the specimen surface. The compressive stress was caused by two effects: (1) the enhanced bonding between the Ni and the SiC substrate due to interdiffusion of SiC into the Ni overlayer and (2) the difference in thermal contraction between Ni and SiC; on cooling, the Ni contracts to a greater extent than the SiC. In order to determine the magnitude of the compressive stress, the size of the critical

(a) (b)

Figure 15. Common flaws found after KrF laser irradiation.

flaw was measured directly from a surface fractograph of each specimen. The average value of K_{Ic} for the as-received SiC was determined using equation (4) (σ_c was the strength determined by 4-point flexure for this calculation) and was found to be ≈ 2.8 MPa-m$^{1/2}$; the shapes of the flaws in the starting material were difficult to fit to a model - as an approximation (the same approximation was made for the laser annealed samples), and for appropriate specimens which did not have extremely complex flaw shapes, the critical flaws were considered as spherical or elliptical shaped cracks or voids. By determining the theoretical fracture strength for each of the laser irradiated specimens (using the K_{Ic} value found for the as-received SiC) using equation (4) and comparing the result to the fracture strength experimentally determined from 4-point bending, σ_f, the residual compressive stress, σ_r, can be calculated as follows:

$$\sigma_r = \sigma_f - \sigma_c \tag{5}$$

Thus, using (a) fracture mechanics techniques to determine the critical flaw size and geometric parameter Y, (b) the average value of K_{Ic} as determined from fractography, and (c) equations (4) and (5), the nominal residual compressive stress determined for the KrF laser irradiated, Ni-coated SiC was $\approx$50 MPa, which accounts for almost the entire strength increase.[21]

A completely opposite effect was noted following ruby laser irradiation. Not only were the critical flaws smaller in size, but most were located significantly below the surface, in contrast to the flaws found in the as-received SiC. These flaws are shown in Figure 16. From the SIMS and X-TEM analysis presented, it is evident that the change in critical flaw size and position was due to the flow of molten Ni into the upper volume of small surface voids during ruby laser irradiation. It is reasoned that the much larger remaining volume of each of the Ni-containing voids acts as a thermal insulator which mitigates heat loss from the Ni for a time sufficient to allow chemical reaction between the metal and the SiC within the void itself. During cooling, the greater contraction of the Ni places a compressive stress in the near surface region of the voids. Thus, these voids are, in a sense, closed off and are prevented from acting as the fracture origins; instead, subsurface flaws become the strength limiting flaws. The compressive stress was determined to be $\approx$56 MPa for these specimens which, although fairly significant, only accounts for about 35% of the strength increase. Therefore, the strength increase occurs primarily as a result of altering the size and/or position of the critical flaw during ruby laser annealing.

CONCLUSIONS

Ion beam mixing of Ni overlayers on α-SiC produced no change in

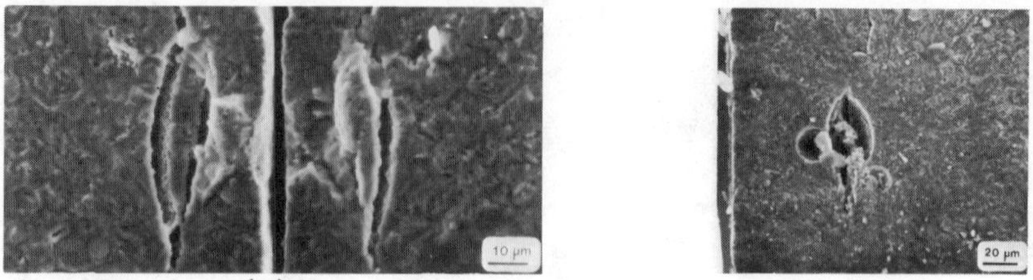

(a) (b)

Figure 16. Common flaws found after ruby laser irradiation.

the fracture strength of the SiC. An amorphous, mixed Ni-SiC region was formed at the surface as a result of ion beam mixing, but this processing method did not cause a change in critical flaw size and/or position; nor did it induce a residual compressive stress at and/or near the specimen surface. In this respect, ion beam mixing proved unsuccessful as a surface modification technique.

Similarly, laser glazing of uncoated SiC caused no significant change in fracture strength. During this process, local surface melting appeared to occur; however, there was no flaw "healing" or blunting to cause a strength increase. The SiC surface was highly faulted due to the extreme temperatures reached during laser glazing. Double irradiation of the SiC damaged the surface further; thermal shock during the second heating appeared to have caused surface cracking. Thus, flaws already present before irradiation became larger and more severe (as evidenced by considerable secondary fracture at the tensile face of the specimens) and the fracture strength was subsequently decreased by ≈20%.

On the other hand, laser surface modification of Ni overlayers on SiC significantly increased the fracture strength of the SiC. Following excimer laser irradiation, the fracture strength was increased by ≈20%. During annealing, sufficiently high temperatures were reached to cause melting of the Ni and the subsequent interdiffusion of SiC into the molten overlayer; thus, the bonding between the Ni layer and the SiC substrate was considerably enhanced. In addition, on cooling, the Ni contracts more than the SiC and, as a result, a residual compressive stress of ≈52 MPa was induced at the surface. The ≈50% increase in fracture strength following ruby laser irradiation was due to a different mechanism, however. During irradiation, the Ni overlayer melted, but the temperatures reached were not high enough to cause a chemical reaction between the Ni coating and the SiC substrate at the interface. The molten Ni flowed into small surface voids, reacted with the walls of the voids (the temperature within the voids was apparently high enough to allow for this), and, on cooling, subsequently produced a compressive stress in the near surface regions of the SiC. Consequently, the size and position of the critical flaws were changed to those having smaller dimensions and located beneath the surface.

REFERENCES

1. S. D. Ferris, H. J. Leamy, and J. M. Poate, eds., *Laser-Solid Interactions and Laser Processing - 1978*, American Institute of Physics, No. 50 (1979).
2. C. W. White and P. S. Peercy, eds., *Laser and Electron Beam Processing of Materials*, Academic Press, New York (1980).
3. B. R. Appleton, *J. Materials for Energy Systems*, 6(3), 1984, p. 200-11.

4. B. R. Appleton in: *Ion Implantation and Beam Processing*, J. S. Williams and J. M. Poate, eds., Academic Press, New York (1984).
5. J. M. Poate and J. W. Mayer, eds., *Laser Annealing of Semiconductors*, Academic Press, New York (1984).
6. J. W. Mayer, B. Y. Tsaur, S. S. Lau, and L. S. Hung, *Nucl. Inst. and Meth.*, 182/183, 1981, p. 1-13.
7. B. Y. Tsaur, Z. L. Liau, and J. W. Mayer, *Appl. Phys. Lett.*, 34(2), 1979.
8. G. C. Farlow, B. R. Appleton, L. A. Boatner, C. J. McHargue, C. W. White, G. J. Clarke, and J. E. E. Baglin, Presented at the Fall Meeting of the Materials Research Society, San Fransisco, CA (1984).
9. C. W. White, G. Farlow, J. Narayan, G. J. Clarke, and J. E. E. Baglin, *Materials Letters*, 2(5A), 1984, p. 367-72.
10. J. Narayan, D. Fathy, O. W. Holland, B. R. Appleton, R. F. Davis, and P. F. Becher, *J. Appl. Phys.*, 56(6), 1984, p. 1577-82.
11. D. Fathy, J. Narayan, O. W. Holland, B. R. Appleton, and R. F. Davis, *Materials Letters*, 2(4B), 1984, p. 324-7.
12. D. Fathy, O. W. Holland, J. Narayan, and B. R. Appleton, *Nucl. Inst. Meth.*, B 7/8, 1985, p. 571-5.
13. C. J. McHargue, G. C. Farlow, C. W. White, J. M. Williams, B. R. Appleton, and H. Naramoto, *Mat. Sci. and Eng.*, 69, 1985, p. 123-7.
14. B. R. Appleton, H. Naramoto, C. W. White, O. W. Holland, C. J. McHargue, G. Farlow, J. Narayan, and J. M. Williams, *Nucl. Inst. Meth.*, B1, 1984, p. 167-75.
15. E. H. Kraft and J. A. Coppola, Presented at the Fifth Army Materials Technology Conference *Ceramics for High Performance Applications - II*, March 25, 1977, Newport, RI.
16. S. G. Seshadri and M. Srinivasan, *J. of Eng. Mat. and Tech.*, 105, 1983, p. 219-23.
17. H. P. Kirchner and R. M. Gruver in: *Fracture Mechanics of Ceramics, Vol. 1*, R. C. Bradt, D. P. H. Hasselman, and F. F. Lange, eds., Plenum Press, New York (1973).
18. Roy W. Rice in: Reference 17.
19. Roy W. Rice in: *Fractography of Ceramic and Metal Failures*, J. J. Mecholsky and S. R. Powell, eds., ASTM STP827, Baltimore (1984).
20. D. B. Marshall, B. R. Lawn, and J. J. Mecholsky, *J. Amer. Cer. Soc.*, 63(5-6), 1980, p. 358-60.
21. K. L. More and J. J. Mecholsky, to be published.

SUB-CRITICAL CRACK GROWTH, SURFACE ENERGY AND FRACTURE TOUGHNESS OF BRITTLE MATERIALS

Daniel Maugis

Equipe de Recherche de Mécanique des Surfaces (CNRS)

LCPC, 58, Bd Lefebvre, 75732 Paris Cédex 15

ABSTRACT

A tentative attempt is proposed to interpret subcritical crack growth in brittle materials without any mention to stress corrosion. Based on theoretical and experimental results on adherence of viscoelastic solids on glass, where the true Griffith criterion ($G = 2\gamma$) was observed, with crack healing for $G < 2\gamma$, and crack propagation for $G > 2\gamma$, it is proposed that subcritical crack growth is a normal mode of propagation in any material. The crack velocity v results from a balance between crack extension force $G - 2\gamma$ and losses at the crack tip, fonction of the crack speed and the intrinsic surface energy γ :

$$G - 2\gamma = 2\gamma\phi_T(v)$$

Crack propagation is stable only on branches of the $\phi(v)$ curve with positive slope. At a critical crack speed v_c, $d\phi/dv$ becomes negative, and the velocity jumps on a second positive branch. This criterion for crack speed unstability which defines the fracture toughness K_c must not be confused with the Griffith criterion.

In region I of subcritical crack growth, $\phi(v)$ has generally the form $\phi(v) \sim v^m$ and one has the Charles equation $v = AK^n$ with $n = 2/m$. Due to the multiplicative factor γ in the second member, reduction of surface energy by adsorption shifts the whole v (K) curves towards lower K - values. When crack velocity increases, the arrival rate of active molecules from gaseous environment or by diffusion through liquid medium is insufficient to maintain equilibrium coverage on nascent surfaces, the reduction of surface energy is less and less, and region 2 arises (Mere viscous drag can also reduce the crack velocity in region 2).

Examination of literature data shows that reduction of surface energy can reach 60 to 70 percents. A representative value for glass could be $\gamma = 0.8\ J.m^{-2}$ in vacum and $\gamma = 0.45\ J.m^{-2}$ in 100 percent relative humidity. Assuming à Langmuir type isotherm with dissociation of water upon adsorption, v-K curves for glass have been reconstructed, which roughly resemble the real curves.

Examples of velocity jumps and stick-slip motions are given for brittle polymers.

INTRODUCTION

This paper is a tentative attempt to generalize to brittle maretials our results concerning kinetics of crack propagation between rigid bodies and elastomers, obtained during the course of a study of adherence of solids |1,2|.

Both thermodynamics and mechanics of contact are in agreement with fracture mechanics |3| to state that the equilibrium area of contact between a rigid punch and an elastic half-space is given by the Griffith equation

$$G = w \tag{1}$$

where G is the strain energy release rate, and

$$w = \gamma_1 + \gamma_2 - \gamma_{12}$$

the Dupré energy of adhesion (γ_1 and γ_2 are the intrinsic surface energies of the two solids, and γ_{12} their interfacial energy.) Analogy with fracture mechanics is complete, non only because the edge of the contact can be considered as an interface crack tip that recedes or advances as the area of contact increases or decreases, but also because stresses and displacement at a distance r of the edge vary as $K_1/\sqrt{2\pi r}$ and $K_1\sqrt{r/2\pi}$ (with K_1 the stress intensity factor) due to attractive molecular forces |4|.

Observation of stable equilibrium radii of contact between glass ball and polyurethane gave values of w ranging from 50 mJ/m^2 to 100 mJ/m^2, depending on room humidity content. These values are in agreement with those expected for Van der Waals attractive forces.

When the applied load increases, the area of contact increases, and the crack recedes (crack healing) ; when the applied load decreases, the area of contact decreases, and the crack advances either to a new equilibrium point or to rupture if the load is sufficiently negative. The following expression has been proposed |1,2| for the kinetics of crack propagation :

$$G - w = w\phi(a_T v) \tag{2}$$

with the following assumptions :

1/ kinetic energy is neglected

2/ the losses are limited to the crack tip zone where the stresses and the strain rates are high, so that gross displacements are elastic and G can still be evaluated by the elastic theory (with relaxed modulus) during kinetic phenomena.

3/ results at various temperatures can be shifted to a reference temperature $T_S = T_g + 50$ (where T_g is the glass transition temperature measured at zero frequency) by using the WLF shift factor |5| a_T :

$$\log a_T = \frac{8.86(T-T_S)}{101.6+T-T_S}$$

so that a master curve is obtained when studying adherence as a function of the reduced parameter $a_T v$ (A given value of $a_T v$ corresponds to either low speed at low temperature or to high speed at high temperature).

4/ viscoelastic drag is proportional to the thermodynamic work of adhesion as suggested by peeling experiments in various liquids |6| or on various substrates |7|. That means that losses only arise if the interface itself is capable of withstanding stress |7|.

Thus, in eq.(2) surface properties (w) and viscoelastic properties (ϕ) are completely decoupled from elastic properties, geometry and loading conditions included in G. The dimensionless function $\phi(a_T v)$ is characteristic of crack propagation in mode I in the material. Once $\phi(a_T v)$ is known, eq.(2) allows to predict any feature such as kinetics of detachment at fixed load, fixed grips or fixed cross head velocity.

Eq.(2) was verified for glass on polyurethane for various geometries where G can be computed (sphere, flat punch, smooth ended spheres, peeling) |1,2,8| by studying crack propagation at fixed load and at various temperatures. As shown on Fig. 1 ϕ is independent of the geometry and varies as $(a_T v)$ 0.6 over about five powers of ten. When plotted as log v vs log K (Fig. 2) the results clearly show the shift of the whole curve, with reduction of the Dupré energy of adhesion by water adsorption, illustrating the multiplicative effect of w in the RHS of eq.(2).

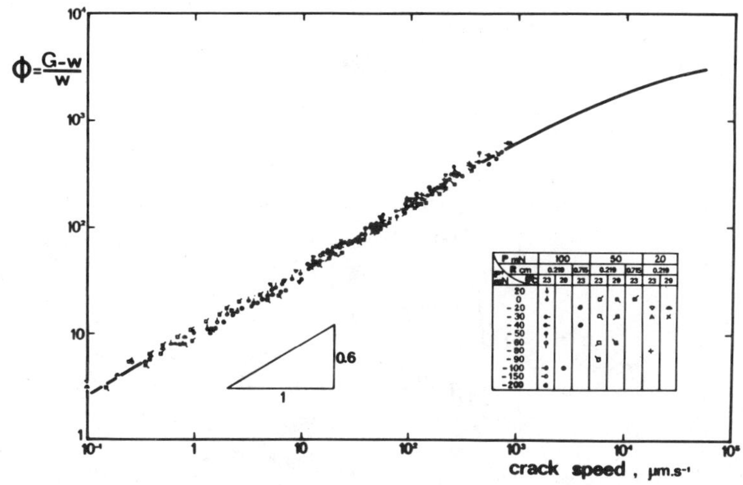

Fig.1 - Reduced crack extension force against velocity for glass-polyurethane systems.

The slope 0.60 of the log G - log v curves was correlated with the viscoelastic properties of the polyurethane. Such a material can be represented by a parabolic Zener model (analog to Cole-Cole model for dielectrics), with the loss modulus E" varying with frequency ω as $\omega^{0.60}$ at low frequency. (For another rubber tested, the slope of the logG - logv curve was 0.30 and the loss modulus varied approximately as $\omega^{0.25}$ at low frequency).

Thus, in the case of a crack between glass and elastomers, the Griffith criterion G = w was observed for equilibrium cracks, with crack healing for G < w, and slow crack propagation for G > w, the kinetics of crack propagation being related to internal losses in the material, at the crack tip. We can wonder whether these results can be generalized to other materials.

THE PROPOSED MODEL

Let us consider in a diagram log G - log v (Fig. 3), the superposition of a curve for brittle fracture with dynamic effects (assuming limiting crack velocity at Rayleigh waves speed C_R), and a curve for internal losses at the crack tip. Let

$$G - 2\gamma = 2\gamma\phi_T(v) \tag{3}$$

the general equation of the curve so obtained, and

$$G - 2\gamma = 2\gamma\alpha(T)v^m \tag{4}$$

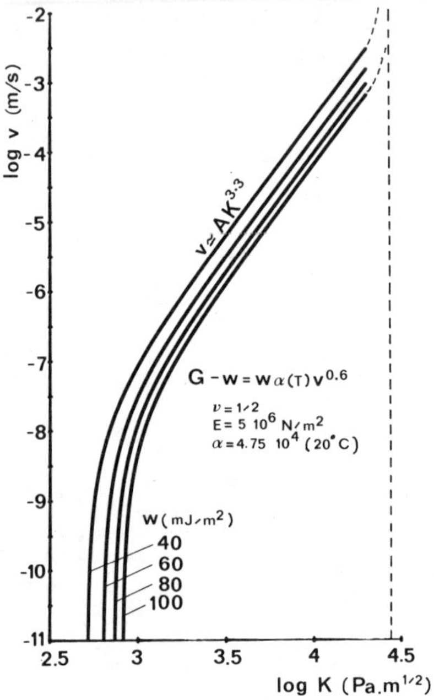

Fig 2 - Results of Fig.1 replotted in log v - log K coordinates

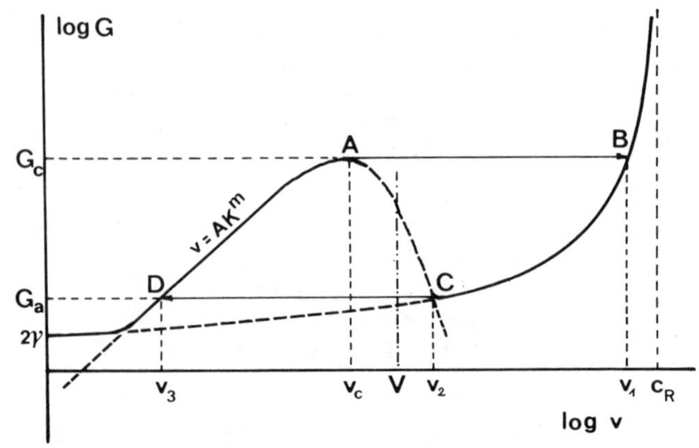

Fig.3 - Schematic diagram of a log G - log v curve showing velocity jumps and stick-slip motion.

its approximation for low velocities. It can be inferred from the fact that loss modulus decreases at high frequencies, that the curve for internal losses decreases above a given velocity v_c (dashed curve).

The negative resistance branch AC cannot be observed for resistance to crack propagation decreases as the velocity increases. Thus, when testing samples with geometry such that G increases as the crack length increases (DCB at fixed load, SEN...) a slow crack propagation is first observed, until $G = G_c$ at which the velocity jumps on the second stable branch (with acoustic emission), and one has

$G < 2\gamma$ crack healing

$G > 2\gamma$ crack propagation

$2\gamma < G < G_c$ slow crack growth

$G > G_c$ fast crack growth

The Griffith criterion $G = 2\gamma$ for crack equilibrium, must not be confused with the criterion $G = G_c$ for crack speed instability and catastrophic failure. It is possible to write this second criterion in a form equivalent to Griffith's one : $G = 2\gamma_f$, where γ_f is the fracture surface energy (γ_f can be thousands times γ for materials with high internal losses), but if it is confused with the true Griffith criterion, the slow crack propagation at $G < 2\gamma_f$ seems in contradiction with thermodynamics. Stress corrosion has to be invoked, but fails to explain subcritical crack growth in vacuum.

For samples whose G or K is independent of the crack length (peel tests, DT, TDCB...), the crack velocity can be monitored, and the velocity is stable on the two branches with positive slope. However, when imposing a mean velocity V in the range $v_c - v_2$ corresponding to the negative resistance branch, a stick-slip cycle ABCD occurs, with periodic apparent crack arrest and acoustic emission at each velocity jump. (A good example is peeling an adhesive tape : crack propagation is continuous at low or high velocity, but is jerky at intermediate speed).

In this model, adsorption of active species at the crack tip, reduces the surface energy γ according to the Gibbs formula

$$\Delta\gamma = -\int_{\mu_0}^{\mu} \Gamma d\mu = - RT \int_{p_o}^{p} \Gamma \frac{dp}{p} \tag{5}$$

where μ is the chemical potential of active species, p its partial pressure, Γ the adsorption in moles per unit area at the chemical potential μ or partial pressure p, and shifts the whole curve along the G or K axis, so that at the same crack velocity one has

$$\eta = \frac{\gamma_1}{\gamma_2} = \frac{G_1}{G_2} = \frac{K_1^2}{K_2^2} \tag{6}$$

The slope of the curve G versus log p allows thus, in principle, to compute the adsorption Γ at the crack tip (fig.4). If the active species can reach the crack tip the velocity v_c, the critical strain energy release rate for catastrophic failure is rediced in the same proportion as the

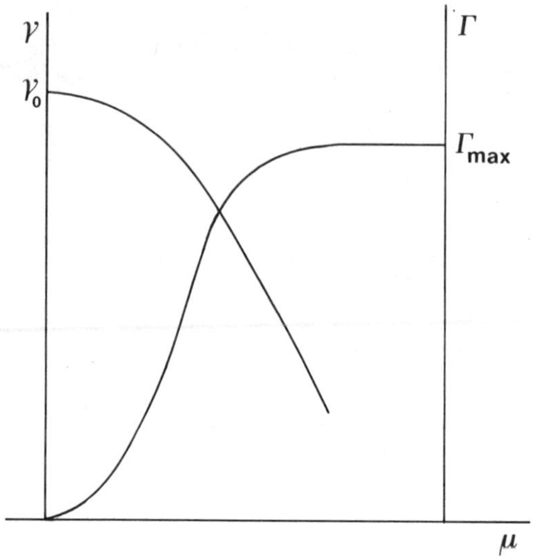

Fig.4 - Gibbs relation between surface energy, adsorption and chemical potential.

surface energy. These embrittlement effects are discussed in more details in |9|.

Another advantage of using eq.(3) is that the reduction of Young modulus with increasing temperature, gives an increasing crack extension force $G - 2\gamma$ which cannot be taken into account when using K, a temperature independent quantity.

Thus, it is proposed that subcritical crack growth is a normal mode of slow crack propagation above the Griffith criterion, in which the crack speed adjust to a stable balance between the energy dissipation rate and the driving force $G - 2\gamma$. It can be observed in vacuum, and on crystalline solids as well as on amorphous ones. The effect of reduction of surface energy by active environment is to reduce the stresses needed to break the bonds, and proportionnally to reduce energy dissipation (plastic work, viscoelasticity or internal losses of any kind) at the crack tip as first suggested by Mc Lean |10|. This by no means, implies that stress corrosion does not exist : besides giving a reduction of surface energy, a surrounding medium can react at crack tipe or give electrical double layers along the crack lips leading to a repulsion as in the DLVO theory.

In the following, we will examine some experimental results on brittle solids to see if they are compatible with this model.

THE STATIC FATIGUE LIMIT K_o AND THE REGION I

When G approaches 2γ (or K approaches K_o), the crack extension force and the crack velocity tend to zero. In a diagram logv - log K, eq.(4) becomes

$$v = \frac{1}{\alpha^{1/m}} \left(\frac{K^2}{K_o^2} - 1 \right)^{1/m} \qquad (7)$$

which accounts for the bending of curves near K_o, and gives

$$v = AK^n \qquad (8)$$

with $n = 2/m$ for $K >> K_o$. Eq.(7) could probably allow for more precise lifetime predictions than eq.(8).

The bending of curves near K_o is clearly visible below 10^{-9} m /s in a number of papers |11-17|, and with less evidence in |17bis| for quartz. The deduced values of K_o are in agreement with expected values for intrinsic surface energies which rarely exceed 2 J/m^2. For soda-lime |11,14| or borosolicate |11| in water, it comes K_o = 0.25 MPa.m$^{1/2}$, i.e. γ = 0.45 J/m^2 (assuming E = 7 10^{10}Pa). For silica glass in water, the value K_o = 0.37 MPa.m$^{1/2}$ (i.e. γ = 0.9 J/m^2) was estimated by Sakaguchi et al |18|. For silicon in ambient air, curves from Bhaduri and Wang |19| show bending near 0.6 MPa.m$^{1/2}$ which, with E = 8.10^{10}Pa, gives γ = 2.3 J/m^2, a rather high value compared with γ = 1.29 J/m^2 given by Miedema |20|. (The value E = 1.2 10^{11} Pa generally found in the literature would give a better agreement). However, it must be recognized that static fatigue limit is not always evident, and in particular, has never been observed in geological materials |21|. For PMMA, the experimental results of Marshall et al. |22| on subcritical crack growth at 20°C, can be closely fitted by the equation, |9|

$$\frac{G}{2\gamma} - 1 = 800\ v^{0.17}$$

(assuming γ = 1 J/m^2) and the bending of curves would be observed below 10^{-20} m/s (Note that the power 0.17 is in agreement with the value 0.17 for the slope of the loss modulus versus frequency |23| at low frequency).

In the proposed model, the action of an active, noncorrosive medium, is simply to reduce surface energy and shift the v-K curves along the K-axis, according to eq.(6). Such shifts can be found in |14, 18, 24 - 26| for glasses, and |27, 28| for polymers. Estimates of the ratio η give 0.6 for soda-lime between 0.017 % and 100 % RH |14|, 0.5 for commercial glass between octanol with 15 percent humidity and water |24|, 0.6 for soda-lime in butyl alcohol solutions with different quantities of dissolved water |25|, 0.7 for silica glass between 0.08 % RH and water |18|, 0.65 for vitreous silica between formamide and water |26|, 0.7 for quartz between 68 % RH and water |17bis|, 0.2 for polyethylene between detergent and water |27|, 0.5 for epoxy resins between 55 % RH and water |28|. Taking the surface energy in 100 percent humidity as γ_1 = 0.45 J/m^2 for glass, and γ_1 = 0.9 J/m^2 for silica, the variation of surface energy $\Delta\gamma$ by water adsorption are respectively 0.30, 0.45, 0.30 and 0.40 J/m^2.

These reductions in surface energy must be in agreement with the Gibbs adsorption formula (5). Such determination of $\Delta\gamma$ on glass powder by Schultz and Simon |29|, or quartz powder by Boyd and Livingston |30| gave respectively $\Delta\gamma$ = 0.23 J/m^2 and $\Delta\gamma$ = 0.25 J/m^2 in saturated vapor, which is of similar order of magnitude. Similarly Parks |31| estimates the reduction of surface energy of quartz between the beginning of water adsorption and immersion to be 0.15 - 0.35 J/m^2, with a ratio η reaching 0.2, 0.3.

Fig. 5 shows, for v = 10^{-7} m/s (region I), the ratios K^2/K^2max = $\gamma/\gamma max$ taken for Fig 1-3 of ref. |25| for soda-lime silicate glass in gaseous nitrogen (25°C), ethyl and butyl alcohols of various relative

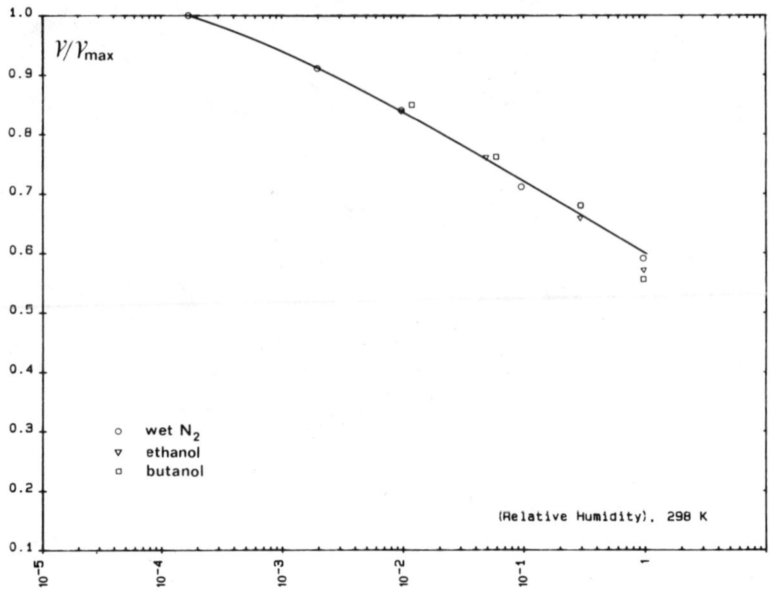

Fig.5 - Ratio $\gamma/\gamma max$ against relative humidity for soda-lime glass

humidity content, where K_{max} is the largest value at that speed in each figure. Taking γ = 0.45 J/m^2 for 100 % RH, the ratio γ/γ_{max} = 0.6 gives γ_{max}= 0.75 J/m^2 at 0.017 % RH. From the final slope

$$\frac{d\gamma}{d\ell np} = \frac{\gamma max}{2.3}\ \frac{d(\gamma/\gamma max)}{dlogp} = -\ 3.9.10^{-2}J.m^{-2}$$

it comes Γ = 1.58 10^{-5} moles m^{-2} = 9.5 10^{14} molecules /cm^2 which is of the proper order of magnitude for surface adsorption. Assuming a Langmuir isotherm, with dissociation of water molecules upon adsorption, as it is generally thought |32|, (at least for low relative humidity), one has for the coverage

$$\theta = \frac{\Gamma}{\Gamma_{max}} = \frac{\sqrt{br}}{1+\sqrt{br}} \tag{9}$$

where b is a numerical constant and r the relative humidity. Putting this expression for Γ in the Gibbs equation (5) and integrating, it comes

$$\Delta\gamma = -\ 2\Gamma_{max}\ RT\ \ell n\ (1 + \sqrt{br}) \tag{10}$$

As the final slope in Fig.5 is approximately constant, the computed adsorption Γ can be taken as Γ_{max} and one has

$$\gamma = \gamma_o - 7.82\ 10^{-2}\ \ell n\ (1+\sqrt{br}) \tag{10^{bis}}$$

The full line in Fig.5 given by

$$\frac{\gamma}{\gamma max} = \frac{\gamma_o - 7.82\ 10^{-2}\ell n\ (1+\sqrt{br})}{\gamma o - 7.82\ 10^{-2}\ell n\ (1+\sqrt{1.7\ 10^{-4}b})}$$

and adjusted with γ_o = 0.8 J.m^{-2} and b = 7580 closely represents the results.

However, the v-K curves are not always parallel, in particular when the pH of the solution is varied |33,34|. This could be due to the repulsion of double layers as suggested by Parks |31|, increasing K (see Appendix) but Wiederhorn and Johnson |35| failed to observe any ζ - potential effect.

Below K_o the crack must heal, according to the Griffith criterion. Such crack healing in brittle materials has been studied in a number of papers, as reported by Hockey |36|. Wiederhorn and Townsend |37| have shown that recovery of strength during healing was approximately 80 % for crack formed by shock. Observation of brittle crystalline solids in transmission electron microscope |36, 38, 39| have revealed interfacial misfit dislocations which form when crack interfaces rebound in imperfect registry, and no evidence for crack tip plasticity. However, the phenomenon of crack healing can be obscurred by corrosive effect of the environment. Recently, Ito and Tomozawa |40-42| have shown that crack blunting of high silica glass in water and silicic acid solutions occurs by dissolution at the crack lips and precipitation at the crack tip because solubility is a function of the curvature of the surface. This irreversible change of the crack shape compared with that of the elastic solution probably occurs only near K_o and impedes both crack healing and crack repropagation upon reloading. For borosilicate glass in acidic solutions Wiederhorn and Johnson |33| observed that crack motion ceased abruptly at 0.404 $MPa.m^{1/2}$, and that 0.486 $MPa.m^{1/2}$ was required to restart the crack. Tear marks were left on the surface at the point of crack arrest suggesting crack blunting. The same observations were done by Michalske |43| for aging soda-lime in water. These results could explain the well known fact that glass left in air or immersed in water or acid in absence of stress (aging) tends to gain strength |44-49|, and confirm the ancient remark by Wiederhorn |50| that water strengthens glass at stresses below that of the static limit and weakens it at stresses greater than that of the static fatigue limit.

The idea that K_o corresponds to intrinsic surface energy is not new, and was proposed by Hasselman et al |51| and Rice |52|. Previously, Shand |53,54| noted that γ deduced from static fatigue limit of glass in air approaches the value of γ extrapolated from high temperature measurements. Recently, Michalske and Freiman |26| proposed a molecular interpretation of stress corrosion in silica, suggesting that environments which enhance stress corrosion are composed of molecular groups with electron donor sites at one end and proton donor sites at the other, like H_2O, NH_3, hydrazine and formamide. This is not in contradiction with reduction of surface energy, and reflects the quantum mechanical aspect of modern theories of adsorption. Note that Parikh |55| in his study of effect of atmosphere on surface energy of soda-lime glass in the range of 500° to 700° C, has shown that H_2O, NH_3 and SO_2 were the more efficient for reduction of surface energy. Note also that gases like NH_3 and SO_2 are known to change the surface tension σ of glass as well as its surface energy γ (see |56,57|). These two quantities are related by the Shuttleworth equation |58|

$$\sigma = \gamma + A \frac{d\gamma}{dA} \tag{11}$$

where A stands for surface area. Whereas γ can only decrease by adsorption, σ can either decrease or increase |57, 59, 60| giving extension or contraction, according to the Bangham formula |61|.

$$\frac{\Delta L}{L} = - \lambda \Delta\sigma \tag{12}$$

(where their $\Delta\gamma$ is replaced by $\Delta\sigma$ |57|). A correlation was found between the expansion of porous glass by adsorption and its mechanical strength |62|, but the role of surface stresses induced by variation of surface tension, is still an open question in fracture mechanics.

REGION II, VISCOUS DRAG AND LIMITED RATE OF TRANSPORT

In the preceding section it was argued that lowering the surface energy by equilibrium adsorption lowers K_o and shifts the whole v-K curve. The toughness K_c should also be reduced in the same proportion. It is the case when active species segregate from the bulk to interface giving integranular brittleness as discussed in |9| (Such shifts of K_c with the whole v-K curve is observed for intergranular slow crack propagation in $Al_2O_3 - ZrO_2$ ceramics when changing the Y_2O_3 content from 1 mole % to 3 moles % |63|). However, in a fluid environment the reduction of surface energy can be limited by the rate of transport of active species to the crack tip, and the crack velocity can be reduced by the viscous drag of the fluid itself. As the fluid and the active species comes from both the crack mouth and the crack side, a thickness effect can be expected in region II, contrarily to region I.

Viscous effects are well known in adhesion. For example, the pull-off force between two discs of radius R separated by a liquid of thickness h is a function of the time t for separation

$$F = \frac{2\pi R^2 \gamma \cos\theta}{h} + \frac{3\pi\eta R^4}{4h^2}\frac{1}{t}$$

where γ and η are the surface energy and the viscosity of the liquid and θ the wetting angle. The first term is the quasistatic adherence force (Laplace law) the second is the rheological term derived from the Stefan law for viscous flow. Similarly the pull-off force between rigid sphere and plane with inserted oil film is strongly dependent of the time for separation and of temperature through the viscosity |64|. These effects are limited by cavitation, so that at high speed the crack tip can be completely separated from the environment. Experiments by Carré and Schultz |65| on the peeling of elastomers on aluminium in oils of similar surface energy but of various viscosities clearly show that one returns to the curve for peeling in air at critical velicities depending on oil viscosity.

Experiments on propagation of cracks filled by water have been performed by Michalske and Frechette |66, 67| on glass. Their results are schematically shown on Fig 6a. For crack velocities below 3.10^{-2} m/s, water facilitates crack propagation compared to air. At about 3.10^{-2} m/s (equivelocity point, point K) wetted and non-wetted cracks require the same stress intensity factor, whereas at velocity above 3.10^{-2} m/s the viscosity hinders crack propagation until cavitation occurs at $v_{cv} = 4.10^{-2}$ m/s and crack velocity instantaneously jumps to its values on the dry curve. A scarp is formed on the fracture surface at the point where the velocity jumps from 4.10^{-2} m/s to 2.10^{-1} m/s, and the appearance of fine cavitation bubbles at velocities slightly lower than v_{cv} explains the formation of hackles.

Hydrodynamic phenomena and conditions for cavitation in crack filled with liquids has been discussed in several papers |65, 67-70|. The expression derived by Michalske and Frechette |67| gives good correlation with measured critical velocities for cavitation. It could be anticipated that for Newtonian viscosity the crack velocity should be proportionnal to the driving force $G-2\gamma$, so that log G - log v and log K - log v curves would respectively display a slope 1 and 1/2 in region II. Such a slope is apparent in peeling experiment in oils |65|.

When the crack velocity increases, a point is reached at which diffusion of active species will be too low to continue feeding the crack tip fast enough to maintain equilibrium adsorption. The reduction in surface

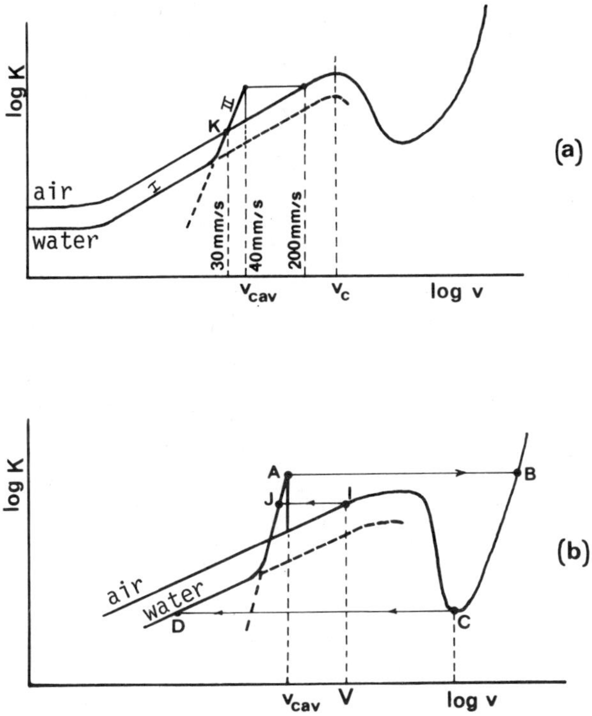

Fig.6 - Schematic diagram for viscous drag and cavitation.

energy becomes less and less as the crack velocity increases. If seems that two cases must be distinguished according as the active species are in a gaseous state or diffuse through a viscous liquid.

The knees in v-K curves in presence of water vapour were first observed by Shand |53| and by Wiederhorn |71| who showed that the height of the "plateau" increased from 10^{-7} to 10^{-4} m/s with the humidity constant, and pointed out the influence of the limited rate of water transport at the crack tip. Varner and Frechette |72| observed in SEN specimen, 1 or 2 mm thick, that the velocity in region II was too great for water vapor to be transported to the central portion of the crack, so that the movement was stepwise in the center and continuous on the edges. Hibino et al |73| showed that for silica glass the transition I-II occurs at constant stress intensity factor K_T = 0.61 $MPa.m^{1/2}$ independent of temperature and humidity, and at a velocity v_T depending of the partial vapor pressure at the power 0.57. The slope of the curves in region II was between 3.4 and 4 at 20°C instead of about 40 in region I.

Let us perform a crude calculation to evaluate the transition velocity v_T for glass in humid air. Consider a crack front of length L, where the local vapor pressure is p, advancing at the velocity v. During the time τ, a new surface $Lv\tau$ is created on one crack side. The arrival of molecules per unit area of surface is $p\tau/\sqrt{2\pi mkT}$, where m is the mass of the molecules. They are immediately adsorbed, and the coverage is

$$\theta = \frac{\Gamma}{\Gamma_{max}} = \frac{p\tau}{\Gamma_{max}\sqrt{2\pi mkT}}$$

Letting $a_o^2 = 1/\Gamma_{max}$ the area per atom of substrate, and taking $\tau = a_o/v$ the time to expose a new row of atoms, it comes

$$\theta = \frac{pa_o^3}{v\sqrt{2\pi mkT}} = r\,\frac{a_o^3 p_{sat}}{v\sqrt{2\pi mkT}} \qquad (13)$$

The minimum velocity to maintain the equilibrium coverage given by eq.(9) (resulting from a balance between adsorption and desorption rates) is thus :

$$v_T = \frac{a_o^3 p_{sat}}{b\sqrt{2\pi mkT}} \sqrt{br}\,(1 + \sqrt{br}) \qquad (14)$$

For specimens of same thickness and same crack length (same local pressure p), the transition velocity increases with the vapour pressure at a power between 0.5 and 1. Fig. 7 shows v-K curves for soda-lime in wet nitrogen at 25°C constructed from the point A ($K = 0.45$ MPa.m$^{1/2}$, $v = 10^{-7}$ m/s) and the slope $n = 20.5$ taken from fig 3 of ref |14|. In region I, curves are shifted according to the Langmuir isotherm, eq.(10bis) with $\gamma_o = 0.8$ J.m^{-2} and $b = 7580$. The transition velocities are computed from eq.(14) with $a_o = 3\ 10^{-10}$m, a mean value for the distance Si - Si deduced from radial distribution functions |74|. In region II the curves are drawn by computing the pressure p' which would give at equilibrium the same coverage as that given by eq.(13) (assuming local thermodynamic equilibrium), and putting this value in eq.(10bis). The agreement for v_T and K_T is not too bad, but the change in slope at the transition is very sharp compared to experimental results.

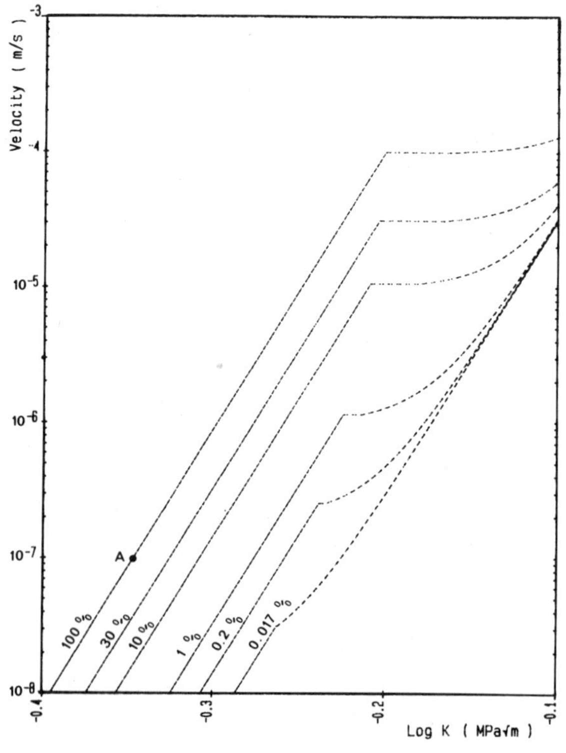

Fig.7 - v-K curves for soda-lime glass in wet nitrogen constructed from point A and slope n = 20.5, assuming Langmuir - type adsorption.

For cracks filled with a viscous liquid, the effect of the limited rate of transport of active species to the crack tip can be difficult to distinguish from a mere viscous drag effect |75, 76|. Only if region II appears for a liquid and not for a more viscous one |12, 77, 78|, or by adding active species in an inert liquid (e.g comprox in water for polyethylene |27|), or if an effect of the concentration c of active species occurs |24, 25|, a limitation of the rate of transport can be affirmed. A correlation was found between the transition velocity v_T and the reciprocal of the viscosity |27, 75, 76|, and more precisely with ratio c/η |25|. For SEN specimens of soda-lime immersed in humid alkanes Quackenbush and Frechette |79| observed that in region II the crack front has a reversed curvature with the water-free central part retarded in comparison with the edges feeded with water diffusing through alkane from the nearest sample boundary. As in cavitation experiments by Michalske and Frechette |67| a wedge separates the area which failed in humid and in water free alkane, during crack acceleration through region II. Due to the redistribution of stresses along this mixed crack front, depending on velocity and sample thickness, prediction of slopes of v-K curves in region II will be certainly difficult. Nevertheless a rough estimation of transition velocities can be made for thick samples when edge effects are negligible.

Let us consider an active species of concentration c (moles/m^3) diffusing through a liquid of viscosity η. The relative "humidity" will be defined as $r = c/c_{max}$, to take into account a limited solubility (for pure water $c_{max} = 5.56\ 10^4$ moles/m^3, for water in butyl alcohol $c_{max} = 9.2\ 10^3$ moles/m^3 |25|). The arrival rate of molecules at the surface can be computed in two different ways |80| according as a depletion layer is feeded or not by convection and stirring. In absence of any stirring one has

$$\frac{dn}{dt} = \left(\frac{D}{\pi t}\right)^{1/2} cN \tag{15}$$

where N is the Avogadro number and D the coefficient of diffusion given by the Stokes-Einstein formula :

$$D = \frac{kT}{6\pi\eta\rho} \tag{16}$$

with ρ the radius of the diffusing molecule ($\rho = 3.10^{-10}$m for water). For a constant depletion layer of thickness δ, one has

$$\frac{dn}{dt} = \frac{D}{\delta}\ cN \tag{17}$$

We will make use of eq.(17), for eq.(15) gives transition velocities too strongly dependent on concentration. The number of molecules reaching the unit area of surface during the time τ is

$$n = \frac{kTcN\tau}{6\pi\eta\rho\delta}$$

Taking $\tau = a_o/v$ and $\Gamma_{max} = a_o^{-2}$ as above, it comes

$$\theta = \frac{RTa_o^3\ c}{6\pi\eta\rho v\delta} \tag{18}$$

The minimum velocity to maintain the equilibrium coverage given by eq.(9) is thus

$$v_T = \frac{RTa_o^3}{6\pi\rho\delta}\ \frac{c}{\eta}\ \frac{1 + \sqrt{br}}{\sqrt{br}} \tag{19}$$

or

$$v_T = \frac{0.016kT}{6\pi\rho\delta} \frac{c}{\eta} \frac{1 + \sqrt{br}}{\sqrt{br}} \tag{20}$$

with c expressed in moles/ℓ (or kg moles/m^3). Since the term $(1 + \sqrt{br})/\sqrt{br}$ varies between 1.01 at 100 % RH and 1.11 at 1 % RH, eq.(20) is equivalent to eq.(12) of Wiederhorn et al |25| giving the correlation between v_T and c/η. (Ignoring the use of an order of chemical reation in |71| and taking it equal to 1, their calculation is in its essence the same as above for $\theta = 1$).

For 99 % ethylalcohol ($c = 9.556$ moles/ℓ, $r = 10^{-2}$, $\eta = 1.1\ 10^{-3}$ Pa.s) the observed transition velocity is $1.46\ 10^{-5}$ m/s, hence the thickness $\delta = 4.5.\ 10^{-7}$m. With this value the transition velocities at 20°C are

$$v_T = 2.6\ 10^{-8}\ \frac{c}{\eta} \frac{1 + \sqrt{br}}{\sqrt{br}} \tag{21}$$

which, with b = 7580 as computed above, is exactly the line plotted on Fig.10 of ref |25|.

For $v > v_T$, it is still possible to compute a fictitious "humidity" r' giving the same coverage θ that given by eq.(18), and by putting this value of r' in eq.(10) to have the surface energy γ and draw the v-K curves in region II. The results are very similar to fig. 7, with sharp transitions at v_T, and K_T varying in a similar manner.

VELOCITY JUMP AND STICK-SLIP

There is no firm evidence for velocity jump and stick-slip in glass but these phenomena are well documented for brittle polymers. Whether velocity jump or stick-slip occurs depends on the stability of the system , which in turn depends on the geometry tested and the stiffness of the measuring apparatus |9|.

Fast propagation along the second positive branch has been studied on PMMA |81-83|, epoxy resins |84, 85| and polyesters |85|, and the stick-slip cycle ABCD (Fig. 3), first proposed by Williams, Radon and Turner |82|, was clearly observed by Kobayaski and Dally |85| on epoxy resins. Stick-slip motion range was observed between two stable velocity ranges on elastomeric adhesives |86|. It was observed by lowering the cross-head velocity in epoxy resins |87| and by increasing the cross head velocity in PMMA |22|.

Velocity unstabilities were early associated with decrease in crack resistance with increasing velocity |88,89|. For polymers, various reasons have been invoked : molecular relaxation |90|, transition in craze formation |83|, isothermal to adiabatic transition causing thermal softening |91|. However, the very sensitive influence of curing agents, the low velocities at which stick-slip can be observed, the absence of craze in epoxy resins, and the absence of thermal softening effect near running crack tips in PMMA |92| (despite the high temperature rises measured by Fuller et al. |93|), and the coincidence of power index for crack propagation and for loss modulus, leads to prefer the hypothesis of a decrease of internal friction. Such negative strain rate sensitivity exist also in metallic alloys, when dislocations are teared off from their solute atmosphere giving rise to jerky flow characteristic of the Portevin - Le Chatelier effect |94, 95| .

CONCLUSION

The enhancement of subcritical crack growth by reduction of surface energy seems to give a reasonable alternative to stress corrosion. Although not explaining all the experimental results, the proposed model has nevertheless the advantage of giving a mnemonic and simple picture of various facets of fracture mechanics : static fatigue limit, subcritical crack growth, fracture toughness, velocity jump, stick-slip, acoustic emission and embrittlement.

Appendix: Repulsion of double layers

Repulsion of double layers forms the basis of the DLVO theory of stability of colloids |32, 96, 97|. Between two plates immersed in an electrolyte (n_o ions per unit of volume) the pressure is

$$p = 2n_o kT \left(\cosh \frac{e\phi_d}{kT} - 1\right)$$

where ϕ_d is the midpoint potential, and e the electron charge. Calculation of ϕ_d as a function of the surface potential ϕ_o and the distance 2d between the plates involves elliptic integrals, but can be simplified when $e\phi_d / kT$ is small, i.e when d is large compared with the Debye length

$$\lambda_D = \frac{1}{K} = \left(\frac{\varepsilon kT}{2n_o e^2}\right)^{1/2}$$

where ε $\varepsilon_r\varepsilon_o$ is the dialectric constant. (For monovalent electrolyte λ_D varies between 10^{-8} m and 3 10^{-10} m when the concentration varies between 10^{-3} mole/ℓ and 1 mole/ℓ). The result is

$$p = 64 n_o kT\gamma^2 e^{-2Kd} \tag{A1}$$

where

$$\gamma = \left|\frac{e^{y_o/2} - 1}{e^{y_o/2} + 1}\right|$$

with $y_o = e\phi_o/kT$

This pressure acts inside the crack, and increases the value of K. Let

$$K_{tot} = K + K_{el} \tag{A2}$$

and consider a crack of length c (from x = o to x = c). K_{el} can be computed from the Barenblatt formula |98|

$$K_{el} = 2 \left(\frac{c}{\pi}\right)^{1/2} \int_o^c \frac{p(b)db}{\sqrt{c^2-b^2}} \tag{A3}$$

where p(b), given by eq.(A1), varies with the half crack opening displacement

$$d = \frac{4K_{tot}}{E} \sqrt{\frac{c - b}{2\pi}} \qquad (A4)$$

Assuming the pressure localized at the crack tip, as cohesion forces in the Barenblatt theory, $\sqrt{c^2-b^2} \simeq \sqrt{c}\sqrt{c-b}$, and integrating from o to infinity gives

$$K_{el} = \frac{\alpha^2}{K_{tot}} \qquad (A5)$$

with

$$\alpha^2 = 32 n_o kT\gamma^2 E\lambda_D$$

The true stress intensity factor is thus

$$K_{tot} = \frac{1}{2}\,(K + \sqrt{K^2 + 4\alpha^2}) \qquad (A6)$$

and the ratio of velocities is

$$\frac{v}{v_o} = \left(\frac{K_{tot}}{K}\right)^n = \left(\frac{1 + \sqrt{1 + (2\alpha/K)^2}}{2}\right)^n \qquad (A7)$$

Let us take a numerical example with glass ($E = 7.10^{10}$ Pa) in 2N NaOH ($n_o = 1.2\ 10^{27}\ m^{-3}$), and take $\gamma = 1$ as an upper bound. It comes $\alpha = 0.05$ MPa$\sqrt{m}$, and for K = 0.5 MPa$\sqrt{m}$ the ratio $v/v_o = 1.20$. This value is negligible compared with the ratio 1000 between 2N HCl (n = 20) and 2N NaOH observed by Atkinson and Meredith |34|.

Note however that for $d \ll \lambda_D$, the pressure varies as d^{-2} |96|, which gives an infinite value for K_{el}. The problem of repulsion of double layers inside cracks therefore must be examined more carefully. Note also that for quartz, for which the point of zero charge (pzc) is reached at pH around 2 |99|, the crack velocities are lowest in acidic solutions. It would be interesting to know if for rutile TiO_2, for which the pzc is reached around pH 6 |99|, crack velocities are of the same order of magnitude for acidic and basic solutions.

ACKNOWLEDGEMENTS

The author is indebted to GRECO 46 and DRET for financial support.

REFERENCES

1. D. Maugis, M. Barquins, J.Phys. D : Appl. Phys. 11 : 1989 (1978)
2. D. Maugis, M. Barquins, in : "Adhesion and adsorption of polymers", part A, p. 203, L.H. Lee, ed, Plenum Press, New-York (1980)
3. D. Maugis, in "Unilateral problems in structural analysis", CISM courses and lectures, G. Del Piero, F. Maceri, ed, Springer-Verlag, Berlin (1985)
4. M. Barquins, D. Maugis, J. Mec. Theor. Appl., 1 : 333 (1982)
5. J.D. Ferry, "Viscoelastic properties of polymers", J. Wiley, New York (1970)
6. A.N. Gent, J. Schultz, J. Adhesion, 3 : 281 (1972)
7. E.H. Andrews, A.J. Kinloch, Proc. Roy. Soc. A, 332 : 385 (1973)

8. D. Maugis, M. Barquins, J. Phys. D : Appl. Phys, 16 : 1843 (1983)
9. D. Maugis, J. Mater. Sci (in press)
10. D. Mc Lean, "Grain boundaries in metals", Clarendon Press, Oxford (1957).
11. S.M. Wiederhorn, L.H. Bolz, J. Am. Cerom. Soc., 53 : 543 (1970)
12. A.G. Evans, J. Mat. Sci., 7 : 1137 (1972)
13. D.P. Williams, A.G. Evans, J. Test. Eval., 1 : 264 (1973)
14. S.M. Wiederhorn, in "Fracture mechanics of ceramics", vol 2, R.C. Bradt, D.P.H. Hasselman, F.F. Lange, ed, Plenum Press, New York (1974) p. 613
15. A.G. Evans, Int. J. Fracture, 10 : 251 (1974)
16. G.W. Weidmann, G. Holloway, Phys. Chem. Glasses, 15 : 116 (1974)
17. C.J. Simmons, S.W. Freiman, J. Non-Cryst. Sol. 38 : 503 (1980)
17bis B.K. Atkinsow, Pure appl. Geophys, 117 : 1011 (1979)
18. S. Sakagushi, Y. Sawaki, Y. Abe, T. Kawasaki, J. Mat. Sci, 17 : 2878 (1982)
19. S.B. Bhaduri, F.F.Y. Wang, in "Fracture mechanics of ceramics", vol 5, R.C. Bradt, A.G. Evans, D.P.H. Hasselman, F.F. Lange, ed., Plenum Press, New York (1983) p. 327
20. A.R. Miedema, Z. Metallkde, 69 : 287 (1978)
21. B.K. Atkinson, J. Geophys. Res., 89 : 4077 (1984)
22. G.P. Marshall, L.H. Coutts, J.G. Williams, J. Mat. Sci., 9 : 1409 (1974)
23. A.M. Hill, L.A. Dissado, J. Mat. Sci, 19 : 1576 (1984)
24. S.W. Freiman, J. Am. Ceram Soc., 57 : 350 (1974)
25. S.M. Wiederhorn, S.W. Freiman, E.R. Fuller, C.J. Simmons, J. Mater. Sci., 17 : 3460 (1982)
26. T.A. Michalske, S.W. Freiman, J. Am. Ceram. Soc., 66 : 284 (1983)
27. M.K.V. Chan, J.G. Williams, Polymer, 24 : 234 (1983)
28. E.J. Rippling, S. Mostovoy, C. Bersch, J. Adhesion, 3 : 145 (1971)
29. J. Schultz, H. Simon, Verres refract., 34 : 192 (1980)
30. G.E. Boyd, H.K. Livingston, J. Am. Ceram. Soc., 64 : 2383 (1942)
31. G.A. Parks, J. Geophys. Res., 89 : 3997 (1984)
32. J. Fripiat, J. Chaussidon, A.Jelli, "Chimie physique des phenomenes de surface, applications aux oxydes et aux silicates" Masson, Paris (1971)
33. S.M. Wiederhorn, H. Johnson, J. Am. Ceram. Soc., 56 : 192 (1973)
34. B.H. Atkinson, P.G. Meredith, Tectonophysics, 77 : 1 (1981)
35. S.M. Wiederhorn, H. Johnson, J. Am. Ceram. Soc., 58 : 342 (1975)
36. B.J. Hockey in "Fracture mechanics of ceramics", vol 6, R.C. Bradt, A.G. Evans, D.P.H. Hasselman, F.F. Lange, ed, Plenum Press, New York (1983), p. 637
37. S.M. Wiederhorn, P.R. Townsend, J. Am. Ceram. Soc., 53 : 486 (1970)
38. B.R. Lawn, B.J. Hockey, S.M. Wiederhorn, J. Mat. Sci., 15 : 1207 (1980)
39. B.J. Hockey, B.R. Lawn, J. Mat. Sci., 10 : 1275 (1975)
40. S. Ito, M. Tomozawa, J. Am. Ceram. Soc., 65 : 368 (1982)
41. Y. Bando, S. Ito, M. Tomazawa, J. Am. Ceram. Soc., 67 : C 36 (1984)
42. M. Tomozawa, Y. Bando, S. Ito, J. Am. Ceram. Soc. 67 : C 254 (1984)
43. T.A. Michalske, in "Fracture mechanics of ceramics", vol 5, R.C. Bradt, A.G. Evans, D.P.H. Hasselman, F.F. Lange, ed. Plenum Press, New York (1983), p. 277
44. A.J. Holland, W.E.S. Turner, J. Soc. Glass Technol., 21 : 383 (1937)
45. T.C. Baker, F.W. Preston, J. Appl. Phys., 17 : 179 (1946)
46. R.E. Mould, J. Am. Ceram. Soc., 43 : 160 (1960)
47. B.A. Proctor, Phys. Chem. Glasses, 3 : 7 (1962)
48. R.H. Doremus, E.K. Pavelchek, J. Appl. Phys., 46 : 4096 (1975)
49. T.P. Dabbs, B.R. Lawn, J. Am. Ceram. Soc., 65 : C37 (1982)
50. S. Wiederhorn, in "Environment sensitive mechanical behavior" A.R.C. Westwood, N.S. Stoloff, ed., Gordon and Breach, New-York (1966)
51. D.P. Hasselman, D.A. Krohn; R.C. Bradt, J.A. Coppola in "Fracture mechanics of ceramics", vol 2, R.C. Bradt, D.P.H. Hasselman, F.F. Lange, ed. Plenum Press, New York (1974), p. 749
52. J.R. Rice, J. Mech. Phys. Solids, 26 : 61 (1978)

53. E.B. Shand, J. Am. Ceram. Soc., 44 : 21 (1961)
54. E.B. Shand, J. Am. Ceram. Soc., 48 : 43 (1965)
55. N.M. Parikh, J. Am. Ceram. Soc., 41 : 18 (1958)
56. P.J. Sereda, R.F. Feldman, in "The solid-gas interface", E.A. Flood ed, Dekker, New York (1967)
57. D. Maugis, Cah. du Gr. Franç. de Rheologie, 5 : 209 (1980)
58. R. Shuttleworth, Proc. Phys. Soc. A, 63 : 444 (1950)
59. J.C. Eriksson, Surface Sci., 14 : 221 (1969)
60. P.R. Couchman, D.H. Everett, W.A. Jesser, J. Colloid Interf. Sci., 52 : 410 (1975)
61. D.H. Bangham, N. Fakhoury, A.F. Mohamed, Proc. Roy. Soc. A, 147 : 152 (1934)
62. Y. Oka, J.M. Wahl, M. Tomozawa, J. Am. Ceram. Soc., 64 : 456 (1981)
63. P. Becher, J. Am. Ceram. Soc., 66 : 485 (1983)
64. J.S. Mc Farlane, D. Tabor, Proc. Roy. Soc. A 202 : 91 (1950)
65. A. Carré, J. Schultz, J. Adhesion (in press)
66. T.A. Michalke, J.R. Varner, V.D. Frechette in "Fracture mechanics of Ceramics", vol. 4, R.C. Bradt, D.P.H. Hasselman, F.F. Lange, ed, Plenum Press, New York (1978)
67. T.A. Michalske, V.D. Frechette, J. Am. Ceram. Soc., 63 : 603 (1980)
68. N. Perrone, H. Liebowitz, in "Proc. 1st conf. on fracture, Sendai 1965", T. Yokobori, T. Kawasaki, J.C. Swedlow, ed. p 2065 (1966)
69. J. Newman, W.H. Smyrz, Met. Trans., 5 : 469 (1974)
70. S.M. Wiederhorn, in "Fracture mechanics of ceramics", vol. 4, R.C. Bradt, D.P.H. Hasselman, F.F. Lange, ed, Plenum Press, New York (1978)
71. S.M. Wiederhorn, J. Am. Ceram. Soc, 50 : 407 (1967)
72. J.R. Varner, V.D. Frechette, J. Appl. Phys., 42 : 1983 (1971)
73. Y. Hibino, S. Sakaguchi, Y. Tajima, J. Am. Ceram. Soc., 67 : 64 (1984)
74. J. Zarzycki, "Les verres et l'état vitreux", Masson, Paris (1982)
75. J.G. Williams, G.P. Marshall, Proc. Roy. Soc., A 342 : 55 (1975)
76. R.P. Kambour, A.F. Yee, Polym. Eng. Sci., 21 : 218 (1981)
77. S.W. Freiman, J. Am. Ceram. Soc., 58 : 339 (1975)
78. P. Chantikul, B.R. Lawn, H. Richter, S.W. Freiman, J. Am. Ceram. Soc., 66 : 515 (1983)
79. C.L. Quackenbush, V.D. Frechette, J. Am. Ceram. Soc., 61 : 402 (1980)
80. J.T. Davies, E.K. Rideal, "Interfacial phenomena", Academic Press, New York (1961) p. 165.
81. A.B.J. Clark, G.R. Irwin, Exp. Mech. p 321 (June 1966)
82 J.G. Williams, J.C. Radon, C.E. Turner, Polym. Eng. Sci., p 130 (1968)
83. A.K. Green, P.L. Pratt, Eng. Fract. Mech., 6 : 71 (1974)
84. S. Mostovoy, P.B. Crosley, E.J. Ripling, ASTM STP 601. p. 234 (1976)
85. K. Kobayaski, J.W. Dally, in "Fast fracture and crack arrest", ASTM, STP 627, p. 257 (1977)
86. D.W. Aubrey, M. Sherriff, J. Polym. Sci., 18 : 2597 (1980)
87. A.J. Young, P.W.R. Beaumont, J. Mater. Sci., 11 : 776 (1976)
88. G. Gurney, J. Hunt, Proc. Roy. Soc., A299 : 509 (1967)
89. D.H. Kaelble, J. Colloid Sci., 19 : 413 (1964)
90. F.A. Johnson, J.C. Radon, Eng. Fract. Mech., 4 : 555 (1972)
91. J.G. Williams, Int. J. Fract. 8 : 393 (1972)
92. K. Takahashi, S. Hyodo, J. Macromol. Sci. Phys.,B19 : 695 (1981)
93. K.N.G. Fuller, P.G. Fox, J.E. Field, Proc. Roy. Soc., A341 : 537 (1974)
94. P. Penning, Acta Metall. 20 : 1169 (1972)
95. L.P. Kubin, Y. Estrin, Acta Metall., 33 : 397 (1985)
96. A.W. Adamson, "Physical chemistry of surfaces", Wiley, p. 204 (1976)
97. B. Chu, "Molecular forces", Wiley, p. 56 (1967)
98. G.I. Barenblatt, Adv. Appl. Mech., 7 : 55 (1962)
99. H.C. Li, P.L. de Bruyn, Surface Sci., 5 : 203 (1966)

MICROSTRUCTURAL EFFECTS ON THE SUBCRITICAL CRACK GROWTH IN POLYCRYSTALLINE β"-ALUMINA

J. P. Singh,* A. V. Virkar, D. K. Shetty,† and R. S. Gordon

University of Utah
Salt Lake City, Utah 84112

ABSTRACT

Results are presented for the subcritical crack growth behavior of polycrystalline β"-alumina in water as a function of microstructure. A load relaxation technique using double cantilever-beam-type specimens and dynamic fatigue experiments using rectangular bar- and ring-type specimens were utilized to evaluate subcritical crack growth parameter N. The value of N ranged from 64-96 in the load relaxation experiments and from 26-76 in the dynamic fatigue experiments. The N values obtained from the two techniques for identical microstructures indicated no general agreement. This discrepancy is discussed on the basis of the spatial variation in material properties and dynamic effects of crack propagation.

INTRODUCTION

With the increasing use of technical ceramics as structural materials, the achievement of long lifetime for these ceramics is very important. Since the time-dependent failure in a corrosive environment is caused by subcritical growth of the pre-existing flaws, prediction of the lifetime of a brittle ceramic will be greatly influenced by the nature of subcritical crack growth. The nature of subcritical crack growth is usually characterized by the parameter, N, given by the well-known equation[1]

$$V = A\,(K_I/K_{IC})^N, \tag{1}$$

where V is the crack velocity, K_I is the stress intensity factor, K_{IC} is the critical stress intensity factor, and A is a constant. Two very commonly used methods for evaluating N are the load relaxation technique[2] using double-torsion- or double-cantilever-beam-type[3] specimens and the dynamic fatigue experiments[4-6] using bar- or ring-type specimens. In the load relaxation experiments, the crack velocity V is measured as a function of stress intensity factor K_I and the data are used to evaluate N from Eq. (1). In the dynamic fatigue experiments, the fracture stress σ_F

*Now with Argonne National Laboratory, Argonne, IL 60439.
†Now with Battelle Columbus Laboratories, Columbus, OH 43201.

is measured as a function of loading rate $\dot{\sigma}$. The slope of the straight line on a log-log plot of σ_F versus $\dot{\sigma}$ is used to determine the value of N from Eq. (2):[7]

$$\sigma_F = C\,(\dot{\sigma})^{\frac{1}{N+1}}, \tag{2}$$

where C is a constant. However, the results obtained from these two techniques do not always agree. For example, Davidge et al.[4] obtained an N value of 21.5 for 95% alumina by the dynamic fatigue experiments. N values for 95% alumina ranged from 30 to 60 when obtained from K-V diagrams resulting from double-torsion specimens. A similar discrepancy was observed in our laboratory in a preliminary study on β"-alumina specimens. According to Davidge et al.,[4] this discrepancy is due to the difference in the crack size relative to the microstructure in the two types of experiments. In a dynamic fatigue experiment where a small inherent crack may be contained inside a single grain, a lower fracture surface energy is required for crack propagation than in a double-cantilever beam or a double-torsion-type specimen where the initial precracks are large. In a study of the influence of microstructure on crack propagation in ZnSe, Freiman et al.[8] also suggest that the fracture originating from a small flaw contained in a large grain may be controlled by the single-crystal fracture surface energy and not by the polycrystalline surface energy. Kirchner et al.[9] also used the spatial variation of material properties to evaluate the crack growth from small flaws in alumina. However, it is not obvious why the two N values should differ based on the difference of fracture surface energy. Land and Mendiratta[10] as well as Huang et al.[11] explained this discrepancy by taking into account the spatial variation of N and critical stress intensity factor K_{IC}. In this paper, the possibility of crack arrest in the polycrystalline region after it becomes critical due to subcritical crack growth in the region of smaller N and fracture surface energy γ is taken into account. At this point, depending upon the crack size/grain size ratio, the crack may propagate catastrophically or it may stop in a region of higher γ[12] and N and be followed by further subcritical crack growth. The apparent value of N will depend upon which of the two paths the crack follows.

The purpose of this paper is to present the results of a systematic study on slow crack growth behavior of polycrystalline β"-alumina using load relaxation and dynamic fatigue techniques. The difference in the N values obtained from the two techniques are discussed in light of a quantitative model based on the dependence of the crack propagation behavior on the crack size/grain size ratio.

EXPERIMENTAL PROCEDURE

The specimens for both the dynamic fatigue and the load relaxation experiments were prepared from β"-alumina powder of the composition 8.8 wt % Na_2O, 0.75 wt % Li_2O, and 90.45 wt % Al_2O_3. The details of sample preparation are explained in Ref. 13.

The tube and bar specimens were first preformed after which they were isostatically pressed at 55,000 psi. The specimens were then bisque fired at ~900°C for an hour to burn off any volatiles that may have been picked up during the storage. The bisque-fired specimens were heat treated at predetermined temperatures (1400-1600°C) in platinum encapsulation and powder packs.[13] The details of heat treatment are given in Ref. 13.

Dynamic fatigue experiments consisted of measuring fracture stress as a function of loading rate using bar- and ring-type specimens. The experiments were conducted in deionized water at room temperature (~23°C). Bar-type specimens of ~4.6 x 1.0 x 0.25 cm were used to measure the fracture stress in four-point bending with a load span of 1.9 cm and a support span of 3.5 cm. The fracture stress, σ_f, was calculated using the formula

$$\sigma_f = \frac{3}{2} P_f \frac{(L - a)}{bt^2} , \tag{3}$$

where P_f is the fracture load, b is the specimen width, t is the specimen thickness, a is the loading span, and L is the support span.

The ring-type specimens were cut from sintered β"-alumina tubes. The wall thickness of the tube was approximately 1 mm. The fracture stress, σ_f, of the ring in the diametral test was calculated using the formula[14]

$$\sigma_f = \frac{P_f K (a/b)}{\pi bL} , \tag{4}$$

where P_f is the breaking load, a and b are inner and outer radii of the tube, respectively, L is the thickness of the ring, and K(a/b) is a function of a/b such that it rapidly increases with a/b beyond a/b ~0.8.

Loading rate ($\dot{\sigma}$) in the dynamic fatigue experiments was varied from ~0.1 to ~55 MN/m^2s. Very small loading rates were obtained using a rubber cork in series with the loading system on a universal testing machine.

RESULTS AND DISCUSSION

Results of dynamic fatigue experiments are shown in Figs. 1 and 2 which are the plots of fracture stress versus loading rate on a log-log scale. Least squares analysis was utilized to fit straight lines through the data points. The values of the subcritical crack growth parameter N were determined from the slopes of these straight lines using Eq. (2). Figure 1(a) represents the results of the diametral tests for β"-alumina tubes sintered at 1597°C for 5 min and annealed at 1400°C for 2 h in a box furnace. The N value, as obtained from the slope of the strength plot, was 26. Each data point represents the average of nine specimens. Results of the four-point-bending experiment for bar specimens with identical heat treatment are shown in Fig. 1(b) with an N value equal to 16.

The results for four-point-bending experiments for bar specimens sintered at 1600°C for 1 h and annealed at 1400°C for 2 h in a tube furnace are shown in Fig. 2. The N value as obtained from the slope of strength plot was found to be 76. Each data point in the four-point-bending experiment represents the average of five specimens.

The N values obtained from the load relaxation experiments using double-cantilever-type specimens with identical heat treatments as those for specimens in Figs. 1(a) and 2 are 96 and 64, respectively. The corresponding microstructures for these specimiens are shown in Figs. 3 and 4. Figure 3 shows the microstructure of β"-alumina sintered at 1597°C for 5 min and annealed at 1400°C for 2 h. The microstructure is duplex in which large grains are embedded in a fine-grained matrix. The corresponding microstructure for β"-alumina sintered at 1600°C for 1 h and annealed at 1400°C for 2 h is relatively uniform as shown in Fig. 4.

(a)

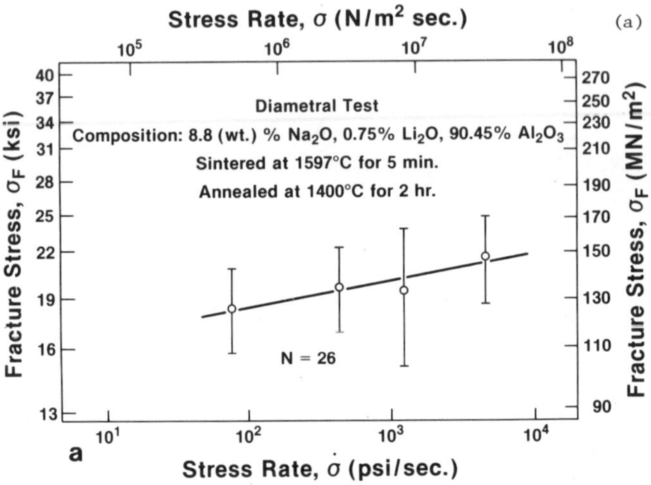

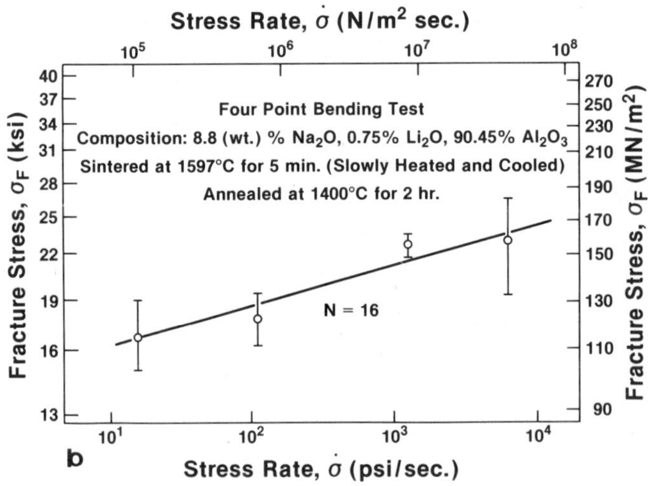

Fig. 1. Strength-loading rate data for β"-alumina (sintered at 1597°C for 5 min and annealed at 1400°C for 2 h) in water at room temperature obtained from (a) the diametral test and (b) the four-point-bend test. Subcritical crack growth parameter N was determined from the slope of the straight line plot using Eq. (2).

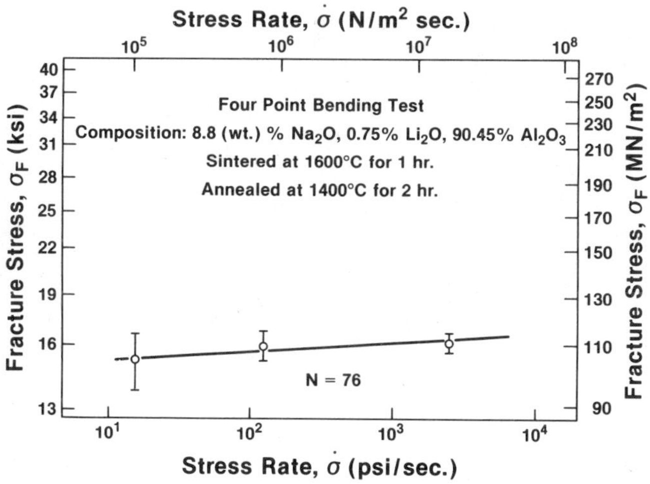

Fig. 2. Strength-loading rate data for β"-alumina (sintered at 1600°C for 1 h and annealed at 1400°C for 2 h) in water at room temperature obtained from the four-point-bending test. Subcritical crack growth parameter N was determined from the slope of the straight line plot using Eq. (2).

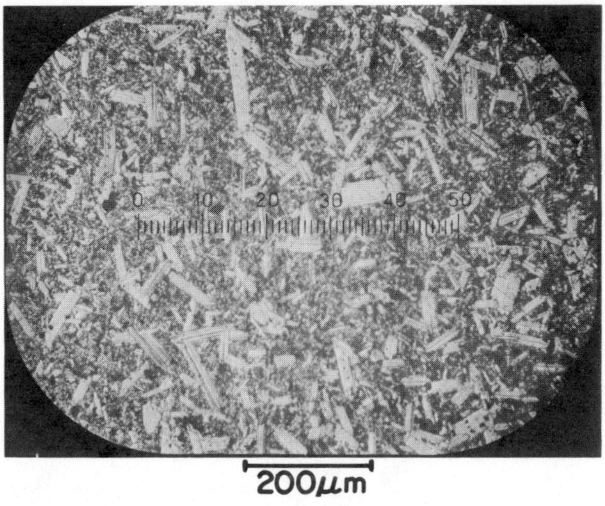

Fig. 3. Microstructure of β"-alumina sintered at 1597°C for 5 min and annealed at 1400°C for 2 h.

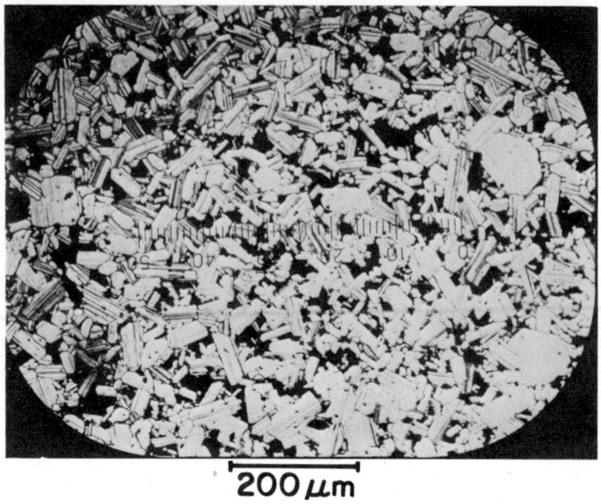

Fig. 4. Microstructure of β"-alumina sintered at 1600°C for 1 h and annealed at 1400°C for 2 h.

A comparison of N values obtained from the stress relaxation experiments and the dynamic fatigue experiments is shown in Table 1. These values indicate that the difference in N values obtained from the dynamic fatigue and stress relaxation is small for specimen type B, whereas the difference is large for specimen type A. A statistical analysis was used to find out that for 90% confidence interval, the N value (26) obtained from the dynamic fatigue experiments would be in the range of 15 to 90. These statistics were based on a normal distribution of σ_f for a given $\dot{\sigma}$ value. A consistent trend of a lower[4] N value obtained from the dynamic fatigue experiments led to a belief that there may be a fundamental reason in addition to a statistical variation for this discrepancy. As indicated in the introduction, it is proposed that a part of this discrepancy may be due to the spatial variation of the material properties (N, K_{IC}). For stress relaxation experiments, the propagation of a large precrack is controlled by the polycrystalline material properties and the resulting value of N corresponds to polycrystalline value. On the other hand, in the case of dynamic fatigue experiments where the inherent cracks are small, the crack propagation may not be controlled by polycrystalline material properties alone.[12] In this case, a small inherent crack subcritically grows with increasing load until the stress intensity factor at the crack tip becomes critical, at which time dynamic crack propagation[12] ensues. The effect of loading rate is to change the amount of subcritical crack growth before the onset of the dynamic crack propagation. For a large grained material, the inherent crack is generally contained inside a single grain and the initial subcritical crack growth is controlled by single-crystal material properties. With increasing load, the crack grows subcritically inside the grain before the onset of dynamic crack propagation. As the crack propagates dynamically, it enters the polycrystalline region of higher fracture surface energy which offers higher resistance to crack propagation. However, the specimen has already gathered a certain amount of kinetic energy due to the laterally moving mass of the body while the crack is moving through the single-crystal region. The resistance to crack growth in polycrystalline region will tend to slow down the crack and the crack may arrest when all kinetic energy is converted to surface and strain energy. If kinetic energy is larger than required to create new fracture surfaces, the crack will not

Table 1. Comparison of the Values of Subcritical Crack Growth Parameter N Obtained from Dynamic Fatigue Experiments and from Stress Relaxation Experiments

Specimen Type	Heat Treatment	Value of N: Dynamic Fatigue Experiment	Value of N: Stress Relaxation Experiment
A	Sintered at 1597°C for 5 min followed by annealing at 1400°C for 2 h	26	96
B	Sintered at 1600°C for 1 h followed by annealing at 1400°C for 2 h	76	64

arrest inside the body and a complete fracture will result. In the first case, the crack arrest in the polycrystalline region will be followed by further subcritical crack growth until the critical stress intensity factor corresponding to the polycrystalline region is reached at the crack tip when complete fracture will occur. The fracture stress in this case will be affected by the single-crystal and the polycrystalline values of material properties as well as by the crack size/grain size ratio. In the latter case in which the crack does not stop in the polycrystalline region, the fracture stress will be controlled by the single-crystal material properties. From the above discussion, it is clear that the N value obtained from the slope of the fracture stress versus loading rate plot in a dynamic fatigue experiment may be a function of single-crystal and polycrystalline material properties as well as crack size/grain size ratio. These concepts were used to develop an analytical model for the calculation of fracture stress as a function of loading rate as described in the appendix. For the numerical calculations, the following values of the specimen size and the various parameters were used:

Specimen width (B) = specimen thickness (W) = 5×10^{-3} m;
specimen length (L) = 5×10^{-2} m, A = 10^{-1} m/s;
Poisson's ratio (ν) = 0.25;
single-crystal value for critical stress intensity factor ($K_{IC(s)}$) = 5×10^{5} N/m$^{3/2}$;
polycrystalline value for critical stress intensity factor ($K_{IC(p)}$) = 11.2×10^{5} N/m$^{3/2}$;
initial flaw size (Co) = 10 μm;
subcritical crack growth parameter N_s corresponding to single-crystal region = 30; and
N_p corresponding to polycrystalline region = 40.

Using the analytical technique described in the appendix, fracture stress versus loading rate plots were constructed for various grain sizes. Figure 5 represents the plots of strength versus loading rate calculated for grain sizes 50 and 160 μm. These grain sizes represent the large grains in which the inherent cracks are contained. The N values obtained from the slopes of these lines are indicated in the figure. From these plots (and several others not shown here), it is clear that the effective values of N in dynamic fatigue may range from N_S (single-crystal value) to N_P (polycrystalline value) and even higher. For large grain size (e.g., D = 160 μm), fracture stress is controlled by subcritical

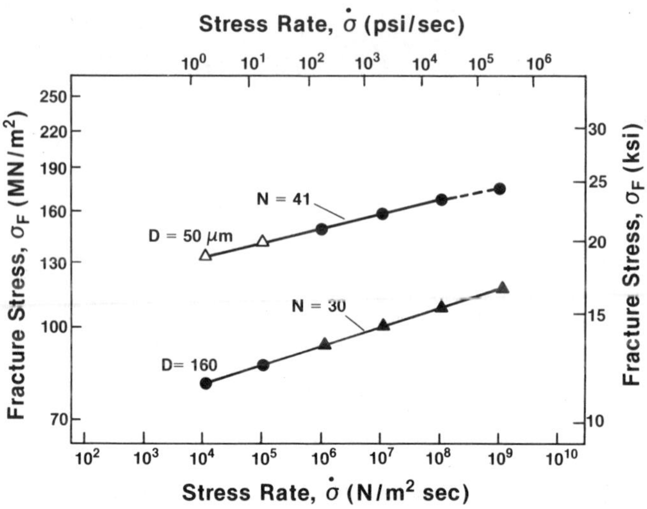

Fig. 5. Strength versus loading plots calculated for grain size D = 50 and 160 μm. The effective values of the subcritical crack growth parameter N were determined from the slopes of the straight line plots using Eq. (2). ● - slow crack growth in both single and polycrystalline regimes; ▲ - slow crack growth in single crystal region followed by a catastrophic failure; and Δ - slow crack growth in both single and polycrystalline regimes without dynamic fracture.

crack growth in a single-crystal region for most of the loading rates. Thus, the calculated N value equals the single-crystal value N_S. On the other hand, for intermediate small grain size (e.g., 50 μm), the fracture stress is controlled by subcritical crack growth in both single-crystal and polycrystalline regions as well as by the dynamic crack propagation. In this case, the calculated value of N approximately equals N_P (the polycrystalline value).

A closer look at the microstructure of the β"-alumina in Figs. 3 and 4 clearly indicates that specimen set A has large-grained duplex microstructures whereas the grain structure of specimen set B is relatively small and uniform. Thus, for a given initial flaw size, the dynamic crack propagation will influence the crack growth in two sets of specimens differently. For specimen set A with large grains, the crack size may be much smaller than the grain size. Hence, a subcritical crack growth in the single-crystal region will probably control the fracture stress for most of the stress rates. Consequently, the fracture stress versus stress rate plot for specimen set A may be similar to a plot for grain size D = 160 μm in Fig. 5 where the apparent N value corresponds to a single-crystal value. Obviously, the difference between this N value and the one obtained from the stress relaxatin experiment is large since the N value in the stress relaxation experiment corresponds to a polycrystalline value. However, for specimen set B with intermediate grain sizes, the effects of dynamic crack propagation may be predominant and the fracture stress versus stress rate plot may be similar to one for D = 50 μm in Fig. 5. In this case, the apparent N value is approximately equal to the polycrystalline value. Consequently, the difference between the N values obtained from the dynamic fatigue and stress relaxation experiments is minimum for specimen set B.

It is to be noted that the exact evaluation of the fracture stress as a function of stress rate was not made for β"-alumina specimens with different microstructures due to the unavailability of several materials and crack parameters. However, a qualitative agreement between the experimental results and analytical predictions for the variation of N with grain size suggests that N values obtained from the dynamic fatigue and stress relaxation experiments may or may not agree depending upon the crack size/grain size ratio and the nature of the spatial variation of the material properties. Therefore, caution must be exercised in using the value of N for lifetime prediction.

ACKNOWLEDGMENT

This work was supported by the National Science Foundation (RANN), Contract No. C805 and NSF Contract No. DMR 76-04110.

REFERENCES

1. S. M. Wiederhorn, p. 613 in: "Fracture Mechanics of Ceramics," Vol. 2, ed. R. C. Bradt, D. P. H. Hasselman, and F. F. Lange, Plenum Press, New York, 1979.
2. A. G. Evans, J. Mater. Sci. 7, 1137 (1972).
3. A. V. Virkar and R. S. Gordon, J. Am. Ceram. Soc. 59 [1-2], 68 (1976).
4. R. W. Davidge, J. R. McLaren, and G. Tappin, J. Mater. Sci., 8[12], 1699 (1973).
5. R. S. Gordon, "Fabrication and Characterization of β"-Alumina Electrolytes for Ford Sodium Sulfur Battery," Annual Technical Report for the Period July 1973-July 1974, prepared under subcontract to Ford Motor Co. under the support of the NSF (RANN) Contract NSF-C805.
6. J. E. Ritter, Jr. and C. L. Sherburne, J. Am. Ceram. Soc., 54[12], 601 (1971).
7. S. M. Wiederhorn, p. 6136 in Fracture Mechanics of Ceramics, Vol. 2, ed. R. C. Bradt, D. P. H. Hasselman, and F. F. Lange, Plenum Press, New York, 1974.
8. S. W. Freiman, J. J. Mecholsky, Jr., and R. W. Rice, J. Am. Ceram. Soc., 54[9-10], 406 (1975).
9. H. P. Kirchner and J. M. Ragosta, J. Am. Ceram. Soc., 63[9-10], 490 (1980).
10. P. L. Land and M. G. Mendiratta, J. Mater. Sci. 12[7], 1421 (1977).
11. J. L. Huang, D. K. Shetty, and A. V. Virkar, unpublished work.
12. J. P. Singh, A. V. Virkar, D. K. Shetty, and R. S. Gordon, J. Am. Ceram. Soc., 62[3-4], 179 (1979).
13. J. P. Singh, "Fracture Studies in Ceramics with a Particular Emphasis on β"-Alumina," Ph.D. dissertation, University of Utah, 1979.
14. E. A. Ripperberger and N. Davids, Trans. Am. Soc. Civ. Engr., Paper No. 2308, p. 619 (1948).
15. I. N. Sneddon and M. Lowengrub, "Crack Problems in the Classical Theory of Elasticity," John Wiley & Sons, Inc., New York, 1969.

APPENDIX

The present analysis of flexural strength as a function of loading rate is based on strength-grain size model by Singh et al.[12] A rectangular bar specimen of a polycrystalline ceramic loaded in three-point-bending mode in a corrosive environment is considered for this analysis. The specimen has a semicircular crack of radius C_o contained inside a single grain on the tensile surface. It is assumed that subcritical crack growth parameter N, critical stress intensity factor K_{IC}, and fracture surface energy γ depend on crack size (C)/grain size (D) ratio. For analytical convenience, the values of N, K_{IC}, (and γ) were taken to be N_S, $K_{IC(S)}$, and γ_S corresponding to the single-crystal region ($C/D \leq 1/2$). The corresponding polycrystalline values were N_P, $K_{IC(P)}$, (and γ_P) for $C/D > 1/2$. The specimen is loaded at a constant rate $\dot{P}$. The corresponding stress rate $\dot{\sigma}$ in the outer fiber is given by

$$\dot{\sigma} = \frac{3}{2} \frac{L}{BW^2} \dot{P} , \tag{A-1}$$

where L, B, and W are the dimensions shown in Fig. A-1. The stress σ at any time t is given by $\dot{\sigma}t$. The inherent crack of length C_o subcritically grows during loading and its length at any time t is given by C. The corresponding stress intensity factor K_I is given by[15]

$$K_I = 2\sqrt{\frac{1-\nu^2}{\pi}}\, C\, \sigma , \tag{A-2}$$

where ν is Poisson's ratio. For $P = \dot{P}t$ ($\sigma = \dot{\sigma}t$), Eqs. (A-1) and (A-2) result in

$$K_I = Y\sqrt{C}\, P = Y\sqrt{C}\, \dot{P}t, \tag{A-3}$$

where

$$Y = 3\sqrt{\frac{1-\nu^2}{\pi}} \frac{L}{BW^2} .$$

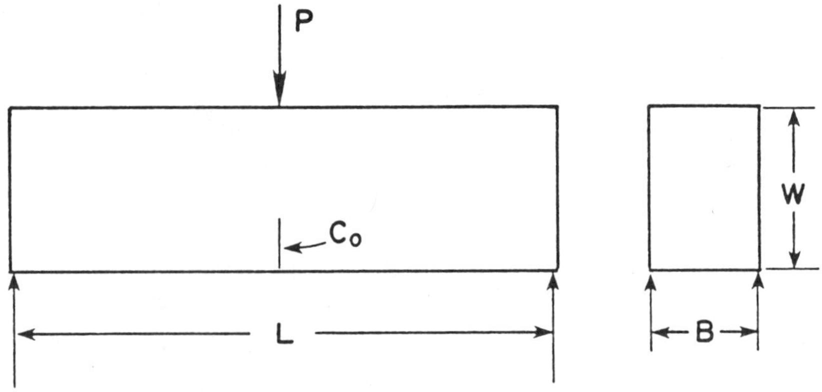

Fig. A-1. Schematic of a three-point-bend specimen.

For a specimen being loaded in a corrosive environment, the crack growth occurs much before the critical stress intensity factor K_{IC} is reached at the crack tip. The crack velocity V is given by[1]

$$V = A\,(K_I/K_{IC})^N, \tag{A-4}$$

where A is a constant and other terms have been defined earlier. Using $V = dC/dt$, from Eqs. (A-3) and (A-4), it can be shown that

$$\frac{dC}{dt} = \alpha\,\dot{P}^N C^{N/2} t^N, \tag{A-5}$$

where $\alpha = AY^N/K_{IC}$. Upon integration of Eq. (A-5) between time t_o, and t, one obtains

$$C_1^{-\frac{N}{2}+1} - C_o^{-\frac{N}{2}+1} = \frac{\alpha\,\dot{P}^N}{(N+1)}\left[t_1^{N+1} - t_o^{N+1}\right], \tag{A-6}$$

where C_o and C_1 are crack lengths at times t_o and t_1, respectively. Thus, for a given initial flaw size C_o, the time $(t_1 - t_o)$ required for it to grow to C_1 can be calculated by Eq (A-6). Hence, at the end of a small subcritical crack growth of $\Delta C = C_1 - C_o$, the load P_1 is given by $P_1 = P_o + \dot{P}\,(t_1 - t_o)$. With a known load P_1 (or stress σ_1), and the crack length C_1, the stress intensity factor K_{I1} can be calculated at the end of time t_1. In general, for the ith increment in crack length:

$$C_i^{-\frac{N}{2}+1} - C_{i-1}^{-\frac{N}{2}+1} = \frac{\alpha\,\dot{P}^N}{N+1}\left[t_i^{N+1} - t_{i-1}^{N+1}\right], \tag{A-7}$$

or

$$t_i = \left[\frac{(N+1)}{\alpha\,\dot{P}^N}\left\{C_i^{-\frac{N}{2}+1} - C_{i-1}^{-\frac{N}{2}+1}\right\} + t_{i-1}^{N+1}\right]^{\frac{1}{N+1}}. \tag{A-8}$$

The crack length C_i, load P_i, and stress intensity factor K_{Ii} after the ith increment in crack length is given by $Co+i\Delta C$, $P_o+\dot{P}\,(t_i - t_o)$, and $Y P_i \sqrt{C_i}$, respectively. After i^{th} iteration, the calculated value of K_{Ii} can be compared with K_{IC} to obtain the criticality condition. The increase in K_{Ii} with time can follow either of the following two paths:

(1) The stress intensity factor K_{Ii} can become equal to $K_{IC(S)}$ with the corresponding crack length C_i being less than D/2 (Fig. A-2). At this point, the crack becomes unstable and a dynamic crack propagation[12] would either cause a complete fracture or the crack would finally arrest at C_2 in the polycrystalline surface energy region. In the latter case, the crack arrest is followed by another subcritical crack growth in the polycrystalline region to C_3 where K_{Ii} equals $K_{IC(P)}$ and a complete fracture occurs. In the former case, the fracture stress is given by the stress corresponding to crack length C_1, whereas in the latter case, it is given by the stress corresponding to crack length C_3.

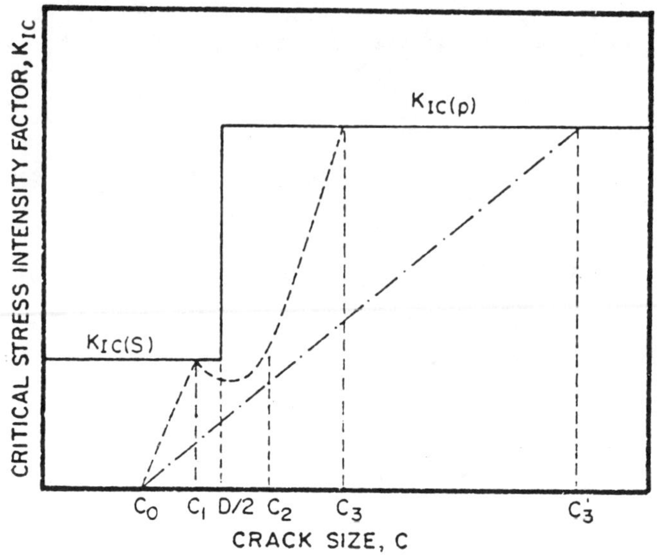

Fig. A-2. Schematic showing the crack path during subcritical crack growth.

(2) If the stress intensity factor K_{Ii} does not equal $K_{IC(S)}$ before the corresponding crack length, C_i equals D/2, the dynamic crack propagation does not take place in the single-crystal region and the subcritical crack growth continues until K_{Ii} equals $K_{IC(P)}$ where the crack length C_i equals C_3' (Fig. A-2). At this point, catastrophic fracture occurs and the fracture stress is given by the stress corresponding to the crack length C_3'. The K_{Ii} versus C_i path is shown by the chain line in Fig. A-2.

Using this mathematical approach, fracture stress was calculated for different loading rates and straight line plots of fracture stress versus loading rate were obtained for different grain sizes. The slopes of these straight line plots were used to calculate the effective values of subcritical crack growth parameter N.

CRACK GROWTH IN SINGLE-CRYSTAL SILICON*

C. P. Chen and M. H. Leipold

Jet Propulsion Laboratory
California Institute of Technology
Pasadena, CA 91109

ABSTRACT

Crack growth in single-crystal silicon at room temperature in air was evaluated by double torsion (DT) load-relaxation method and monitored by acoustic emission (AE) technique.

Both DT and AE method indicated lack of subcritical crack growth in silicon. At the critical stress intensity factor, the crack front was found to be jumping several times in a "mirror" region and then followed by fast crack growth in a "hackle" region. Hackle marks were found to be associated with plastic deformation at the tip of the fast moving crack. No dislocation etch pits were found in the "mirror" region, in which crack growth may result from interatomic bonds broken at the crack tip under stress without any plastic defomation. Acoustic emission appears to be spontaneously generated from both interatomic bonds broken and dislocation generation at the moving crack tip during the crack growth in single-crystal silicon.

1. INTRODUCTION

The physics of crack growth in brittle ceramics has been developed quite rapidly in the past 30 years or so, long after the Griffith's original work[1]. Efforts have been made to provide models that allow us to understand better on the fundamental mechanisms of crack growth and fracture of ceramics.

One theory is based on the phenomenon of the chemical interaction at the strained crack-tip bonds and has been developed rather extensively for silica and water molecules from the environment. In this theory, crack growth in this system and similar brittle ceramics can be described by the crack velocity (V) and the stress intensity factor (K) for a given system (microstructure and corrosive species). Each segment of the typical K-V curve[2] can be described by a different rate controlling mechanism. In general, there are three principal regions: Region I in

*One phase of Flat-Plate Solar Array Project sponsored by the Department of Energy through an agreement with NASA.

which the rate of chemical reaction at the crack tip controls crack motion; Region II in which the crack velocity is essentially constant and crack motion is controlled by diffusion of corrosive species to the crack tip; Region III in which the crack velocity is independent of chemical species and increases very rapidly with increase in K. At the onset of Region III, K is close to K_{IC}, the critical stress intensity factor for crack propagation in the absence of slow crack growth. This model has been received rather extensive evaluation on glasses and ionic ceramics (e.g., Ref. 2-5). Crack growth in these materials is controlled by environmental enhanced stress-corrosion process.

Another model for crack propagation and brittle fracture of solids to illustrate the phenomenon of "lattice trapping" at crack tip has been discussed by Fuller and Thomson[6]. This model proposes a crack that is stably trapped in the atomic lattice structure. The mechanism of crack growth in material is accomplished from interatomic bonds broken at the crack tip by thermal activation over the lattice-trapping energy barriers; and by quantum tunneling through the barrier.

Both stress-corrosion and lattice trapping models consider that the materials are ideally brittle, i.e., there are no zones of plastic deformation at propagation crack tip. Nevertheless, it is believed that some plastic flow occurs at the crack tip during the crystal fracture even for very brittle materials at room temperature, such as silicon. A crude criterion to determine the presence of ductility in brittle frature was first proposed by Kelly et al[7]. This criterion was further developed by Rice and Thomson[8] to determine the condition of ductile fracture versus brittle fracture. These models were developed for brittle fracture in crystals in which dislocations are spontaneously generated from an atomically sharp cleavage crack.

Crystalline silicon in its reponse to mechanical stress behaves as classic brittle material and fails by extension of pre-existing flaws. It is also an important semiconductor material for photovoltaic and microelectronic applications. Single crystal silicon exhibits cleavage cracking in the {111} plane[9]. Subcritical crack growth in silicon can reduce the long term mechanical integrity of the material and the electrical performance of the device.

There has been controversy on the existence of subcritical crack growth in crystalline silicon. Chen and Knapp[10] performed delayed-fracture tests of single-crystal silicon under several liquid solutions. Their results showed no indication of delayed fracture. Bhaduri and Wang[11], however, studied slow crack growth of single crystal silicon under the influence of corrosive etchant and electrical fields. Their results indicated that corrosive etchant or applied electric field can induce subcritical crack growth in silicon. The tendency of stress corrosion slow crack growth in single crystal silicon tested in ambient air was observed. Recently, Chen and Leipold[12] evaluated the effect of stress rate and proof testing of silicon wafers in ambient air. Dynamic fatigue test results suggested that subcritical crack growth does not occur in crystalline silicon. More understanding of the crack growth behavior in crystalline silicon is desirable.

Acoustic emission (AE) techniques have been used to evaluate the fracture process of brittle ceramics[13-16] and glasses[17]. AE study of the crack propagation in silicon can provide additional information on the mechanism of crack growth in crystalline solids.

The purpose of this paper is to present the results of the evaluation of crack growth behavior in single-crystal silicon in ambient air at room temperature monitored by acoustic emission technique. Attempt was made to resolve the question of subcritical crack growth and to determine the mechanism of crack growth process in single-crystal silicon.

2. EXPERIMENTAL PROCEDURE

A. Test Method

The double torsion load relaxation method[18] was used to measure crack velocity while simultaneous AE measurements were made on the specimens. The double torsion (DT) test is shown schematically in Fig. 1. Since the specimens used for these tests were precracked but ungrooved, the stress instensity factor, K_I, of the specimen is given[18] by

$$K_I = PWm \left[\frac{3}{Wd^4(1-\nu)\xi}\right]^{1/2} \tag{1}$$

where P is the load

ν is the Poisson's ratio

Wm, W and d are defined in Fig. 1 for a typical specimen and loading configuration.

ξ is the specimen thickness correction factor. A simplified expression for ξ in terms of a thickness ratio, t=2d/W, which is accurate to better than 0.1 percent, is given by,

$$\xi = 1 - 0.6302t + 1.20te^{-\frac{\pi}{t}} \tag{2}$$

For the constant displacement-load relaxation method, the crack growth velocity, V, is related[18] to the instantaneous load and the corresponding load relaxation rate (dP/dt) as

$$V = -\frac{\left(P_{i,f}\right)\left(\alpha_{i,f}\right)}{P^2}\left(\frac{dP}{dt}\right) \tag{3}$$

where

P_i = initial load
α_i = initial crack length
P_f = final load, and
α_f = final crack length.

Thus, by measuring the instantaneous load and corresponding dP/dt, and the crack length before and after a load relaxation experiment, V can be determined over a range of K_I from a single experiment.

These tests were carried out at room temperature in ambient air.

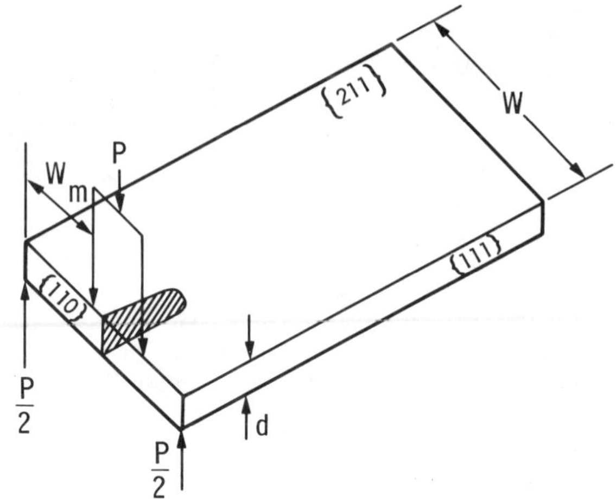

Fig. 1. Schematic showing the double torsion test of silicon wafer specimen

B. Specimens

The specimens were rectangular wafers, approximately 50 x 100 mm and 0.76 mm in thickness. The dimensions of each test sample were measured individually by a micrometer. The specimens were cut from a Czochralski single crystal silicon+. The silicon material was boron doped, p-type with resistivity ranging from 0.8 to 2.0 ohm-cm. The surface orientations of the specimens are shown in Fig. 1. These surface orientations were examined by the Laue technique and were found to be accurate within 5°. All the surfaces were in as-cut condition.

Since single crystal silicon has a minimum fracture toughness[9] on the {111} lattice plane, cleavage crack propagation in {111} is usually found in silicon. Each specimen was precracked to various length from one end surface as shown in Fig. 1 by pushing a wedge into a notch. Therefore, this initial crack is along a {111} cleavage plane of each specimen. The initial crack length was measured by a stereo microscope. Acoustic emission measurement of the crack propagation along this cleavage plane was evaluated.

C. Acoustic Emission Measurements

Acoustic emission pulses were detected and analyzed by a computerized multichannel AE data acquisition system.** A six transducer*** array was applied to the test specimen to establish the location of the propagating crack. The location of the transducers are shown in Fig. 2. Transducers 1 to 4 formed a square array while the transducers 5 and 6 formed a linear array. Epoxy glue was used to cement transducers in

+Supplied by Applied Solar Energy Corp., City of Industry, CA
**System Model No. 8000, Dunegan Corp., Mission Viejo, CA
***Model S9223M3 miniature AE sensors, Dunegan Corp.

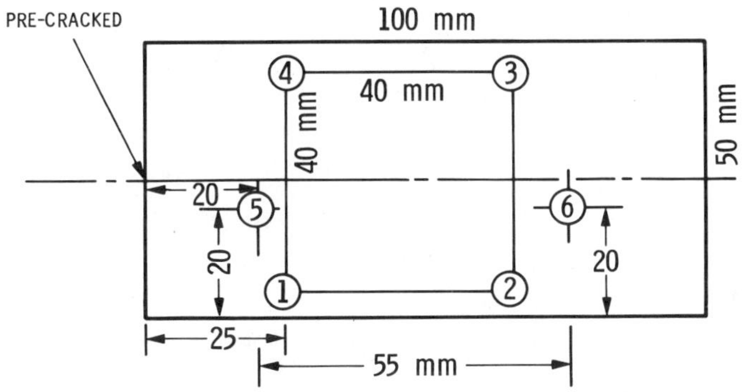

Fig. 2. Location of the transducers.

position on the specimen. The output of each sensor was amplified by a preamplifier (40 dB gain) before going into the data acquisition system.

The legitimate AE events are expected to be detected by both sensor array sets. The data analysis discussed here will concentrate on the signals recorded by square array, while the linear array data is used for confirmation. Before testing, the accuracy of the source locations was calibrated using lead pencil breaking which simulated cracking events.

The loading conditions and AE measurements were recorded on a floppy disc of the computer in the AE system which can also be used to perform on-site analysis as well as the post test analysis.

During the post test analyses, filtering of any amplitude level can be applied to the computer-stored AE data. Therefore, the undesirable noise from the AE tests can be eliminated. Filter levels here are total and thus include reduction of the 40 dB preamplifier gain.

3. RESULTS AND DISCUSSION

A. Double Torsion Test

Constant crosshead rate was applied to the double torsion load relaxation test of the specimen. For conventional ceramics (e.g., $A\ell_2O_3$), when the stress intensity factor is sufficient that the crack starts moving rapidly, a large load relaxation is usually observed as a function of time after the crosshead movement has been arrested. The crack growth rate is then determined from the relaxation curve.

A typical load versus time and deflection versus time curves from the strip chart recorder for a silicon test are shown in Fig. 3. The deflection was measured by an extensometer under the sample, directly underneath the inner loading point. The load application was stopped at the first appearance of a load decrease. In Fig. 3, the load versus time curve reveals that the crosshead was stopped after two peak loadings. The occurrence of these peak loadings resulted from the repeated process of crack front jumping and load dropping before the load application was

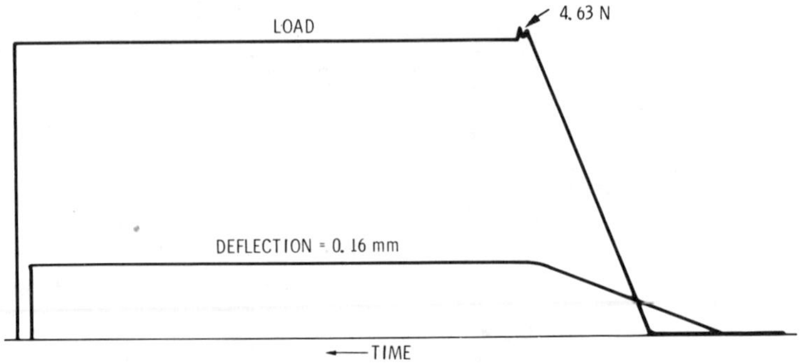

Fig. 3. Typical load versus time and deflection versus time curves from the strip chart recorder for a silicon test.

stopped. When the crosshead motion was arrested, the load held constant. Since no load relaxation during load holding period was found, the slow crack growth velocity from Eq. (3) cannot be determined. Therefore, Regions I and II of K-V curve of silicon cannot be observed. This suggested that no load relaxation and no subcritical crack growth can be observed. The load holding times were usually 5 to 10 minutes. The specimen was unloaded and removed from the test jig for crack length measurement using a stereo microscope and to take an X-ray topography photo. This test process can be repeated on a specimen until the crack is propagated through the specimen.

The critical stress intensity factor (K_{IC}, fracture toughness) was calculated by equation (1). The measured K_{IC} values in {111} cleavage plane as a function of crack length is shown in Fig. 4. As shown in Fig. 4, an increase in K_{IC} at short crack lengths can be expected[18]. The measured K_{IC} within the valid range[18] $0.18 < a/L < 0.78$ is averaged to be 1 $MN/m^{3/2}$. This value is greater than those measured using indentation-induced-flaw technique[9], because of the presence of residual stresses at the indentation site.

B. Acoustic Emission Behavior

Considerable noise was usually detected by the AE measurement. During the post test analyses several amplitude levels of filtering were applied to the data during computer play-back. It was found that AE peak amplitudes greater than 60 dB occurred along the line of cleavage crack propagation in the silicon sample. These peak AE amplitudes were found only when the stress intensity factor reached the critical value.

The typical AE count rate versus time curve and the locations of the AE events after the 60 dB filter are shown in Fig. 5(b). The dots in Fig. 5(a) indicate the location of AE events. The squares containing dots indicate event clusters. The location of each event as a function of time is well defined. Therefore, the crack extensions of each DT crack propagation test can be determined. Fig. 6 shows the crack extensions detected by AE method of 5 progressive DT tests of a silicon specimen. The crack extension and crack size determined from AE method was

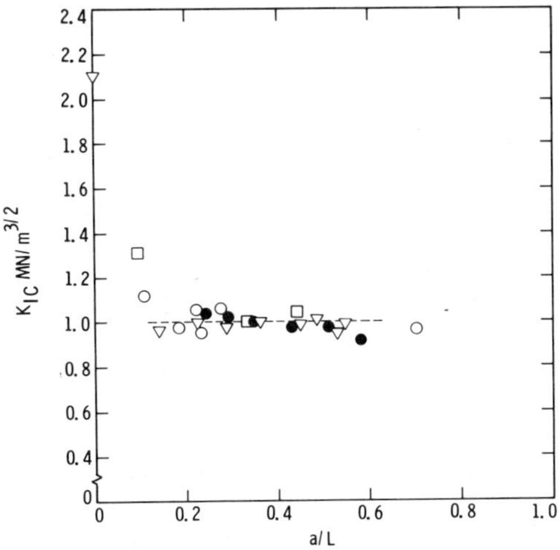

Fig. 4. Measured K_{IC} in {111} cleavage plane of single crystal silicon as a function of crack length (symbols indicated different samples)

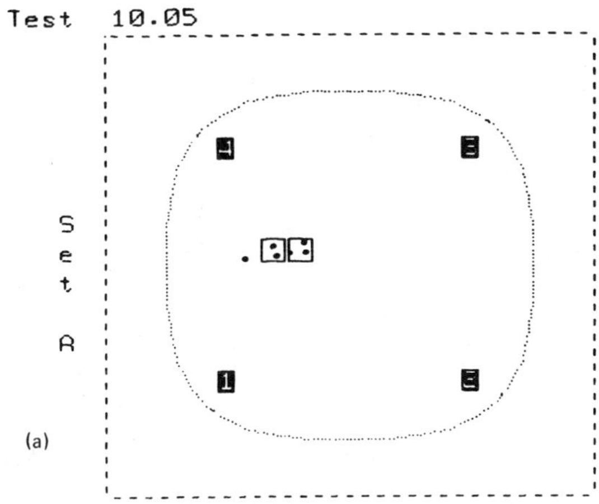

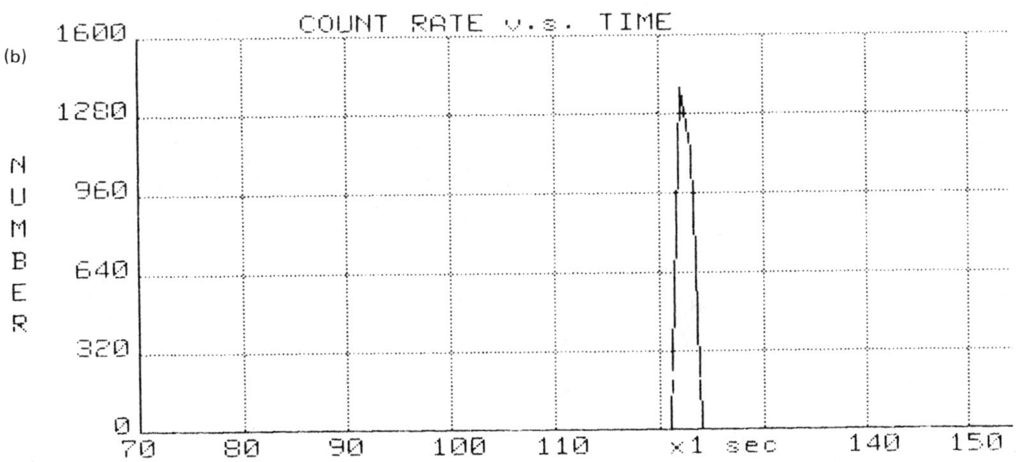

Fig. 5. (a) Locations of the AE events and (b) the count rate versus time of a typical AE test of single crystal silicon. 60 dB filter was applied.

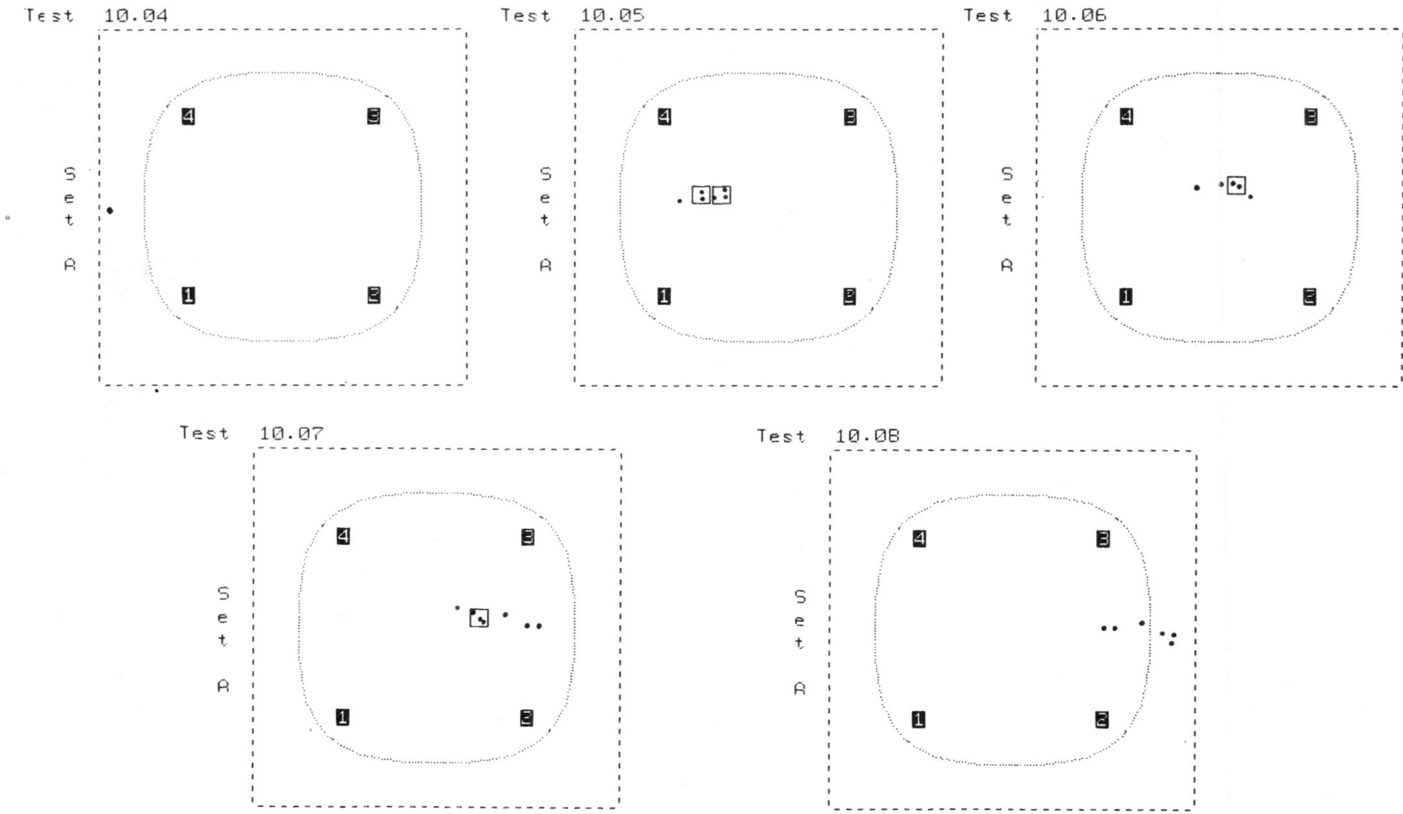

Fig. 6. The detected crack locations by AE method of 5 progressive DT tests of a silicon specimen

found to be longer than that from DT load relaxation tests. This results from the fact that the crack front is not prependicular to the specimen surface. These results were correlated with the fracture surface evaluation. This will be discussed later. Fig. 5(b) indicates the AE count rate versus time of a typical DT test (e.g., Test No. 10.05). By using the load versus time curve in Fig. 3 and the AE event vs time Fig. 5(b), it was found that the maximum count rate occurred at the maximum loading at which the stress intensity factor at the crack front was critical and crack front jumped. As shown in Fig. 5(b), no acoustic emission events were detected during loading before reaching the critical stress intensity factor nor during the load relaxation holding period. This result also suggests no subcritical crack growth in silicon. A more detailed acoustic emission behavior of crack propagation in silicon was reported in Ref. 19.

C. Crack Growth Behavior

The crack growth on the {111} cleavage plane of each silicon specimen was examined by X-ray topography between crack extensions and AE monitoring. Typical X-ray topograhic photos showing crack extension in a sample from "a" to "b" of a DT load relaxation test are given in Fig. 7(a) and (b), respectively. A small zone suggesting a local strain field surrounded the crack tip arresting point was usually found on the X-ray topography. Note that the zone suggesting local strain at the crack tip arresting point "a" remained in position, as shown in Fig. 7(b), when the crack tip moved further to point "b". This observation suggested that plastic deformation occurred while the crack was stopped.

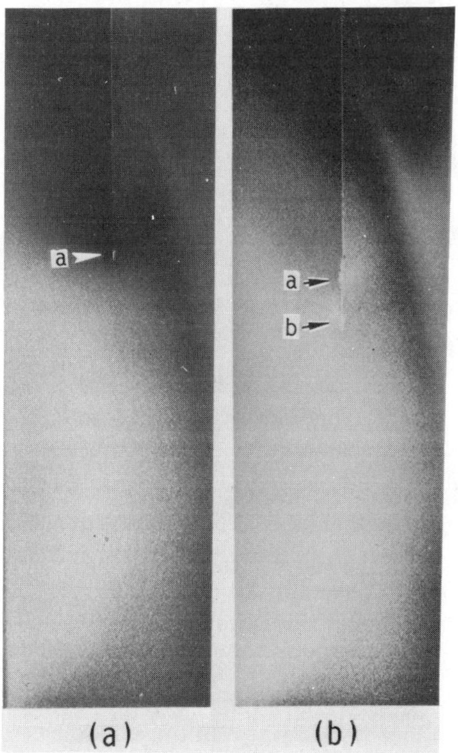

Fig. 7. Typical X-ray topography showing crack extension in a sample from point "a" to "b" of a double torsion test.

The fracture surface of each tested specimen was examined by using scanning electron microscopy (SEM). The crack front configuration of each DT load relaxation test was identified and measured. A typical SEM micrograph of the fracture surface at the area of crack tip arresting point is shown in Fig. 8. Several crack front markings can be found in each DT load relaxation test. These crack front markings indicate multiple crack arresting. As seen in Fig. 8, numerous hackle marks, normal to the moving crack front configuration, can be seen after a "mirror" surface with several crack front markings. It appears that at the critical stress intensity factor, the crack jumped several times before the fast crack propagation took place. Hackle marks are known to be the result of crack branching during the high velocity crack propagation. The mechanism of hackle mark formation has been received much consideration, (e.g., Ref. 20 and 21). The formation of the hackle marks in single crystal brittle materials requires additional evaluation. From the SEM micrographs, the distance between the leading end and the trailing end of the crack front configuration was found to be in reasonable agreement with the overlapping distance of AE measured crack length of each crack growth DT tests. It can be concluded that the AE generation accompanied the motion of the entire crack front. AE technique appears to be a useful tool for monitoring macrocrack propagation in brittle ceramics.

Analysis of the sample top surfaces and fracture surfaces were made using Sirtl etching and dislocation etch pit counting. Analysis of the as-cut top surface was somewhat confused by the presence of sawing damage, but there was a tendency for the dislocation density to be higher in the vicinity of the crack arrest points. On the fracture surface, the lack of surface flatness as a result of analysing the as-fractured surfaces made focussing the entire surface, and hence good photomicrographs, difficult. In spite of this difficulty, the dislocation etch pit arrays in hackle marks near the crack front marks are shown in Fig. 9 and near the sample surface in Fig. 10. It can be concluded that dislocations are

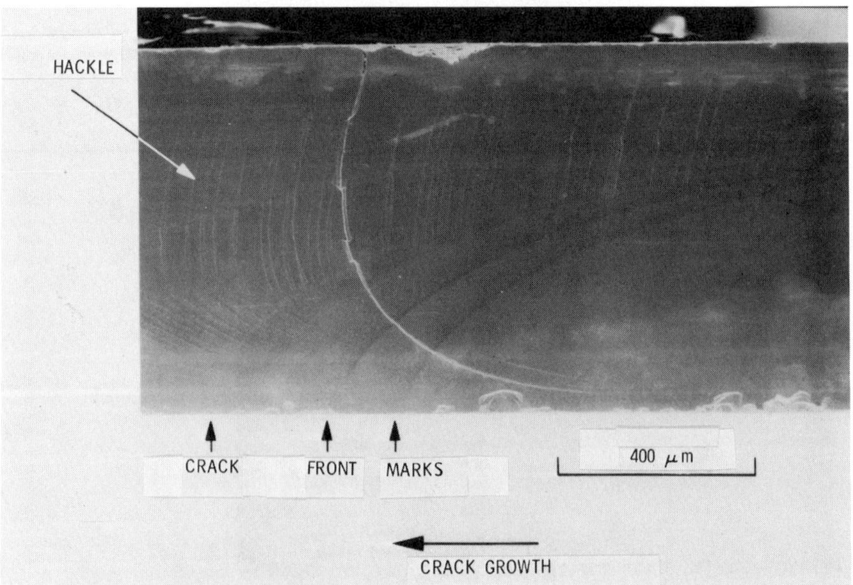

Fig. 8. A typical SEM micrograph of a cleavage fracture of single-crystal silicon tested by double torsion method. Tensile surface is at the bottom.

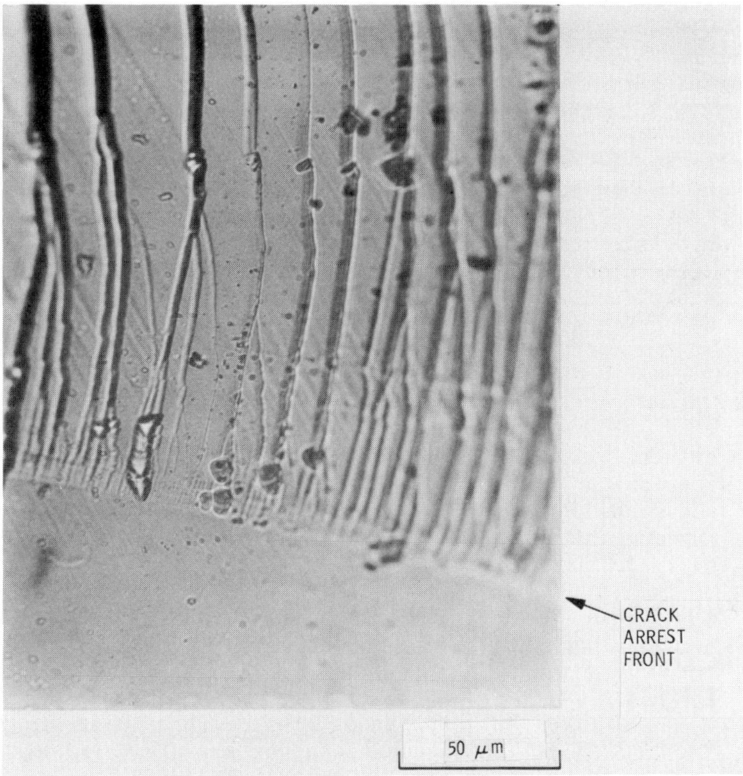

Fig. 9. Dislocation etch pit arrays in hackle marks near crack front mark.

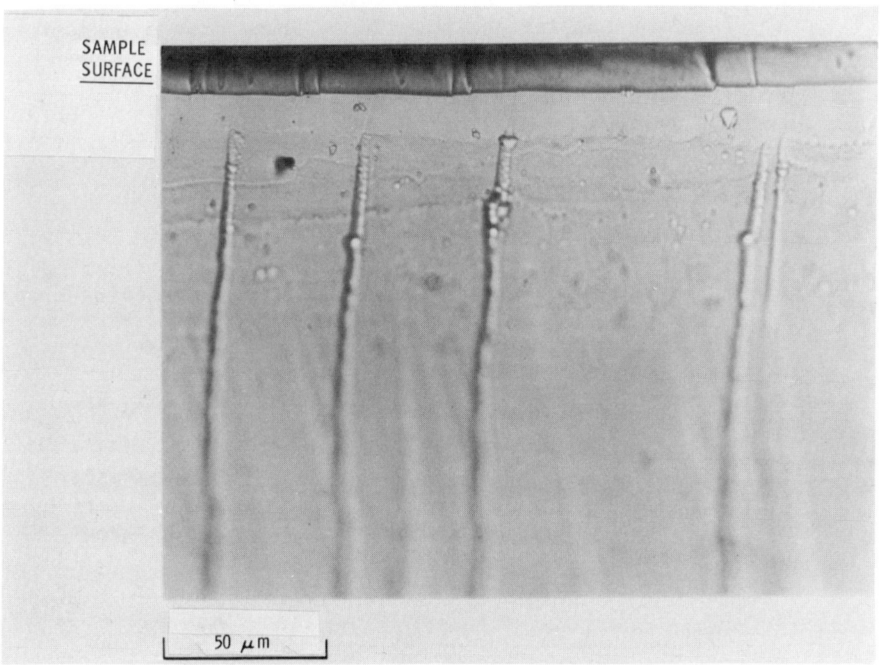

Fig. 10. Dislocation etch pit arrays in hackle marks near the sample surface which is compressive surface under double torsion test.

generated at the crack tip of a fast moving crack front in silicon crystal tested at room temperature in ambient air. The Rice and Thomson[8] model appears to be more applicable to describe the fast crack growth in silicon. However, dislocation etch pits were not observed on the crack front markings and in the "mirror" surface area of crack front. In this region the crack growth in silicon is believed to be moving at slower speed than that in the hackle marks area. The crack growth mechanism in this region may be due to the interatomic bonds "broken" at crack tip by thermally activated process as described by lattice trapping theory[6] but without any dislocation generation. Acoustic emission appears to be spontaneously generated from both the crack front moving more slowly (no dislocation generation) and hackle mark formation (dislocation generation) during the crack growth in single-crystal silicon.

CONCLUSIONS

1. The critical stress intensity factor for cleavage crack propagation in {111} of single crystal silicon was measured to be 1.0 $MNm^{-3/2}$ by double torsion (DT) load-relaxation methods. Crack propagation by jumping was found at the critical stress intensity factor.

2. No load relaxation during load holding period of DT test was found, slow crack growth was not detected by DT load-relaxation methods.

3. Acoustic emission peak amplitudes greater than 60 dB were found at the critical stress intensity factor for cleavage crack propagation in {111} of single crystal silicon.

4. No acoustic emission event was detected during loading before reaching the critical stress intensity factor nor during the load relaxation holding period. These observations suggest no subcritical crack growth in silicon.

5. Fracture surface examination revealed the process of crack growth in silicon involved crack front jumping several times in a "mirror" region, followed by a fast crack growth "hackle" region.

6. Hackle marks on the fracture surface were associated with dislocation etch pit arrays. Hackle mark formation appears to be from the plastic deformation at the tip of a fast moving crack.

7. Dislocation etch pits were not observed on the crack front markings or in the flat area of "mirror" surface. In this region, crack growth may result from interatomic bond breakage at crack tip.

8. Acoustic emission appears to be spontaneously generated from crack front jumping (interatomic bonds broken) and hackle mark formation (dislocation generation) during the crack growth in single-crystal silicon.

REFERENCES

1. A. A. Griffith, Phil. Trans. Roy. Soc. (London) A221:163 (1921).
2. S. M. Wiederhorn, "Environmental Stress Corrosion Cracking of Glass," National Bureau of Standards, NBS Report 10-565, April 1971.

3. S. M. Wiederhorn and L. H. Bolz, J. Am. Ceram. Soc., 56(10):543-548 (1970).
4. A. G. Evans, Int'l Journal of Fracture, 10(2):251-259, (1974).
5. T. A. Michalske and S. W. Freiman, J. Am. Ceram. Soc., 66(4):4-8 (1983).
6. E. R. Fuller, Jr. and R. M. Thomson, Fracture Mechanics of Ceramics, Vol. 4:507-548, R. C. Bradt, D. P. H. Hasselman and F. F. Lange, ed., Plenum, New York, 1978.
7. A. Kelly, W. R. Tyson and A. H. Cottrell, Phil. Mag. 15:567-586 (1967).
8. J. R. Rice and R. Thomson, Phil. Mag. 29:73-97 (1974).
9. C. P. Chen and M. H. Leipold, Am. Ceram. Soc. Bulletin, 59(4):469-472 (1980).
10. T. J. Chen and W. J. Knapp, J. Am. Ceram. Soc. 63(3-4):225-226 (1980).
11. S. B. Bhaduri and F. F. Y. Wang, Fracture Mechanics of Ceramics, Vol. 5:327-336, R. C. Bradt, A. G. Evans, D. P. H. Hasselman and F. F. Lange, ed., Plenum,New York, 1983.
12. C. P. Chen and M. H. Leipold, J. Am. Ceram. Soc., 68(2):C54-55 (1985).
13. L. J. Graham and G. A. Alers, Fracture Mechanics of Ceramics, Vol. 1: 175-188, R. C. Bradt, D. P. H. Hasselman, and F. F. Lange, ed., Plenum, New York, 1974.
14. M. J. Noone and R. L. Mehan, ibid., pp. 201-229 (1974).
15. A. G. Evans, M. Linzer, and L. R. Russell, Materials Sci. Eng. 15: 253-61 (1974).
16. J. S. Nadeau, J. Am. Ceram. Soc., 64, [10]:585-590 (1981).
17. K. P. McKinney, H. H. Chaskelis and S. W. Freiman, J. Am. Ceram. Soc. 59, [7-8]:369-370 (1976).
18. B. J. Pletka, E. R. Fuller, Jr., and B. G. Koepke, Fracture Mechanics Applied to Brittle Materials, S. W. Freiman, ed., ASTM STP 678:19-37 (1979).
19. C. P. Chen and S-Y. S. Hsu, Review of Progress in Quantitative Nondestructive Evaluation, Vol. 4B:689-699, D. O. Thompson and D. E. Chimenti, ed., Plenum, New York, 1985.
20. V. D. Frechette, Introduction to Glass Science, Pye, Stevens and LaCourse, ed., pp. 433-450, Plenum, New York, 1972.
21. J. J. Mecholsky and S. W. Freiman, Fracture Mechanics Applied to Brittle Materials, S. W. Freiman, ed., ASTM, STP678, pp. 136-150, 1979.

A FRACTURE MECHANICS AND NON-DESTRUCTIVE EVALUATION INVESTIGATION OF THE SUBCRITICAL-FRACTURE PROCESS IN ROCK

Peter L. Swanson*

Center for Materials Science
National Bureau of Standards
Gaithersburg, MD 20899

ABSTRACT

Results of subcritical-fracture experiments performed in air on five different rock types are presented. A non-unique relationship between calculated stress-intensity factor and crack velocity suggests violations of the assumptions made when using conventional fracture testing techniques. To obtain a better understanding of fracture propagation in rock without regard to single-parameter characterizations of the fracture process several non-destructive testing techniques were employed in conjunction with subcritical crack-growth experiments on Westerly granite. The size and shape of the fracture process zone is estimated from (1) 2-dimensional surface-deformation maps obtained using holographic interferometry, (2) the location of acoustic-emission events, and (3) the spatial distribution of ultrasonic-wave attenuation. By focusing attention in the areas shown to be of importance in the NDT experiments, in-situ microscope observations are used to identify several micromechanisms of fracture.

INTRODUCTION

Time- and environment-sensitive tensile crack propagation is being recognized as a problem of practical significance in the geosciences. A few noteworthy areas of concern are geotechnical-structure stability, energy-resource extraction, nuclear- and toxic-waste disposal and earthquake prediction. As preexisting fractures are ubiquitous in geologic settings, large-crack fracture mechanics methods have been used to try to characterize the fracture properties of rock.

This paper provides a synopsis of an investigation into the subcritical-fracture process in rocks using fracture mechanics and non-destructive testing (NDT) techniques. The initial objective was to characterize slow crack growth in rocks under various environmental conditions relevant to the earth's crust using stress-intensity

*Majority of work performed while author was at the Cooperative Institute for Research in Environmental Sciences (CIRES) and Department of Geological Sciences, University of Colorado, Boulder, CO.

factor/crack velocity (K-v) diagrams. This objective presupposes the applicability of linear elastic fracture mechanics (LEFM) concepts and techniques. Results from a series of subcritical-fracture experiments on five different rock types are presented and show that the relationship between calculated crack velocity and calculated stress-intensity factor is non-unique. The apparent non-uniqueness is suggested to stem from violation of several assumptions needed to apply the theory (both double-torsion testing and LEFM in general) to the experiments.

Having recognized the inapplicability of the elastic theory to the experiments, the investigation was given new direction. The new objectives were to (1) establish the magnitude of the deviation of rock-fracture behavior from ideal or "small-scale yielding" conditions and (2) identify the specific micromechanisms responsible for providing resistance to fracture in these materials. These data are needed for the design of "valid" fracture mechanics tests and for providing fundamental insight into the rock-fracture process.

To address the first objective without regard to the second we focus on the question, "What is the size and shape of the fracture process zone ?". The term "fracture process zone" is used to denote the volume within which fracture energy is dissipated. As visualization of fractures is very difficult in opaque polycrystalline materials, optical microscopy was not thought to be a sufficiently quantitative tool to shed adequate light on this subject. Instead, NDT techniques were used, allowing quantitative physical measurements to be made during the propagation of fractures. The NDT techniques include: (1) measurement of two-dimensional surface-displacement maps using holographic interferometry, (2) location of acoustic-emission events using a laboratory version of earthquake-location techniques, and (3) mapping the spatial distribution of ultrasonic-wave attenuation in the vicinity of extending fractures. Comparison of NDT measurements made on glass, the LEFM ideal material, and rock provides an estimate of the size scale of the material-breakdown zone in rock if we consider the inelastic process zone in glass to be acceptably miniscule.

Two approaches were used to identify the operative micromechanisms of fracture in rock. In the first approach we contrast the extent of the fracture process zone, measured in the above-described fashion, with what would be expected from a popular microcrack process-zone model (critical-principal-normal-tensile-stress model, CPNTS). The CPNTS model is apparently the only model to date which actually makes a prediction of the process-zone shape based on a specified mechanism. Most other microcrack process-zone models specify the zone size, shape or microcrack density without any direct physical basis. In the second approach in-situ optical microscopy observations were made of the fracturing process, focusing in regions shown to be of importance by the NDT studies. Observed micromechanisms of fracture, which were recorded on video tape and 35-mm film, are briefly described.

As single-phase polycrystalline and multiphase ceramics are "simple" cousins to the generally more complex multiphase rocks, it was thought that the Fracture Mechanics of Ceramics volumes would be an appropriate medium for communicating to the ceramic-fracture audience some of the findings reported herein. An example of some more recent work performed by the author on a wide variety of ceramics is also included and offered as evidence that similar micromechanisms of fracture also operate in these materials.

EXPERIMENTAL PROCEDURES

Rock Descriptions

Mineralogical descriptions of the rock types investigated are not given because the paper does not touch upon, in detail, aspects of the work related to petrography. Several statements concerning igneous rocks in general, however, are provided as background. Igneous rocks form by solidification of a melt at or below the earth's surface. Various cooling rates give rise to a range of grain sizes and in this study they vary from micrometers up to a millimeter, with a mixture of both end-member sizes present in at least one case. Each rock type tested is composed of several different phases (typically 3-4 major minerals) with a wide range in physical properties. This variety exists not only between phases but also within each mineral, as the anisotropy in single-crystal fracture energy, thermal expansion, and elastic modulus can be extreme. Such diversity can lead to extensive microcracking when the rocks have cooled and are brought to the earth's surface. For example, in some rock types microcracks exist along as much as 90% of the grain boundaries.[1]

Fracture Mechanics Techniques

Precracked double-torsion specimens of rock (1.5-5x25x150 mm) and glass (1x25x75 mm soda-lime silicate microscope slides and 5x100x230 mm float-glass samples) were used in both the K-v determinations and the holography and acoustic-emission location experiments.

In relaxation tests,[2] the samples were rapidly loaded to a predetermined force level and the load-point displacement was subsequently fixed. During the relaxation, the force-gage output was sampled at approximately 25 times per second and recorded by an automatic data-acquisition system. Five consecutive readings were averaged and compared with the previous average. When the average differed by a specified amount, the voltage and time were stored and later recorded on magnetic tape. This procedure yielded approximately equal force decrements for the duration of the typically 45-minute tests.

Calculation of the velocity requires the determination of the slope of the force-relaxation curve. For ease in determining the slope between discrete data points that show some scatter, curve-fitting techniques were employed. Sixth-degree polynomials were found to adequately represent the log force/log time data and were used in the determination of the relaxation-curve slopes. The choice of a sixth-degree polynomial was arbitrary and was selected for its satisfactory fit to both a straight line and the major trends in tri-modal relaxation data observed in glass under atmospheric conditions[3,4]. Three to four relaxation tests were typically performed on each specimen.

The specimens were given a low-temperature vacuum-bakeout treatment for environment standardization and stored and tested under equilibrium laboratory conditions. Temperature and relative humidity were monitored throughout the tests and varied less than 1% in any single experiment.

Wedge-loaded double-cantilever beam (WLDCB) specimens of Westerly granite measuring 16-25x78x200 mm were used in the ultrasonic-wave propagation experiments. Tool-steel endcaps were epoxied to the ends of DCB specimens and load was applied to dowel pins protruding from the endcaps via a 30-degree hollow wedge. Force and crack-opening displacement were plotted on an x-y recorder. Crack-guide grooves were not employed.

Holographic Interferometry

Holography is sometimes loosely referred to as lensless 3-dimensional laser photography. An ordinary hologram, when reilluminated by laser light, yields a virtual 3-dimensional image that is indistinguishable from the original illuminated object. Double-exposure holography, or holographic interferometry, can be thought of as a 3-dimensional version of the Michelson interferometer in which specularly reflecting surfaces are not needed. If, as a result of specimen deformation, an increment of surface displacement takes place between two exposures, then upon reillumination of the developed hologram, two virtual images optically interfere to produce a fringe pattern of alternating light and dark bands. The fringes can be interpreted as a contour map of the relative surface displacements that took place between exposures when directions of viewing and illumination are coincident (CIV holography[5]). The contour interval is one-half the wavelength of the laser light (approximately 0.3 microns - He-Ne). Surfaces of the glass specimens were given a light coat of white spray paint for reflectance.

Acoustic-Emission Detection and Location

Acoustic-emission (AE) events were detected by amplifying voltage signals from an array of piezoelectric transducers (1 MHz PZT-5A) which were bonded to the fracture specimens. Waveforms were synchronously recorded at 50 nanoseconds per point on multi-channel digital transient recorders with pretrigger memory and stored for later analysis.

AE events were graphically located in two dimensions using a simplified version of a more general 3-dimensional source-location routine developed by Sondergeld and Estey[6]. Assuming a constant isotropic elastic-wave velocity in the thin-plate samples, relative arrival-time isochrons were constructed for each acoustic-transducer pair. Each isochron delineates possible positions of AE-source locations for a single event and is generated from knowledge of the transducer locations and measured time delay between wave arrivals. Mutual intersection of isochrons, determined for multiple transducer-arrays, yields the AE source location. In the examples given below, two transducers were bonded to opposite ends of double-torsion specimens so as to provide good axial control of the source locations. AE events were assumed to originate at the intersection of the isochron and the sidegroove or specimen midplane. No attempt was made to differentiate event locations in the thickness direction. The suitability of this two-dimensional approach was found to be sufficient by location verification using a four-transducer array.

Ultrasonic-Wave Measurements

Three primary modes of ultrasonic probing were used (Figure 1). In each mode, ultrasonic waves were propagated perpendicular to the direction of fracture extension. Both transient surface (1) and body (compressional) waves (2) were recorded as the transducers were moved along a traverse parallel to the crack. Measurements of surface waves (with body-wave components) were also recorded along a perpendicular traverse (3).

In both parallel traverses ((1) and (2)) waveform measurements were taken at 1-2 mm intervals near the crack-tip zone and 2-5 mm intervals at other positions. The location of each measurement is referred to the distance between the centerline of the source-receiver pair and the starter-notch tip (Figure 1). At each level of applied load waveforms were typically recorded over a 150-mm length of the sample. A similar waveform-recording schedule was used in the perpendicular traverse (3).

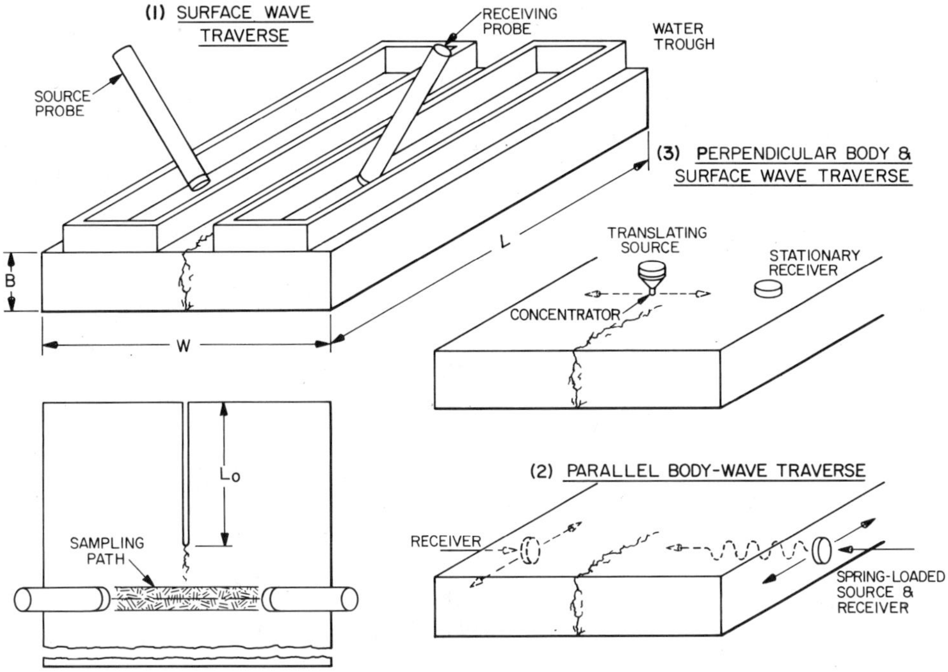

Figure 1. Ultrasonic-wave propagation modes used to delineate fracture process zone size and shape.

Transmitter and receiver probes were constructed using 1-MHz PZT-5A transducers. Waveforms were recorded as described in the AE-measurement section.

Further details concerning specimen preparation, NDT techniques, and general experimental procedures can be found elsewhere[3,4,7].

EXPERIMENTAL RESULTS

Double-Torsion Relaxation Tests on Five Rock Types

K-v curves determined from over 50 relaxation tests on five different rock types are shown in Figure 2. It is quite clear from the scatter in Figure 2 that we cannot immediately utilize these data for construction of useful failure-prediction diagrams. In particular, the data suggest a non-unique relationship between K_I and v. Seemingly consistent K-v data are produced, although with considerable scatter, when the same initial K_I value is used to begin each relaxation test. The clustering of the curves for Westerly granite and Ralston intrusive (Figures 2A and 2E) bears not on the uniformity of the material but on the narrow range of K_I values used to begin those tests. Recent publications by other investigators show that they too suffer similar difficulties in producing repeatable measurements with synthetic polycrystalline materials tested by the relaxation method. Most of the investigators reporting this behavior in single-phase polycrystals[8,9,10] have been able to identify the source of the difficulties as an interaction between the crack and side-groove walls. Several studies[10,11,12] have found that wide side-grooves reduce and narrow side-grooves increase the crack/side-groove wall interaction (also observed in the author's holography studies). Momentary "trapping" of the crack can occur because of the

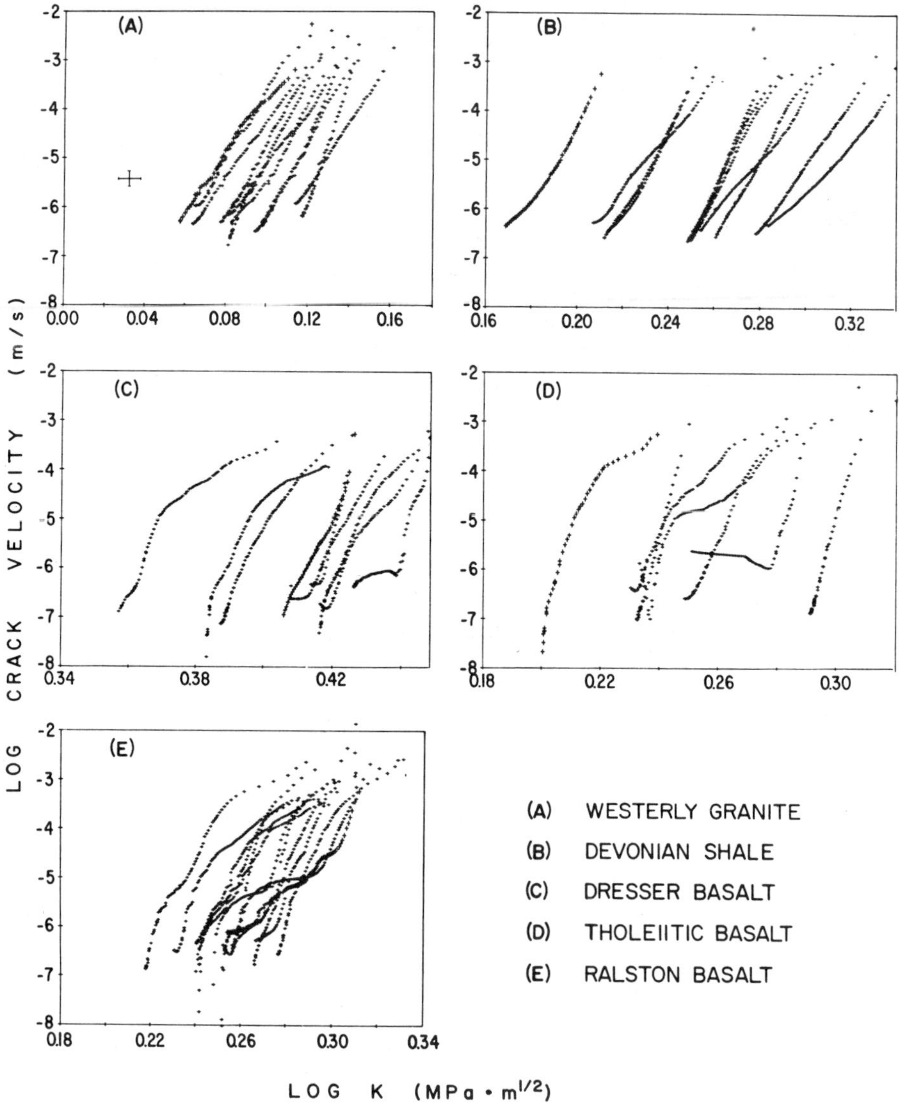

Figure 2. Crack velocity/stress-intensity factor relations determined for five rock types using relaxation form of double-torsion testing. Peculiar hooks at low-velocity ends of several curves are artifacts of the data-reduction process.

influence of the side-groove geometry on the crack-tip stress field. This varying stress field across the groove, unaccounted for in the expression[13] relating the applied load to K_I, produces K-v curves with increased scatter that are offset on the stress-intensity axis. Another possibly significant contribution to the scatter in the data may arise from changes in the resistance to fracture with crack extension (R-curve behavior).

Elimination of the guide in alumina[10], silicon-nitride[11], and glass[3,4] eliminates much of the scatter in the slope and position of K-v curves. With an accurately aligned loading fixture the crack can be confined, in homogeneous isotropic materials, to the center of the sample. Since rocks do not typically display such uniformity, we are restricted to using some form of guiding groove in order to conduct this type of experiment. A series of relaxation tests were conducted using samples with various guide-groove geometries (sizes and shapes) to see if the scatter could be reduced. Multiple relaxation tests on individual specimens, however, showed similar scatter regardless of guide-groove geometry[3,4].

As shown by the NDT experiments described below, the non-uniqueness of the relationship between calculated K_I and v can largely be attributed to violations of the small-scale deformation assumption of LEFM.

Holographic Interferometry

A comparison of typical relative surface-displacement fields observed for the compressive sides of glass and granite double-torsion specimens during relaxation tests is shown in Figure 3. Crack extension is from left to right for a fraction of a millimeter (as estimated from relaxation data) under room conditions at an apparent velocity of approximately 10^{-5} m/s. Overall, the relative surface-displacement patterns are remarkably similar. Closer examination, however, reveals consistent differences. For example, the spatial extent of the difference in the fringe patterns of Figure 3 is evidenced by the fringe spacing encountered as one traverses the crack plane toward the unfractured material, beginning at the central fringe or position of greatest relative surface displacement. In the glass there is a rapid monotonic decrease in fringe density indicating a steep relative-displacement gradient. Open-ended fringes on the right side of the sample are due to the close proximity of the specimen end. Note also that the curvature of each fringe is continuous at its intersection with the central plane. In the granite specimen, the fringe spacing does not monotonically increase away from the central fringe, nor is the fringe curvature continuous in this region. The axial distance over which these differences exist between the fringe patterns in glass and granite is of the order of 10 mm. Significant differences in fringe patterns in the direction perpendicular to the crack plane are not evident in the tests of Figure 3 and other similar experiments.

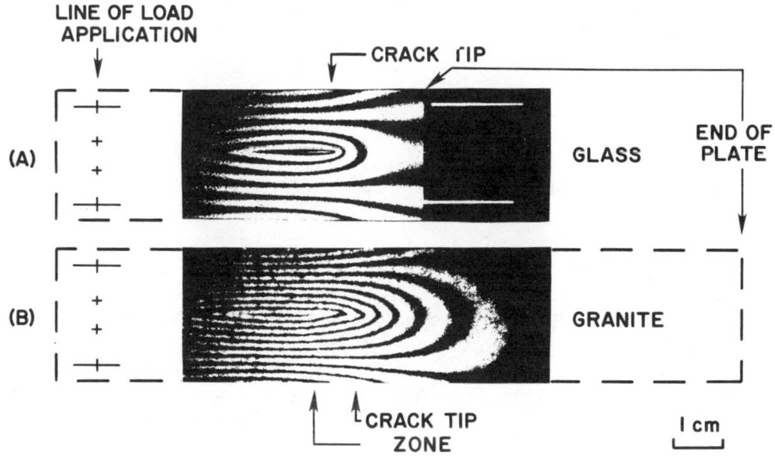

Figure 3. Double-exposure holograms showing relative surface-displacement field associated with subcritical fracture in glass and granite.

Other holography observations of rock fracture include: (1) evidence for discontinuous slip along the "previously-fractured" crack trace (locally asymmetric crack-opening displacement) at great distances (tens of mm) from the crack tip in rocks with long and tortuous cracks, (2) the occurrence of time-dependent rebound of sample surfaces after rapid unloading of double-torsion specimens, and (3) after cessation of rebound, application of water to the fracture trace leads to reactivation of time-dependent rebound.

Acoustic-Emission Detection and Location

Figure 4 shows acoustic-emission locations observed in a double-torsion relaxation test on Westerly granite at an apparent crack velocity ranging from 10^{-2} to 10^{-5} m/s. Propagation was started from a 55-mm precrack. Its length and the length of the "visually-defined" crack-tip zone (Figure 4) were estimated from observations through a 60x microscope. Prior to analyzing the AE results, the crack-tip zone was identified as the furthest extent at which surface damage could be seen yet did not appear to be extending continuously from the obviously-fractured crack trace. Of the approximately 400 events captured by the oscilloscopes after the start of the relaxation, 27 were of sufficient magnitude to be locatable and are numbered in chronological order. A significant number of the located events were not associated with what was later identified as the best estimate of the final crack-tip region. AE events were located ahead of and behind the visually-defined macrocrack tip zone along the apparently "previously-fractured" crack plane.

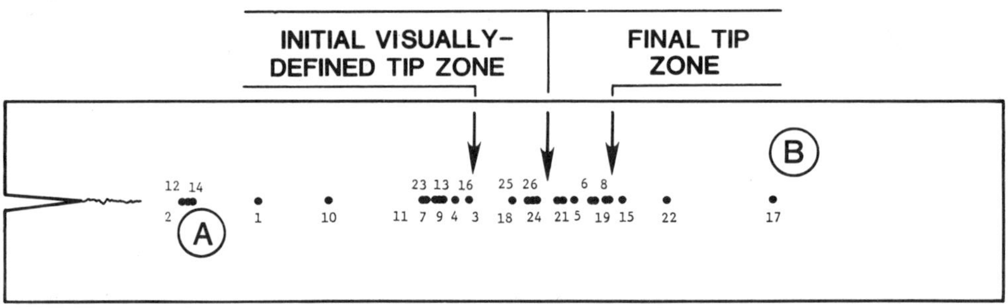

Figure 4. Acoustic-emission event locations associated with a relaxation test on rock sample. Fracture-tip zone estimated using 60x reflective-light microscope.

Acoustic-emission waveforms were also recorded (Figure 5) as another precracked sample was loaded at a constant displacement rate (20% apparent K_{IC}/minute) to a load corresponding to that necessary to propagate the crack at approximately 10^{-7} m/s. Moments after completion of loading, the sample was unloaded at the same rate. Figure 5 shows the event locations for the complete load cycle. Continued cycling is accompanied by continued acoustic activity although at a decreasing rate.

Similar AE-event distributions were found for Ralston basalt in this study and for Murata basalt and Oshima granite in another study by Sano[14].

AE events were also detected in float-glass double-torsion specimens which had a crack guide scribed part way down the tensile side of the specimen. After subcritically propagating the crack into the unscribed

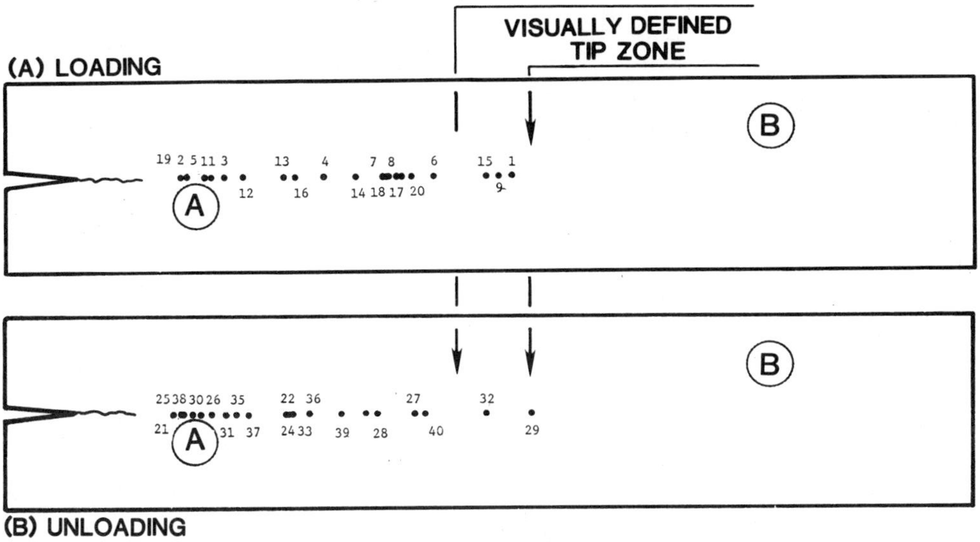

Figure 5. Acoustic-emission event locations observed while loading and unloading double-torsion rock specimen. Maximum load corresponds to that associated with crack velocity of 10^{-7} m/s.

material under fixed-grip conditions, AE were located and were found to emanate, not from the crack tip, but from the previously-fractured, scribed, portion of the crack trace. Microscopic examination of the AE source locations revealed several large 3-dimensional fracture-surface features (lances[15]) created by the initial interaction of the main crack with the scribe damage. The fracture lances (hackle markings) showed evidence of microcrack damage, i.e. loose fractured particles of glass. Large lance-like features which had no significant broken particles associated with them also were not associated with large AE-event locations. It should be emphasized that the AE events located in the glass are linked to microcrack damage on the tensile (not compressive) side of the specimen. AE events were not detected unless a scribed crack guide was present.

Ultrasonics

Figure 6 shows amplitudes of the first peak in the surface-wave arrival for Westerly granite as a function of distance from the starter-notch tip. Measurements were taken on a parallel traverse ((1) in Figure 1) at two different applied loads. Amplitudes are normalized to measurements obtained at near-zero load. Deviation from unity represents the percentage decrease in amplitude observed upon loading the sample to the middle of the non-linear region (triangles) and the peak (squares) of the force-displacement curve. It should be noted that the measurement locations are referred to the distance from the source/receiver-pair centerline to the starter-notch tip. The actual location or definition of the crack tip is a subject for discussion based upon the ultrasonic measurements. Data points at negative values of distance reflect the finite width of the ultrasonic-wave sampling path and any diffracted-wave energy present.

After stopping the advancement of the wedge, significant time-dependent force decay and crack-opening displacement occurred as shown in the inset of Figure 6. Ultrasonic measurements were taken after approximate equilibrium was achieved, typically in less than one hour.

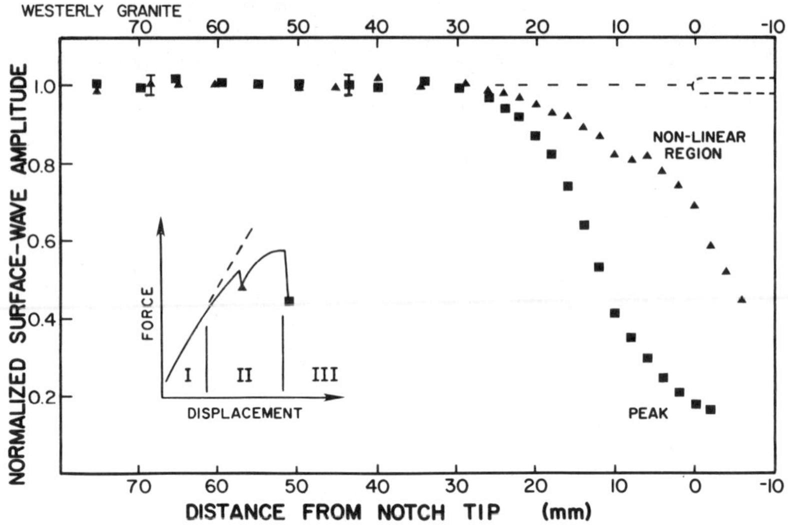

Figure 6. Surface-wave amplitudes versus measurement position for data obtained in the middle of the non-linear region (triangles) and peak (squares) of the force-displacement curve. Data are normalized to amplitudes observed at near-zero load.

According to the investigators originally applying the WLDCB technique to rocks[16], microcracking surrounding the starter-notch tip gives rise to the non-linearity in the force-displacement record. Propagation of the macrocrack by coalescence of the microcracks is interpreted as starting at the peak of the force-displacement curve, where the microcrack process zone is thought to be in a highly developed state. The ultrasonic measurements reported below were obtained at both the middle of the non-linear region and peak of the force-displacement curve.

Several significant features are apparent in Figure 6. First, the length of the process zone can be estimated from the length of the partial-transmission zone. From either Hoagland et al.'s interpretation of force-displacement curves or direct comparison with experiments on glass, the process-zone length is of the order of 20 mm. For different specimens, the initial partial-transmission-zone length was found to vary considerably (14 to 40 mm), with results for surface waves in general agreement with those for body waves. Apparently the force and displacement measurements do not consistently correspond to identical states of damage. Secondly, the position at which the first-peak amplitudes begin to decrease (herein termed the "pinning point") does not significantly penetrate further into the specimen upon loading from the middle of the non-linear region to the peak of the force-displacement curve. At points closer to the notch, however, the amplitudes decrease significantly.

On several occasions ultrasonic-wave amplitudes were observed to increase upon reducing the applied load. This observation is interpreted as evidence for the partial closing of crack surfaces. The fact that it does not always occur or occurs to varying degrees is not surprising given the rough angular nature of the fracture surfaces and production of loose specimen debris.

After loading past the peak in the force-displacement curve, the "pinning point" surges further into the specimen in a discontinuous fashion. Each discontinuous advance is preceded by a steepening of the amplitude roll-off curve (similar to the transition shown in Figure 6) to a form more like that found for a sharp crack tip in glass. In experiments on glass, the amplitude distribution remains unchanged as load is varied from 5 to 99 % of the load required for fracture extension. At the point of extension, the amplitude roll-off curve keeps the same shape and merely translates with the crack tip. The behavior displayed by the granite is markedly different. The amplitude versus distance curve continually steepens as the load or displacement is increased, right up to the extension of the pinning-point position. Also, a gradual steepening of the amplitude roll-off sequence has sometimes been observed when reloading precracked samples. These observations imply that the separation of both preexisting and newly-created fracture surfaces in rock is a gradual process. Based on this interpretation, a single value of load or crack-closure stress, beyond which the surfaces are completely separated or free of contact, was not observed, as has been deduced[17] on the basis of force-displacement curves alone.

The ultrasonic measurements made up to, and beyond, the peak of the force-displacement curve suggest that fracture extension in Westerly-granite WLDCB specimens is a discontinuous process. In these experiments, fractures did not incrementally propagate over intervals equal to the size of the microstructural elements (0.75 mm). Instead, the distance that the pinning point extends (or the incremental increase in length of the partial-transmission zone) appears to be a measure of the spatial distribution of fracture-sensitive material properties, at least in the non-linear region of the force-displacement curve. The increment length in the post-peak region is sensitive to both the local material properties and the test-machine stiffness. In Westerly granite this post-peak propagation distance varied from approximately 5 mm to the remainder of the specimen length (tens of centimeters).

The width of the zone of partial ultrasonic transmission was measured by observing the location at which amplitudes change as transducer probes were moved across the fracture trace on a perpendicular traverse ((3) in Figure 1). Figure 7 schematically shows the receiver and source locations. Note the different horizontal and vertical length scales in the plan view. A perpendicular scan was made 35 mm from the notch tip after a 40-mm zone of partial transmission was created in another granite specimen. The sample is loaded near to the post-peak position. At source positions close to the receiver the amplitudes are observed to be fairly constant, with the slight decrease due to geometric spreading barely detectable in the noise of the measurement. Near the middle of the scan a relatively abrupt decrease in amplitude occurs and remains nearly constant for the remainder of the traverse. A similar pattern is observed for scan lines located all along the partial-transmission zone. The zone width over which the amplitude decreases varies from one to at most four millimeters and is usually closer to two millimeters (1-2 grains). Toward the end of the sample containing the notch the decreases in amplitude are the largest and most abrupt. As the scan lines are moved toward the fracture terminus, the magnitude of the amplitude decrease gets smaller. The width over which the amplitude decreased seemed to be closer to the 3-4 mm dimension but since the amplitude decrease is much smaller, the uncertainty in the measurement is much greater and this observation must be weighed accordingly. Shadowing and focusing effects, which can be associated with a dilatant or an anomalous-velocity zone[18], were not observed in any of the wave-propagation modes.

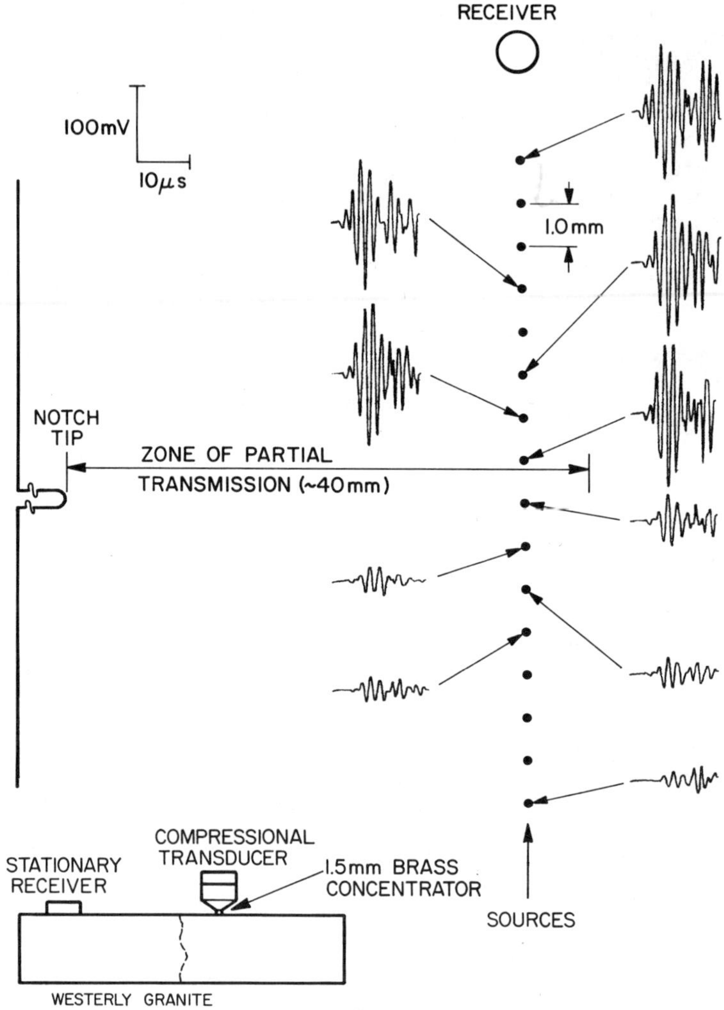

Figure 7. Measurement of the partial-transmission-zone width using perpendicular scan ((3) in Figure 1). Note different scales.

DISCUSSION

Estimate of Fracture Process-Zone Size

All three NDT techniques have provided evidence of the size-scale over which the behavior of stable fractures in rock deviates from behavior observed in glass. Size estimates based on the spatial distribution of holographically-measured surface displacements, acoustic-emission locations, and the spatial variation of ultrasonic-wave transmission indicate that the process zone in Westerly granite has dimensions exceeding 10 mm along the direction of propagation under the various test conditions employed.

In order to characterize the fracture process using LEFM the inelastic fracture process zone must be contained within the $r^{-1/2}$ stress-singularity-dominated region of the stress field (r is distance from tip of crack of length a). In other words, the zone in which fracture energy is consumed must be restricted to a region where the K_I term dominates the elastic field. In general, K_I characterizes the magnitude of the stress field for $r/a << 1$. From an engineering point of view, process-zone dimensions of $r/a < 2\%$ have empirically been found to be acceptable for a one-parameter characterization of fracture through years of experience in metals testing[19]. Based on the 2%-of-crack-length process-zone-size limitation, a one-parameter (K) characterization of fracture in Westerly granite would necessitate, at a minimum, the use of meter-sized (homogeneous) specimens.

Comparison of Experimental Results with the CPNTS Microcrack Process-Zone Model

The NDT experiments were designed to detect a process zone consisting of a cloud of microcracks, a portrayal frequently described in the ceramics and rock mechanics literature. According to these models [20,21,22,23], microcracks form in the vicinity of macroscopic fractures in response to a critical-principal-normal-tensile stress (CPNTS). As the models are based upon continuum principles, the microcracks are considered to be distributed throughout a volume large in comparison to the microstructure so that they do not interact with each other. The theory predicts a zone form having a double-lobe shape similar to a kidney. As approximated by the linear crack-tip stress equations, the zone length (measured in the direction of propagation) is less than one-half its width.

All three NDT techniques show that the deviation of the behavior of fractures in granite from the behavior displayed by the unscribed glass occurs in a zone elongated parallel to the direction of propagation, quite thin compared to its length. These estimates of the process-zone size and shape are made without identifying the specific micromechanisms of fracture responsible for the NDT observations. The observations do not appear, however, to support the shape predicted by the CPNTS model.

It may be argued that a microcrack process zone with an initial kidney shape would appear elongate after propagating over a length equal to a few process-zone widths. As a counter argument, the ultrasonic-wave measurements have consistently shown that an elongate shape is present soon after the onset of non-linearity in the force-displacement relation, indicating that this zone initially forms with an elongate shape. Its width also appears to be restricted to 1-2 grain dimensions where continuum-mechanics concepts are inappropriate for describing the local stress conditions appropriate to microcracking. Failure to find evidence of the CPNTS-style microcrack zone is not proof of its non-existence. Instead, the techniques employed have shown that there are other mechanisms of fracture-energy consumption which appear to dominate.

Frictional Interference Between Interlocking Crack Surfaces

A most conspicuous result of the NDT experiments is the observation of certain phenomena which can be interpreted as evidence of crack-plane friction or frictional-resistance to crack opening due to interlocking fracture surfaces. This mechanism has received comparatively little attention as an important feature of fracture-energy consumption in brittle materials, undoubtedly because most observations of fractures are made post-mortem instead of in-situ. Its existence in the rock fracture experiments is documented in part by (1) the occurrence of acoustic emissions along (apparently) previously-fractured crack surfaces during fracture extension as well as during unloading and reloading (recall that AE can actually be produced behind the crack tip along rough previously-fractured surfaces as was shown in the experiments on scribed glass), (2) holographic observations of both discontinuous crack-plane slip and time-dependent rebound following rapid unloading of specimens, and (3) the observed narrow, elongate, zone of partial ultrasonic transmission that, unlike the experiments on glass, can be quite sensitive to applied load all the way up to the fracture propagation load. Specimens which are fractured in a stable fashion also frequently remain in one piece even though fracture damage is clearly visible along the entire length of the sample surface. Similar observations and references to crack-plane friction have been made by a number of investigators of ceramic fracture [24,25,26,27].

Microcracking is one way in which the resistance to frictional interlocking can be overcome. Indirect evidence for friction-induced microcracking, in addition to the AE experiments, comes from the significant amounts of specimen debris, or small loose fractured particles, which are always present on the sample-support structure and fracture surfaces following rock fracture testing. An exaggerated example of the proposed fracture-surface-geometry-related frictional interlocking is shown schematically in Figure 8. Elimination of frictional contact is possible only by parallel translation of the two surfaces. Separation of these surfaces in a crack-like fashion results in an increasing resistance to further propagation, friction-induced microcracking and rising R-curve behavior. Frictional resistance to separation would seem to be guaranteed over some length of fracture by the combination of (1) local shearing displacements associated with the tortuous, predominantly grain-boundary, fracture and (2) immediate distortion of the new surfaces due to the release of stored elastic strain energy (when present).

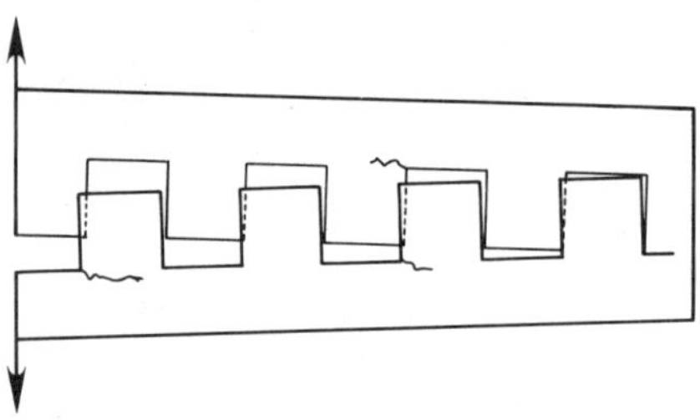

Figure 8. Schematic illustrating how fracture-surface geometry can lead to frictional contact and microcracking between locally-separated crack surfaces.

Real-Time Microscope Observations of the Fracture Process

To provide direct evidence of the proposed geometry-related friction-induced microcracking mechanism operating in-situ, a series of tests on miniature WLDCB Westerly-granite and Ralston-basalt specimens (3-6x20-30x60 mm) were performed while observing the exterior sample surfaces through a 40-400x reflective-light microscope. Sample surfaces were polished with successively finer grit sizes down to 0.25 microns. Notched samples were loaded directly on a microscope stage by a screw-driven 80-degree steel wedge.

Figure 9 presents several instances of transgranular microcracking induced by crack-plane friction. In each case, the primary fracture initially follows a grain-boundary path from left to right. Frictional contact accompanying local mode-II displacements results in transgranular fracture of adjacent grains. Complete separation of particles by friction-induced microcracking or "pull-out" was observed taking place at

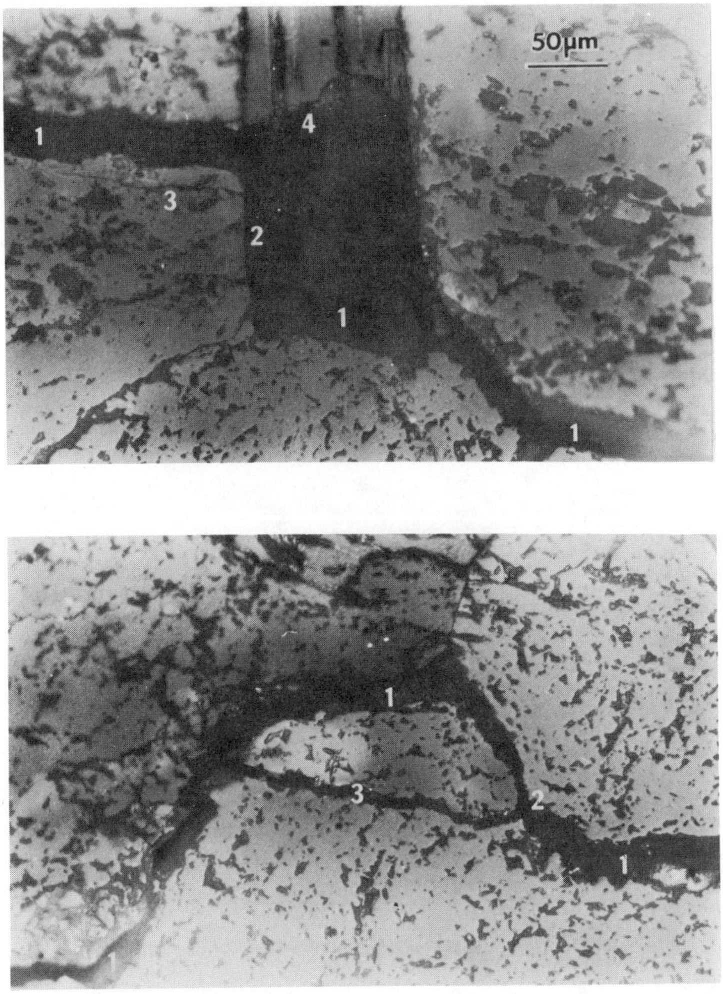

Figure 9. In-situ observations of friction-induced microcracking associated with macrofracture propagation in WLDCB granite specimens Macrocrack propagation from left to right. (1) Normal displacements, (2) shear displacements, (3) and (4) friction-induced microcracks.

positions distributed along the fracture trace. Observed sizes of separated particles range from the grain scale (0.75 mm) down to the resolution limit of the microscope (microns). A video camera attached to the microscope was used to record fractures in the process of extending[28]. The video documents the existence of this energy-dissipating mechanism operating as far as several centimeters behind the furthermost visible extent of fracture damage. Observed microcrack damage was restricted to 1-2 grain dimensions on either side of well-developed fractures. Continuum-style, distributed, microcracking was not apparent.

The in-situ microscope observations are consistent with the NDT results. In particular, the frictional interlocking mechanism provides a reasonable explanation for the elongate distribution and load dependence of AE, holographically-observed time-dependent crack closure and the load-dependence and spatial distribution of partial ultrasonic-wave transmission. Frictional interlocking of crack faces, however, is not the only mechanism which can be used to describe the NDT results. A thin zone of partial decohesion (Figure 10) or elongate zone of grain-localized microcracking that leaves "bridges" of intact material behind "crack tips" will produce similar effects. Such continuous bridges of material across what was later to become a traction-free crack segment were observed in the in-situ optical-microscopy study of the rocks, particularly when the fracture intersects and segments a readily-cleavable mineral.

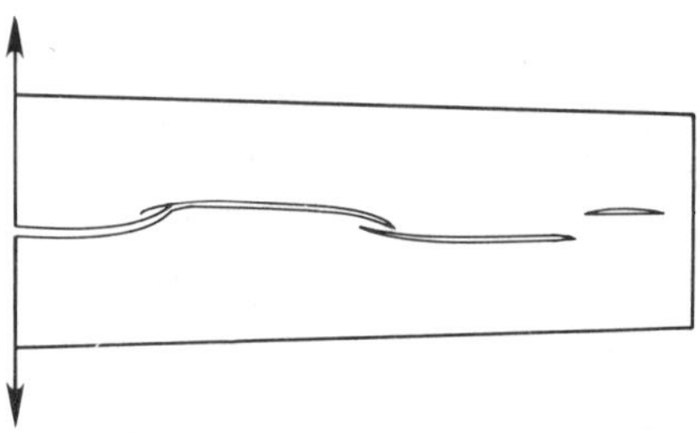

Figure 10. Partial decohesion, or grain-localized microcracking, leaves intact "bridges" of material behind the "crack tip". This mechanism is also capable of producing elongate zones of partial ultrasonic-wave transmission and AE event locations.

In translucent single-phase ceramics and glass ceramics, microscopic observation of fractures is less difficult and instances of both friction-induced microcracking and incomplete decohesion have been observed. From surface observations alone, however, it is not always clear which mechanism is involved. For example, a sequence of three photographs of a crack propagating in a coarse-grained sample of alumina (Coors - Vistal) is shown in Figure 10. As the relatively well-defined crack tip propagates out of the photographs to the right, damage continues to occur at significant distances behind the visually-defined tip along the crack flanks. Similar crack-flank damage has been observed occurring over considerable lengths (tens to thousands of micrometers) in other aluminas and several glass ceramics with particle sizes ranging from several to many tens of micrometers. The nature of these crack-flank processes and their role in providing toughness, rising fracture resistance (R-curves), and resistance to subcritical fracture in ceramics are currently under study at NBS (e.g. Fairbanks et al., this volume).

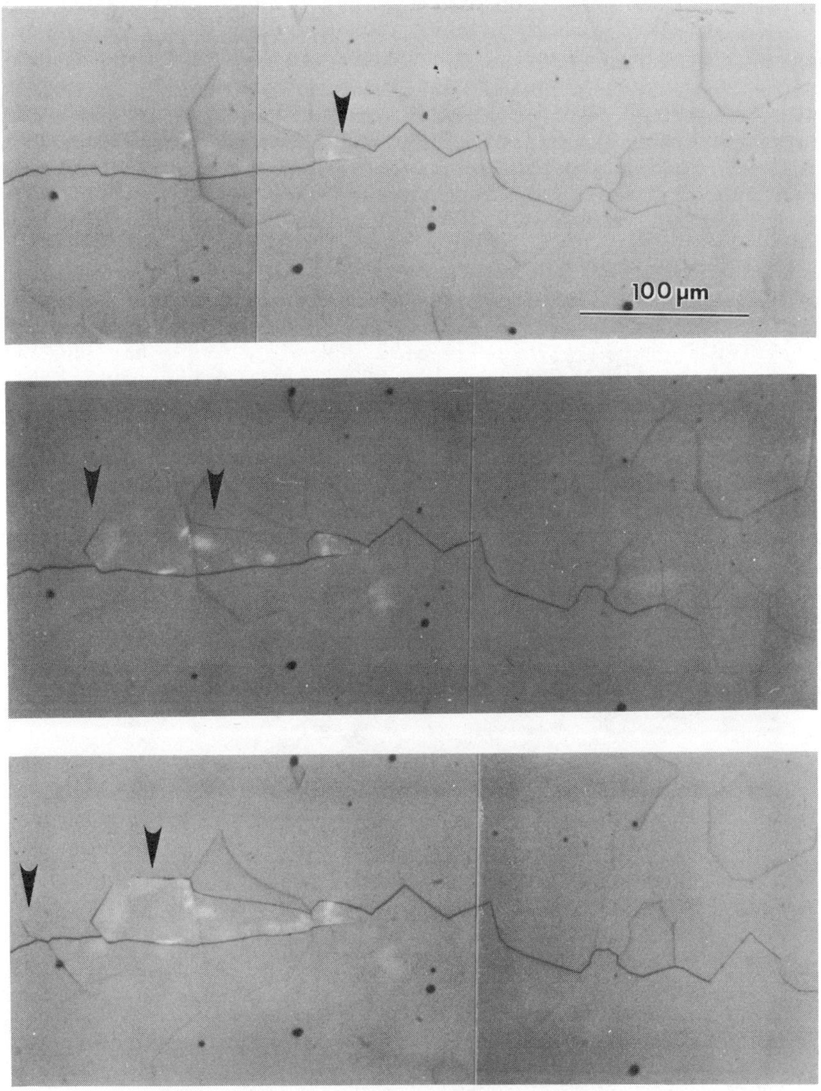

Figure 11. Sequence of photos illustrating microfracture damage occurring behind relatively well-defined crack tip, propagating from left to right, in alumina (Coors-Vistal).

SUMMARY AND CONCLUSIONS

The size and shape of the fracture process zone in Westerly granite double-torsion and wedge-loaded double-cantilever beam specimens has been estimated from holographically-measured surface displacements, acoustic-emission locations and the spatial distribution of ultrasonic-wave attenuation. Due to the large dimensions of the process-zone length (14-40 mm) and width (1-2 grain dimensions), direct application of LEFM to the experiments is inappropriate.

The standard interpretation of force and displacement measurements in terms of a minutely-propagating macrofracture accompanied by a cloud of distributed microcracks appears incorrect for the fracture experiments on Westerly granite. A schematic illustrating the author's understanding of the fracture propagation process in this particular rock type is offered in Figure 12. It is based on the interpretation of the NDT experiments and the in-situ optical-microscopy study.

Fractures appear to form as gradually separating surfaces having a longitudinal extent, X_o (see Figure 12), of the order of centimeters, many times larger than the grain size. X_o terminates at a pinning point or barrier to further separation. Partial decohesion or grain-localized microcracking and friction-induced microcracking take place all along X_o as load and displacement are increased. With continued loading, increasingly-severe damage propagates along X_o until enough stress has been transferred to the pinning point to cause it to fail. Following this, a second zone X_o' is formed in which another progression of damage propagation, similar to the first, takes place. Discontinuous propagation on this scale (up tens of mm)

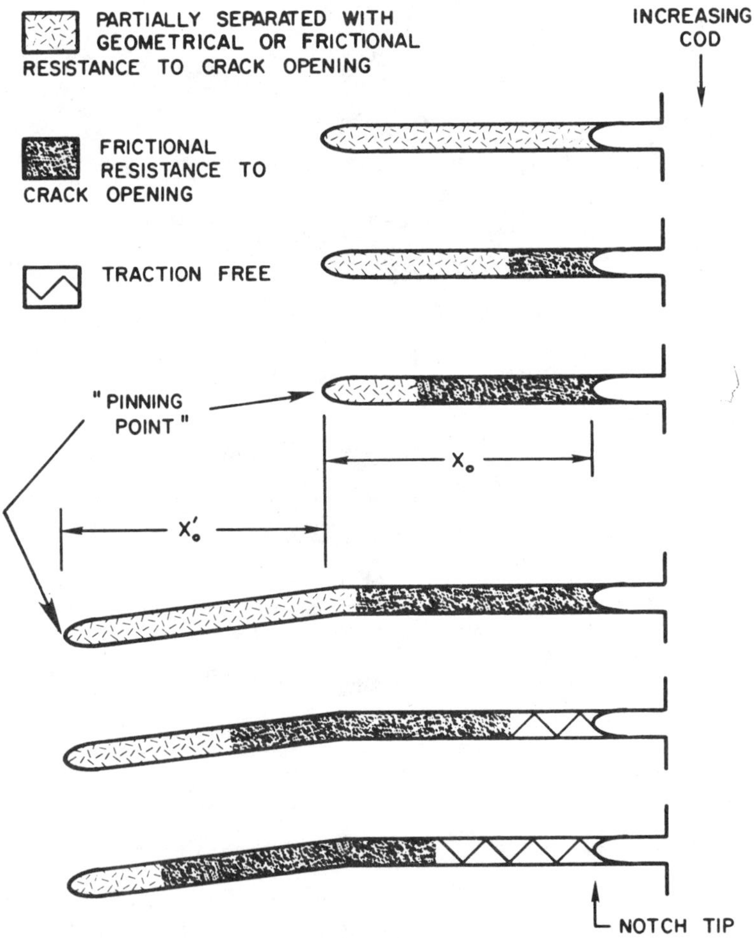

Figure 12. Schematic illustration of the fracture propagation process in WLDCB specimens of Westerly granite. Based on interpretation of NDT experiments and in-situ microscopy observations.

was not observed in double-torsion specimens where the process zone is constrained by the presence of large stress gradients.

A comparison of the observed process-zone shape with that expected on the basis of a critical-principal-normal-tensile-stress criterion shows that the zone is much more elongate in the crack propagation direction than predicted by the continuum-based microcracking model alone. Friction-induced microcracking due to the interlocking of "previously-fractured" crack surfaces and partial decohesion due to segmentation of the crack plane are two mechanisms partially responsible for the elongate nature of the process zone. Both micromechanisms of fracture in Westerly granite were documented by in-situ observations of the fracture process using optical microscopy. Preliminary experiments on alumina ceramics and glass ceramics suggest that these materials also display similar crack-flank process-zone mechanisms.

REFERENCES

1. E.S. Sprunt and W. F. Brace, Int. J. Rock Mech. & Min. Sci. 11:139, (1974).
2. A.G. Evans, J. Mat. Sci., 7:1137, (1972).
3. P.L. Swanson, J. Geophys. Res., 89:4137, (1984a).
4. P.L. Swanson, PhD dissertation, Univ. of Colorado, (1984b).
5. H. Spetzler, C. Scholz, and C.J. Lu, Pure Appl. Geophys., 112:571, (1974).
6. C.H. Sondergeld and L.H. Estey, J. Geophys. Res., 86:2915, (1981).
7. P.L. Swanson and H.A. Spetzler, Proc. 25th U.S. Nat. Symp. on Rock Mechanics, 25:67, (1984).
8. J.G. Bruce, W.W. Gerberich, and B.G. Koepke, in: "Fracture Mechanics of Ceramics", R.C. Bradt, D.P.H. Hasselman, and F.F Lange, eds., 4:687, Plenum, NY, (1978).
9. M.K. Ferber and S.D. Brown, J. Am. Ceram. Soc., 63:424, (1980).
10. R.F. Pabst and J. Weick, J. Mat. Sci., 16:836, (1981).
11. C.G. Annis and J.S. Cargill, in: "Fracture Mechanics of Ceramics", R.C. Bradt, D.P.H. Hasselman, and F.F Lange, eds., 4:737, Plenum, NY, (1978).
12. B.J. Pletka, E.R. Fuller, Jr., and B.G. Koepke, in: "Fracture Mechanics Applied to Brittle Materials", ASTM STP 678, p.19, (1979).
13. E.R. Fuller, Jr., in: "Fracture Mechanics Applied to Brittle Materials", ASTM STP 678, p.3, (1979).
14. O. Sano, Int. J. Rock Mech. Min. Sci., 11:18, (1981).
15. E. Sommer, Eng. Frac. Mech., 1:539, (1969).
16. R.G. Hoagland, G.T. Hahn, and A.R. Rosenfield, Rock Mech., 5:77, (1973).
17. R.A. Schmidt and T.J. Lutz, in: "Fracture Mechanics Applied to Brittle Materials", ASTM STP 678, p. 166, (1979).
18. H. Spetzler, G.A. Sobolev, C.H. Sondergeld, B.G. Salov, and A. Kolstov, J. Geophys. Res., 86:1070, (1981).
19. J.F. Knott, "Fundamentals of Fracture Mechanics", Butterworth, Boston, (1976).
20. R.G. Hoagland, J.D. Embury, and D.J. Green, Scrip. Met., 9:907, (1975).
21. A.G. Evans, A.H. Heuer, and D.L. Porter, Proc. 4th Int. Conf. on Fracture, 4:529, Waterloo, Canada, (1977).
22. R.A. Schmidt, Proc. 21st U.S. Nat. Symp. Rock Mech., 21:581, (1980).
23. A.G. Evans and K.T. Faber, J. Am. Ceram. Soc., 67:255, (1984).
24. F.F. Lange, J. Am. Ceram. Soc., 56:518, (1973).
25. F.W. Kleinlein and H. Hubner, Proc. 4th Int. Conf. on Fracture, 3:883, Waterloo, Canada (1977).
26. H. Hubner and W. Jillek, J. Mat. Sci., 12:117, (1977).
27. R. Knehans and R. Steinbrech, J. Mat. Sci. Lett., 1:327, (1982).
28. P.L. Swanson, B.J. Douglas, and H. Spetzler, EOS, Trans. Am. Geophys. Union, 65:1081, (1984).

STATIC FATIGUE RESISTANCE OF HOT PRESSED SILICON NITRIDE

George D. Quinn

U. S. Army Materials and Mechanics Research Center
Watertown, Massachusetts 02172

ABSTRACT

The static fatigue behavior of magnesia doped hot pressed silicon nitride is characterized by stress temperature regimes wherein failure occurs from slow crack growth from preexisting flaws, or alternatively, creep fracture. The concept of a fracture mechanism map will be applied as originally proposed by Gandhi and Ashby. Flexural stress rupture experiments up to 15,000 hours duration were performed with both artificially (Knoop) and naturally flawed specimens. Two empirical approaches were used to define the slow crack growth versus creep fracture regimes.

INTRODUCTION

This paper is a sequel to an earlier paper[1] which unsuccessfully attempted to correlate life prediction parameters for hot pressed silicon nitride. This new paper reinterprets the earlier results and presents extensive new data.

Static fatigue in high performance ceramics can occur by several mechanisms including stress corrosion, creep fracture, slow crack growth and surface pitting.[2] Hot pressed silicon nitride (HPSN), in particular the NC 132 form★, has received considerable attention as a model material for static fatigue studies. This is in part due to its very consistent behavior and the reproducibility of stress rupture experiments.[3,4] Two mechanisms of static fatigue at elevated temperature have been identified for NC 132 HPSN: slow crack growth and creep fracture.[4,5] Similar observations on another HPSN have been reported by Grathwohl.[6]

Slow crack growth (SCG) is defined in this paper as slow growth from preexisting flaws such as machining damage, inclusions, or artificially implanted Knoop flaws. Most previous studies have focused on this mechanism since it will predominate for many stress-temperature conditions. On the other hand, attention has more recently focused upon creep fracture. Under certain conditions, creep deformations can blunt preexisting flaws. Excessive creep deformation may lead to microcrack nucleation, intergranular

★Norton Company, Worcester, Mass.

growth, and finally, coalescence ultimately leading to fracture. Although both have intergranular crack growth stages, the key distinctions are that SCG emanates from preexisting flaws, with creep deformation and microcracking confined to the crack tip. Creep fracture on the other hand has bulk deformation and microcracking and may not necessarily involve preexisting defects.

The regimes where each mechanism is predominant can best be represented with fracture mechanism maps as suggested by Gandhi and Ashby.[7] Preliminary maps for NC 132 HPSN have been presented previously.[4,5]

EXPERIMENTAL PROCEDURE

All experiments were performed in flexure with specimens cut from two plates of NC 132 HPSN identified as billets A and P. Extensive results with billet A have been presented in Reference 1. Additional experiments have been performed with new specimens from billet A and billet P. Both were fabricated from the same powder lot and were pressed at the same time. Preliminary experiments with P specimens indicated there was no significant difference in failure times between A and P specimens.* Additional material characterization information is in Reference 1.

Four point flexure was performed with hot pressed silicon carbide fixtures with spans of 0.75 and 1.5 inches. Loads were applied with fixed knife edges or cylinders of 1/8 inch diameter. All experiments were in air. Specimens were either tested in the as-machined state, or with an artificial flaw implanted in them. In the latter case, a Knoop indentation was applied to a polished surface, which created not only the hardness indent, but a subsurface semielliptical crack. The purpose was to create a flaw large enough to guarantee it was the severest in the specimen, and to precisely control its size and location. If it was large enough, fast fracture would always originate at the artificial flaw. No provision was made to remove the residual indentation stresses, since it was originally assumed they would be annealed during furnace heating (an unavoidable step) or during a 5 minute soak prior to load application. Additional experimental procedure details have been previously reported.[1] In this report, to help discern trends, failure times less than .001 hours are occasionally shown, but these cannot be considered accurate since the loading sequence was about .001 hours.

Creep deformations were not measured during the experiments, but final deformations were. The midspan deflection was measured relative to the inner flexure span loading points. The curvature in the inner span is constant. If is is assumed the material creeps similarly in tension and compression and that strain is linearly related to the distance from the neutral axis, then a simple formula can be used to evaluate strain at failure. Steady state creep rates can be estimated by dividing the strain by time to failure. Of course, such oversimplified analysis ignores primary, tertiary or viscoelastic strains. It further assumes the applied stress is constant. These creep estimates may be incorporated in a future paper, but there are serious shortcomings with such analyses as will be discussed below.

RESULTS: AS-MACHINED SPECIMENS

Figures 1 and 2 show on a semilog stress versus time-to-failure plot,

*Unlike results between billets A and C as reported in Reference 1.

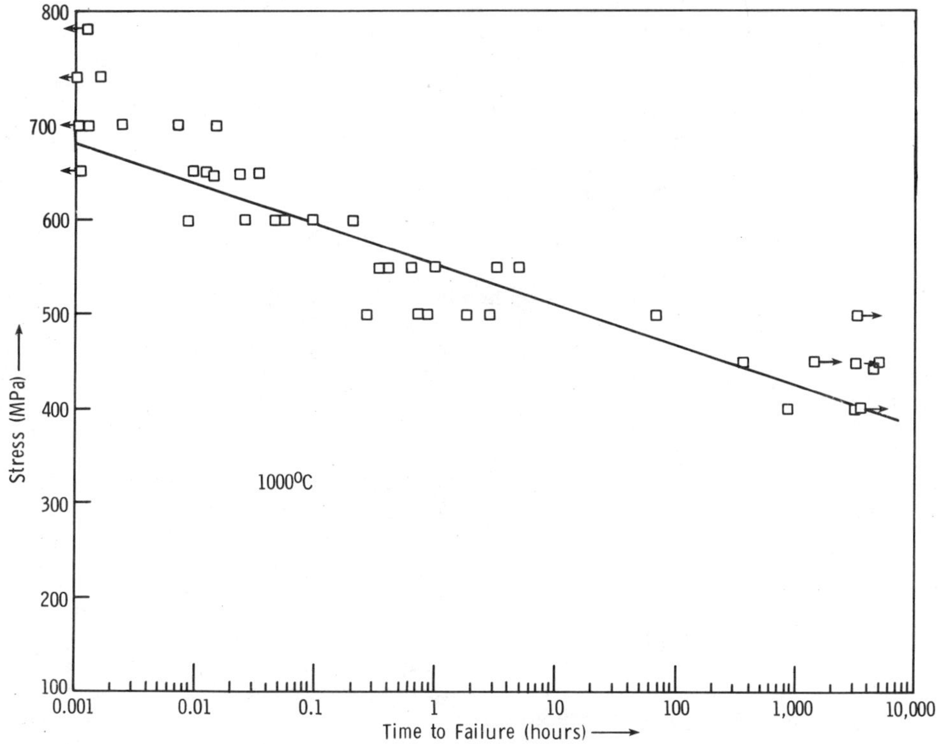

Figure 1. Natural flaw flexural stress rupture results at 1000°C.

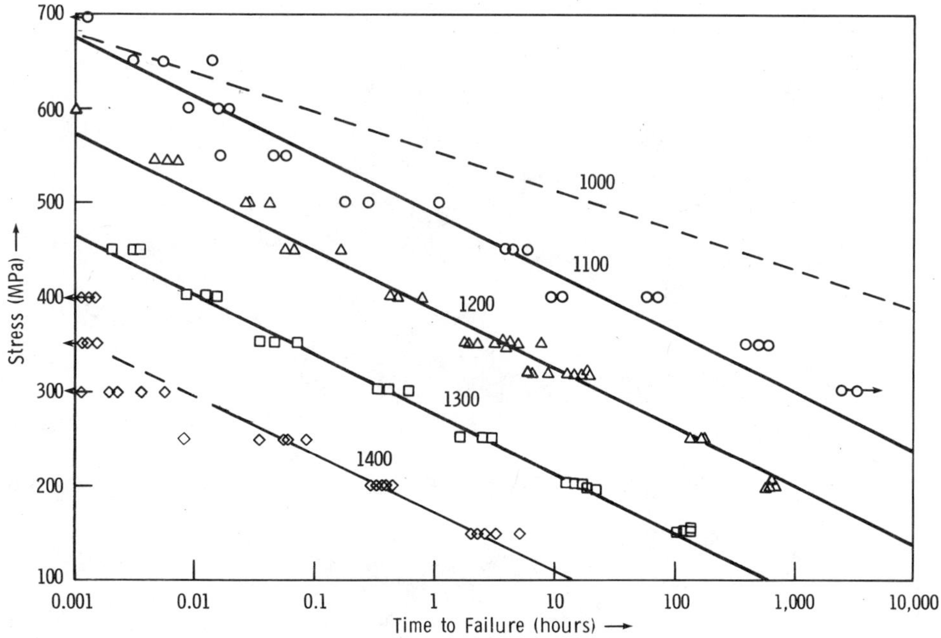

Figure 2. Natural flaw flexural stress rupture results at 1100-1400°C.

the stress rupture results for as-machined specimens. Each point represents an individual specimen outcome. Points with arrows are either specimens that failed on loading or specimens that survived intact. New data is shown both at 1000 and 1400°C. A least squares fitted line is now possible at 1000°C (whereas insufficient data was available in earlier publications [4,5]). At intermediate lifetimes, the 1400°C results seem to follow the 1100 to 1300°C trends. Indeed, much of the data over the 1100 to 1400°C range is well modeled by the relationship:

$$t = 3.09 \times 10^{-21} \exp(180/RT) \exp(-.037\ S) \qquad (1)$$

where t is the time-to-failure in hours, R is the gas constant, T the absolute temperature, and S the applied elastic stress. The apparent activation energy is 180 kcal/g mole. Relationship (1) was used to draw the lines in Figure 2. The short time-to-failure data at 300 MPa, 1400°C was not used in the least squares curve fitting since it appeared to be influenced by the loading sequence.

For comparison, a log-log representation of the data is given in Figure 3. The slopes of the lines can be interpreted in terms of N, the slow crack growth exponent. The values of N derived from curve fitting in Figure 3 are: 29.6, 19.5, 12.8, 9.9, and 9.2 for 1000, 1100, 1200, 1300, and 1400°C respectively.

Figure 4 shows an alternative way of depicting these results. In this instance, each point is derived by picking a temperature and time condition, then reading a stress level from the curves in Figures 1 and 2. The dotted lines in Figure 4 represent reasonable extrapolations. The reason for presenting the stress rupture data in this format will be evident shortly.

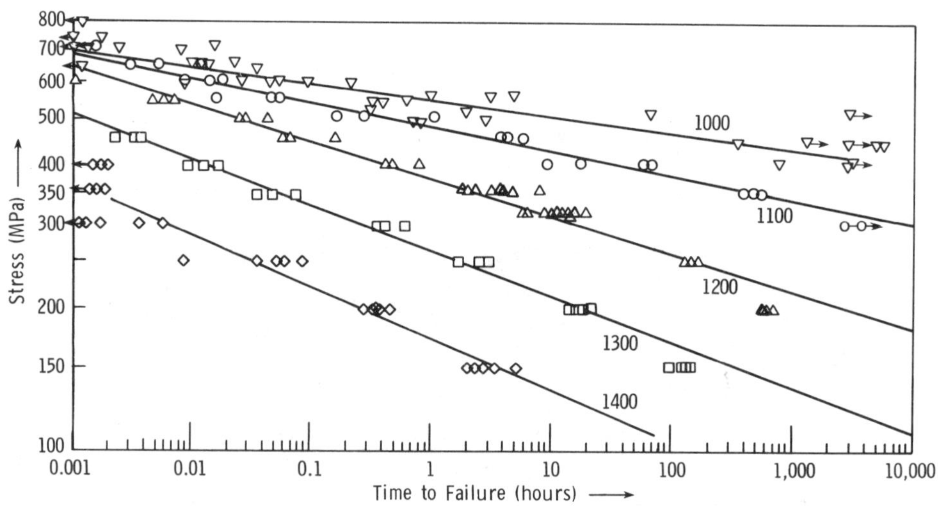

Figure 3. Natural flaw stress rupture results at 1000 to 1400°C on a log-log format.

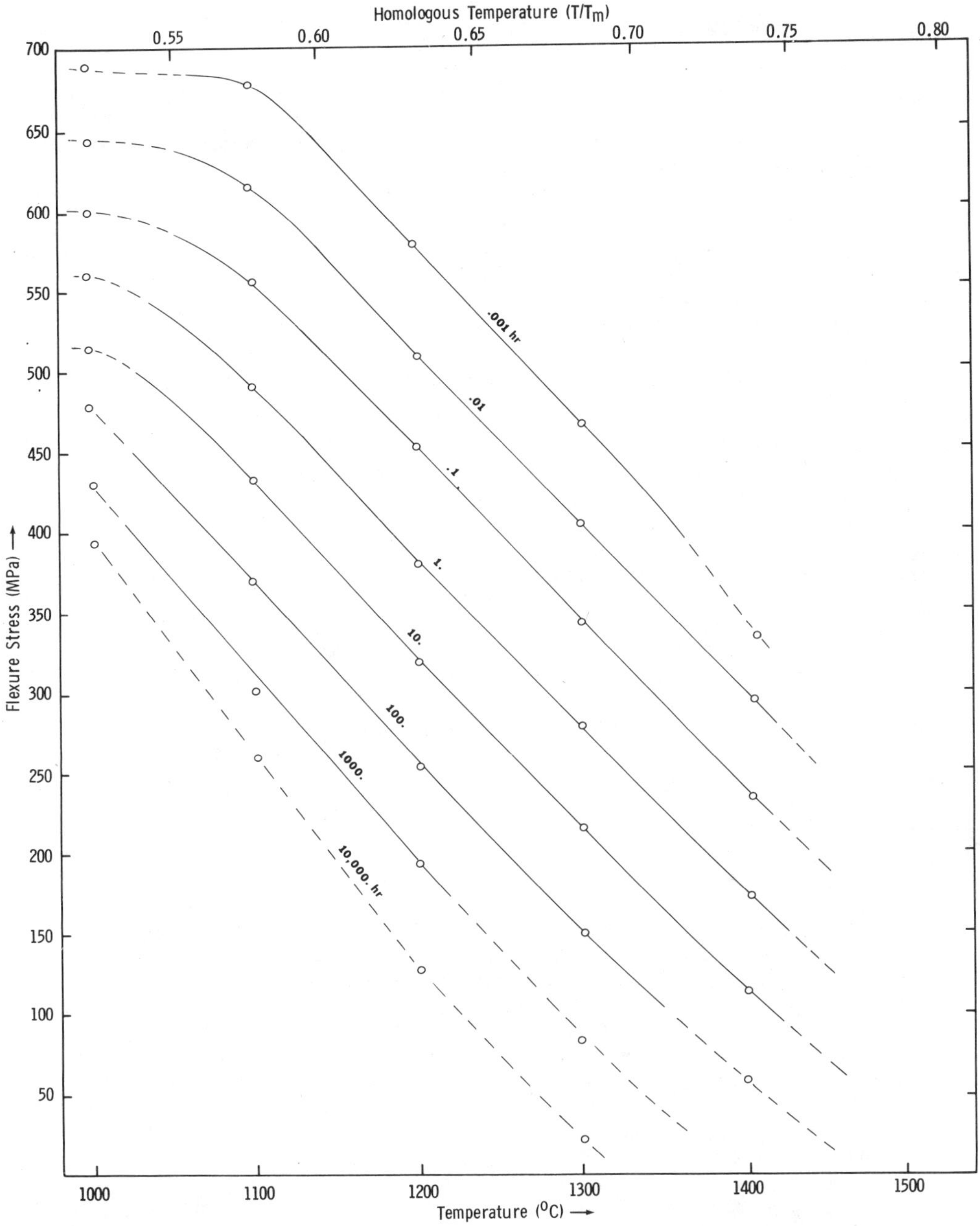

Figure 4. Natural flaw stress rupture. The hollow circles were interpolated from Figures 1 and 2. Loci of constant failure time in hours are shown. Dotted lines are reasonable extrapolations.

The stresses shown in the Figures were computed on the basis of the elastic formulation which is appropriate for conditions where creep relaxation is negligible; that is, for low temperatures, or very short failure times at higher temperatures. The effect of creep relaxation is to diminish with time, the magnitude of the tensile stresses acting in the bend specimen. Therefore, the stresses shown are all upper limits. Specimens that failed under conditions where creep relaxation occured would have "effective" or "time averaged" stresses of less magnitude. Thus, the slopes of the stress rupture lines are upper limits and the strength degradations could be even steeper than shown on the plots.

A number of investigators have attempted to estimate these stress relaxations and account for them.[8-11] Difficulties arise because secondary creep is nonlinear with time and creep in tension and compression can be significantly different.[9,10] Viscoelastic effects can be present,[12-14] which will obscure efforts to measure steady state rates. Arons and Tien, with meticulous effort on tensile specimens, observed that steady state was not achieved in HPSN in hundreds of hours.[12] Oxidation which is simultaneously occuring during stress rupture tests, can cause significant changes in the secondary phase composition and structure in the material.[15,16] These changes may lead to gradients in structure through a flexure specimen which will in turn alter the local constitutive equations. This has been demonstrated for reaction bonded silicon nitride by Grathwohl and Thummler.[17] Recently, some attention has been given to the effects of creep microcracking upon flexural specimen compliance.[18,19] Frictional constraints of fixture knife edges and contact point tangency shifts of flexural specimens that have crept more than a few tenths of a percent strain have been completely ignored. All of the above phenomena are significant and lead one to wonder whether detailed analyses of flexural creep data are warranted. For these reasons, a detailed creep relaxation analysis was not performed in this study, and all stresses were left in terms of effective elastic stress.

RESULTS: ARTIFICIALLY FLAWED SPECIMENS

Figure 5 shows the results for artificially flawed specimens. An indentation force of 15.7 N (1.6 kg) was used on all specimens. This load created a semielliptical surface flaw of 50 microns depth which limited the room temperature fast fracture strength to 370 MPa. This confirms the dominance of the artificial flaws over the natural flaws that ordinarily limit strength to 500-1000 MPa.[1] The 370 MPa strength level is marked with a bar in Figure 5. Each labeled point in the figure corresponds to a stress temperature condition at which stress rupture experiments were performed. The points are labeled by a ratio which gives the number of specimens that did fail from the artificial flaw, to the total number tested. The number in parenthesis is the geometric mean time to failure. For example, at 266 MPa and 1200°C, of seven specimens tested, six broke at the artificial flaw and the geometric mean time to failure was 60 hours.* Figure 6 shows a fracture surface with SCG from an artificial flaw. Figure 5 contains numerous additional results compared to previously presented versions in Reference 4 and 5, and includes a failure at 9,100 hours (1150°C, 200 MPa).

Loci of constant failure time are also shown in Figure 5. Over much of the stress-temperature field, failure did occur as the result of SCG extension from the artificial flaws. This was confirmed by careful fractography. Alternatively, there is a regime where creep fracture was dominant since creep deformation became significant and specimens did not fail from

*Except at 1150°C, 266MPa, where the median is given instead because of two runouts.

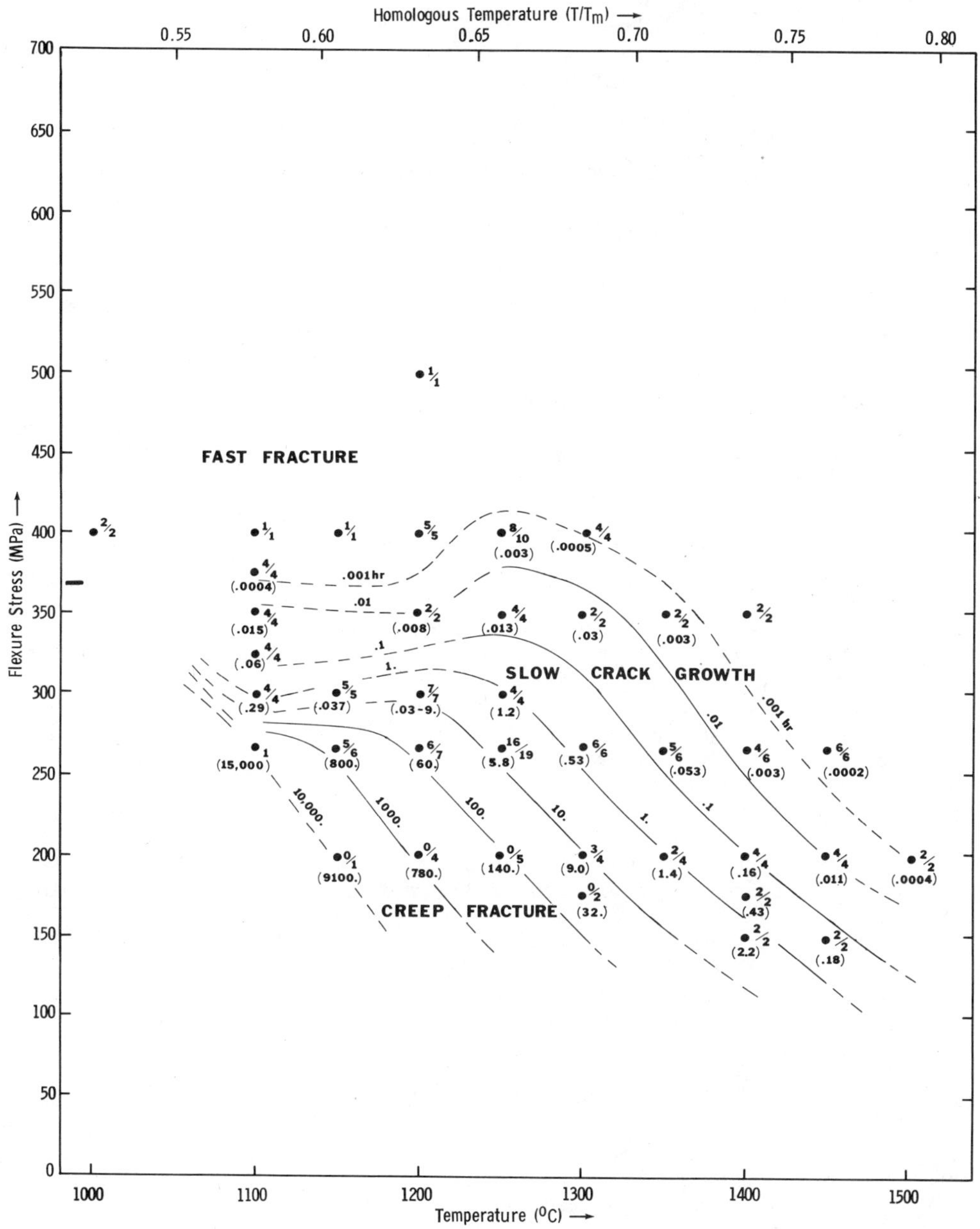

Figure 5. Fracture map for flexural stress rupture specimens containing 1.6 kg Knoop flaws. The symbols are explained in the text.

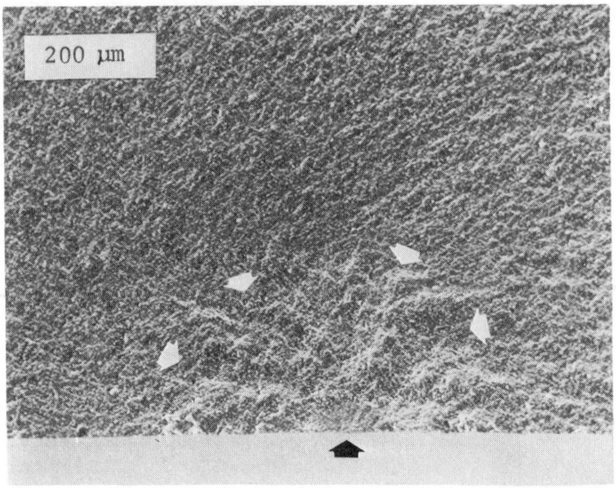

Figure 6. Fracture surface of a specimen which failed at 10 seconds at 1250°C, 400 MPa. The SCG zone is highlighted by white arrows; the original 1.6 kg flaw by a black arrow.

the artificial flaw. Thus, fractographic examination was able to delineate the mechanism of failure for the artificially flawed specimens, unlike the case for naturally flawed specimens. Of special interest are two trials at 266 MPa and 1100°C. One experiment was regrettably terminated at 10,000 hours, but the other was allowed to continue until failure at 14,941 hours. The fracture surface had several tiers including a portion which had crack growth from the artificial flaw, but on another level had gross creep cracking. The specimen had considerable permanent strain (about 1.7 %).

There are a number of ways to represent these results. Figure 7 was prepared to assist in interpolation of lines of constant failure time in Figure 5. Figure 7 emphasizes the strong effect residual stresses (and their relief) have upon failure time for stresses greater than 300 MPa and temperatures less than 1200°C. Lines of constant stress (MPa) are shown. Each solid point is the geometric mean time-to-failure of between 2 and 19 experiments. Hollow points are individual datums. Scatter is generally low except at 1200°C, 300 MPa, where a dichotomy of results exist (arrowed line). As stated before, the times less than .001 hour are not accurate, but are shown in order to detect the trends.

In the one instance where a large number (19) of experiments were performed (266 MPa, 1250°C), the log of time-to-failure was normally distributed. The geometric mean and median thus coincide. The geometric mean was 5.7 hours and the standard deviation of the log time-to-failure was .13.

COMBINED NATURAL AND ARTIFICIAL FLAW RESULTS

A different analysis of the above data can better delineate the SCG and creep fracture regimes. Figure 8 combines the constant failure time loci of Figures 4 and 5. The 10,000 hour loci are tentative and based upon ex-

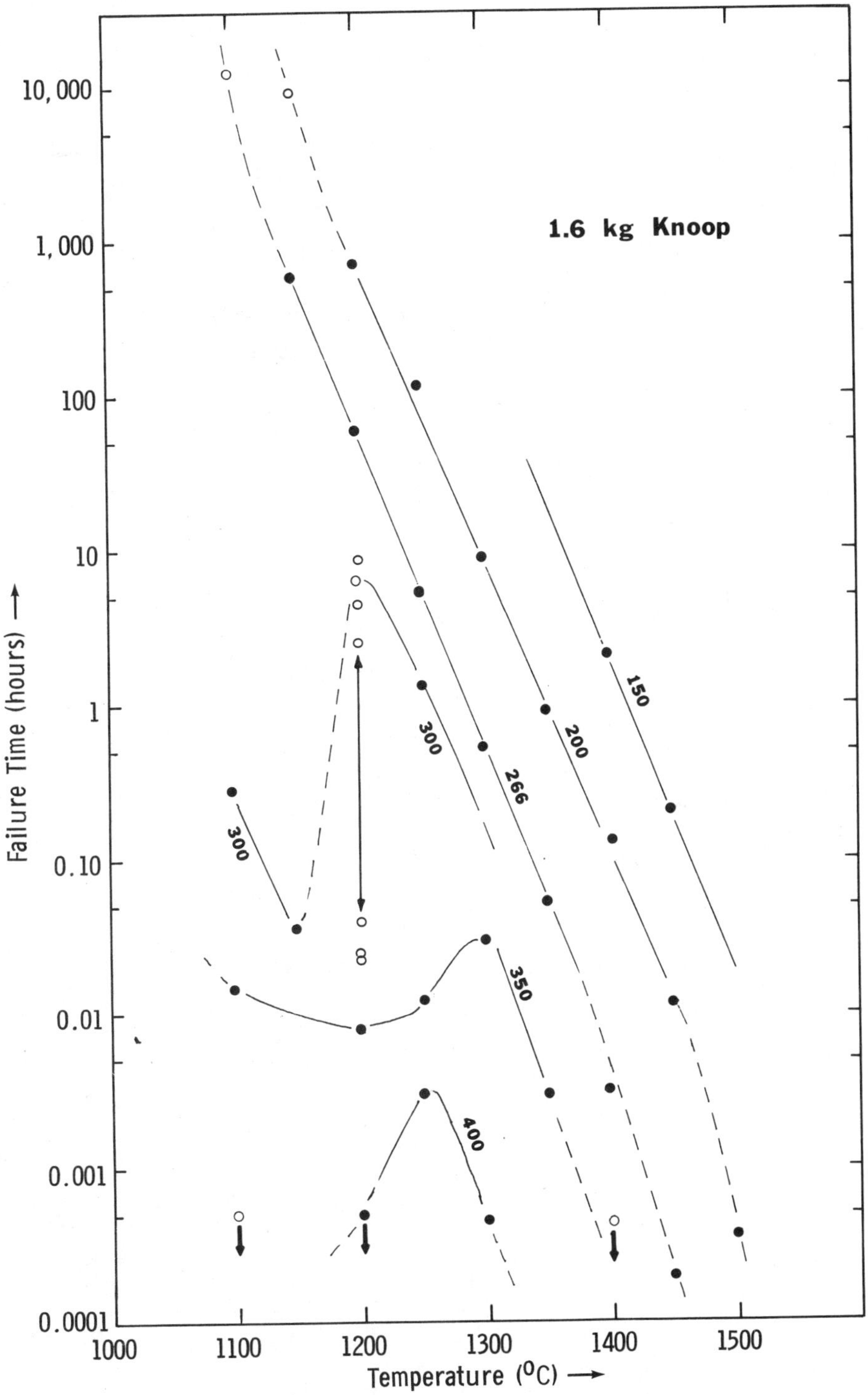

Figure 7. The results of Figure 5 in a different format.

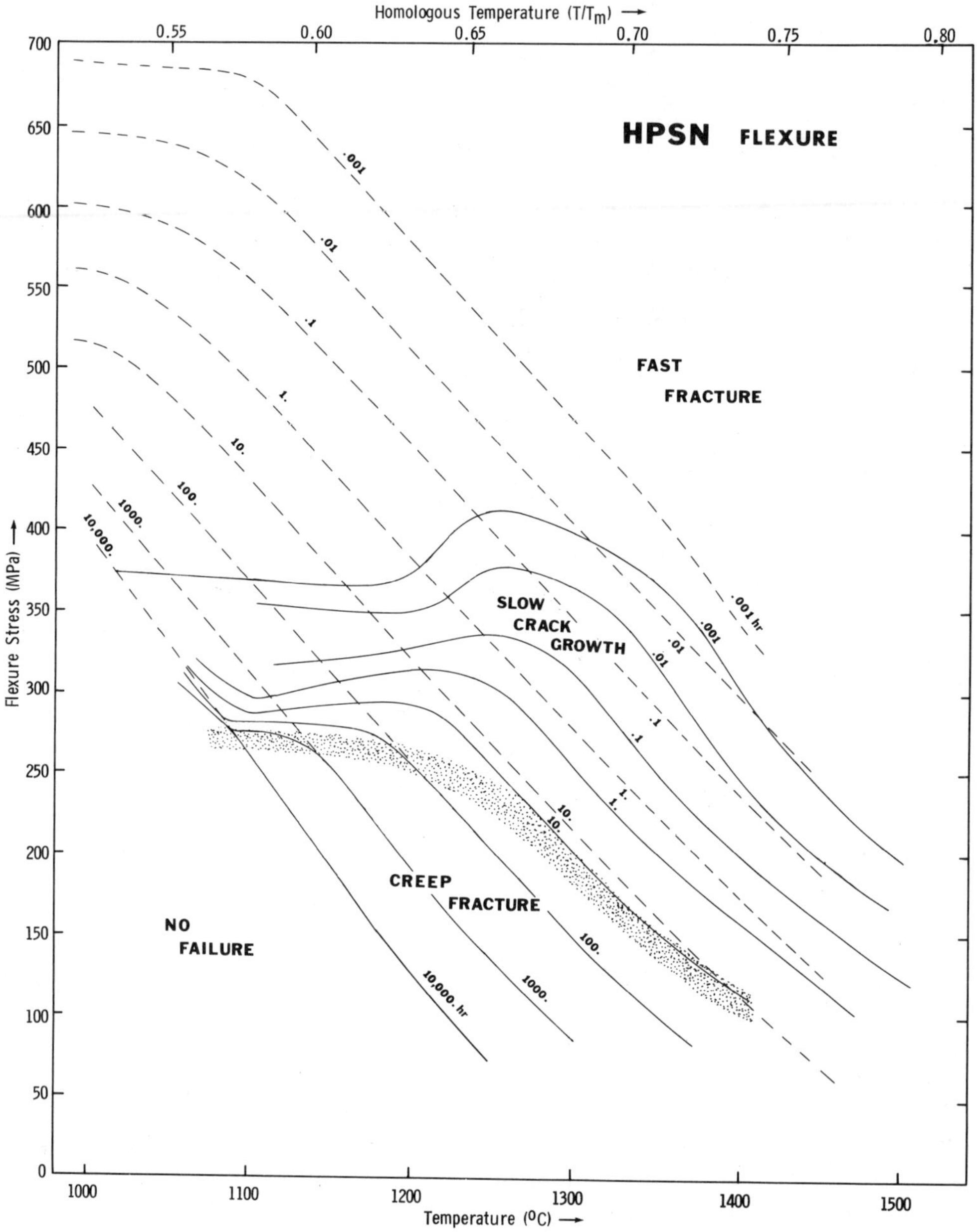

Figure 8. Combined natural (dotted) and 1.6 kg artificial (solid) flaw stress rupture data. The coincidence of the loci of failure time for both specimen types, determines the creep fracture regime.

trapolation. In the creep fracture regime, the overlap of the natural and artificial flaw lines confirms that the presence of a large artificial flaw had no impact whatsoever upon the failure time. In the slow crack growth regime, for a given stress-temperature condition, the artificial flawed specimen always failed in less time. The shaded zone is the boundary between the SCG and creep fracture regimes.

Reference 16 has supportative findings in that five flexure specimens at 1300°C and 196 MPa did not break from the 2.6 kg Knoop flaws implanted in them. Times-to-failure were appreciably different (shorter), but testing in Reference 16 was in vacuum. An earlier study similarly observed a substantial decrease in stress rupture lifetime and strain in inert atmosphere compared to an air environment.[20]

The apparent activation energies for the two mechanisms can be assessed in Figure 9. At 200 and 266 MPa, the data encompases both creep fracture and SCG conditions. There is no significant change in slope in either line, implying the same activation energy applies to both mechanisms. This undoubtedly is the activation energy for grain boundary sliding processes which account for both SCG and creep. The apparent energies are 211 and 216 kcal/g mole which are in reasonable agreement with the 180 kcal/g mole estimated from the natural flaw experiments.

The shaded boundary which delineates the creep fracture and SCG regions levels off at about 260 MPa. It is possible that this level bears some relationship to the 1.6 kg indentation load and the fast fracture level of 370 MPa. It is conceivable that a different indentation load would change the fast fracture strength which in turn could shift the apparent mechanism boundary. To test this hypothesis, limited additional experiments were performed with 1.0 and 3.5 kg indented specimens. Insufficient experiments were done to fill out an entire map such as Figure 5 and attention was focused on experiments near the boundary of Figure 8.

The 3.5 kg indentations reduced the fast fracture strength to 290 MPa and did increase the propensity for failure to occur due to SCG from the very large (85 micron) initial defects. The creep fracture upper stress boundary was estimated to be about 220 MPa. The 1.0 kg indentations only limited the fast fracture strength to 445 MPa. The upper edge of the boundary shifted upwards to 270-280 MPa. Thus,the severity of the initial flaws does seem to have an influence on the exact boundary loaction: the severer the initial flaws, the more likely SCG can cause failure. This is unfortunate because generalization of the fracture map may be more difficult.

DISCUSSION

Creep mechanism maps are well known, but fracture maps are relatively recent. Gandhi and Ashby[7] presented a preliminary diagram for an earlier vintage HPSN.* Their terminology for the SCG regime was "brittle intergranular fracture 3". They used normalized temperature and stress axes. The temperature axis was a homologous temperature, with the T_m for HPSN set at 1900°C. Figures 5 and 8 of this paper include such a normalized axis. The normalized stress axis used by Gandhi and Ashby is not shown since it ignores differences such as flexure versus tension testing, and artificial versus natural flaw. The only tension stress rupture data available on NC 132 HPSN has been generated by Govila[21] and while it shows similar trends it cannot yet be analytically related to flexure data.[4] This is probably due to creep relaxation interfering with the stress distributions in flexure,

*HS 130 Grade, Norton Company, Worcester, Mass; or Joseph Lucas Co., West Midlands, U.K.

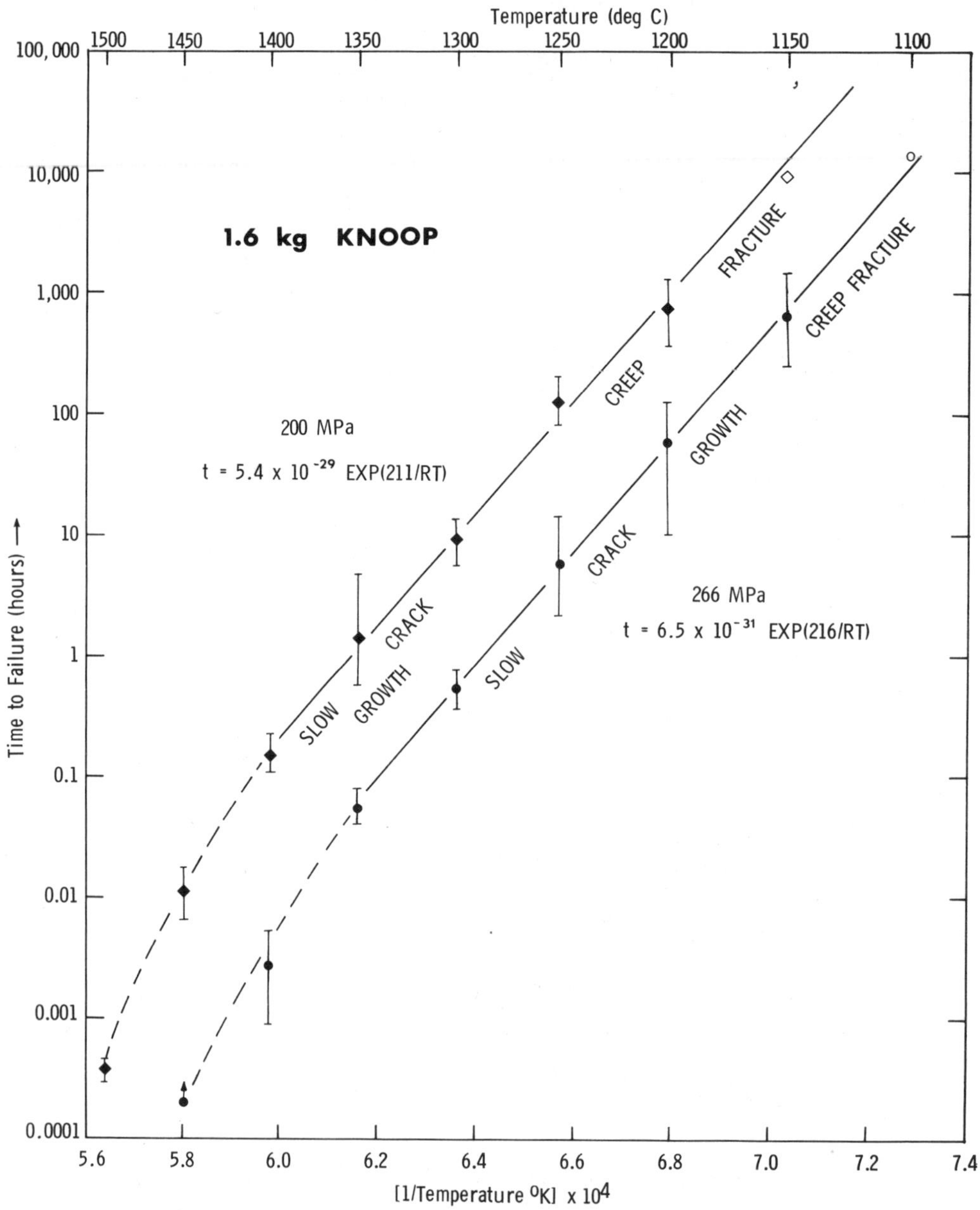

Figure 9. Failure time versus inverse temperature for 1.6 kg indented specimens. The deviation bars are limits.

whereas no such problem exists in tension. Furthermore, the flexure specimens in Govila's study likely failed due to SCG. The tension specimens may have failed in creep fracture.

The analysis of the creep deformations in this study is ongoing as discussed earlier, but some preliminary observations are possible. The apparent strains in the creep fracture regime are of the order of 1 percent or more. There appears to be no coincidence between the creep fracture boundary and strain at failure. There may be a correspondence of the right hand portion of the creep fracture boundary to an apparent strain rate of 2 x 10^{-7}/sec. Of course these statements must be considered in light of the coarse assumptions used in the estimation of strain and strain rate.

Most life prediction analyses to date have interpreted stress rupture results in terms of SCG from preexisting flaws. A power function relating crack velocity and stress intensity:

$$V = A K_I^n \qquad (2)$$

where V is crack velocity, A and n are constants, K_I is stress intensity; is often used to develop equations for time-to-failure. The equation can be rewriten in terms of flaw size, stress and other terms and integrated to yield an equation for failure time. Equation 2 is a convenient form, but is empirical and as yet has no theoretical foundation. Indeed, it may not be relevant to instances where static fatigue failure is due to creep fracture. Life prediction equations for SCG from preexisting flaws and creep fracture will surely differ. Initial flaw size is a variable for the former, but may be irrelevant to the latter. Creep rate and total strain at failure will be important parameters to predict lifetime under creep fracture conditions.

Preoxidation or special heat treatments can affect lifetimes and whether specimens break from artificial flaws.[1,3,16] Flaw healing and deep pit formation, with or without applied stress, can significantly alter strength and lifetime as well.[16,22] Thus, although much of the fracture map region shows "no failure," that phrase is used in the context of no failure in continuous stress rupture loading in air. Significant weakening (or strengthening) could occur nonetheless as shown in Reference 22. Furthermore, at low stresses between 1200 and 1300°C, although fracture may not occur, the material may not be suitable for structures because of excessive deformation. In no instance did strength loss due to pitting or corrosion "catch-up" with or exceed either SCG or creep fracture.

The maps presented in this paper for HPSN are the culmination of years of testing and are revised compared to earlier versions. No additional HPSN testing is planned.

Multiple mechanisms of static fatigue exist in other high performance ceramics as well. SCG and alternatively, a stress corrosion phenomena which acts upon surface connected porosity, cause failure in sintered silicon carbide.[2,23-25] SCG and creep fracture have been identified in sintered silicon nitrides.[26,27]

CONCLUSION

Two mechanisms of static fatigue failure at elevated temperature exist for NC 132 hot pressed silicon nitride: creep fracture and slow crack growth. Both are related to the creep resistance of the material. The difference is a matter of scale. In the former, creep deformations are confined to the immediate vicinity of a crack tip and lead to crack extension. Alternately, if there is significant deformation throughout the material, creep fracture will result.

REFERENCES

1. G. D. Quinn and J. B. Quinn, in: "Fracture Mechanics of Ceramics 6," R. Bradt, A. Evans, D. Hasselman, and F. Lange, ed., Plenum Press, 603-36 (1983).
2. G. D. Quinn, Ceram. Eng. Sci. Proc., 3 (1-2) 77-98 (1982).
3. G. D. Quinn and L. Swank, J. Amer. Ceram. Soc., 66:C31 (1983).
4. G. D. Quinn, in:"Methods for Assessing the Structural Reliability of Brittle Materials," S. Freiman and C. Hudson, eds., American Society of Testing and Materials STP 844 177-93 (1984).
5. G. D. Quinn, Ceram. Eng. Sci. Proc., 5 (7-8) 596-602 (1984).
6. G. Grathwohl, in: "Deformation of Ceramic Materials II," R. Tressler and R. Bradt, ed., Plenum Press, 573-587 (1984).
7. C. Gandhi and M. Ashby, Acta Met., 27:1565-1602 (1979).
8. G. W. Hollenberg, G. R. Terwilliger, and R. S. Gordon, J. Amer. Ceram. Soc., 54 (4) 196-9 (1971).
9. C. A. Anderson, D. P. Wei, and R. Kossowsky, in: "Deformation of Ceramic Materials," R. Bradt, R. Tressler, eds., Plenum Press, 383-98 (1975).
10. P. K. Talty and R. A. Dirks, J. Mat. Sci., 13:580-6 (1978).
11. T. Fett and D. Munz, in: "Methods for Assessing the Structural Reliability of Brittle Materials," S. Freiman and C. Hudson, eds., American Society of Testing and Materials STP 844 154-76 (1984).
12. R. M. Arons and J. K. Tien, J. Mat. Sci., 15:2046-58 (1980).
13. F. F. Lange, B. I. Davis, and D. R. Clarke, J. Mat. Sci., 15:601-610 (1980).
14. F. F. Lange, D. R. Clarke, and B. I. Davis, Ibid., 611-5.
15. F. F. Lange, B. I. Davis, and D. R. Clarke, Ibid., 616-8.
16. G. Das, M. G. Mendiratta, and G. R. Cornish, J. Mat. Sci., 17:2486-94 (1982).
17. G. Grathwohl and F. Thummler, J. Mat. Sci., 13:1177-86 (1978).
18. A. Venkateswaran and D. P. H. Hasselman, J. Mat. Sci., 16:1627-32 (1981).
19. D. P. H. Hasselman, A. Venkateswaran, and Y. Tree, in: "Fracture Mechanics of Ceramics 5," R. Bradt, A. Evans, D. Hasselman, and F. Lange, ed., Plenum Press, 639-57 (1983).
20. R. Kossowsky, D. Miller, and E. Diaz, J. Mat. Sci., 10 (6) 983-97 (1975).
21. R. K. Govila, J. Amer. Ceram. Soc., 65 (1) 15-21 (1982).
22. N. Tighe and S. Wiederhorn, in:"Fracture Mechanics of Ceramics 5," R. Bradt, A. Evans, D. Hasselman, and F. Lange, ed., Plenum Press, 403-24 (1983).
23. G. D. Quinn and R. N. Katz, J. Amer. Ceram. Soc., 63 (1-2) 117-9 (1980).
24. G. D. Quinn, "Stress Rupture of Sintered Alpha Silicon Cardide," U. S. Army AMMRC Technical Report TN 81-4 (1981).
25. M. Slavin and G. D. Quinn, "Mechanical Property Evaluation at Elevated Temperature for Beta Silicon Carbide," U. S. Army AMMRC Technical Report, To be published, 1985.
26. G. D. Quinn, "Static Fatigue of a Sintered Silicon Nitride," U. S. Army AMMRC Technical Report TR 84-40 (1984).
27. I. Oda, M. Matsui, and T. Soma, in: "Progress in Nitrogen Ceramics," F. Riley, ed., Martinus-Nijhoff Publ. Co. Boston, Mass., 501-6 (1983).

TIME DEPENDENT FAILURE OF CERAMIC MATERIALS IN SUSTAINED AND FATIGUE LOADING

A.S. Krausz*, and K. Krausz

Department od Mechanics Engeneering
University of Ottawa
Ottawa, Canada K1N 6N5

ABSTRACT

The physical principles of time dependent failure are reviewed and their application to environment assisted sustained and fatigue loading of ceramic materials presented.

Time dependence is the consequence of thermal activation: the associated rearrangement of the atomic bonds in crack propagation and in environment induced chemical reactions are controlled by the same physical processes. Their common physical principles are described by the constitutive equations of fracture that express rigorously the crack velocity as

$$v = f \quad \text{(external and internal constraints)}.$$

The fracture constitutive equation expresses the effects of the macroscopic design quantities (load, etc.) and of the material characteristics quantitatively and specifies the test methods in exact terms.

Fracture kinetics analyses were carried out for various glasses subject to stress corrosion cracking and the corresponding fracture constitutive equations were established. These are extended to the description of corrosion fatigue of ceramics in which the Elber-type rebonding is understood as a thermally activated process, leading to frequency, wave-form, and temperature dependence in agreement with experimental results. Conclusions have shown that a probabilistic behavior is associated with these processes. The corresponding analyses indicate that threshold effects may be observed with a statistical variance that can lead to failure within the order of a few months when the conventional design expects 30 or more years of life time.

INTRODUCTION

Many industrial uses of ceramic materials require rational descriptions of the progress of crack growth: require the establishment of rigorous, physically based fracture constitutive equations. Such equations have to represent quantitatively the effects of

(a) the external conditions; the geometry of the component and the crack, the spatial and time distribution of the load, the distribution and time variation of the temperature, and the chemical environment,

(b) the internal, material, characteristics; the microstructure and its changes during crack propagation. The fracture constitutive equation can be formally represented as

v = f (load, geometry, temperature, chemical environment, material composition and microstructure, and their time variation).

This is, of course, a formidable task. To aim for it, however, is so imperative that the efforts are worthwhile.

The very character of a rationally based approach requires that it should be firmly rooted in the fundamental concepts of physics that define the behavior of solid aggregates. Statistical mechanics provides this basis[1]. In the following the principles of the fracture constitutive equations are described as applicable to ceramic materials. Realizing that statistical mechanics is not necessarily a component of the common background in ceramic engineering, the report starts with a brief discussion of the essential characteristics of the physical process of crack propagation. Using this conceptual framework a mathematical formulation will be presented, leading to a constitutive equation that describes subcritical crack propagation in a sufficiently practical form so that it can be used in engineering applications.

THE PHYSICAL PROCESS OF TIME DEPENDENT CRACK PROPAGATION

The bewildering complexity of microstructural and loading condition combinations can be brought under conceptually simple control (and, as will be shown in the next section, to qualitative terms as well) by noting that

(i) all materials are composed of atoms with interactions of the same character, their differences being quantitative;

(ii) crack propagation is a sequence of atomic bond breaking steps and is, therefore, controlled by energy conditions.

The understanding that results from this conceptual framework is complete, physically firmly based, and leads to a quantitative description.

Consider first the appearance of a crack at the microscale, Figure 1. The circles represent the atoms; for simplicity, regularly spaced identical circles are shown, but it is understood that they represent the variety in kind and distribution of the atoms that correspond to the actual ceramic material. However, the simplification will not affect the conclusions in any manner. Notice in the figure that the atoms are interconnected by the atomic interaction forces. Again, a simplified picture is presented in which only nearest neighbor interactions are formally represented by springs between atoms. The visual model is then valid for the description of the most important features of time dependent crack propagation.

When tensile load is applied to the model, the interatomic distances increase simultaneously and the springs are stretched. If the load is high enough, the already extended springs in front of the crack tip become overextended and break: the crack propagates by one or more interatomic distances. This is the process of dynamic crack propagation. When the load is less than the critical value, the crack propagation is slow, subcritical, time dependent.

Consider that the load applied to the model is not sufficiently high to break the springs (atomic bonds). Under this effect only, the system remains in a stable condition and no crack propagation could occur. To understand the physical process of subcritical, time and temperature dependent crack propagation, the effects of the always present atomic vibrations have to be considered. It is well recognized that the atoms are not at a standstill at the positions shown in Figure 1: they do a perpetual vibrational motion about this point, with a frequency of about 10^{12} sec^{-1}, and with randomly varying amplitudes and directions[2]. The randomness is essential to the process. The amplitude is related to the interatomic distance and is the atomic level equivalent of the heat content: the higher the amplitude the greater is the energy associated with the atom. This energy, being of a randomly varying quantity, is the physical reality of the macroscopic concept of heat content: the higher the temperature the larger is the average amplitude. Now, when the vibrational movement is superposed on the bond-stretching imposed by the load it is clear that the atomic distance increases during a particular half cycle of the vibration. If the randomly varying amplitude happens to be large enough, this superposed effect breaks the bond and the crack propagates by one atomic distance. The rate of subcritical crack propagation is, therefore, determined by the applied load and by the frequency at which large amplitudes occur, which, in turn, depends on the temperature.

Atomic bonds are defined by energy conditions, by the energy needed to break the bond. This atomic interaction energy is represented in Figure 2 for a pair of atoms. The character of the atomic interaction energy vs. interatomic distance is typical for all materials, only quantitative differences exist[3]. It is obvious that to break the bond the interatomic energy has to be supplied to the crack tip. This is then a critical value, the interatomic energy, the strength of the ceramic material expressed at the atomic level. Because this energy depends on the interacting atom types and their spatial configuration, it represents the chemical composition and the atomic configuration of the material: it represents quantitatively the microstructure, including structural defects. The crack will propagate when the work supplied by the load, W, and the thermal energy of the instant, ΔG, is equal to the bond energy, $\Delta G^{\dagger}$, that corresponds to the microstructure

$$\Delta G^{\dagger} = W + \Delta G. \tag{1}$$

Rigorous theoretical statistical mechanics considerations have shown that the frequency of bond breaking is expressed as[4]

$$k = \frac{kT}{h} \exp(- \frac{\Delta G}{kT}) \tag{2}$$

In Equation (2) k is the elementary rate constant, the average frequency of bond breaking (having the dimension of $time^{-1}$); k and h is the Boltzmann and Planck's constant respectively, both are universal constants, independent of the material and external constraints; T is the absolute temperature. A very important conclusion follows from Equation (2): the frequency, that is, the rate of bond breaking is fully controlled by $\Delta G/T$; the effects of all external constraints and the microstructure are fully and quantitatively expressed by the apparent activation energy, ΔG. A major simplification of the complex fracture process is recognized by the application of the statistical concept of rate theory that led to Equation (2).

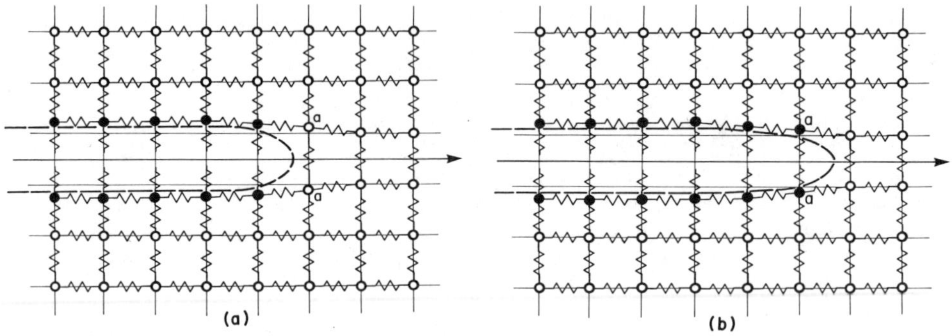

Fig. 1. In crack propatation an elementary step of thermal activation occurs when the atomic bond in front of the crack tip breaks and the crack moves from the position indicated in (a) by one atomic distance to the right as shown in (b). The intact springs represent intact atomic bonds where atomic interactions exist, and the broken bonds are shown by the broken springs.

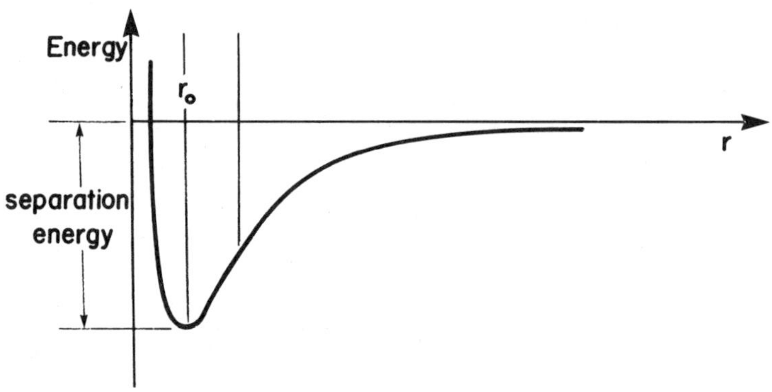

Fig. 2. The schematic representation of the typical interatomic energy vs. interatomic distance relation. The minimum energy corresponds to the equilibrium atomic distance r_o.

Because Equation (2) has a firm foundation on basic physical theories and consequently is a rigorously derived quantitative expression, it is eminently qualified for the development of precisely defined testing methods, for the analyses of experimental results, and for the design of life time conditions. To do so, it is to be noticed that ΔG is an internal quantity, the average heat content needed to break bonds: it is not, by itself, a directly measurable quantity. To render it into a convenient, and practical form, Equation (1) is substituted into (2)

$$k = \frac{kT}{h} \exp(- \frac{\Delta G^{\dagger} - W}{kT}) \tag{3}$$

All quantities are now in the form that is directly measurable, or can be defined by the designer for engineering application. The values of the universal constants are: $k = 1.38 \times 10^{-23}$ J K^{-1}, and $h = 6.6 \times 10^{-34}$J sec. The temperature is either measured, calculated, or specified. The bond energy represents — expresses — quantitatively the state of microstructure. As a first approximation, atomic bond energy values listed in physics and chemistry handbooks can be used. It is to be realized that while the theory of interatomic bond strength is absolutely well established, indeed it is one of the fundamentally valid theories of physics (quantum mechanics), the mathematical processing is beyond the present capacity of calculation for anything but the simplest bonds. The so-called activation energy, $\Delta G^{\dagger}$, is, therefore, a quantity that has to be measured. As such, it represents the qualities of the microstructure that are pertinent to the understanding of crack propagation. The fact that it has to be measured should not lead to the notion that Equation (3) is an empirical expression: it was derived rigorously from physical principles and is defined fully in terms of exactly defined physical quantities. Material properties usually must be measured. To illustrate: elastic stress analyses are carried out with a rigorous theory, and certainly not with empirical expressions. But, the elastic moduli (Young's modulus, Poisson's ratio) have to be measured because these are material characteristics. Exactly identical considerations are valid for Equation (3). At the same time, the measured value of $\Delta G^{\dagger}$ provides a quantitative identification of the microstructure and its evolution during fracture.

The work, W, originates from the external load and from the change of the stored elastic energy: it is the mechanical energy contributed by these two mechanical sources to the bond breaking step. The work depends on the load; the geometry of the component; the elastic properties; displacement boundary conditions; and — because of its effects on the stress and strain field — on the defect structure, most notably cavity formation, and grain and anisotropy conditions in the vicinity of the crack tip. For brittle, linearly elastic ceramics the mechanical energy can be calculated — in principle. A fracture mechanics consideration leads immediately to the practical conclusion that the work should be expressed as the function of the stress intensity factor, K, or of the crack driving force, $\mathcal{G}$, $(\mathcal{G} = K^2/E)$:

$$W = W_1(K) \quad \text{or} \quad W = W_2(\mathcal{G}) \ . \tag{4}$$

For nonlinearly elastic ceramics the corresponding quantity is, of course, the J-integral. A convenient, and mathematically rational, approximation of the functional form for Equations (4) is

$$W = \alpha_K K \quad \text{or} \quad W = \alpha_{\mathcal{G}} \mathcal{G},$$

where the work factor α can be calculated from stress and strain analysis or can be measured when the material microstructural conditions of the ceramic have significant effects on the work associated with crack propagation.

The description of the crack propagation step rate is now complete: to express the crack velocity one more aspect of the process has to be considered.

Fracture Kinetics

It is a generally valid notion that crack propagation is a complex of elementary steps: no process is so simple that it would consist of only simple repetitions of the step rates defined by a single rate constant.

All crack propagation processes are associated with various mechanisms. As an example, recall that it was shown that cracks typically propagate by forming double kinks and spreading sideways afterwards[5,6,7]. These are controlled by two different kinds of bond breaking resistances — two different G values — and work contributions. Also, the progress of cracks through a string of cavities and their adjoining ligaments is associated with a combination of cavity formation and crack propagation rate constants. Two other particularly conspicuous examples are stress corrosion cracking and corrosion fatigue[8,9]. Both are intimately tied with the chemical degradation of the ceramic. The chemical reactions themselves are always composed of several interrelated steps; the breaking of the corroded bonds is a further addition to the complexity. (It is to be noted that the chemical reaction steps are also always described by the Equation (2) type expressions[10]. Even from these few examples it is obvious that the rate controlling steps are complex indeed. Again, as in the previous section, the recognition of the theoretical foundations of these processes and the application of the theory leads to an ordering of the confusing varieties: leads to the conclusion that all components can be fully represented by a single kinetics expression. This expression gives the residual rate of crack propagation steps, k_r, the fracture kinetics equation for a system of two mechanisms as[11,12]

$$k_r = \frac{k_{b_1} k_{b_2} - k_{h_1} k_{h_2}}{k_{b_1} k_{b_2} + k_{h_1} k_{h_2}} , \tag{5}$$

where the subscripts 1 and 2 identify the two step-types and the subscripts b and h refer to breaking and healing steps as explained in the Appendix. Extension of the fracture constitutive equation to more than two types of steps follows directly in the pattern given by Equation (5) and the references.

Equation (5) is a typical representation of the general expression of fracture kinetics that is valid for the description of any time and temperature dependent, subcritical crack propagation process. It is an important observation that often many of the step-types have only negligible effects on the residual rate. In fact, Equation (5) often reduces to a single term so that

$$k_r = k_i .$$

A typical example is shown in Figure 3. The individual k_i expressions depend on the external constraints and on the microstructure: their combination also reflects the effects of these two.

<u>The Fracture Constitutive Equation.</u> To establish the fracture constitutive equation of time and temperature dependent propagation as a rational, physically rigorous theoretical expression, the concepts presented in the two previous sections are used.

The velocity, v, of crack propagation is clearly the function of the frequency of the atomic bond breaking steps and the distance over which the crack travels at each event

$$v = n\lambda\, k_r , \tag{6}$$

where n is a small integer, λ is the interatomic distance, and k_r is

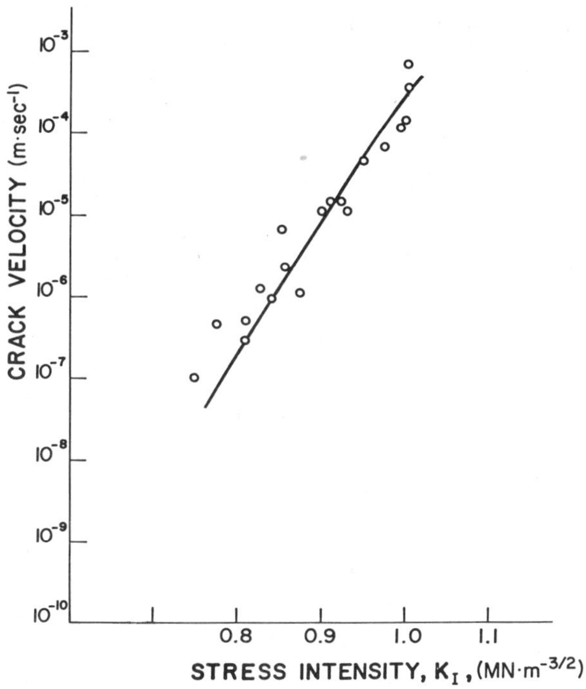

Fig. 3. Porcelain, tested in water at room temperature[11]. The curve was calculated with the reduced form of Eq. (5), that is, with the equivalent of Eq. (6).

defined by Equation (5). The constitutive equation is then a logical, and unique assembly of the elementary rate constants k_i. Because each rate constant is rigorously defined as the function of external constraints and the microstructure and so is their kinetic combination leading to k_r, the constitutive equation is theoretically rigorous, unique, and represents fully all tests and design characteristics.

CONCLUSIONS

The complex processes of time and temperature dependent subcritical crack propagation can be rationalized to obtain a generally valid, theoretically founded description of the crack propagation velocity as the function of the external constraints and the internal, microstructural, characteristics. The corresponding fracture constitutive equation is a unique, physically rigorous quantitative expression.

It is an important conclusion that the constitutive equation is shown to be fully defined by two elements only: (1) the thermal energy needed to break the atomic bonds, ΔG; (2) by the kinetics combination of the elementary rate constants defined uniquely by ΔG.

Some aspects of the foregoing deliberations were presented as separate building blocks before[2,7,8,11]: here the coherent theory is elaborated. Further conclusions and applications have been submitted elsewhere.

APPENDIX

It is a fundamental concept of statistical mechanics that at the atomic level all processes occur in both directions[1,4]: not only bond breaking, but bond healing steps take place as well. The macroscopic observation records the net results of these two. Accordingly, cracks grow on the macroscopic scale when the breaking rate is greater than that of healing. Experiments have shown that the reverse can also be true and cracks can shrink[13,14].

The description of the average healing rate is very similar to the breaking rate. The healing rate constant is[4]

$$\begin{aligned} k_h &= \frac{kT}{h} \exp\left(- \frac{\Delta G_h}{kT}\right) \\ &= \frac{kT}{h} \exp\left(- \frac{\Delta G_h^{\dagger} + W_h}{kT}\right) , \end{aligned}$$

where the terms have the meaning similar to that for the bond breaking rate constant

$$\begin{aligned} {}_b &= \frac{kT}{h} \exp\left(- \frac{\Delta G_b}{kT}\right) \\ &= \frac{kT}{h} \exp\left(- \frac{\Delta G_b^{\dagger} - W_b}{kT}\right) . \end{aligned}$$

Note that the sign of the work terms is different. Accordingly, ΔG_h is usually (but not always!) greater than ΔG_b, and consequently

$$k_h < k_b .$$

REFERENCES

1. T.L. Hill, "Statistical Mechanics", McGraw-Hill, N.Y. (1956).
2. A.S. Krausz and K. Krausz, <u>Canadian Metall. Quarterly</u> 23: 107 (1984).
3. K. Laidler, "Chemical Kinetics", McGraw-Hill, N.Y. (1965).
4. A.S. Krausz and H. Eyring, "Deformation Kinetics", Wiley-Interscience, N.Y. (1975).
5. B.R. Lawn and T.R. Wilshaw, "Fracture of Brittle Solids", Cambridge University Press, Cambridge (1975).
6. A.S. Krausz, <u>J. Appl. Phys.</u> 49: 3774 (1978).
7. A.S. Krausz and K. Krausz in "Deformation of Ceramics II", p. 655, R.E. Tessler and R.C. Bradt, eds., Plenum, N.Y. (1984).
8. K. Krausz and A.S. Krausz in "Deformation of Ceramics II", p. 547, R.E. Tessler and R.C. Bradt, eds., Plenum, N.Y. (1984).
9. K. Krausz and A.S. Krausz in "Fatigue 84", C.J. Beevers, ed., U. of Birmingham, Birmingham, U.K. (1984).
10. S. Glasstone, K.J. Laidler and H. Eyring, "The Theory of Rate Processes", McGraw-Hill, N.Y., (1941).
11. A.S. Krausz, <u>Int. J. Frac.</u>, 14:5 (1978).
12. A.S. Krausz, <u>Eng. Frac. Mech.</u>, 11:33 (1979).
13. E. Orowan, <u>Z. Physik</u>, 82:235 (1933).
14. S.M. Wiederhorn and P.R. Townsend, <u>J. Am. Ceram. Soc.</u>, 53:486 (1970).

THE ROLE OF SURFACE FORCES IN FRACTURE

D. R. Clarke*, B. R. Lawn** and D. H. Roach**

*Thomas J. Watson Research Center, IBM, Yorktown Heights, NY 10598
**National Bureau Of Standards, Gaithersburg, MD 20899

ABSTRACT

The role of surface forces in fracture is considered by incorporating the contribution of the adhesive forces between the crack walls into the net mechanical driving force. It is shown that attractive surface forces lead to a threshold stress intensity below which a crack will heal. In addition, the surface forces modify the kinetics of slow crack growth. Crack growth may provide an alternative experimental method for measuring fundamental surface forces.

I. INTRODUCTION

The central notion behind the present work is that surfaces forces acting between the walls of a crack behind the tip contribute to the overall balance of forces acting to drive the crack. These forces must be taken into account in any comprehensive description of fracture. It is convenient then to consider a crack to be under the influence of two distinct driving forces, one a crack advance force due to the applied stress and the other a crack closure force

From an historical perspective, the neglect of surface forces when considering the fracture process is understandable. In the majority of fracture configurations considered in the past emphasis has been placed on the conditions for crack growth, in particular unstable crack growth. Under such conditions the crack driving force provided by the applied stress swamps any contributions from surface forces. In some fracture configurations, for instance in the de-adhesion of an interface, where surface forces might be expected to exert a dominant influence, the fracture mechanics have not been well characterized: this despite the extensive work in the last fifteen years (1-5) on the adhesion of spheres to flat surfaces (a geometry that has a direct correspondence, for elasticity analysis, to that of a mode I crack, as was pointed out in 1939 by Westergaard, 6). Perhaps more importantly, however, was that until the last decade or so there were few direct measurements of the surface forces, especially down to distances of the order of atomic spacings. The best that could be done was to resort to values of the empirically determined van der Waals forces from the colloid science literature, values that different workers had difficulty reproducing to within an order of magnitude. However, the incredibly delicate experiments using crossed mica cylinders first carried out by Tabor and Winterton (7) and subsequently developed to a higher level of sophistication by Israelachvili, with Tabor (8) and with a variety of co-workers in Australia (9-12), have transformed our understanding of surface forces by providing precise, direct measurements. The theory of surface forces has also been significantly advanced in the last two decades. In particular, the work of the Russian group under Lifshitz (13) extended the quantum electrodynamics treatment of long range Van der Waals-dispersion forces to the pertinent case of parallel bodies separated by an intermediary phase. In addition, a number of groups (14-16) have demonstrated how these theories could be applied to problems in cell biology, wetting phenomena and colloid science.

As a result of these advances we are now in a position to consider the role that surface forces might have on the fracture process. We might anticipate surface force effects to be significant in view of the remarkably small crack opening displacements that characterize brittle solids, eg figure 1. Our purpose in this paper is to explore these effects. We shall be focussing on van der Waals forces because of their amenability to analytical description, but the underlying principles are expected to embrace other types of surface forces.

II. FORCES ACTING ON A CRACK

Our focus in this section is on the forces that may act on a crack and which must be taken into account in establishing a force balance at the crack tip. We deal specifically with physical forces as distinct from any chemical forces that may occur at the crack-tip itself as a result of bond-rupturing reactions. For the purposes of calculating stress intensity factors and relating to elasticity equations later we use continuum descriptions of the force distributions. We also assume that a fluid phase wets the crack along the walls and is connected to an external reservoir. These assumptions allow us to describe the surface force distributions in terms of interfacial stresses and, ultimately, of interfacial energies.

The principal force that can be applied to the crack is, of course, the applied stress intensity, an experimental variable dependent only on the loading geometry and load itself. The surface forces augment this force. Derjaguin (17) refers to the net effect of the surface forces as a disjoining pressure, Π. We shall simply regard this pressure as a distributed normal stress, $\Pi(x)$, at the interface, but (contrary to convention in colloid chemistry), adopting a positive sign to designate a closing stress.

A number of specific force contributions (17,18) to the total disjoining pressure can be identified: the van der Waals dispersion force, Π_D; electrical double layer forces, Π_{EDL}; forces due to hydrogen bonding in the phase between the crack walls, Π_{HB} ; forces due to preferential solute adsorption on the crack walls, Π_{AD} ; electrostatic forces, Π_E; forces due to the structure of the liquid, Π_{ST}. Since these forces are independent of one another, their individual contributions are additive and can be expressed as:

$$\Pi = \Pi_D + \Pi_{EDL} + \Pi_{HB} + \Pi_{AD} + \Pi_E + \Pi_{ST} \qquad [1]$$

Little is known about these terms with the exception of the dispersion and possibly also electrical double layer forces. These two figure importantly in the well-known DLVO theory (17,19,20) of colloidal stability, but as indicated earlier even these were not experimentally accessible until the crossed cylinder measurements. Also, the region of interest to the colloid science community was the long range, out to distances of microns rather than those pertinent to fracture here, namely less than 100 nm. The theoretical framework of Dzyaloshinskii et al. (13) and the subsequent work of Parsegian, Ninham and Gingell (14-16) enables precise calculation of the dispersion force to be made from optical constants of the materials. Specifically, the van der Waals dispersion force for two slabs of a given phase separated by a second phase of thickness, d, can be expressed as:

$$\Pi_D = \frac{\hbar}{16\pi^2 d^3} \int_0^\infty \left[\frac{\varepsilon_\alpha(i\zeta) - \varepsilon_\beta(i\zeta)}{\varepsilon_\alpha(i\zeta) + \varepsilon_\beta(i\zeta)}\right]^2 d\zeta$$

where $\varepsilon(i\zeta)$ are the dielectric permeabilities of the phases involved and ζ is a dummy variable. This may be re-written as:

$$\Pi_D = \frac{H}{6\pi d^3} \qquad [2]$$

where H is conventionally referred to as the Hamaker "constant". This equation has the form long familiar to the colloid science community. The difference is that the Hamaker "constant" is now a predictable quantity dependent only on the dispersion properties of the phases. The value of the Hamaker constant for most ceramic systems is of the order of $1 - 100\,kT$, namely $10^{-21} - 10^{-19}\,J$. We may note that if we define the van der Waals interaction energy as the work, $2\gamma_D$, to separate the two walls from an initial separation d_o to ∞ we obtain:

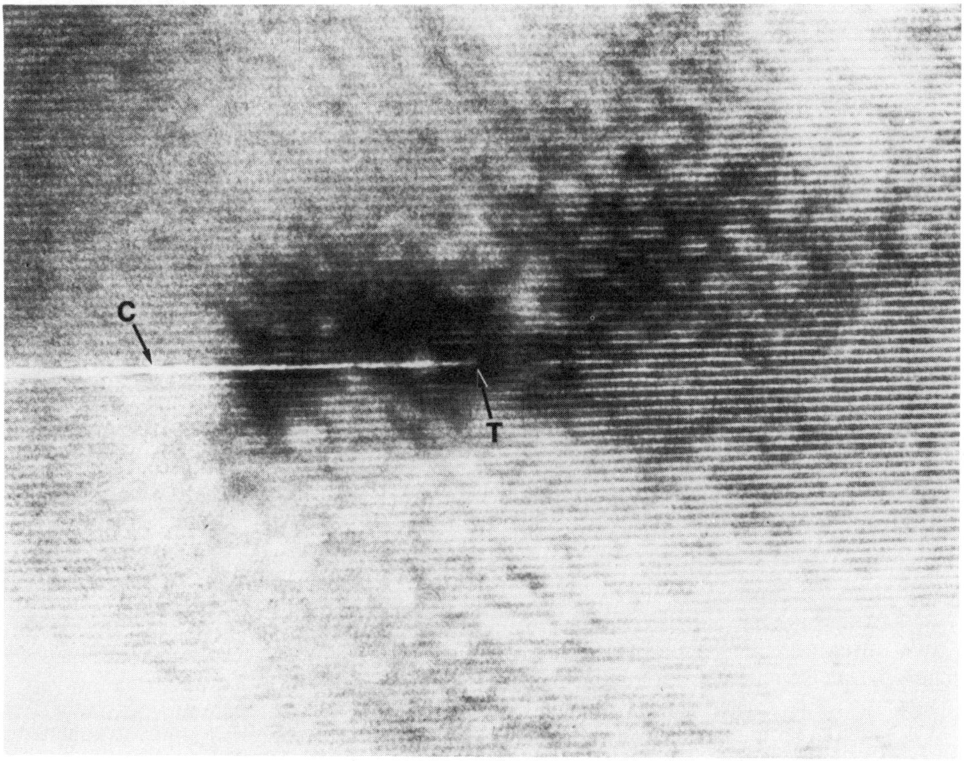

Figure 1. A high resolution transmission electron micrograph of a crack-tip region in a magnesium Sialon material. The crack, C, is in a loaded state as evidenced by the diffraction contrast around the crack-tip, T. The lattice fringes correspond to the lattice spacing of the material.

$$2\gamma_D = \frac{H}{12\pi d_o^2} \qquad [3]$$

III. FORCE BALANCE AND NET CRACK DRIVING FORCE

Given the form of the forces acting on a crack the next step is to calculate the effect they have on the crack driving force. This can perhaps be most usefully done through the sequence of thought experiments sketched schematically in figure 2 . Following the spirit of elasticity theory we introduce a narrow slit of length, 2c, by cutting the bonds across the slit. In the second step, a uniform external stress, σ_x , is applied normal to the plane of the slit, and this has the effect of opening the slit to form a crack. From elasticity theory[1] the opening of the crack can be represented (within the confines of the linear, continuum approximation) as an elliptical cavity:

$$u(x) = \frac{2\,K}{E\sqrt{\pi c}}\sqrt{c^2 - x^2} \qquad [4]$$

where E is Young's modulus and K is the stress intensity factor. In the third step the interfacial forces are applied to the walls of the crack. The final form of the crack shape will be considered later but one pertinent feature may be singled out here. In general, the magnitude of $\Pi(u)$ is greater the smaller

---

Plane stress solutions; plane strain solutions can be obtained by replacing E by $\frac{E}{(1-\nu^2)}$, where ν is Poisson's ratio.

the crack opening, u, so the effect of the surface forces will be more strongly felt as we approach the tip.

It is possible to use a Green's function approach to calculate the details of both the crack opening displacement and the net crack driving force (21). For an embedded penny crack the crack opening displacement is modified by the presence of the surface forces to:

$$u(x) = \frac{4}{\pi E}\int_x^c \frac{s}{\sqrt{s^2 - x^2}}\int_0^s [\sigma_\infty - \Pi(t)]\,\frac{t}{\sqrt{s^2 - t^2}}\,dt\,ds \qquad [5]$$

where s and t are dummy coordinates. In turn the net stress intensity factor is given by

$$K = \frac{2}{\sqrt{\pi c}}\int_0^c [\sigma_\infty - \Pi(x)]\,\frac{x\,dx}{\sqrt{c^2 - x^2}} \qquad [6]$$

Our principal interest here is in the solution of equation 6. This equation can be re-expressed as the difference between a crack advance driving force and a crack closure driving force, viz:

$$K = K_\infty - K^\Pi \qquad [7]$$

Unfortunately, equation 5 is a nonlinear integral equation, so a general expression for the crack closure driving force K^Π is not possible at present. In the face of such difficulties we seek approximate solutions, as follows:

The first level of approximation is to consider that the surface force has a constant magnitude, Π, and acts over a finite range, $u^\times$. This approximation is in the spirit of the Barenblatt cohesive zone (22). We define a threshold stress intensity factor, $K_\infty = K^\Pi_{TH}$ at $K = 0$ in equation 7, whence equation 6 reduces to:

$$K^\Pi_{TH} = \sqrt{E\,\Pi\,u^\times} = \sqrt{E\gamma_i} \qquad [8]$$

where γ_i defines the interfacial energy.

The next level of approximation is to assume that the crack shape is determined by the net stress intensity factor, as in equation 4. The shape fixes the magnitude of the interfacial forces. Then equations 4, 6 and 7 may be combined to give, for any general stress-separation function, $\Pi(u)$,

$$K = K_\infty - \frac{E\,\gamma_i}{K} \qquad [9]$$

where $\gamma_i = \int_0^{u^\times}\Pi(u)\,du$. The threshold condition is $K = K_\infty = K_{TH}$ which gives:

$$K_{TH} = 2\sqrt{E\,\gamma_i} \qquad [10]$$

The third level of approximation is a first order perturbation solution in which it is assumed (as in the thought experiment of figure 2) that the surface forces perturb the shape imposed by the applied stress. Thus far a solution has only been obtained for the van der Waals force function equation 2 (putting $d = 2u$). This solution is achieved by substituting equations 2 and 4 into 5 and then in turn into equation 6. The details of the calculations are beyond the scope of this contribution and will be published elsewhere (Clarke and Liniger, 23). They lead to an exact expression for the threshold stress intensity given by:

$$K^D_{TH} = \sqrt{\frac{HE}{12\pi d_o^2}} = \sqrt{2\,E\,\gamma_D} \qquad [11]$$

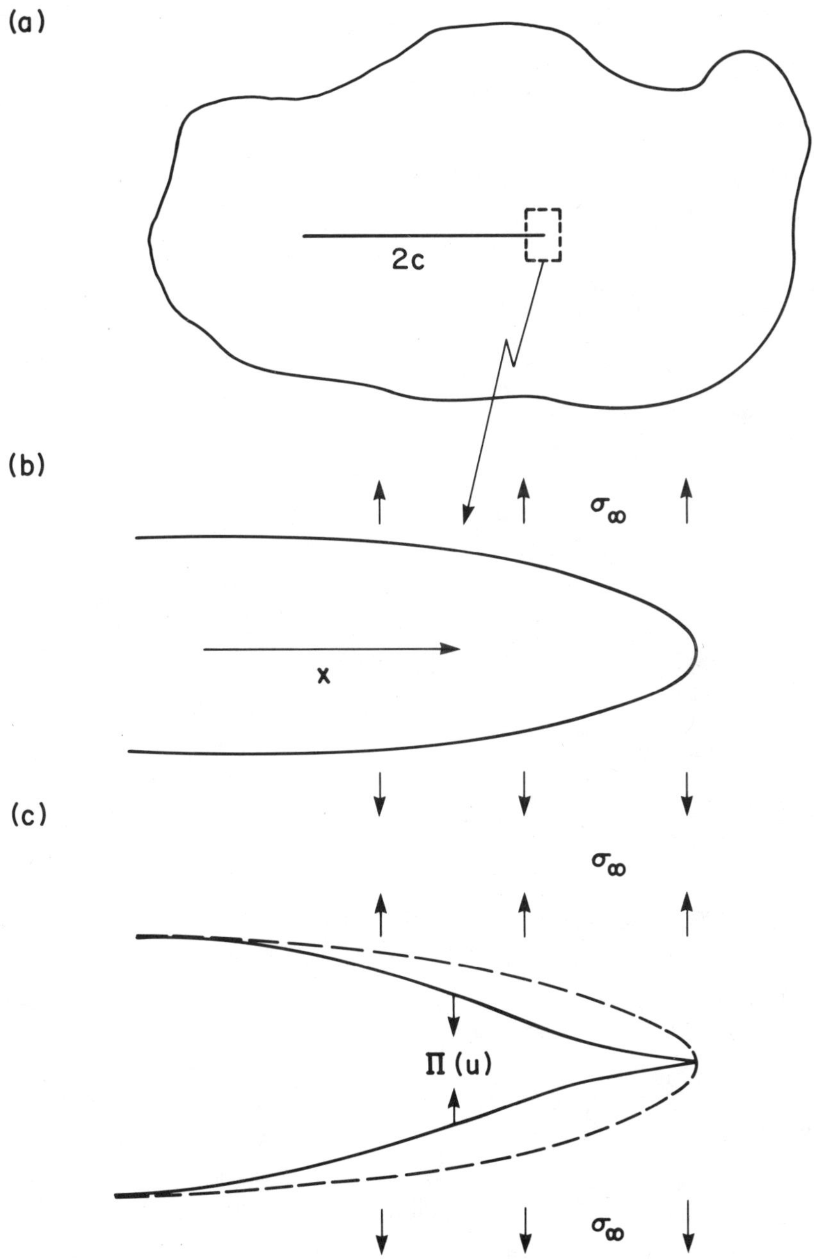

Figure 2. Schematic diagram of the operations used to evaluate the displacements and stress intensity factors for a crack subject to an applied stress and surface forces between the walls of the crack.

In this region the solution to the displacement equation 5 shows that the crack walls "pinch down" into a cusp at the tip (figure 2c), in the manner of a Barenblatt cohesive zone, marking the onset of crack closure and retraction.

There is an implied equivalence between the surface force terms and corresponding interfacial energies in equations 8, 10 and 11 which bears further elaboration here. In the classic Griffith description (24) the condition for crack equilibrium is met when the rate of dissipation of mechanical energy just balances the rate of increase of surface (or interface) energy. In terms of fracture mechanics parlance this balance is expressible as:

$$K_c = \sqrt{2 E \gamma} \qquad [12]$$

It will be seen that if we identify γ in equation 12 with γ_D in equation 11 our third solution above represents a true Griffith equilibrium configuration, $K_c = K_{Th}^D$, in the sense that at $K > K_c$ the crack will advance and at $K < K_c$ will retract. We shall explore the implications of this further in the following section. Note also that the identity between equations 11 and 12 means that the first and second approximations in equations 8 and 10 must respectively underestimate and overestimate the threshold stress intensity factor.

Typically, van der Waals interaction energies fall in the range 10 to 100 $mJ.m^{-2}$. More exact estimates, using published values of Hamaker constants in equation 3, are listed in Table I.

TABLE I

DISPERSION FORCE CONTRIBUTION TO INTERFACIAL ENERGY

Material System	Hamaker Constant 10^{-21} J	Interfacial Energy $mJ.m^{-2}$
Mica-Air-Mica	100	48
Mica-Water-Mica	22	10
Silica-Air-Silica	65	31
Silica-Water-Silica	8.3	4
Quartz-Air-Quartz	88	42
Quartz-Water-Quartz	17	8
Alumina-Air-Alumina	156	74
Alumina-Water-Alumina	53	25

IV. CRACK HEALING IN RELATION TO THRESHOLD CONFIGURATIONS

The physical significance of the threshold stress intensity factor is that it corresponds to the driving force below which a crack will retract and heal. There is ample evidence in the literature that healing can occur in highly brittle solids, even in the presence of chemical environments (25). Perhaps the most extensive review of the healing literature is given in an article by Case et al.(26) Particularly dramatic pictorial evidence for crack reversibility is shown by the beautiful electron microscopy studies of Hockey (27).

Although crack healing in glass (25) and alumina (27) is well documented, one of the best materials in which to observe the healing effect is mica, by virtue of its near molecularly smooth cleavage

(28). We have recently begun to use this material in an investigation of healing. Figure 3 is an example of muscovite mica cleaved in air. In figure 3a a wedge has been inserted into a flake, causing a crack to propagate to an equilibrium position. The open crack is rendered visible by virtue of the Fizeau fringe pattern. Figure 3b shows the same field after partial withdrawal of the wedge. The original front has clearly retracted. A spurious particle has become trapped at the interface in this second frame, leaving a residual penny crack within the otherwise totally re-adhered region. On re-inserting the wedge the configuration in figure 3a is once more attained. This experiment can be carried out in different environments, to determine the chemical dependence of the interfacial energy (28). Thus cyclic loading experiments in the threshold region have the potential for providing quantitative information on fundamental surface forces.

In most of our discussions here it has been tacitly assumed that the interfacial stresses will be attractive. This need not be so. There are indications that many of the force types in equation 1 will be repulsive. Electrical double layer forces are of this kind. A net repulsive force corresponds to a negative interfacial energy. It is evident from equations 8,10 and 11 that the threshold stress intensity factor for this case becomes imaginary. One can show (e.g. from equation 9) in such a case that a net force will persist on the crack even at zero externally applied loading.

V. SURFACE FORCES AND SLOW CRACK GROWTH

The influence of surface forces are expected to extend beyond equilibrium fracture properties, to the well known phenomenon of chemically-assisted, rate-dependent crack growth. Quantitatively, one characterizes this phenomenon by a crack velocity function $v(K)$, which depends on the material-environment system but is always monotonically increasing. Most of the velocity studies have been carried out on silicate glasses, although the rate effect has been demonstrated in a wide range of brittle materials. The relevance of such studies to surface forces is that the velocity functions appear to exhibit zero velocity cut-offs at low K values (29) suggestive of exactly the kind of threshold behavior described above. The question arises: if surface forces are indeed responsible for the threshold at $K = K_{TH}$, what effects do they have on the $v(K)$ function at $K > K_{TH}$?

The key to the answer to this question lies in the realization that it is the net stress intensity factor K at the crack tip which determines the velocity function, whereas in an experiment we actually measure the applied stress intensity factor K_∞; K and K_∞ are of course connected via K^{II} in equation 7 where K^{II} may itself be a function of K. Thus given an intrinsic velocity function $v(K)$, we may solve for $v(K_\infty)$ for any specified surface force state (ie K_{TH}). Calculations of this kind have been made in the spirit of the approximations embodied in equation 9, and are presented elsewhere (30). It becomes apparent that, in addition to the imposition of a threshold, surface forces can manifest themselves strongly as shifts in the entire crack velocity curve. Such shifts have been widely reported in the glass fracture literature, particularly as a result of variations in environmental solution pH (Wiederhorn and Johnson, 31).

It is of some interest to investigate the threshold itself a little more quantitatively. Recent data from indentation studies (32) give $K_{TH} \simeq 0.1\ MPa.m^{1/2}$ for soda-lime glass in water. Taking $E = 70\ GPa$ we evaluate $\gamma_i \simeq 70\ mJ.m^{-2}$ This value is considerably larger than that due solely to van der Waals dispersion forces (Table I-- fused silica data). This suggests that we must consider the other force terms in equation 1. For aqueous environments we might pay particular attention to the hydrogen bonding and structural forces. Water is a strongly associated liquid and appears to have a long-range structure, quite unlike that of other liquids. The long range structure has been measured directly in the sensitive crossed cylinder experiments of Pashley and Israelachvili (10). Furthermore, the chemical interaction between the silicates and the water not only alters the glass but also alters the properties of the water itself (33). Perhaps most striking is the fact that silicates and water react to form what was once termed "anomalous water", a phase having anomalous transport properties relative to those of water itself (17). Although initial claims that anomalous water may constitute a new liquid phase of water were subsequently discounted, the experiments substantiated the picture of some structural interaction between the liquid phase and the adjacent silica surface. The experiments in which the effect was particularly marked were performed with the water trapped in narrow capillaries of silica, a geometry not dissimilar to that of a crack interface. In considering these possibilities it is well to be aware of the extremely small crack wall displacements which typify highly brittle solids (eg. figure 1), a confinement which can only be enhanced by the Barenblatt-type cusping re-

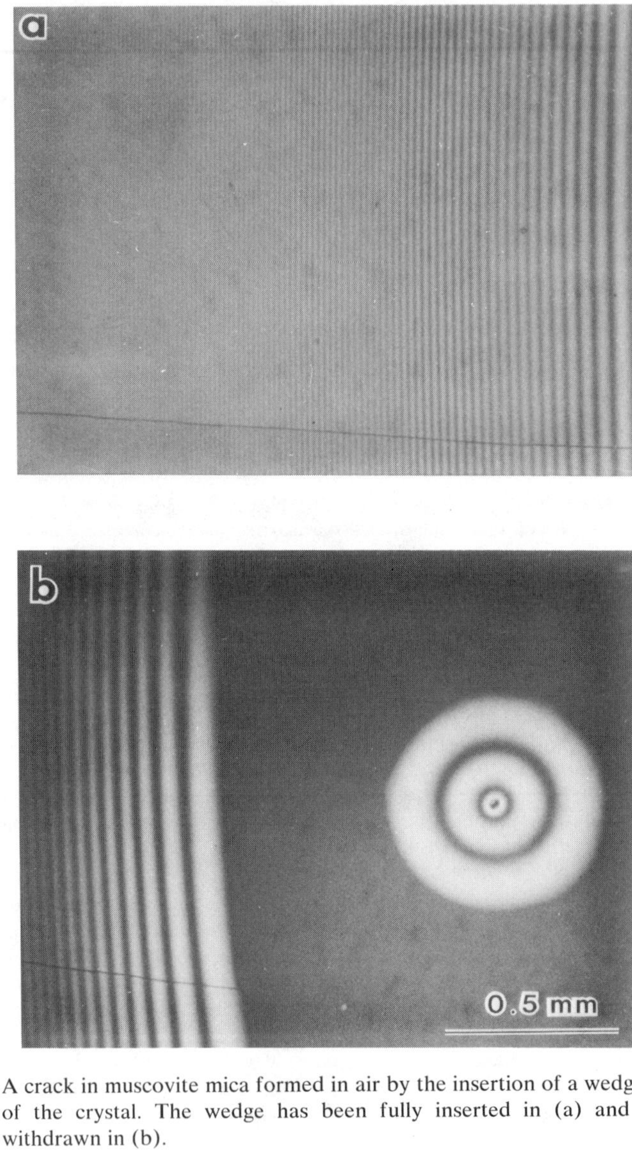

Figure 3. A crack in muscovite mica formed in air by the insertion of a wedge into the edge of the crystal. The wedge has been fully inserted in (a) and then partially withdrawn in (b).

ferred to earlier. In this picture it may be more proper to regard the interface in the near-tip region as akin to a grain boundary rather than an open cavity (18). Such constriction could well determine in a steric manner those ions which can enter the critical reaction zone at the crack tip (Michalske and Bunker, 34).

VI. CLOSING REMARKS

We have mentioned the crossed mica cylinder experiments as the only existing direct way of measuring surfaces forces down to distances close to the interatomic spacing. Unfortunately, as presently implemented, this method is restricted to mica (or other such conformable materials having atomically flat surfaces). The recognition that surface forces may have a predictable effect on fracture, and in particular on a definable threshold behavior, opens up an entirely fresh avenue for measuring surface forces, namely via the fracture process itself. The study of fracture has been preoccupied with characterizing the conditions under which a material might fail. Now, fracture might well prove to be a useful tool for measuring some of the fundamental forces acting in nature, having direct relevance to fields as disparate as colloid science, wetting phenomenon, ceramic processing and cell biology. It is anticipated that fracture in brittle materials such as sapphire, mica and glass may prove to be especially suitable for such surface force measurements.

In summary we would emphasize the following points:

1. Surface forces acting between the walls of a crack, but behind the crack tip itself, provide a crack driving force which supplements that originating from an externally applied stress.

2. For the specific case in which the surface forces cause the opposing walls of a crack to be attracted to each other, the additional driving force acts as a crack closure force. When the crack advance force is balanced by the crack closure force, a condition of zero net driving force on the crack exists. This condition is a measure of the surface forces acting on the crack. The threshold stress intensity is also pertinent to kinetically controlled slow crack growth since it is the value at which the crack growth rate will go to zero.

3. The specific case of van der Waals dispersion force acting on a crack is shown to lead to a threshold stress intensity whose magnitude is proportional to the square root of the Hamaker constant for the crack system.

4. For surface forces that are attractive, it is indicated that in the immediate vicinity of the crack-tip the crack is "pinched" down. This effect is more marked the closer the applied stress intensity is to the threshold.

5. For those cases in which the surface forces are repulsive in nature, no threshold stress intensity will exist, but rather the crack will exhibit nonzero crack velocity at zero applied driving force.

6. Particularly exciting is the prospect of using the fracture process itself, at small crack driving forces, to measure surface forces directly.

ACKNOWLEDGEMENTS

It is a pleasure to thank S. M. Wiederhorn for many constructive discussions during the development of the concepts in this paper.

REFERENCES

1. D. Tabor, J. Colloid Interface Sci., 58:2 (1977)
2. K. L. Johnson, K. Kendall and A. D. Roberts, Proc. Roy. Soc. A324: 301 (1971)
3. B. V. Derjaguin, V. M. Muller and Yu. P. Toporov, J. Colloid Interface Sci., 53:314 (1975)
4. V. M. Muller, V. S. Yushchenko and B. V. Derjaguin, J. Colloid Interface Sci., 77:91 (1980).
5. M. D. Pashley, Colloids and Surfaces, 12:69 (1984)
6. H. M. Westergaard, Trans. Am. Soc. Mech. Engrs. 61:A49 (1939)

7. D. Tabor and R. H. S. Winterton, Proc. Roy. Soc. A312:435 (1969)
8. J. N. Israelachvili and D. Tabor, Proc. Roy. Soc. A331:19 (1972).
9. J. N. Israelachvili and G. E. Adams, J. Chem. Soc., Faraday Trans. I. 74:975 (1978).
10. R. M. Pashley and J. N. Israelachvili, J. Colloid Interface Sci., 101:511, (1985).
11. R. G. Horn and J. N. Israelachvili, Chem. Phys. Letts, 71: 192 (1980).
12. H. K. Christenson, R. G. Horn and J. N. Israelachvili, J. Colloid Interface Sci., 88:79, (1982).
13. I. E. Dzyaloshinskii, E. M. Lifshitz and L. P. Pitaevskii, Adv. Phys., 10:165 (1961).
14. in J. Mahanty and B.W. Ninham, "Dispersion Forces", Academic Press, New York (1976)
15. V. A. Parsegian and B. W. Ninham, Nature, 224:1197 (1969).
16. J. N. Israelachvili, "Intermolecular And Surface Forces", Academic Press, Orlando (1985).
17. in A. W. Adamson,"Physical Chemistry Of Surfaces", 4th Edition, Wiley, New York (1982)
18. D. R. Clarke, J. Am. Ceram. Soc., submitted.
19. B. V. Derjaguin and L. Landau, Acta Phys. Chem. USSR 14:633 (1941)
20. E. J. W. Verwey and J. Th. G. Overbeek, "Theory Of The Stability Of Lyophobic Colloids", Elsevier, Amsterdam (1948).
21. I. N. Sneddon and M. Lowengrub, "Crack Problems In The Classical Theory Of Elasticity", Wiley, New York (1969).
22. G. I. Barenblatt, Adv. Appl. Mechanics, 7:55 (1962)
23. D. R. Clarke and W. Liniger, to be published.
24. A. A. Griffith, Phil. Trans. Roy. Soc., A221:163 (1920)
25. B. Stavrinidis and D. G. Holloway, Phys. Chem. Glasses, 24:19 (1983)
26. E. D. Case, J. R. Smyth and O. Hunter, in " Fracture Mechanics Of Ceramics" vol. 5, ed. Bradt, Evans, Hasselman and Lange, p507, Plenum Press, New York, (1983)
27. B. J. Hockey, in "Fracture Mechanics Of Ceramics", Vol. 6, ed. by Bradt, Evans, Hasselman and Lange, p637, Plenum Press, New York, (1983).
28. A. I. Bailey and S. M. Kay, Proc. Roy. Soc., A301:47 (1967).
29. T. Michalske, in "Fracture Mechanics Of Ceramics" vol. 5, ed. Bradt, Evans, Hasselman and Lange, p277, Plenum Press, New York, (1983).
30. B. R. Lawn, Appl. Phys. Letts., In Press.
31. S. M. Wiederhorn and H. Johnson, J. Am. Ceram. Soc., 56:192 (1967).
32. B. R. Lawn, K. Jakus and A. C. Gonzalez, J. Am. Ceram. Soc., 68:25 (1985).
33. R. K. Iler, "The Chemistry Of Silica", Wiley, New York, (1979).
34. T. A. Michalske and B. Bunker, unpublished work.

SLOW CRACK GROWTH IN SINTERED SILICON NITRIDE

P.K. Khandelwal, J. Chang, and P.W. Heitman

Allison Gas Turbine Division
General Motors Corporation
P.O. Box 420
Indianapolis, IN 46206

ABSTRACT

The strength and crack growth characteristics of a sintered silicon nitride were studied at 1000°C. Fractographic analysis of material failing in dynamic fatigue revealed the presence of slow crack growth (SCG) at stressing rates below 6 ksi/min. This material can sustain a 40 ksi flexural stress at 1000°C for 400 hr or more but is susceptible to both slow crack growth and creep deformation at higher stress levels. The crack velocity exponent (N) determined both from dynamic and static fatigue experiments lies in a range from 13 to 22. The subcritical crack growth and creep behavior at 1000°C is primarily controlled by the deformation of an intergranular glassy phase.

INTRODUCTION

Sintered silicon nitride materials are currently being considered for use in hot flow-path components of gas turbine engines, because of their good thermal shock and oxidation resistance, as well as strength at high temperatures. Injection molded sintered silicon nitride (6% Y_2O_3, 2% Al_2O_3) turbine blades have been developed.[1] This material possesses the necessary fast fracture strength up to 1250°C for turbine blade applications. Blades made from this material were spin-tested-to-failure achieving an average speed of 52,300 rpm. This is equivalent to 142% design speed. Integral gasifier rotors, 4.4 in. in diameter, have also been fabricated by injection molding. These rotors were HIP-processed and spin-tested-to-failure achieving an average failure speed ρf 91,040 rpm[2]. This is 105% of design speed. Silicon nitride, however, is often susceptible to slow crack growth and creep at elevated temperatures. Both of these failure processes are time, temperature, and stress dependent and affect lifetime predictions.[3,4,5,6,7] In this paper we present current results of a study in which dynamic and static fatigue behavior of the specific sintered silicon nitride composition used in the fabrication of the ceramic blades and rotors is assessed.

The failure of ceramic materials can occur from the slow growth of preexisting flaws under applied stress until the critical size for spontaneous crack propagation and fracture is reached. The nature of the slow crack growth behavior of ceramic materials is generally characterized by the parameter N given by the equation[8]

$$V = A(K_I/K_{IC})^N \quad (1)$$

where V is the crack velocity, K_I is the stress intensity factor, K_{IC} is the critical stress intensity factor, and A is a constant. In static fatigue experiments, a material is subjected to constant applied stress to determine the time-of-failure (t_f). From equation (1) it can be derived that the time-to-failure (t_f) under constant stress[8] is

$$t_f = B\sigma_i^{N-2} \sigma_a^{-N} \quad (2)$$

where $B = 2/(AY^2 (N--2) K_{IC}^{N-2})$, A and N are material and environment constants, σ_i = fracture strength in inert environment, and K_{IC} = critical stress intensity factor. In equation (1), B and σ_i are material constants in a given environment. Therefore, for many engineering applications, equation (2) reduces to

$$t_f = (\text{Const})* \sigma_a^{-N} \quad (3)$$

The slope of the log σ_a versus log t_f plot gives the crack velocity exponent (N) which reflects the resistance of the material to subcritical crack growth. Lower values of N indicate a material is susceptible to slow crack growth and vice versa.

Dynamic fatigue data is obtained by measuring fracture strength (σ_f) at several stressing rates ($\dot{\sigma}$). From equation (1) it can be shown that the fracture strength of ceramics is related to stressing rates as follows:

$$\sigma_f^{N+1} = B\ (N+1)\ \sigma_i^{N-2}\ \dot{\sigma} \quad (4)$$

where B and N are fatigue constants as in equation (2). For a given material and environment, $B(N+1)\ \sigma_i^{N-2}$ is constant. Therefore,

$$\sigma_f^{N+1} = (\text{const})\ \dot{\sigma} \quad (5)$$

The value of the crack velocity exponent (N) can be determined from the slope of log $\dot{\sigma}$ versus log σ_f regressed plot.

In practice, however, the inert strength distribution of a material (σ_i) is not generally known. By assuming that the initial inert strength distribution is equivalent to the initial flaw distribution and is given by the Weibull function

$$\ln(\ln(1/1-F)) = m \ln \frac{\sigma_i}{\sigma_o} \quad (6)$$

Jakus and co-workers[9] rewrote the equation (2) for static fatigue as

$$(\ln t_f + N \ln \sigma_a) = C + (N-2)/m\ (\ln(\ln(1/1-F))) \quad (7)$$

where $C = [\ln B + (N-2) \ln \sigma_o]$ and used iterative trivariant regression analysis until convergence of N is obtained. Similarly, combining equations (4) and (6), the dynamic fatigue equation can be written as

$$[\ln \sigma_f + \frac{1}{N+1} \ln \dot{\sigma}] = \frac{1}{N+1} [D + (N-2)/m\ (\ln(\ln(1/1-F)))] \quad (8)$$

where $D = \ln B + \ln(N+1) + (N-2) \ln \sigma_o$ and the converged value of N is obtained.

MATERIAL

The sintered silicon nitride* examined in this study is a commercially available material containing 6% Y_2O_3 and 2% Al_2O_3. The material is designated as AY6-405* and was received during the time period 1980-82. The starting ceramic powder is mixed with a specially developed organic binder and injection molded to produce near net shaped bars. Green bars are thermally treated to remove the organic binder and subsequently sintered in a nitrogen atmosphere to full density. The modulus-of-rupture (MOR) specimens were injection molded individually. Thereafter, MOR bars were longitudinally machined to measure 2 x 0.25 x 0.125 in. with chamfered edges. Surface finish was 20 rms or better.

EXPERIMENTAL PROCEDURE

Fast fracture, dynamic fatigue (variable stressing rate), and static fatigue (constant stress) behavior of the material were evaluated in four-point flexure. Specimens were tested in 1/4 point loading with 1.5 in. outer span and 0.75 in. inner span. The test temperature was 1000°C (1832°F).

Figure 1 shows the experimental arrangement. The load train consists of a mullite loading ram, siliconized silicon carbide (NC-430) four-point fixture, and mullite tube which rests on a water-cooling jacket. Temperature is monitored by two thermocouples--one rests very close to the bar in the inner-shell of the fixture and the second is attached to the side of the furnace housing. Specimens are soaked for 15 minutes at a temperature prior to loading. Fast fracture testing was conducted at a cross-lead rate of 0.020 in./min. The range of stress-rates for dynamic fatigue tests extended from 0.6 ksi/min to 600 ksi/min. However, the actual stress rates were derived from the load displacement curves of each specimen.

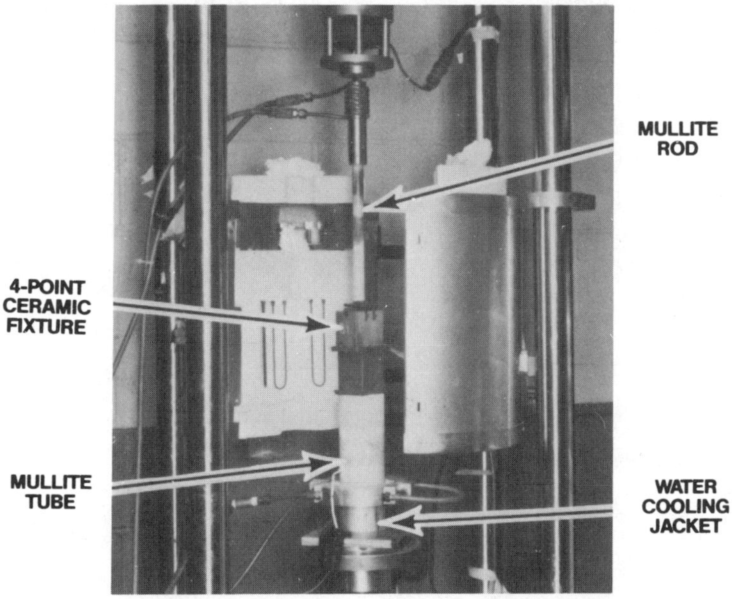

Fig 1. High temperature flexural testing of ceramic materials.

*AY6-405 (6.34% Y_2O_3, 1.92% Al_2O_3, 600 ppm Fe)
GTE Laboratories, Waltham, Massachusetts

Figure 2 shows the static fatigue (or stress rupture) test system. This arrangement can accommodate up to three specimens at a time. It consists of a pressure transducer which measures the input air pressure (P), three air cylinders which are attached to the incoming air and the loading rams, three ceramic fixture, box furnace, a LVDT to measure the deflection of the center bar and three time meters to measure the elapsed time for each specimen. All the valves are closed prior to starting a test and air-pressure calculated for a given stress on the bar. Three specimens are placed individually in three ceramic fixtures inside the furnace. The temperature of the furnace is increased slowly to the test temperature. The LVDT is then raised and brought in contact with the bar under spring load. The air pressure is set at pressure (P) which is magnified to ΠP by air cylinders and applied on the flexural bars by the mullite loading rams to a given stress. The broken specimens are left in the furnace until all three have fractured. Time-to-failure of all three specimens is noted from the meters.

RESULTS AND DISCUSSION

Fast Fracture

Table I shows the measured 1000°C fast fracture strength and Weibull parameters. Included for reference purposes are the room temperature parameters. The average room temperature strength was 91.8 ksi (632.9 MPa) with a standard deviation of 10 ksi (68.9 MPa). The strength controlling flaws observed in these specimens were surface and internal pores, Figure 3. The average strength at 1000°C was found to be 74.5 ksi (513.6 MPa) with a standard deviation of 11.3 ksi (77.9 MPa). Failure at 1000°C was also predominantly controlled by surface and near surface internal pores similar to the type of flaws which controlled fracture at room temperature, Figure 3. No change was observed in the distribution of the strength controlling defects between room temperature and 1000°C. The 19% decrease in the strength from room temperature to 1000°C may be attributed primarily due to the decrease in the viscosity of the intergranular glassy region. It should be noted that both Al_2O_3

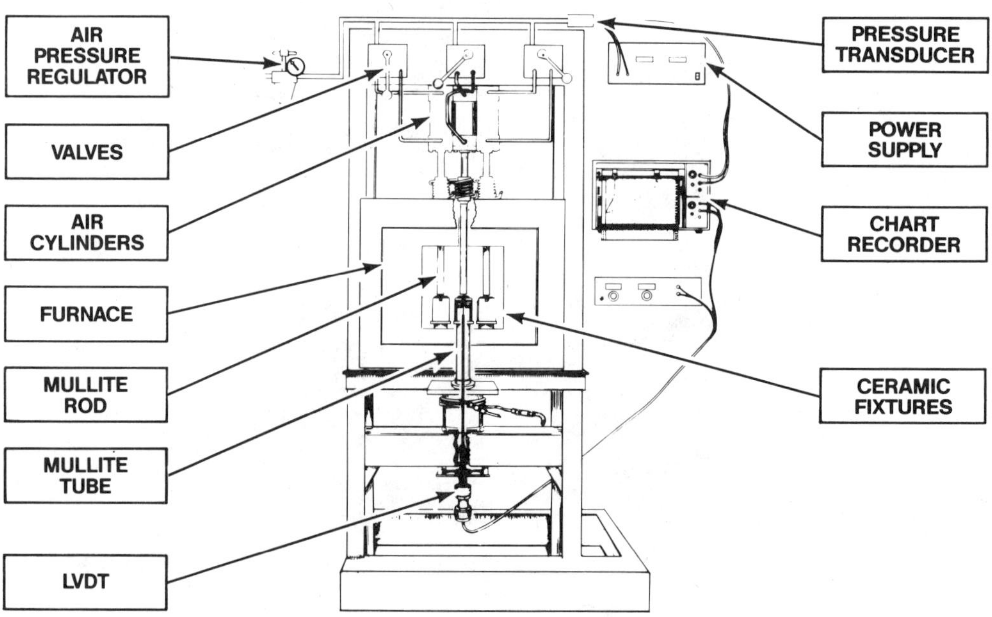

Fig 2. Schematic of static fatigue test system.

Table I. Strength of Sintered Silicon Nitride.

TEST CONDITIONS	NO. OF SPECIMENS	FAST FRACTURE STRENGTH, KSI		WEIBULL PARAMETERS	
		σ_f	STD. DEV.	m	σ_o (KSI)
RM. TEMP.	28	91.8	10.0	9.9	96.4
1000°C	28	74.5	11.3	7.6	79.3

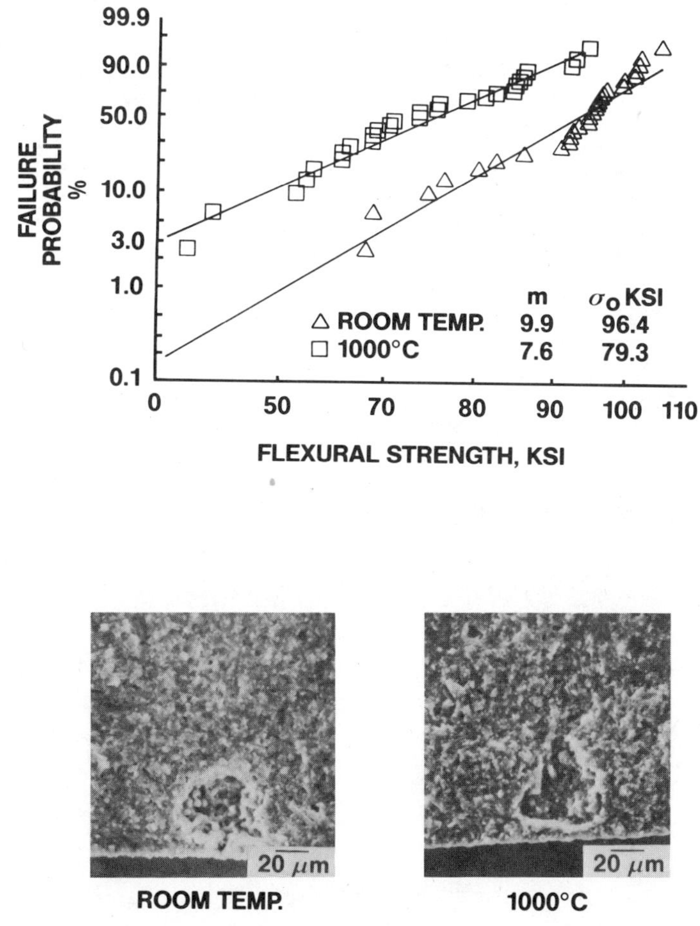

Fig 3. Weibull characteristics of sintered silicon nitride and typical strength controlling defects.

and Fe act as fluxing agents in this composition and significantly affect the viscoelastic behavior at high temperatures. Quackenbush and co-workers[10] have shown that both Al_2O_3 and Fe appreciably decrease the glass transition temperature resulting in substantial degradation in strength of the material. Scanning electron microscopy did not reveal any crack growth regions. Figure 3 shows the least square fit of room temperature and 1000°C strength to two parameter Weibull function. Both the Weibull modulus (m) and characteristic strength (σ_o) decreased from 9.9 and 96.4 ksi at room temperature to 7.6 and 79.3 ksi at 1000°C, respectively. It should be noted that an identical sample population of 28 specimens was used at both room temperature and 1000°C. Also, the standard deviation of the average fracture strength was similar. Therefore, the degradation in Weibull modulus, m, may be significant. The shift in the strength distribution at 1000°C to higher failure probabilities at a given applied stress also indicates that the material has degraded.

Dynamic Fatigue

Twenty-six silicon nitride specimens were subjected to four stressing rates ranging from 0.6 ksi/min. to 600 ksi/min. at 1000°C. Figure 4 shows that the fracture strength degrades with decreasing stressing rate. The crack velocity exponent (N) deduced from both the linear regression and iterative trivariant-regression[9] is about 15. The low value of N is indicative of material susceptibility to slow crack growth (SCG). Failure analysis of fracture surfaces revealed the presence of SCG zones at the lower stress rates (0.6 and 6 ksi/min) while no SCG was observed at higher stress rates (60 and 600 ksi/min). Typical SEM fractographs one each at the four stressing rates is shown in Figure 5. Slow crack growth as a result of intergranular fracture can be readily seen as the high contrast semicircular area in all the specimens at lower stressing rates. At higher stressing rates strength was controlled by isolated pores whereas at lower stressing rates the fracture was initiated from a pore at a pocket of small pores within SCG zone.

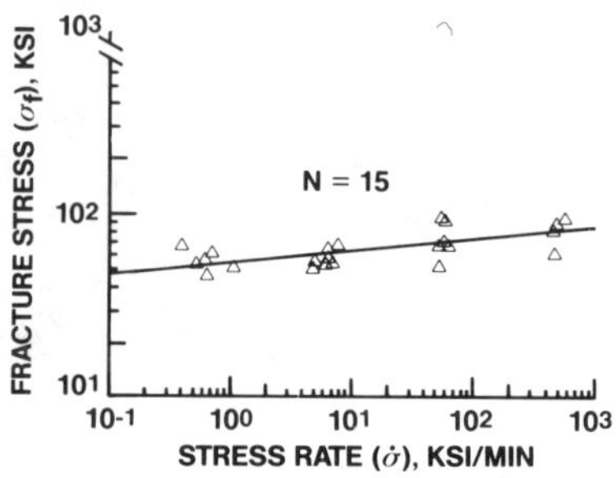

Fig 4. Dynamic fatigue behaviour of sintered silicon nitride at 1000°C.

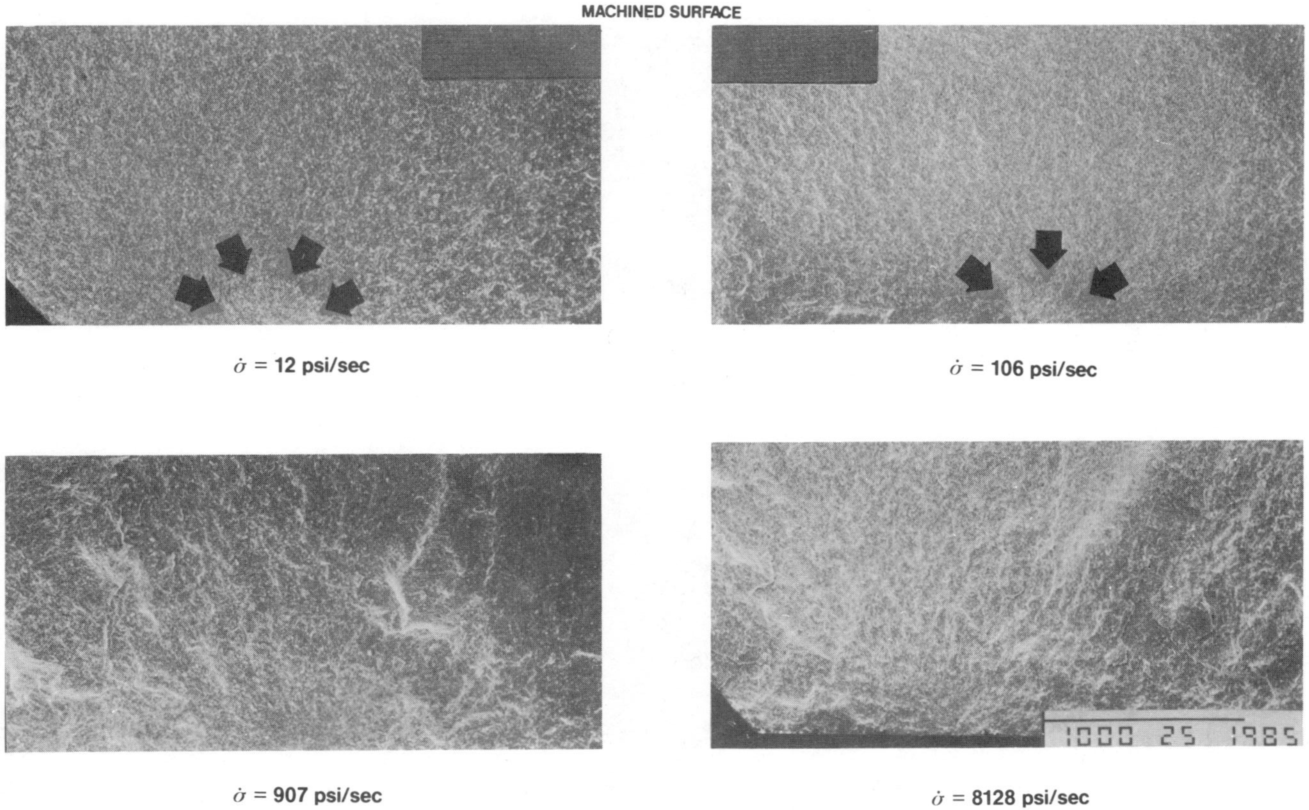

Fig 5. Effect of stress rate on slow crack growth behavior of sintered Si_3N_4 at 1000°C.

Static Fatigue

Twenty-four specimens of sintered silicon nitride were loaded to constant applied stress levels ranging from 40 to 70 ksi to measure the time-of-fracture and creep strain at 1000°C. Results in Figure 6 show that 19 specimens failed in less than 20 hours when subjected to 50 to 70 ksi stress levels with creep strain of up to 0.22%. Five specimens did not fail up to 400 hr when subjected to stresses less than 50 ksi. The crack velocity exponent (N) from the log t_f vs log σ_f was calculated to be 22 whereas iterative trivariate regression technique gave N = 13. The deviation between two values of N is attributed primarily to the effect of small sample population for materials with low value of the Weibull modulus.[11] It is noteworthy, however, that the value of N obtained above is in excellent agreement with the dynamic fatigue value of N = 15. Clearly, similar crack growth parameters can be measured using either technique.

Figure 6 shows that measurable slow crack growth and creep occurs in specimens which fail between 0.2–20 hr when subjected to 50 ksi or more constant applied stress. For example, creep strains of 0.11% in 1.3 hr and 0.22% in 3.6 hr were observed in specimens subjected to 50 ksi stress. Similarily, a specimen failed in 0.27 hr with 0.13% creep strain under 60 ksi applied stress. Fractographic analysis revealed that all the specimens failed from the intergranular fracture of the preexisting subsurface pores. Figure 7 shows a typical example of three specimens which failed at different times when subjected to 60 ksi constant applied stress. Figure 8 shows another example of a specimen which failed in 0.277 hr under 70 ksi applied stress with SCG. Both slow crack growth and creep were measured in these specimens. At high stresses, the viscous grain boundary phases opens up and causes severe internal oxidation by increased oxygen diffusion resulting in grain boundary slid-

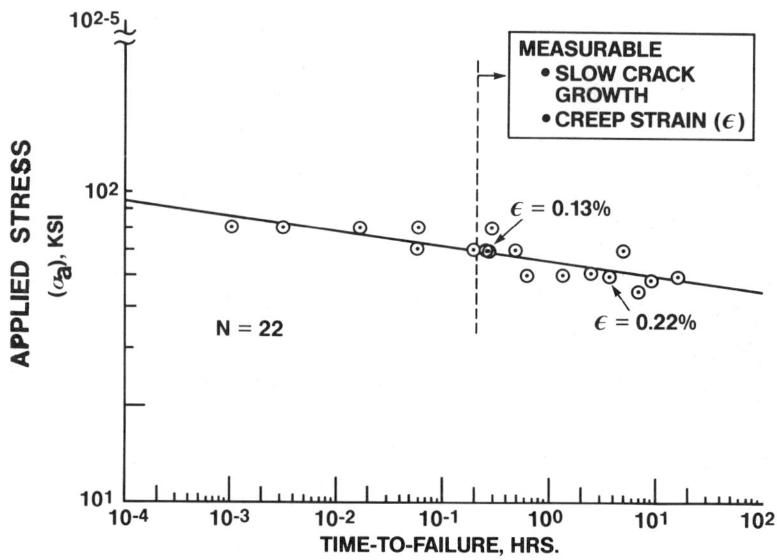

Fig. 6. Static fatigue of sintered silicon nitride at 1000°C.

t_f = 0.242 HR

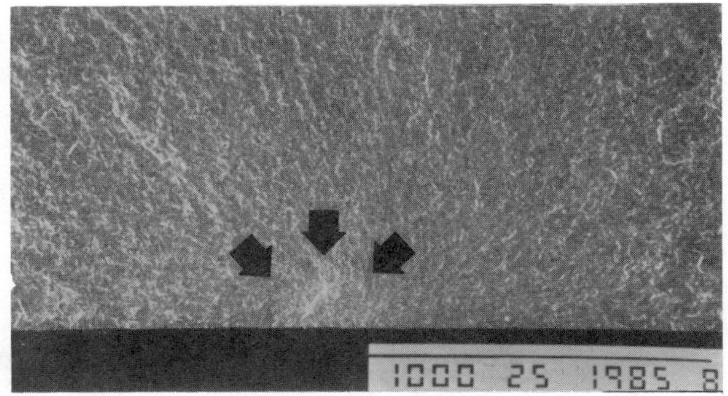

t_f = 0.467 HR

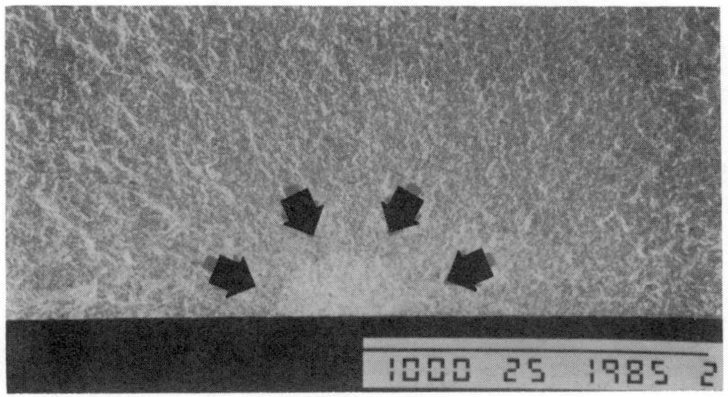

t_f = 4.81 HRS

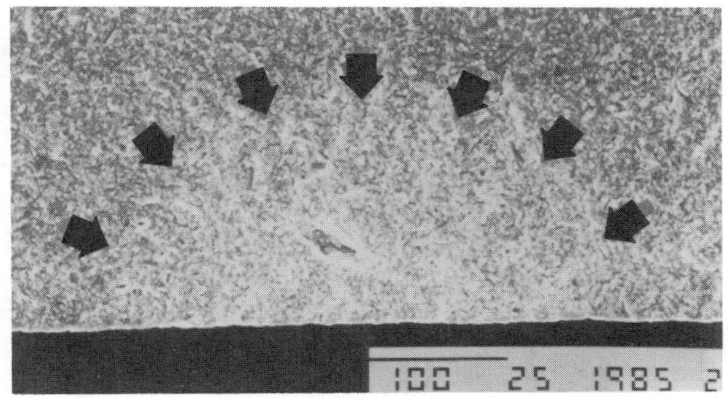

t_f = 4.81 HRS

Fig 7. Static fatigue behavior of sintered silicon nitride at 1000°C and 60 ksi (414 MPa).

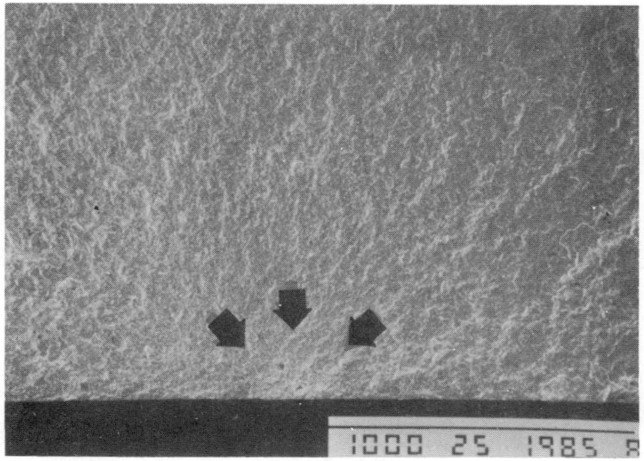

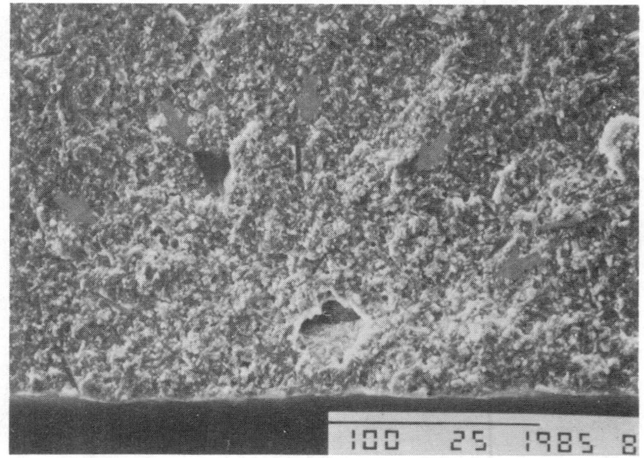

Fig 8. Static fatigue failure of sintered silicon nitride at 1000°C and 70 ksi (483 MPa).

ing, and creep cavitation.[10] Both Lange[3] and Grathwohl[7] have shown that hot pressed silicon nitride with a MgO sintering aid undergoes SCG, stress relaxation, and creep concurrently at high temperatures. A similar phenomenon is observed in the present yttria based material. At and above 50 ksi, failure is predominantly controlled by slow crack growth with simultaneous occurrence of creep due to softening and oxidation of the second phase in the grain boundaries. The present results are also in good agreement with Quinn[5,12] who reported both SCG and creep in this material at 1000°C.

Figure 9 shows that five specimens did not fail up to and after 400 hr when subjected to stresses less than 50 ksi. Three specimens of a similar cold pressed composition* did not fail for 1000 hr when loaded to 30 ksi static stress.[13] The unbroken specimens were subsequently tested without unloading in the fast fracture conditions at 1000°C. The average fast fracture strength of these specimens at 1000°C is 77.4 ksi, which is comparable to 74.5 ksi strength, as shown in Table I. The strength controlling defects were subsurface pores or pockets of small pores similar to flaws that cause fast fracture at 1000°C. Fractographic analysis did not reveal any slow crack growth in these specimens. It appears, therefore, that up to 40 ksi static stress failure does not occur because of crack blunting or healing process. Evidently, there is threshold stress or a "static fatigue limit" up to which this material can successfully perform without degradation. Based on these results, it may be possible to conclude that the present composition may be utilized up to 1000°C for 1000 hr between 30 to 40 ksi static stress.

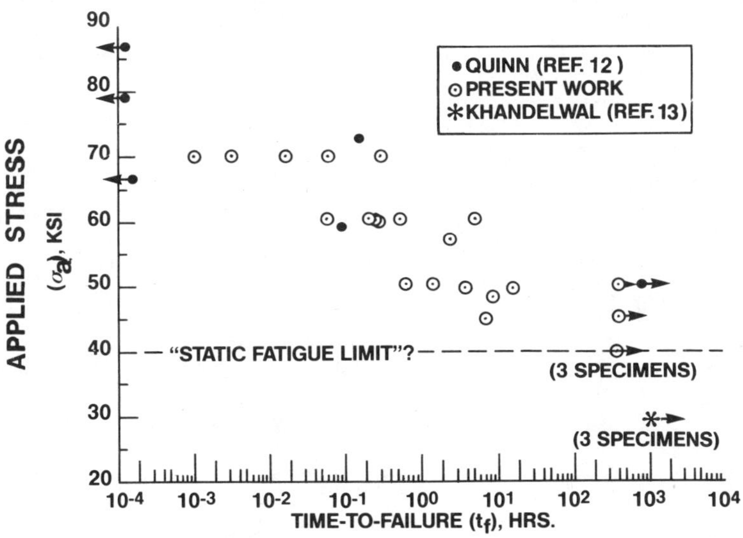

Fig. 9. Static fatigue of sintered silicon nitride at 1000°C.

---

*AY[6] (Al_2O_3 = 2.2%, Y_2O_3 = 5.86%, Fe 190ppm)
GTE Laboratories, Waltham, Massachusetts

SUMMARY

Fast fracture, static, and dynamic fatigue flexural tests were conducted at 1000°C in air on a commercially available sintered silicon nitride. The following observations are made from these tests:

1. The average fast fracture strength decreased from 91.8 ksi at room temperature to 74.5 ksi at 1000°C.

2. Dynamic fatigue experiments revealed the presence of slow crack growth (SCG) at stressing rates lower than 6 ksi/min. The value of the slow crack growth parameter N was 15 was obtained.

3. Both slow crack growth and creep strain were measured in specimens under an applied stress $\geq$ 50 ksi, while no slow crack growth was observed below 30 ksi for 1000 hr and 40 ksi for 400 hr. The value of the slow crack growth parameter N lies between 13 and 22.

4. It may be possible to use the present material up to 1000°C for 1000 hr between 30 to 40 ksi static stress.

ACKNOWLEDGEMENT

This work was supported by the Department of Energy and managed by NASA-Lewis Research Center under Contracts DEN3-17 and DEN3-168.

REFERENCES

1. H. E. Helms, P. W. Heitman, L. C. Lindgren, and S. R. Thrasher, "Ceramic Applications in Turbine Engines," Final Report, NASA CR-174715, October 1984.
2. L. E. Groseclose and R. A. Johnson, ASME Paper 85-GT-205.
3. F. F. Lange, International Journal of Fracture, Vol 12, 1976, pp 739-744.
4. S. H. Knickerbocker, A. Zangvil, and S. D. Brown, Journal of the American Ceramic Society, 68(4)C-99-C-101, 1985.
5. G. D. Quinn, AMMRC TR 84-40, October 1984.
6. N. J. Tighe, S.M. Wiederhorn, T. J. Chuang, and C. L. McDaniel, in Deformation of Ceramic Materials II, edited by R. E. Tressler and R. C. Bradt, 1983, pp 587-604.
7. G. Grathwohl, in Deformation of Ceramic Materials II, edited by R. E. Tressler and R. C. Bradt, 1983, pp 573-586.
8. J. E. Ritter, in Fracture Mechanics of Ceramics, Vol. 5, R. C. Bradt, A. G. Evans, D. P. Hasselman, and F. Lange, eds., Plenum Press, New York, 1983, pp 227-251.
9. K. Jakus, D. C. Coyne, and J. E. Ritter Jr., Journal of Material Science, Vol 13, 1978, pp 2071-2080.
10. C. L. Quackenbush, J. T. Smith, J. T. Neil, and K. W. French, in Progress in Nitrogen Ceramics, edited by F. L. Riley, 1983, pp 669-682.
11. J. E. Ritter, Jr., N. Bandyopadhyay, and K. Jakus, Ceramic Bulletin, Vol. 60, No. 8, 1981, pp 798-806.
12. G. Quinn (private communication), 1985.
13. P. Khandelwal (unpublished).

COMPUTER SIMULATION OF THE GROWTH OF SMALL SURFACE FLAWS IN DYNAMIC FATIGUE

J.M. Jouin, J.L. Chermant, and F. Osterstock

Equipe Matériaux-Microstructure du LA 251, ISMRa-Université
14032 Caen Cedex, France

ABSTRACT

The applicability of toughness models to get a law of flaw growth to instability is presented using computer simulation. The influence of different subcritical crack growth parameters, initial flaw size, local effective critical stress intensity factor and flaw interaction, on the rupture stress, the critical flaw size and the final stress intensity factor is shown. Comparison with experimental results indicates that the local microstructure of the defects and the microstructural inhomogeneities must be taken into account.

INTRODUCTION

Recently Okada and Sines (1)(2) have proposed a toughness model including the individual behaviour of microscopic flaws and their interaction with the aim of giving a law of their growth to instability.

Concerning simulation of flaw growth, the transition from the monocrystalline to the polycrystalline toughness ($\gamma_s^c \rightarrow \gamma_p^c$) has first been introduced by Kirchner and Ragosta (3) in the case of failure due to propagation of small flaws within large grains. Rice (4) proposed a more smooth transition, taking into account a critical flaw size encompassing several crystals, below which the local effective critical stress intensity factor is less than the one usually measured on bulk polycrystals with macrocracks. Okada and Sines used the expression of Rice (4) in combination with an interaction factor proposed by Rooke and Cartwright (5) in order to discuss their time to failure measurements and predictions. Computer simulation using the previously cited expression allowed them to calculate realistic initial flaw sizes and to correlate them with fractographic observations.

The aim of this work is to investigate the applicability of the above models to stressing rate effects. Computer simulation will be used to look for the effect of several subcritical crack growth parameters, local effective critical stress intensity factor and flaw interaction on fracture criteria like rupture stress, critical flaw size and final stress intensity factor. The results will be discussed with respect to general experimental trends observed in fracture mechanical measurements on ceramics.

SIMULATION BACKGROUND

Subcritical crack growth

The model proposed by Okada and Sines (2) takes into account :
- a law describing the growth of each individual flaw,
- the interaction between the growing flaws.

The flaws are randomly distributed on the surface of the specimen and each of them is subjected to subcritical crack growth. On the scale of each defect, i.e. some crystals, the various crystalline orientations, the local surface energy, γ , the residual stresses due to the thermal expansion anisotropy and that of the elastic parameters will strongly affect the instability criterion and the crack growth law. As a consequence the local fracture energy criterion should vary. Several authors (2)(3)(4) have suggested that a smooth and continuous transition from γ_s^c (monocrystal) to γ_p^c (polycrystal) for the fracture energy exists. Thus the effective critical stress intensity factor, K_{eff}^c describing the instability of a flaw, of size a, is proposed to be (2) (see also Fig. 1), for $a < \alpha\bar{D}$:

$$K_{eff}^c = K_p^c[1 + (K_s^c/K_p^c - 1)(1 + \cos \pi a/\alpha\bar{D})/2]$$

with : K_p^c : critical stress intensity factor for the polycrystal,

K_s^c : critical stress intensity factor for the single crystal,

$\bar{D}$: mean grain size,

α : critical ratio for the micro- to macro-rupture transition.

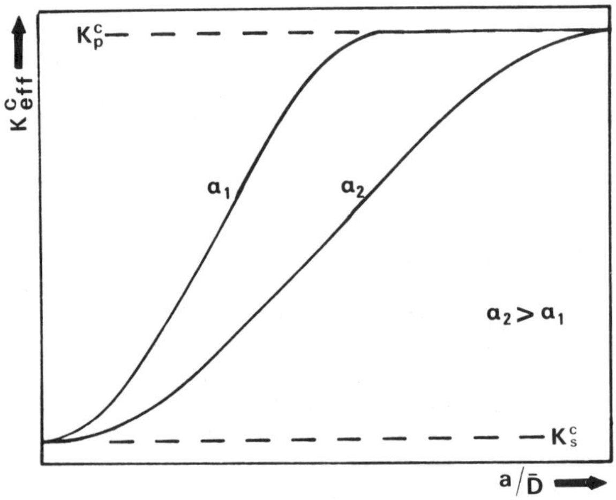

Fig. 1. Schematic transition of the effective fracture criterion K_{eff}^c with the ratio flaw size over grain size for two values of α.

If the flaw size is or becomes larger than $\alpha\bar{D}$, it is supposed to behave like a macrocrack in a bulk polycrystal with a critical stress intensity factor $K_{eff}^c = K_p^c$.

The fact that flaws may be close neighbours and propagate simultaneously must also be taken into account. Investigations (1) of the principal crack propagation modes during time to failure tests at room tempe-

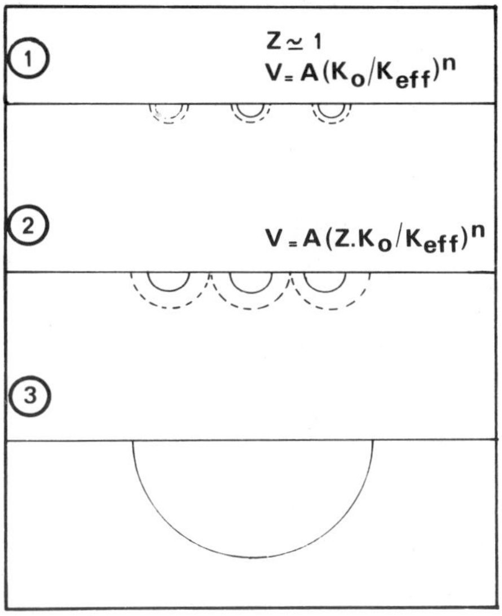

Fig. 2. Schematic sequences of flaw growth and coalescence.

rature in air show that the sequence is as following (Fig.2) :

- the small flaws grow in their own way untill their interaction leads to a modification of their critical stress intensity factor.
- the interaction is described by the parameter $Z = K_I/K_o$ with :
  K_o : the stress intensity factor for an isolated flaw,
  K_I : the stress intensity factor for the set of interacting flaws,
  Z depends thus on the flaw size/flaw separation ratio.
- in the third stage the nearby flaws coalesce creating thus the nucleus for the rupture.

In the scope of this work (computer simulation) we took as typical defects, flat semi-elliptical aligned flaws, e.g. surface cracks produced by machining. The variation of Z with the ratio a/p is shown in figure 3. In the present case rupture will coincide with defect coalescence. Rupture is supposed to be mainly transgranular. This allows to consider only one set of subcritical crack growth parameters (n,A) describing the one phase which is crossed over by the crack. The local crack growth law is thus :

$$V(a) = da/dt = A[K_I/K^c_{eff}]^n = A\left[\frac{Z(a)\sigma Y\sqrt{a}}{K^c_{eff}(a)}\right]^n \qquad [1]$$

Numerical integration method

In the case of dynamic fatigue, the stress, σ, at time, t, is given by :

$$\sigma = \mathring{\sigma}\, t$$

where : $\mathring{\sigma}$ is the stress rate (MPa/sec).

The evaluation of the flaw size at time, t, needs to compute the integral derived from [1] :

$$\int_{a_o}^{a(t)} \left[\frac{K^c_{eff}(a)}{Z(a)\sqrt{a}}\right]^n da = \int_o^t A(Y\mathring{\sigma})^n t^n\, dt = \frac{A(Y\mathring{\sigma})^n t^{n+1}}{n+1}$$

The nature of the integral makes its computation cumbersome. It will thus be solved by calculating the instantaneous growth rates.

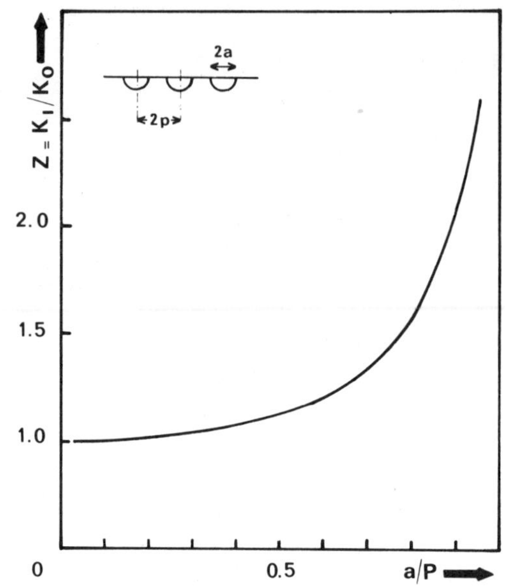

Fig. 3. Variation of interaction factor as a function of the flaw size/flaw separation ratio.

At time t, the critical stress intensity factor is :

$$K_I(t) = Y\ \sigma(t)\ \sqrt{a}\ Z(a)$$

and the growth rate is :

$$V(t) = A[K_I(t)/K^c_{eff}(a)]^n$$

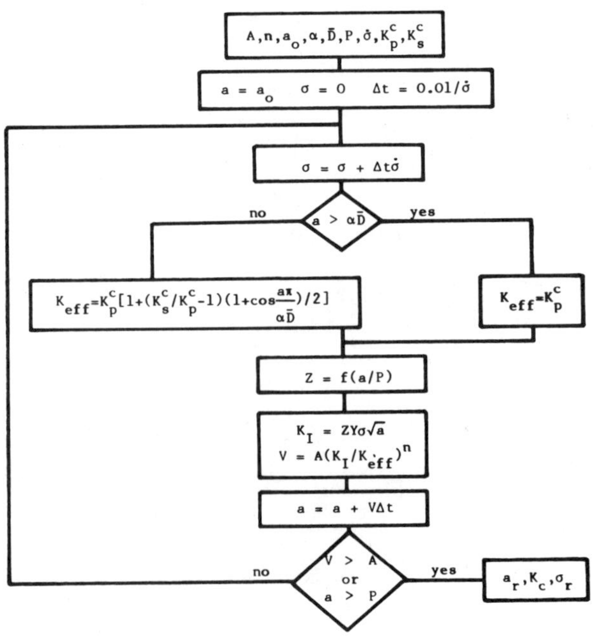

Fig. 4. Flow chart for flaw growth and fracture criteria.

The time step is chosen such that $\Delta\sigma = \dot{\sigma}\,\Delta t = 10^{-2}$ MPa.sec^{-1}. In the next step the defect size becomes :

$$a_{i+1} = a_i + \Delta t\, V_i$$

The flow chart is presented on figure 4. Computation stops either when $K_I = K^c_{eff}$ or when coalescence occurs. The computer gives thus at the instant of rupture the values a, σ , V and K_I used during this last step, i.e., a_r, σ_r and K_c.

Experimental parameters

Commercial alumina is often considered to behave like a model material. Its usual data will be used in our calculations.

The experimental values of Evans (6) for a commercial, 96 % dense, alumina and those of Wiederhorn (7) for the $(10\bar{1}2)$ planes of sapphire chosen by Okada and Sines (2) will also be used here :

$$\bar{D} = 32 \ \mu m,$$
$$K^c_s = 1.7 \text{ MPa } \sqrt{m},$$
$$K^c_p = 5.3 \text{ MPa } \sqrt{m}.$$

On cubic polycrystalline materials (CaF_2, Y_2O_3, $MgAl_2O_4$, ZnSe), Rice et al. (4) have noticed that the $\gamma^c_s \rightarrow \gamma^c_p$ transition is over for values of $\alpha = a/\bar{D}$ ranging from 1 to 6. In further work (8), they investigated among others two types of alumina. One was pressurless sintered ($\bar{D}$ = 8 µm) and the other hot-pressed ($\bar{D}$ = 32 µm). The measured values of α were respectively α = 6 and α = 65. With respect to the largest grains it would vary from 3 to 26. This shows the important role of the microstructure and the processing on the residual internal stresses. For the microstructure we have chosen, a maximum value of α = 10 seems to be best fitted. The value Y = 1.27 used by Okada and Sines will also be used. It corresponds to the value computed by Bansall (9) for tests in which the shape factor had been taken into account.

In order to define the distance, 2p, separating two flaws, following results and considerations must be taken into account. The observations made by Okada and Sines imply that the values of 2p with respect to the grain size are rather small. This is due to the fact that triple point boundaries as well as sintering defects may be looked as crack nuclei. Fractographic analysis reveals that flaws of the same size than the crystals (a_o = 20-40 µm) are rupture initiators. Furthermore for p-values much less than $\alpha\bar{D}$ in the case of an infinite number of flaws, the interaction factor, Z, will contribute alone to the growth law. In a first time a value of p, such that the $\gamma^c_s \rightarrow \gamma^c_p$ transition will rule the crack propagation, will be taken, i.e. p = 160 µm and p = 320 µm.

The K_I-V diagram is mostly obtained experimentally from relaxation or constant K tests on DCB or DT specimens. In view of the numerous results published, a crack propagation rate of da/dt = 10^{-3} m/sec describes rather well the transition from stage I to stage III, although this value seems too high to account properly for stress corrosion rates limited by the flow rate of the chemical species. Neglecting stage II, the crack growth rate when K^c_{eff} is reached will be taken as A = 10^{-3} m/sec.

The value of n = 31 as measured by Wiederhorn (7) will be used as well as other values often met for ceramic materials : say n = 20, 40 and 60.

The highest stressing rate will be the one often used in rupture test : $\dot{\sigma}$ = 10 MPa/sec. The apparent value of n will be determined using values of $\dot{\sigma}$ between 10^{-4} and 10 MPa/sec.

RESULTS

Crack growth rate path

The instantaneous growth rate given by equation [1] is a function of the instantaneous values of the applied stress, flaw size, the interaction factor, Z(a/p), and the effective value of the stress intensity factor, $K^c_{eff}(a, \sigma)$. Since the growth rate depends on the reduced K-value ($\propto ZK/K^c_{eff}$), the changes of Z and K^c_{eff} with flaw size, a, will strongly influence the growth rate. As a consequence a flaw of initial size, a_o, will have its own growth law, which is function of α, n, $\dot{\sigma}$ et P. This growth law is interesting because it will give the values of stress and flaw size at rupture. This graphical law may be looked as the K_I-V diagram related to each flaw. However in order to avoid confusion with K_I-V diagrams obtained with large specimens and cracks, it will be defined simply as the crack growth path. If not specified, the parameters used are :

α = 10,
p = 320 µm,
a_o = 50 µm,
n = 31
$\dot{\sigma}$ = 10 MPa/sec.

In the following we will first investigate the influence of each of these parameters but the distance between flaws, p.

On view of the diagrams given in Fig. 5 through 9 some remarks can be made. Introducing the effective critical stress intensity factor, K^c_{eff},

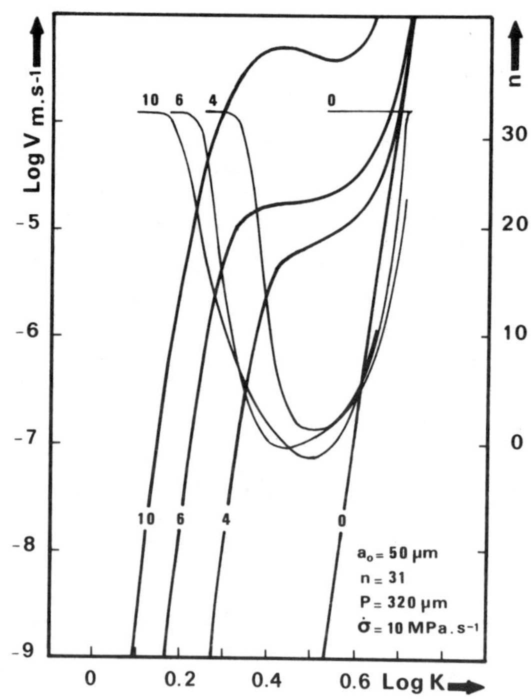

Fig. 5. Flaw growth path as a function of α.

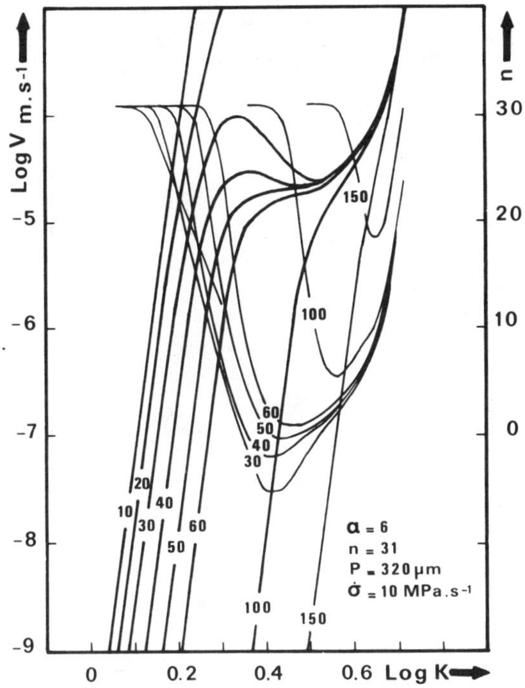

Fig. 6. Flaw growth path as a function of initial flaw size, a_o (α = 6).

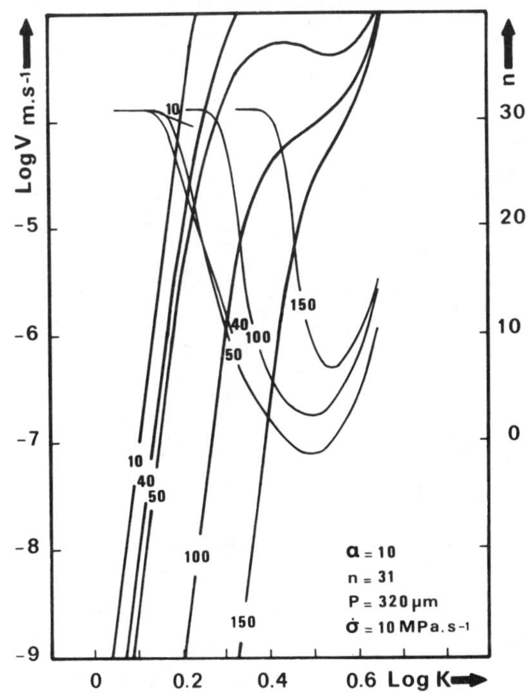

Fig. 7. Flaw growth path as a function of initial flaw size (α = 10).

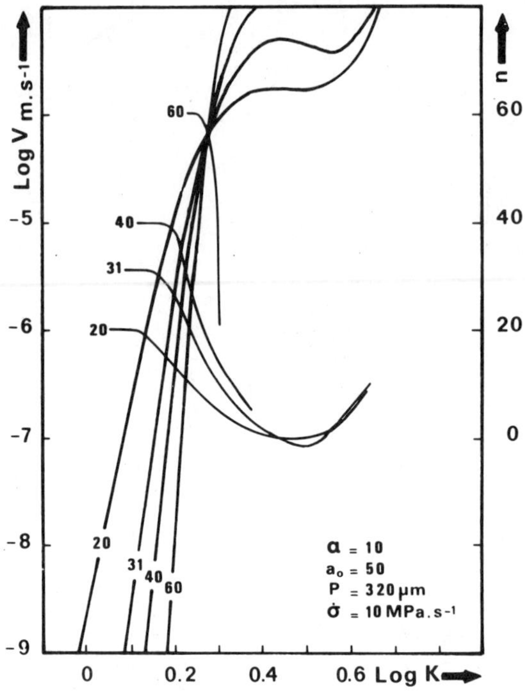

Fig. 8. Flaw growth path as a function of n.

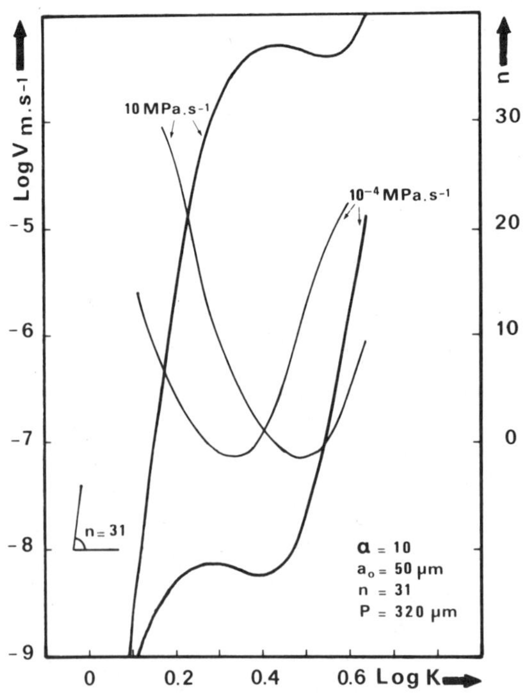

Fig. 9 : Flaw growth path as a function of stress rate, $\dot{\sigma}$.

into the formulation of the crack propagation rate $V = da/dt = A(K/K^c_{eff})^n$ makes the crack growth path to be curved, in opposition to classical K_I-V diagrams, where stage I is described by a straight line. Also the continuous change, with flaw extension, of the relative stress intensity factor introduces a plateau or even a decrease of the crack propagation rate during loading and flaw extension. Furthermore, one observes that the chosen fracture criterion ($da/dt = 10^{-3}$ m/sec) makes the flaw to be critical at K-values less than K^c_p in some conditions.

The critical K-value decreases as α increases and the initial flaw size decreases. This is because α and a_o intervene as a ratio in the expression of K^c_{eff}. For the same reason, large initial flaws tend to behave in a way close to that of a macrocrack.

For high values of n (Fig. 8), the amount of subcritial crack extension is small. The increase in crack propagation rate with $K_I(a, \sigma)$ becomes stronger in a domain of K_I-values closer to K_c. Thus the critical crack velocity and fracture criterion is reached earlier, i.e., the higher n, the lower K_c.

Decreasing the loading rate (Fig. 9) gives much more time to the flaws to reach the size at which the growth rate reaches the plateau or decreases. As a consequence the later is shifted toward lower V = da/dt values. However, the critical value of K is not very affected by this change of parameter in the present cases.

Variation of σ_r with a_o, Z(a) and K^c_{eff}

These effects have been investigated using the following conditions: Z = 1, Z = Z(a) with $K_c = 5.3$ MPa $\sqrt{m}$ and $K_c = K^c_{eff}(a)$. A case without stress corrosion and interaction has also been treated. The results are presented in Fig. 10. They have been calculated using V = A = 6.74 m/sec. That value was also used by Okada and Sines, but appears to be rather high and unrealistic to characterize the end of stage II.

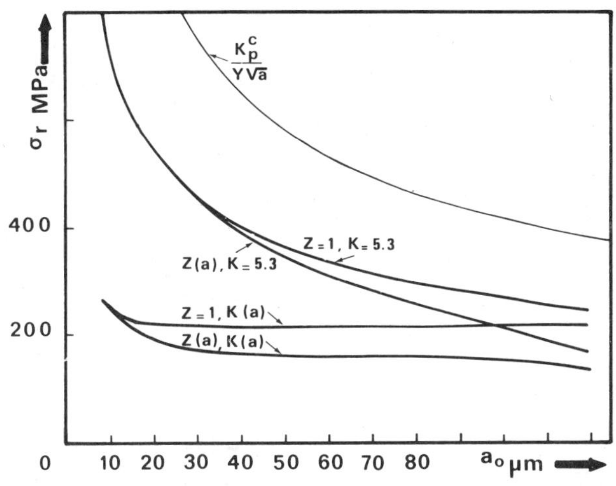

Fig. 10. Rupture stress as a function of initial flaw size for various instability criteria.

Taking no stress corrosion nor K^c_{eff} effects leads to an increases in rupture stress with decreasing initial flaw size according to the $a^{-1/2}$ dependance.

Introducing K^c_{eff}, increases the growth rate of the small flaws ($x10^{10}$) and thus decreases the resulting rupture stress. This effect will be smoothed with increasing initial crack size because the ratio K/K^c_{eff} decreases. Furthermore the shape of the curve $K^c_{eff} = f(a)$ induces a large plateau : the rupture stress remains constant (in the case $Z = 1$ and $K_c = K^c_{eff}(a)$) over a flaw size range from 20 to 120 μm. This observation will be discussed further.

Since the distance between flaws (p = 160 μm) has been chosen to be rather large, it is evident that the effect of introducing Z(a/p) becomes noticeable for large flaw sizes.

Variation of σ_r, a_r and K_c with initial flaw size

In the case where the stress rate ($\mathring{\sigma} = 10$ MPa.sec^{-1}) will not allow much time for the flaws to grow and to behave like macroscopic cracks before failure, the resulting observations are :

- K_c will remain an univocal function of the flaw size at rupture. These two values will determine the rupture stress.
- Fig. 11 through 13 show the variation of a_r, σ_r and K_c with initial flaw size, a_o, ($\alpha = 10$; p = 320 μm) for various values of n.

In fig. 11, one can see that beyond a given initial flaw size, the size at rupture remains constant over a large range. Effectively, as the flaw growth path has shown, beyond this threshold the smallest defect will exhibit some stable crack growth and reach some final size as already mentionned by Okada and Sines (2) in the case of life-time measurements. This is proper to the shape of the $K^c_{eff}(a)$ curve.

A decrease of n firstly increases the value of the a_o-level (it varies from 30 to 65 μm as n increases from 20 to 60), secondly and in the same time increases the plateau-value (constant a_r) slightly from 210 μm to 240 μm. The critical stress intensity factor also depends on the final flaw size, it will thus also be defined by a threshold and a plateau as it is shown in Fig. 12. As a consequence the variation of σ_r with initial flaw size will be similar (Fig. 13).

For flaw size larger than about 200 μm, the interaction between flaws will induce a large decrease of σ_r and an increase in a_r and thus of K_c.

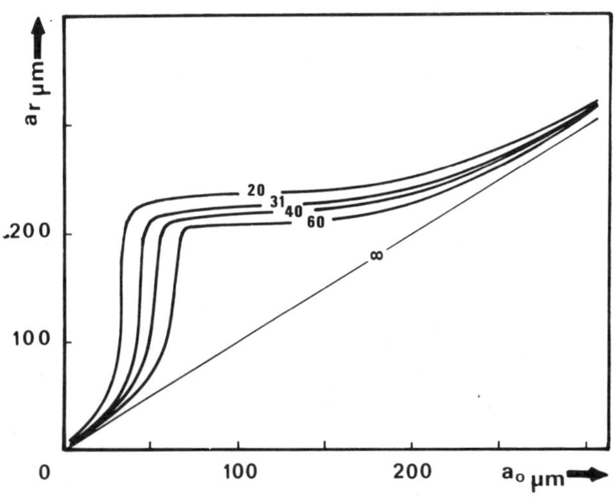

Fig. 11. Variation of final flaw size with initial flaw size for various values of n.

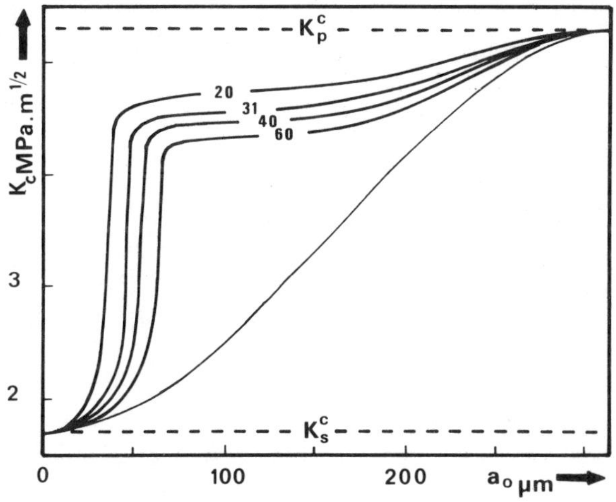

Fig. 12. Variation of K_c with initial flaw size for various values of n.

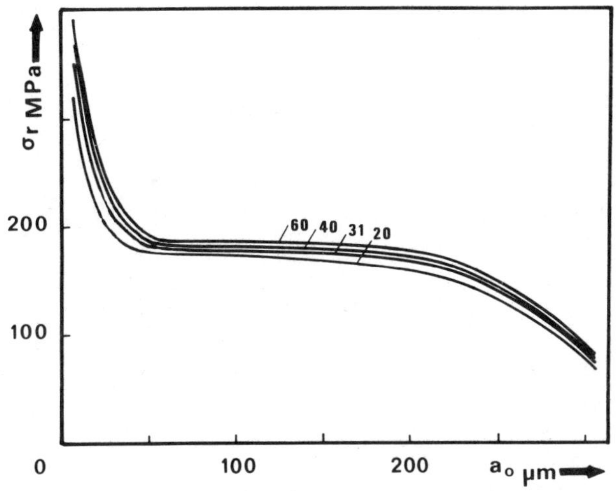

Fig. 13. Variation of the rupture stress, σ_r, with initial flaw size for various values of n.

Effect of stress rate

In the case of a commercial alumina ($\bar{D}$ = 32 µm) it appears reasonnable to keep the values :

$$\alpha = 6 \quad ; \quad p = 160\ \mu m \quad ; \quad n = 31$$

Figures 14 through 17 show respectively the change of K_c, σ_r, a_r and n as a function of the initial flaw size. In comparison with that it was described in the previous paragraph, the new values for α and p lead to much less extended plateaus for each curve.

Figure 16 shows the variation of the rupture stress as a function of stress rate, respectively 10 and 10^{-4} MPa/sec. The general shape is maintained.

From the calculation of the rupture stress made at stress rates of 10 MPa/sec and 10^{-2} MPa/sec or 10^{-4} MPa/sec, the subcritical crack growth parameter, n, can be evaluated using the relation proposed by Evans:

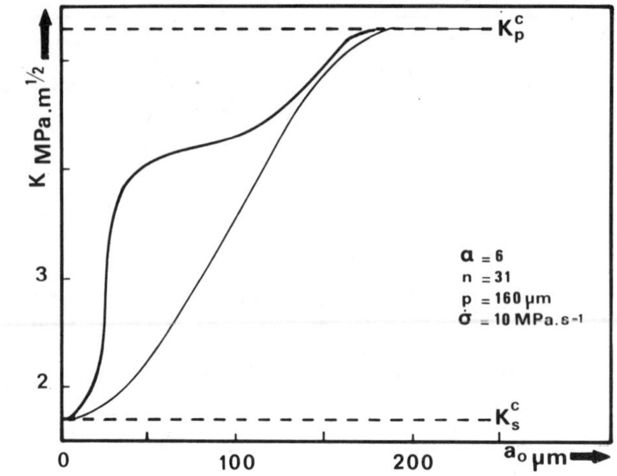

Fig. 14. Effect of reducing the values of α and p on the variation of K_c with initial flaw size (cf. figure 12).

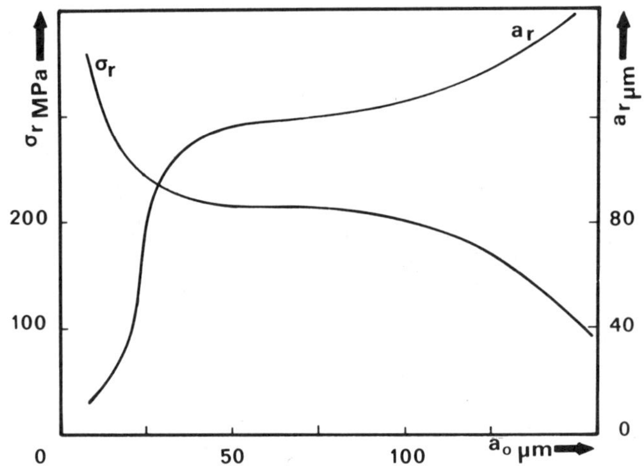

Fig. 15. Effect of reducing the values of α and p on the variation of σ_r and a_r with initial flaw size (cf. figures 11 and 13).

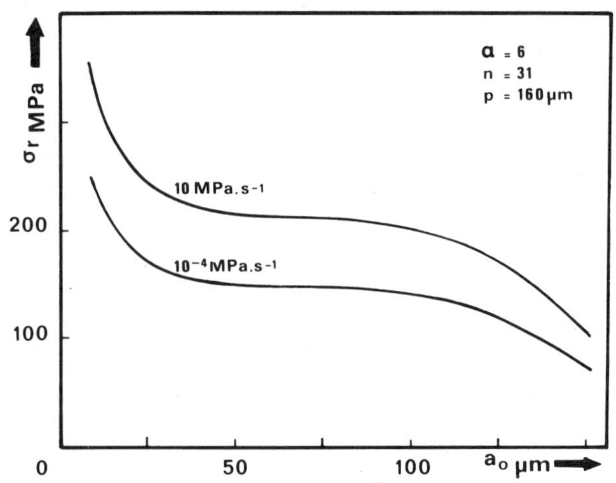

Fig. 16. Variation of σ_r with initial flaw size for two values of stress rates (α = 6, p = 160μm).

$$(\sigma_{r1}/\sigma_{r2}) = (\mathring{\sigma}_1/\mathring{\sigma}_2)^{1/n+1}$$

Its evolution with the initial flaw size is given in Fig. 17 for the two sets of stress rating. One observes that using the $\gamma_s^c \rightarrow \gamma_p^c$ transition, as introduced Okada and Sines (2), as well as the interaction between flaws does not induce any appreciable change of the "real" value of the stress corrosion parameter. The small peak must be related to the threshold values of the $K_c(a)$ curve describing each of the rates described.

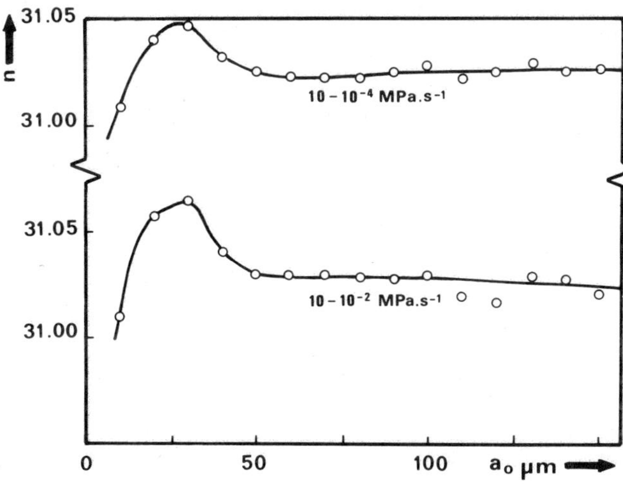

Fig. 17. Influence of initial flaw size and stress rate range on the value of n, as calculated.

DISCUSSION AND CONCLUSION

Some published experimental results support this argumentation. Gruver, Sotter and Kirchner (10), investigated the variation of the fracture stress with flaw character in 96% density Al_2O_3. For some of their rupture tests, they plotted the fracture stress as a function of the surface flaw size determined by taking the average of the measured width and depth. Fracture stress decreases with increasing flaw size but much less than expected from classical fracture mechanics relations, i.e. K_c is constant. The comparison of an ideal $\sigma_r \propto a_r^{-1/2}$ with their data shows effectively that the discrepancy between the expected value of σ_r and that measured increases as the critical flaw size increases. This can be explained by the fact that the $K_c = K_{eff}^c$, function of the flaw size, as used in this paper. The fracture mechanical relationship would not longer be $\sigma_r \propto a_r^{-1/2}$ but $\sigma_{r1}/\sigma_{r2} = \sqrt{a}_{r2}K_{c1}/\sqrt{a}_{r1}K_{c2}$.

There is some scatter in the results of Gruver, Sotter and Kirchner (10) as well as in all the published results of fracture stress measurements of ceramics. Those of Seshadri and Srinivasan (11) on sintered α-SiC are a typical example. They measured also fracture stresses and plotted them as a function of equivalent flaw size, i.e. three dimensional void models (12). The scatter was apparently reduced but remained rather large as long as sintering or machining defects were concerned.

As already noticed, the scatter of the rupture stresses belongs to the fracture testing of ceramics and is in apparent contradiction with the simulation results. The plateaus we have observed have been obtained, each with a given set of (α, A, n) values and the assumption that residual thermal stresses are the same over the whole specimen. The discrepancy of that assumption, although necessary in a first step, and the reality of sintered ceramics have already been pointed out by Kirchner and Ragosta (3). They argued with the number of neighbourgs of a large grain within bulks of different grain sizes. Such a discussion can be extended toward other microstructural inhomogeneities. The residual thermal stresses around a large grain may be much larger than in an assembly of smaller crystals having the same volume. There is also much more probability to find a number of small grains than larger around a large grain. Pores are an other example : the thermal stresses on their border are largely different from those which would exist if it were replaced by several small crystals or one large crystal. Furthermore, concerning pores, the chemical composition of the grain boundaries encompassing it is also likely to be different from the bulk and homogeneous remaining material. All these local inhomogeneities will change the local values of the set (α, A, n). The influence on the rupture stress will be large since it will be these inhomogeneities which will lead to fracture.

In some way, that argumentation is supported by the controlled flaw measurements also made by Seshadri and Srinivasan on the sintered α -SiC (11). They noticed that the K_c-values calculated from rupture stresses of controlled flaws introduced by Knoop indentation are lower than those measured with the classical SENB specimen and a macrocrack. These results are in agreement with those obtained in our laboratory. Even if the residual stresses resulting from indentation are relieved,the K_{CF}-value will be lower than that from a SENB specimen (13)(14).

The second observation from the results of Seshadri and Srinivasan which is worth pointing out is that the σ_r-values from controlled flaw tests remain constant over a range from 120 µm to 240 µm and 140 µm to 280 µm respectively. These results, with rather well defined flaws, are in full agreement with the simulation results. Furthermore the scatter is narrow. This is only in apparent contradiction with the previous discussion about sintering defects and microstructural inhomogeneities. In a classical rupture test - without intentionnally introduced flaw -, the rupture will take place at the later and largest defects and thus the local inhomogeneities will govern the crack growth path. In the case of introducing a hardness flaw there is a very low probability to meet such an inhomogeneity. As a consequence the introduced flaws will all propagate with an almost unique set (α, A, n) and be described by a simulation as we did in this work.

As a conclusion, the results of crack propagation simulation pointed out some results which apparently are in disagreement with experimental results. The discrepancy can be explained by taking into account the local microstructure of as-sintered defects. Simulation of crack propagation should be verified using introduced flaws which have a large probability to meet an homogeneous part of the microstructure.

REFERENCES

1. Okada T., Sines G., Green D., J. Amer. Ceram. Soc., 65 : C64 (1982).
2. Okada T., Sines G., J. Amer. Ceram. Soc., 66 : 719 (1983).
3. Kirchner H.P., Ragosta J.M., J. Amer. Ceram. Soc., 63 : 490 (1980).
4. Rice R.W., Freiman S.W., Mecholsky J.J., J. Amer. Ceram. Soc., 63: 129 (1980).

5. Rooke D.P., Cartwright D.J., Compendium of stress intensity factors, The Hillington, Uxbridge, Middlesex, U.K. (1976) cited by (2).
6. Evans A.G., J. Mat. Sci., 7 : 1137 (1972).
7. Wiederhorn S.M., Fracture Mechanics of Ceramics, edited by Bradt R.C., Hasselman D.P.H. and Lange F.F., Plenum Press, Vol. 2 : 613 (1974).
8. Rice R.W., Pohanka R.C., Mc Donough W.J., J. Amer. Ceram. Soc., 63 : 703 (1980).
9. Bansall G.K., J. Amer. Ceram. Soc. 59 : 87 (1976).
10. Gruver R.M., Sotter W.A., Kirchner H.P., Amer. Ceram. Bull. 55 : 198 (1976).
11. Seshadri S.G., Srinivasan M., J. Amer. Ceram. Soc., 64 : C69 (1981).
12. Baratta F.R., J. Amer. Ceram. Soc, 61 : 490 (1978).
13. Moussa R., Coppolani I., Osterstock F., Proc. Brit. Ceram. Soc., 32 : 237 (1982).
14. Jouin J.M., D.E.A. Université de Caen, Septembre 1983.

UNIAXIAL TENSILE FATIGUE TESTING OF SINTERED SILICON CARBIDE UNDER CYCLIC TEMPERATURE CHANGE

H. Fujita, M. Kawai, H. Takahashi
H. Abe and J. Nakayama

Research and Development Div., Asahi Glass Co., Ltd
kanagawa-ku, Yokohama 221, Japan

ABSTRACT

Uniaxial tensile tests were conducted for sintered SiC up to 1400°C in air under static fatigue, low cycle fatigue, static fatigue with cyclic temperature change and synchronized cyclic change of stress and temperature. In all tests, fatigue characteristics were such that time to failure was controlled only by the accumulation of slow crack growth and that no definite acceleration due to cyclic change of either stress or temperature was observed.

INTRODUCTION

Sintered SiC and Si_3N_4 are currently being considered for use as structural components for gas turbines and other engines[1-5]. However, the history of these non-oxide ceramics to be used as structural components is relatively short and the accumulation of property data, especially data of long term durability is insufficient in comparison to that of metals.

Ceramics, being brittle materials, fracture at defects or flaws and the effects of the size and distribution of flaws have been studied based on Griffith's theory and Weibull statistics[6-11].

Until recently, however, fatigue of ceramics has not been much studied and most of data have been obtained using three points or four points bending specimens which are relatively easy to be tested[12-15]. One of the definite shortcomings of bending test is the small effective volume far smaller than that of actual structural components.

In the present study, long term durability of pressureless sintered silicon carbide was investigated under uniaxial tensile conditions[16-20]. Four kinds of stress and temperature conditions were adopted so that the tests can simulate the actual conditions of ceramic components in use. Test conditions were, static fatigue, low cycle fatigue, static fatigue under cyclic temperature change and thermal fatigue in which cyclic change of temperature and stress were synchronized.

EXPERIMENTAL PROCEDURES

Specimen Preparation

Silicon carbide specimens were fabricated by isostatically pressing and sintering SiC powder with Al_2O_3 additive[21-22]. The sintered blocks were composed of 5-10 μm elongated grains with aspect ratio 5-7 and had a porosity of 4-5%.

The sintered blocks were diamond-machined into the tensile test specimen with circular cross section. The surface of gauge length area, 7mm in diameter and 36mm in length, and shoulders were polished. Prior to testing, specimens were annealed at 1200°C in air for 30 min. to remove the effect of residual stress caused by machining.

Static Fatigue Test

The schematic illustration of a loading system and the details of experimental procedures are shown elsewhere[20]. Static fatigue tests were carried out at 1000°C, 1200°C, 1300°C, 1350°C and 1400°C in air.

Fatigue Tests under Cyclic Change of Stress and Temperatures

Low cycle fatigue test, static fatigue test under cyclic temperature change and thermal fatigue test was conducted using the system which was composed of an Instron type loading machine, a resistant heating furnace and controller which can synchronize the cyclic change of temperature and stress.

The stress was changed between the maximum tensile stress σ_{max} and minimum tensile stress 1/2 σ_{max}.

The temperature cycled either 1100 ± 100°C or 1200 ± 200°C in a period of 5 minutes. The patterns of cyclic change of stress and temperature are given in Fig 1.

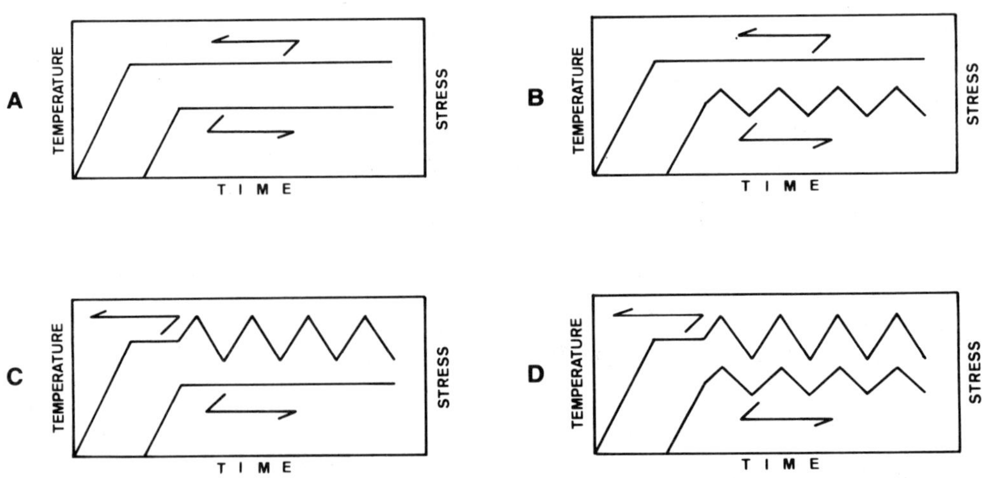

Fig. 1 Patterns of stress and temperature adopted in fatigue tests
A : Static fatigue ; B : Low cycle fatigue
C : Static fatigue with cyclic temperature change
D : Synchronized cyclic change of stress and temperature

RESULTS AND DISCUSSION

Static Fatigue Test

Figure 2 gives the results of static fatigue test up to 1000 hr in air at 1000 - 1400°C. Left sided arrows indicate that specimens fractured at the onset of loading, and right sided ones indicate that the tests were suspended at that time without failure of the specimens.

At 1000 and 1200°C specimens either failed at the onset of loading or survived up to 1000 hr with no indication of fatigue ("On-off" behavior). The level of survival-failure stress depended on temperature, that is, 310, 260 and 210 MPa at r.t., 1000 and 1200°C, respectively. At temperatures over 1300°C, specimens showed delayed fracture, that is, time to failure was dependent on the applied stress.

The fatigues were especially eminent at 1400°C which was the highest temperature adopted in the present study. In terms of slow crack growth (SCG) parameters ($V = A K_I^N$)[23-25], N was infinite at r.t. - 1200°C, 90 at 1350°C and 27 at 1400°C.

Figure 3 shows fractography of the specimens. Most of the specimens tested at r.t. and 1000°C fractured at pores and had relatively rough fracture surfaces. In specimens tested over 1200°C, overall fracture surfaces were relatively flat compared with those of lower temperature specimens. Relatively rough SCG area was observed in fracture surface tested at 1400°C.

The lowering of survival-failure stress level with temperature for low temperature specimens corresponded to the decrease in fracture toughness, K_{IC} as is shown in Fig.4. The Young's modulus, on the other hand, stayed almost constant up to 1200°C (see Fig.4) indicating that the decrease in fracture surface energy caused the decrease in K_{IC} and hence stress level.

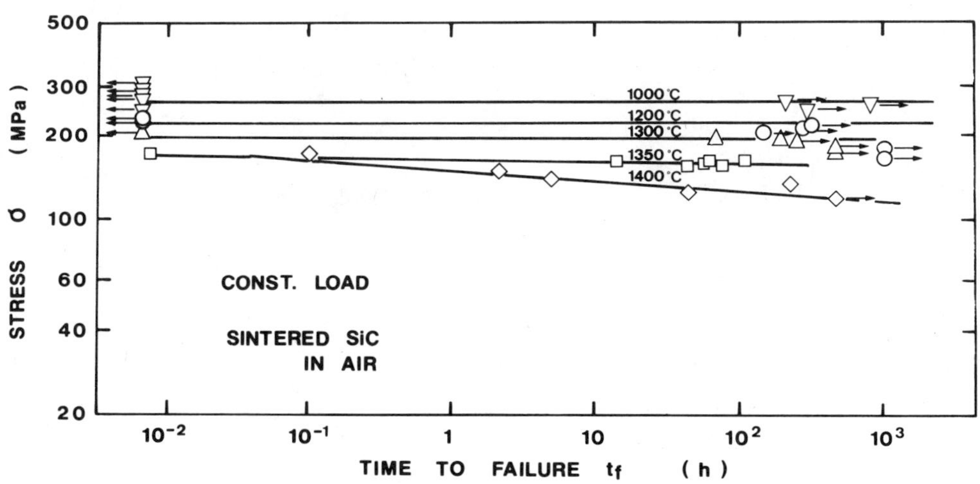

Fig. 2 Results of static fatigue test

Fig. 3 Fracture surfaces of specimens fatigued under constant load

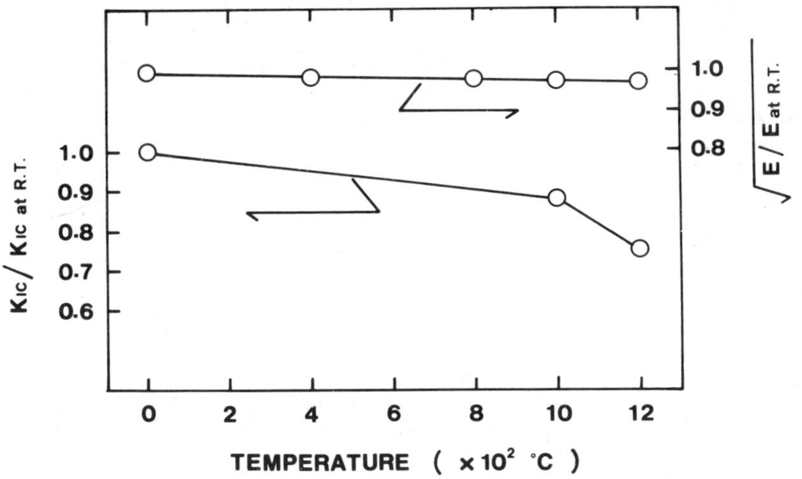

Fig. 4 Changes of fracture toughness (K_{IC}) and Young's modulus (E) as functions of temperature

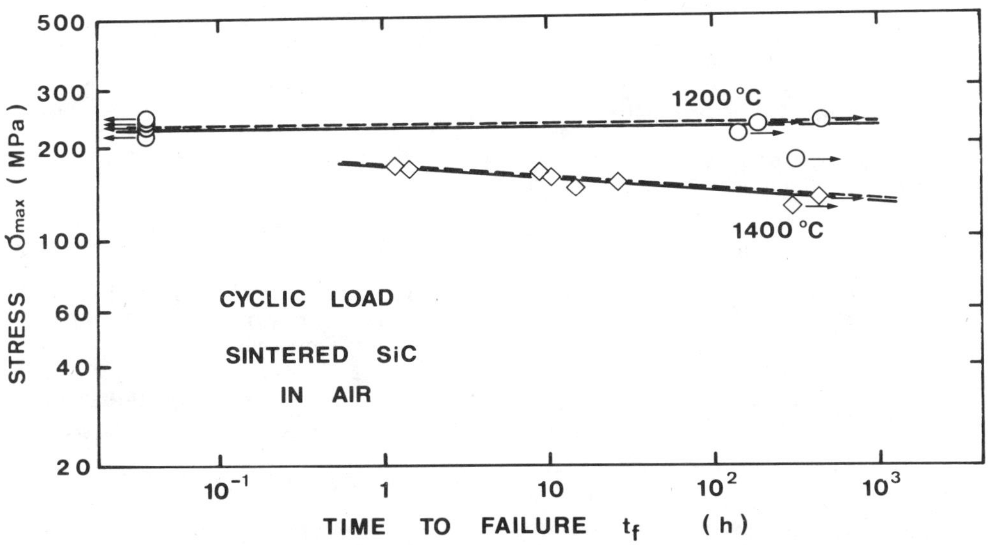

Fig. 5 Results of low cycle fatigue test (estimated lifetime is shown by dotted line)

Low Cycle Fatigue Test

The results of low cycle fatigue test were reported elsewhere[20]. The data are plotted in Fig.5. The results of low cycle fatigue test at 1200 and 1400°C gave the similar trend to the static fatigue test ; "On-off" behavior at 1200°C and delayed fracture at 1400°C.

Lifetime prediction was carried out for low cycle fatigue conditions and the results are plotted as dotted line in Fig.5. The estimated lifetime showed good coincidence with the measured data.

Static Fatigue Test under Cyclic Temperature Change

Figure 6 gives the results of static fatigue test under cyclic temperature change. The temperature cycles were 1000 - 1200°C (1100 ± 100°C) and 1000 - 1400°C (1200 ± 200°C) .

The results of 1000 - 1200°C temperature cycle showed that the survival-failure stress level was the same as that in static fatigue test at 1200°C. The test results of 1000 - 1400°C temperature cycle indicated that delayed fracture was observed and that the slope of the time vs stress plot was similar to that of 1400°C static fatigue test.

The data of the test under cyclic temperature change are discussed as follows. Under temperature change conditions, SCG rate V and fracture toughness K_{IC} change as a function of time. The time to failure was estimated using the SCG data of silicon carbide. If the total length of crack growth before catastrophic failure is Δa, the failure condition will be defined as,

$$a_c(T) < a_i + \Delta a \tag{1}$$

where $a_c(T)$ is the critical crack size at failure which was calculated from K_{IC} and depended on temperature. The initial crack size, a_i , was also calculated from K_{IC} at r.t.. Assuming that the crack growth does not change stress intensity factor, K_{In} , within one particular n-th temperature cycle, the SCG rate $V_n(T)$ is calculated by

$$V_n(T) = A(T)\, K_{In}{}^{N(T)} \tag{2}$$

where T is temperature. Then the crack growth Δa_n in the first half or latter half of n-th temperature cycle is given as

$$\Delta a_n = \int_0^{t_\omega / 2} V_n(T)\, dt \tag{3}$$

where t_ω is the time for one cycle. Figure 7(a) illustrates the relation between temperature T and time t. Figure 7(b) gives the relation between V and T. Using these relations, a function of temperature $V_n(T)$ was converted to a function of time. Hence, $V_n(t)$ was substituted in the equation given above. The time to failure t_f is then determined when the following condition is satisfied.

$$a_c(T) < a_i + \Sigma\, \Delta a_j \tag{4}$$

where j is the number of half temperature cycles, and t_f is calculated by

$$t_f = j \times t_\omega / 2 \tag{5}$$

The results of tensile testing and the calculated time to failure are compared in Fig.6. Both at lower temperature (1100 ± 100°C) and higher temperature (1200 ± 200°C), the results of tensile testing and that of calculation coincided well indicating that the cyclic temperature change did not affect the fatigue mechanism and that the life span of silicon carbide was controlled simply by the accumulation of slow crack growth.

Thermal Fatigue Test

Thermal fatigue test carried out in the present study is the one usually adopted in fatigue testing of metals. In this test, synchronized cyclic change of temperature and stress are repeated until specimens failure.

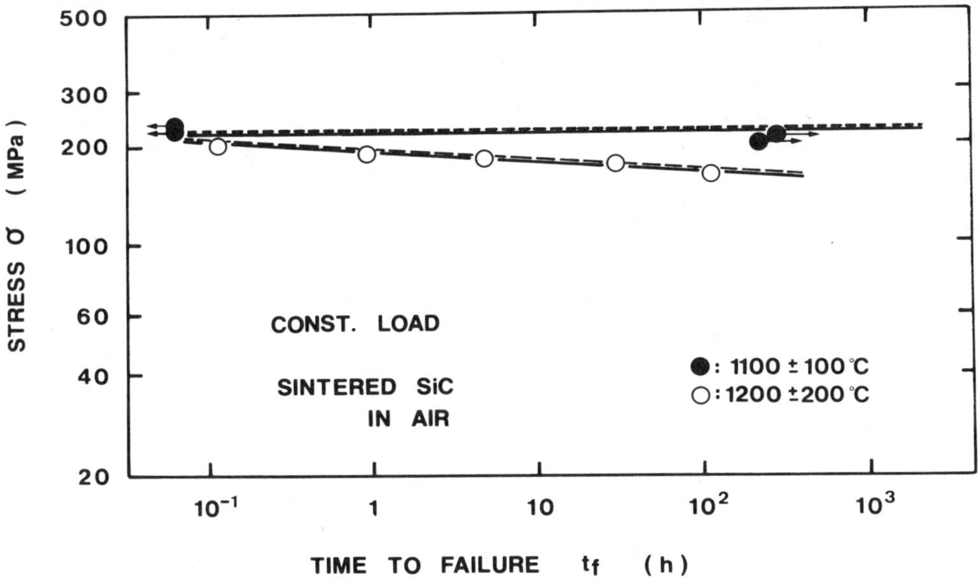

Fig. 6 Results of static fatigue test under cyclic temperature change (estimated lifetime is shown by dotted line)

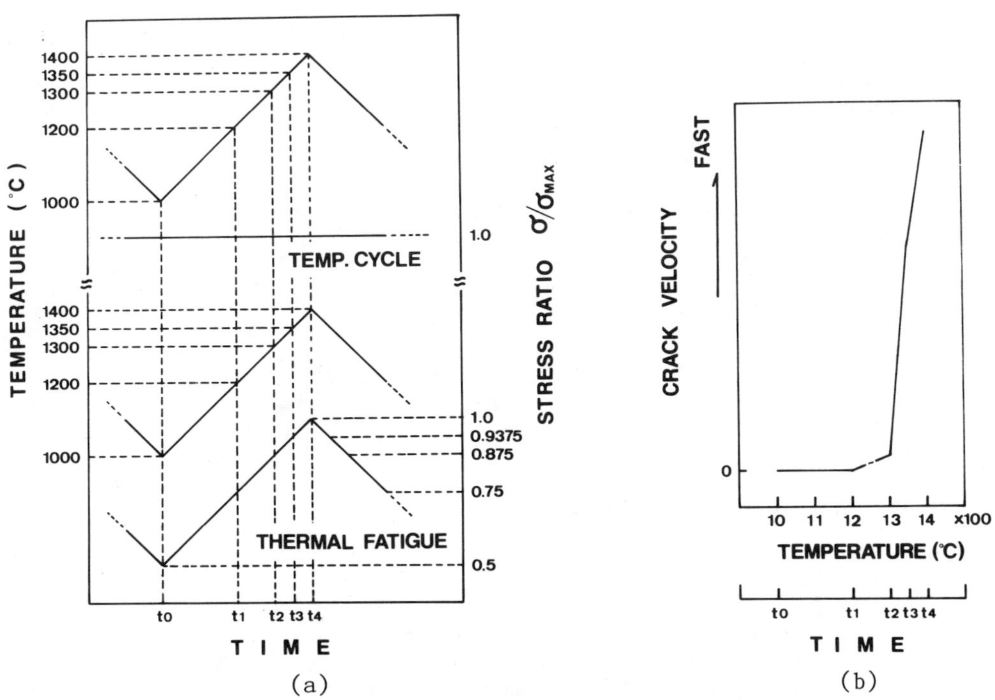

Fig. 7 Relation between temperature T and time t (a) and slow crack growth rate V and t (b)

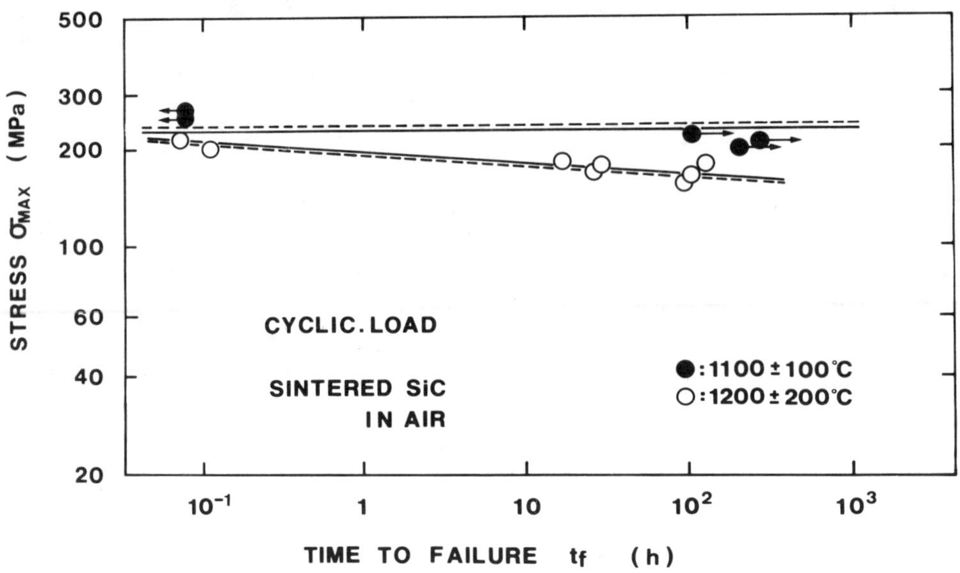

Fig. 8 Results of thermal fatigue test (estimated lifetime is shown by dotted line)

Figure 8 shows the results of thermal fatigue test with the temperature cycles of 1100 ± 100°C and 1200 ± 200°C. Figure 8 indicated that the survival-failure stress level of 1100 ± 100°C thermal fatigue test coincided with that of 1200°C static fatigue test. As for the thermal fatigue test at higher temperature , temperature cycle of 1200 ±200°C, the stress level of fatigue curve was higher than that of low cycle fatigue test at 1400°C.

In order to estimate the effect of cyclic change of temperature and stress on the fatigue characteristics of silicon carbide, time to failure was calculated with the assumption that the life span is controlled only by K_I - V relation.

The principle to estimate the time to failure under thermal fatigue conditions, that is, synchronized application of cyclic stress change and cyclic temperature change, is similar to the one used in the estimation of the time to failure under cyclic temperature change and constant stress.

In this case, stress intensity factor changes as a function of stress σ and the SCG rate $V(T,\sigma)$ is expressed as

$$V_n(T,\sigma) = A(T)\ K_{In}(\sigma)^{N(T)} \tag{6}$$

As both stress and temperature are the function of time, $V_n(T,\sigma)$ is given as a function of time. Crack extension Δa_n in half cycle is then expressed as

$$\Delta a_n = \int_0^{t_\omega/2} V_n(T,\sigma)\ dt \tag{7}$$

In the present condition, the highest temperature and the maximum stress are synchronized, as in Fig.7(a). Then expressing the maximum stress as σ_{max}, the stresses at t1, t2, t3 and t4 are 3/4 σ_{max} , 7/8σ_{max} , 15/16 σ_{max} and σ_{max} , respectively. $V_n(T,\sigma)$ was estimated by the interporation of V_n(t1), V_n(t2), V_n(t3) and V_n(t4).

Finally, time to failure is determined in the same way as the previous calculation.

$$a_c(T,\sigma) < a_i + \Sigma \Delta a_j \quad (8)$$

$$t_f = j \times t_\omega / 2 \quad (9)$$

The results of tensile thermal fatigue test and that of life time prediction are compared in Fig.8. Both at lower temperature, 1100 ± 100°C, and higher temperature , 1200 ± 200°C, the test results and the estimated life span coincided well. It is concluded that the superimposed cyclic temperature change and stress change did not affect the fatigue mechanism and that the life span was controlled simply by the accumulation of slow crack growth.

Figure 9 shows the fractographs of the thermal fatigue test specimens which were tested at 1200 ± 200°C. Some of them (1 and 8 in Fig.9) are similar to the typical fractographs at 1400°C with areas of SCG accompanied with relatively smooth fracture surface. Remarkable difference from other high temperature test specimens were observed in many of the fractographs (2,3,4,5 and 7 in Fig.9) of thermal fatigue test. They showed very rough fracture paths which are commonly observed in fractographs of specimens fractured at lower temperature.

It was noticed that all specimens in this fatigue test fractured at temperatures higher than 1380°C although temperature changed cyclicly between 1000°C and 1400°C, and that the temperature difference at the surface and at the center of the specimen was only 3°C at 1400°C.

In order to clarify the cause of this peculiar feature of fracture surface, silicon carbide specimens were fractured at different stressing rate at 1400°C. The characteristics of the fracture surface differs depending on the stressing rate. Under higher stressing rate the fracture surface was rough, while under lower stressing rate the surface was relatively flat as is common for the specimens fractured in higher temperature fatigue test.

In thermal fatigue test, stress change and temperature change were synchronized. Hence in the process of temperature elevation, stress increase resulted in the increase in K_I and the simultaneous temperature increase caused the decrease in fracture toughness, K_{IC}. Therefore, the rate of increase in K_I / K_{IC} was apparently higher compare with any other test conditions ; static fatigue test, low cycle fatigue test, and static fatigue test under cyclic temperature change. This rapid increase in K_I / K_{IC} per unit time could have caused the characteristic feature of the fractographs of thermal fatigue test.

Transmission electron micrograph of thermal fatigue specimen tested at 1400°C, indicated no sign of dislocation movement or change of stacking fault in SiC crystals. Similar results are reported for Boron-Carbon contained SiC [26].

EPMA analysis of specimens tested at 1400°C with different stress and temperature conditions provided some characteristics. In all specimens tested under static fatigue, low cycle fatigue and thermal fatigue conditions, oxygen rich surface layers were observed. As is shown in Fig.10 indication of Al enrichment was evident especially in those specimens which were tested under cyclic stress and temperature change conditions. Further characterization and the study of the relation between fatigue test conditions and material transport are to be continued.

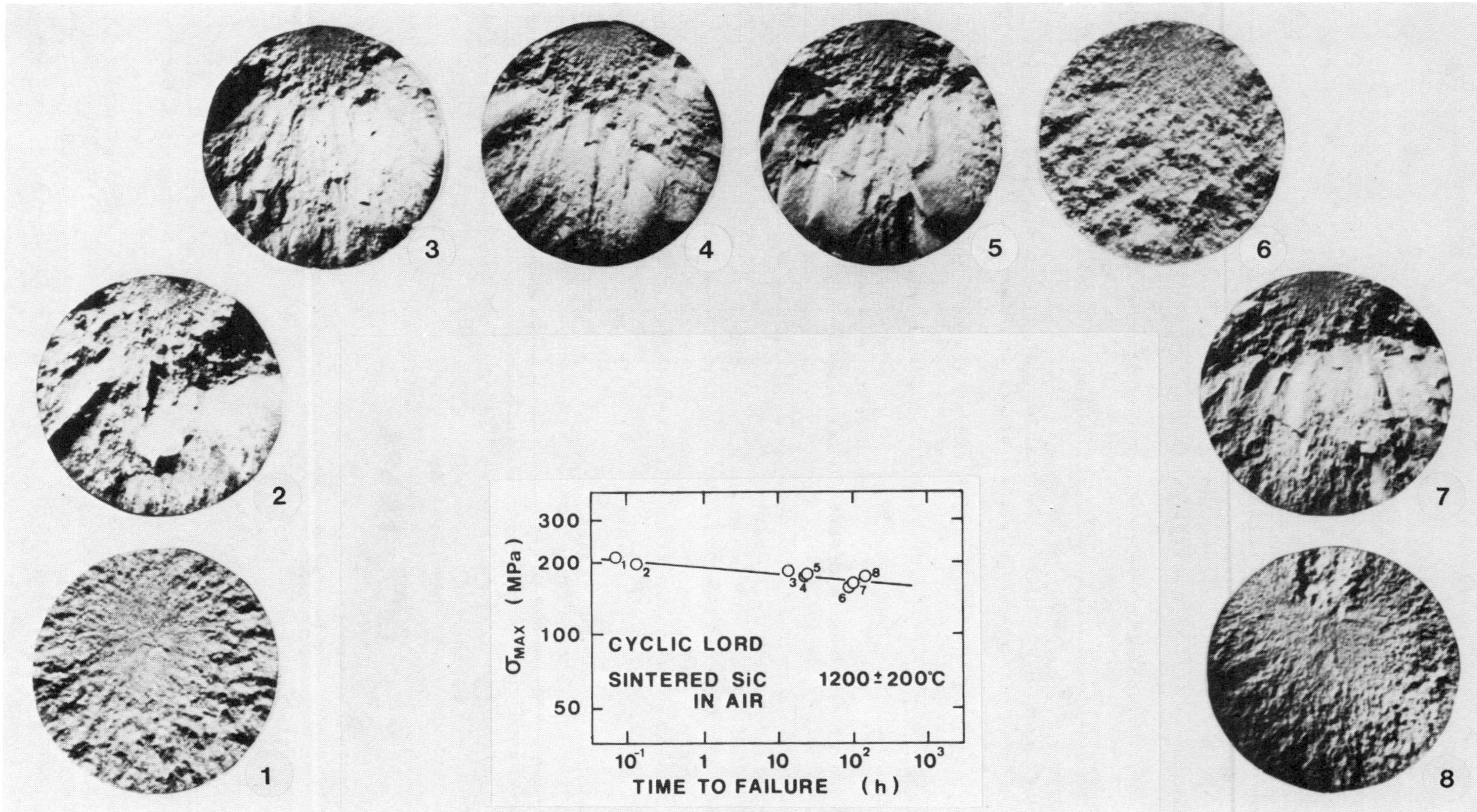

Fig. 9 Fractographs of specimens tested under thermal fatigue condtion

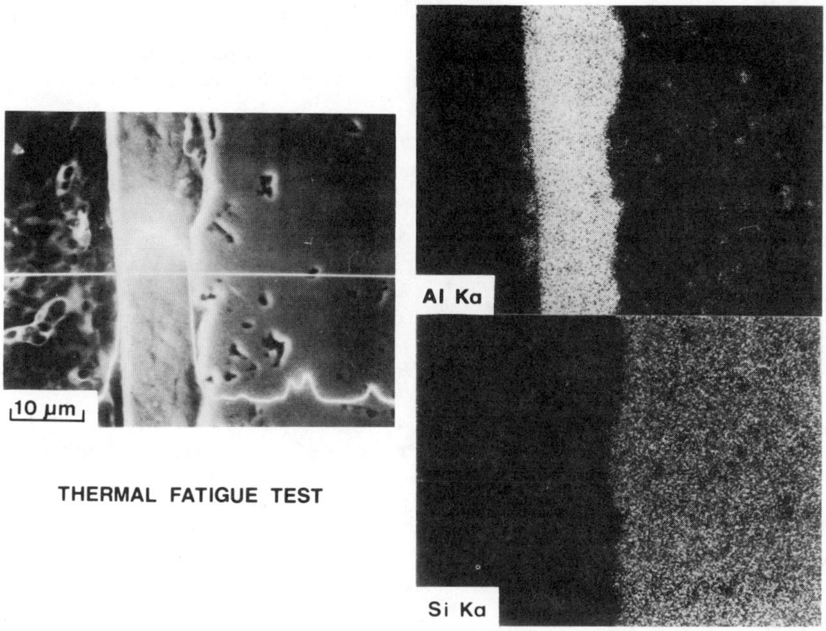

Fig. 10 EPMA analysis of the surface of a specimen tested under thermal fatigue condition (1200 ±200°C)

SUMMARY

Uniaxial tensile tests were conducted for sintered SiC up to 1400°C in air under static fatigue, low cycle fatigue, static fatigue with cyclic temperature change and thermal fatigue conditions.

The results were summarized as follows.

1) Under static fatigue conditions at temperatures lower than 1200°C, no fatigue was observed in alumina containing sintered SiC. Survival-failure stress limit decreased with temperature corresponding to the decrease in K_{IC} . At temperatures over 1300°C, fatigue due to slow crack growth was observed especially at higher temperatures.

2) In all fatigue tests under cyclic stress change, cyclic temperature change and synchronized cyclic change of temperature and stress, fatigue characteristics were such that time to failure was controlled only by the accumulation of slow crack growth and that no definite acceleration due to cyclic change of either stress or temperature was observed within the conditions tested in the present study.

ACKNOWLEDGEMENT

The present study was conducted as part of a programme for the development of an advanced gas turbine power generator sponsored by the Agency of Industrial Science and Technology, the Ministry of International Trade and Industry, Japan. The sponsorship of the Agency is gratefully acknowledged. EPMA study by K. Sakai is appreciated.

REFERENCES

1. A. F. Mclean, "Ceramics for High Performance Applications" edited by J. J. Burke, A. E. Gorum and R. N. Katz, Brook Hill, Chestnut Hill, Mass. 1974, 9-36.
2. R. J. Bratton and D. G. Miller, 689-730 in Ref. 1.
3. R. Kamo, 907-922 in Ref. 1.
4. S. B. Kramer, "DOE Automotive Gas Turbine Modern Technology Engine Program" , Proceedings ,1983 Tokyo International Gas Turbine Congress.
5. W. Bunk, "Overview on the German Program on Ceramic Components for Vehicular Gas Turbines", Proceedings, the First International Symposium on Ceramic Components for Engine, 1983 Japan, edited by S. Sohmiya, E. Kanai and K. Ando.
6. G. R. Irwin, J. Appl. Mech., 29(1962) 651-654.
7. G. K. Bansal, J. Am. Ceram. Soc., Vol.59, No.1-2(1976) 87-88.
8. R. W. Rice, S. W. Freiman, R. C. Pohanka, J. J. Mecholsky, Jr. and C. Cm. Wu, "Fracture Mechanics of Ceramics", Vol.4, Plenum Press, New York (1978), 849-876.
9. W. Weibull, J. Appl. Mech., 18 (1951) 293-297.
10. N. A. Weil and I. M. Daniel, J. Am. Ceram. Soc., Vol.47, No.6 (1964) 268-274.
11. J. Lamon and A. G. Evans, J. Am. Ceram. Soc., Vol.66, No.3 (1983) 177-182.
12. G. W. Hollenberg, G. R. Terwilliger and R. S. Gordon, J. Am. Ceram. Soc., Vol.54, No.4 (1971) 196-199.
13. G. G. Trantina, J. Am. Ceram. Soc., Vol.62, No.7-8 (1979) 377-380.
14. M. Srinivasan, "Elevated Temperature Stress Rupture Pesponse of Sintered Alpha Silicon Carbide", Technical report of The Carborundum Co., presented at the 81st Annual Meeting of the American Ceramic Society, Cincinnati, Ohio, April 27-30, 1979.
15. G. D. Quinn, AMMRC TN 81-4, (1981).
16. R. Kossowsky, D. G. Miller and E. S. Diaz, J. Mater. Sci., 10 (1975) 983-997.
17. F. F. Lange, E. S. Diaz and C. A. Andersson, Am. Ceram. Soc. Bull., Vol.58, No.9 (1979) 845-848.
18. R. K. Govila, J. Am. Ceram. Soc., Vol.65, No.1 (1982) 15-21.
19. R. K. Govila, J. Mater. Sci., 18 (1983) 1967-1976.
20. M. Kawai, H. Fujita, Y. Kanki, H. Abe and J. Nakayama, "Tensile Testing of Sintered Silicon Carbide and Silicon Nitride", in Ref. 5.
21. S. Prochazka, 239-252 in Ref. 1.
22. K. Suzuki, M. Sasaki, "Pressureless Sintereing of Silicon Carbide", Proceedings, Japan - US Seminor on Fundamental Structural Ceramics, (1984) U.S.A..
23. A. G. Evans, J. Mater. Sci. 7 (1972) 1137-1146.
24. S. M. Wiederhorn, 633-663 in Ref. 1.
25. A. G. Evans and F. F. Lange, J. Mater. Sci., 10 (1975) 1659-1664.
26. H. Tanaka and Y. Inomata, J. Ceram. Soc. Japan, Vol.93, No.1 (1985) 55-60.

EFFECT OF SURFACE CORROSION ON GLASS FRACTURE

B. C. Bunker and T. A. Michalske

Sandia National Laboratories
Albuquerque, NM 87185

ABSTRACT

The nature of aqueous corrosion reactions on glass surfaces varies with glass composition and environmental conditions. We have shown how the physical effects of the various surface reactions can be incorporated into simple fracture mechanics models to predict measured crack velocity effects in silicate glasses. Reactions on glass surfaces are divided into four categories:

1) Bond rupture reactions - Crack tip bond rupture reactions in silica are modeled through a combination of theoretical and experimental surface reaction studies. The atomistic fracture model which we developed predicts both the stress dependence and environmental dependence of measured crack growth behavior in vitreous silica.

2) Alteration layer formation - Aqueous corrosion reactions alter the chemical composition and physical properties of glass surfaces. Sodium silicate glasses develop a surface tensile layer when exposed to aqueous environments. This tensile layer adds to the crack tip stress intensity to produce a low velocity plateau in the crack growth curve. Phase separated sodium-borosilicate glasses undergo selective leaching to produce a porous reaction layer. The porous layer around the crack tip reduces the stress which is transferred to the crack tip to produce a fatigue limit.

3) Surface deposition - Preferential deposition of silica at the crack tip can increase the radius of curvature and reduce the crack tip stress concentration. Measured silica deposition rates predict the fatigue limit for soda-lime-silica glass.

4) Intermolecular surface forces - Attractive forces between surfaces decrease the crack tip stress and lead to crack healing. Crack healing measurements for vitreous silica and soda-lime-silica glass can be explained by the formation of hydrogen bonded linkages of surface adsorbed water molecules and electrostatic bonds between anionic surface groups.

These results demonstrate the effect of chemical reaction mechanisms on the long term mechanical behavior of glasses.

*This work performed at Sandia National Laboratories supported by the U.S. Department of Energy under contract number DE-AC04-76DP00789.

INTRODUCTION

It is well known that ceramics and glasses are susceptible to static fatigue which degrades the material's strength with time under load. Since fracture mechanics studies have shown that static fatigue can be attributed to the slow extension of preexisting surface flaws, it is important to determine what factors control the rate at which surface flaws grow to cause catastrophic failure. Controlled crack growth studies in glass have shown that the crack velocity (V) depends both on the applied stress intensity (K_I) and the chemical environment. The measured crack growth behavior is often expressed as a power law relationship of the form $V = AK_I^n$ where A and n are fitting parameters. (Over the small range of stress intensity measured, power law and exponential behavior are indistinguishable). Once a crack growth curve has been established and the representative flaw size determined, fatigue lifetimes can be calculated according to a procedure developed by Wiederhorn[1]. Substituting a relationship for K_I as a function of applied stress S and crack size C into the crack growth relationship and integrating with respect to time yields a time to failure (t_f) given by:

$$t_f = 2K_{Ii}^{2-n}/AS^2Y^2(n-2) \tag{1}$$

where K_{Ii} is the initial stress intensity and Y is a geometric parameter describing the flaw. In practice, crack velocities can be measured as low as 10^{-10} m/s. This data can be used to predict a lifetime on the order of one hour for silica glass in water with a 10 μm initial flaw size. The prediction of more useful lifetimes (10 years) requires an extrapolation of measured crack growth data to velocities of 10^{-15} m/s. The inherent danger in making such an extrapolation is the possibility of a mechanism change which would cause the actual crack velocity to deviate from the measured power law relationship.

Crack growth studies in multicomponent glasses have shown that deviations from power law behavior can occur at low stress intensity (see Fig. 1). In soda-lime-silica glass the slope of the crack growth curve steepens markedly at K_I<0.3 $MPam^{1/2}$. A previous study[2] has suggested that this marks the onset of a fatigue limit below which no crack extension can occur. Alternatively, data obtained for crack growth in binary soda-silica glasses[3] shows a slope decrease at low stress intensity tending toward an apparent plateau in the crack growth curve. This behavior is particularly undesirable since it suggests that the fatigue effect will be present at any applied stress level.

In order to predict the low stress intensity behavior of glasses like vitreous silica, it is necessary to understand what causes the observed deviations from power law behavior in other less durable glasses. Since crack velocities depend on the chemical environment, it suggests that a study of crack tip reactions would be useful in determining the crack growth mechanisms. Unfortunately, it is very difficult to directly observe the crack tip chemistry due to the geometric constraints that the crack imposes. However, if we consider that the crack tip is a region of freshly created surface we can use results of glass surface reaction studies to provide insight into the types of reactions likely to occur.

In this paper we will relate some of the observed crack growth effects to chemical reactions which occur on glass surfaces. After reviewing some results of surface reaction studies, we will show how these chemical phenomena can alter the mechanics of fracture to produce the various effects described in Figure 1.

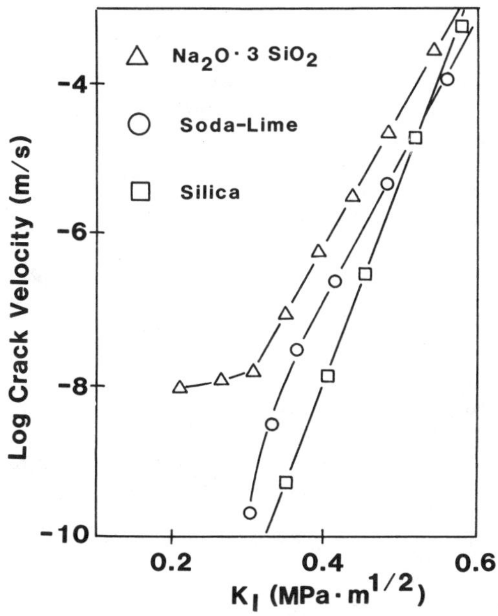

Fig. 1. V vs K_I diagram showing the stress corrosion behavior of three different silicate glasses in water.

REACTIONS ON GLASS SURFACES

A wide range of phenomena can occur when glass surfaces are put in contact with aqueous environments. The specific reactions depend on the glass composition as well as the solution pH. It is not the intent of this paper to explore the details of glass corrosion but rather to present the types of corrosion phenomena which have been suggested in the literature. Four categories of glass surface/aqueous solution interactions will be discussed:

1) Bond rupture reactions
2) Alteration layer formation
3) Surface deposition
4) Intermolecular surface forces

In discussing each category, the implication of the surface reaction to the stress corrosion behavior will be discussed and an example glass composition will be cited.

1) Bond rupture reactions

On an atomic scale, the extension of a crack must involve the rupture of individual chemical bonds. Therefore, a chemical reaction which reduces the strength of those bonds may increase the rate of crack extension. Such a crack tip interaction has been proposed to explain the effect of water on the crack growth rate in silica glass[4]. According to this model water attack at a crack tip bond in silica occurs as a three step process (see Fig. 2). Step (1): A water molecule is adsorbed at the strained crack tip bond. Step (2): The Si-O bond and the O-H bond in the adsorbed water cleave to form two hydrogen bonded silanol groups. Step (3): The remaining hydrogen bond cleaves to leave fracture surfaces covered by silanol groups.

Although this model provides considerable insight into the chemical nature of active stress corrosion species (the molecule must have at least one lone electron pair and a labile proton separated by approx. 0.16 nm), it can not explain why water is so aggressive toward crack tip bonds when adsorption isotherms on silica show that the majority of Si-O surface bonds are hydrophobic[5]. This apparent contradiction can be attributed to the effect of mechanical strain on bond reactivity.

Theoretical calculations and experimental surface reaction studies have been used[6] to determine the effect of Si-O bond strain on its reactivity with chemical species such as water. Molecular orbital (MO) calculations for silica[7-10] suggest that bond angle changes are energetically favored over bond length changes and that decreases in the Si-O-Si angle or the O-Si-O angle can dramatically enhance the acid/base properties of silica thereby making the chemical attack mechanism described in Figure 2 feasible.

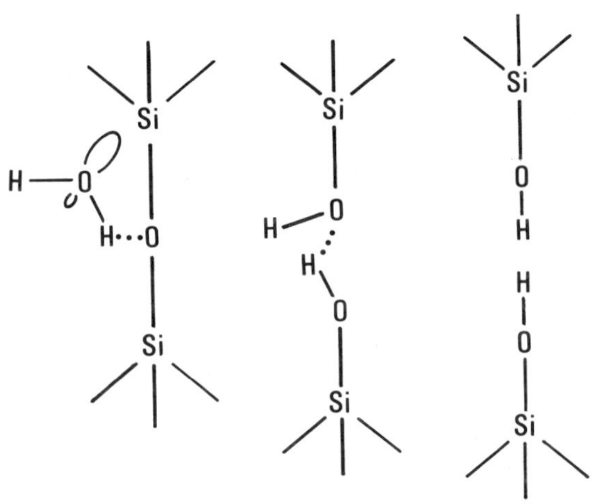

Fig. 2. Chemical reaction proposed by Michalske and Freiman for environmentally enhanced Si-O-Si bond rupture. The strained crack tip bond (1) adsorbs a water molecule, (2) is cleaved by dissociative reaction, and (3) is converted to surface silanol groups.

An experimental confirmation of this strain enhanced reaction mechanism is achieved by studying reactions with strained model compounds such as defect surface sites on silica. Infrared studies[11-13] suggest that edge shared tetrahedral defects are created on silica surfaces dehydroxylated at T > 400°C. These defects possess bond angle strains predicted (by MO calculations) to greatly enhance the acid/base properties of the Si-O bond and make it susceptible to dissociative chemisorption reactions with species like water. Experimental reaction studies[14] show that strained surface defects act as strong acid/base sites which react with chemicals such as water approx. 100,000 times faster than do unstrained siloxane bridges. Some chemicals which react with the defect, and the products of these reactions as determined by IR spectra are listed below:

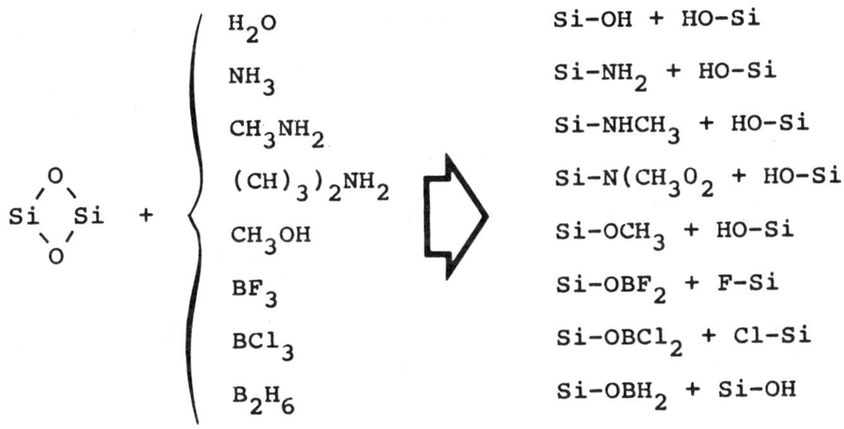

We have shown that chemicals which dissociatively chemisorb on strained surface defects also promote stress corrosion in vitreous silica (see Table I). Alternatively, chemicals which only physisorb to the strained surface defect do not actively promote stress corrosion. This suggests that similar chemical features exist for strained crack tip bonds and strained surface defects.

Using results from MO calculations for silica we have estimated the bond distortions which may occur in a high strain field such as the crack tip and determined the significance of these distortions to bond reactivity. Figure 3 shows the sequence of events expected to occur as a strain free ring in the silica structure is converted to a chemically active crack tip site. The first increment of strain may only require a ring conformation change which will not alter any of the bond angles or distances. The next step involves the extension of the Si-O-Si angles to 180. This distortion results in small energy changes (4kcal/mole) and has little effect on the chemistry of the silica bond. Si-O bond length extension requires a large amount of energy, but has only moderate impact on reactivity and is not expected to be important to the formation of a chemically active site. However, changing the tetrahedral angle has drastic effects on the molecular-orbital configuration and, as shown by the strained surface defect, greatly increases bond reactivity. Once the active site forms, molecules (e.g., H_2O, NH_3) can adsorb and react to cleave the bridging siloxane bond.

We have also shown that the process of active site formation at the crack tip lends itself to a quantitative model for stress corrosion based on

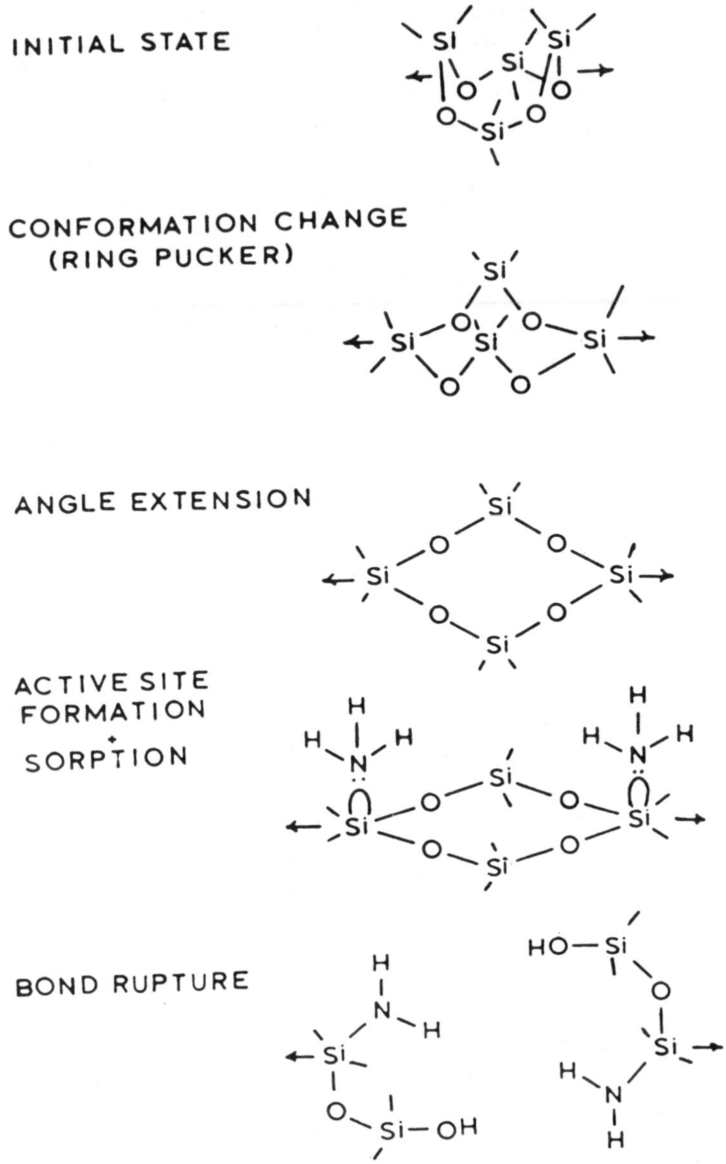

Fig. 3. Stages of silica ring deformation leading to formation of chemically active kink site: (1) initial, (2) conformation change, (3) Si-O-Si angle extension, (4) O-Si-O angle pinching plus adsorption, (5) bond rupture.

kink nucleation and growth (see Fig. 4). According to this model, the crack tip strain field produces some fraction of kink sites (f^*) which are distorted enough to enhance their reactivity. The rate at which these sites are cleaved is dependent upon the fraction of active sites which have adsorbed molecules (f^A) multiplied by the rate of the dissociative chemisorption reaction (k). As one kink site is cleaved, the strain is

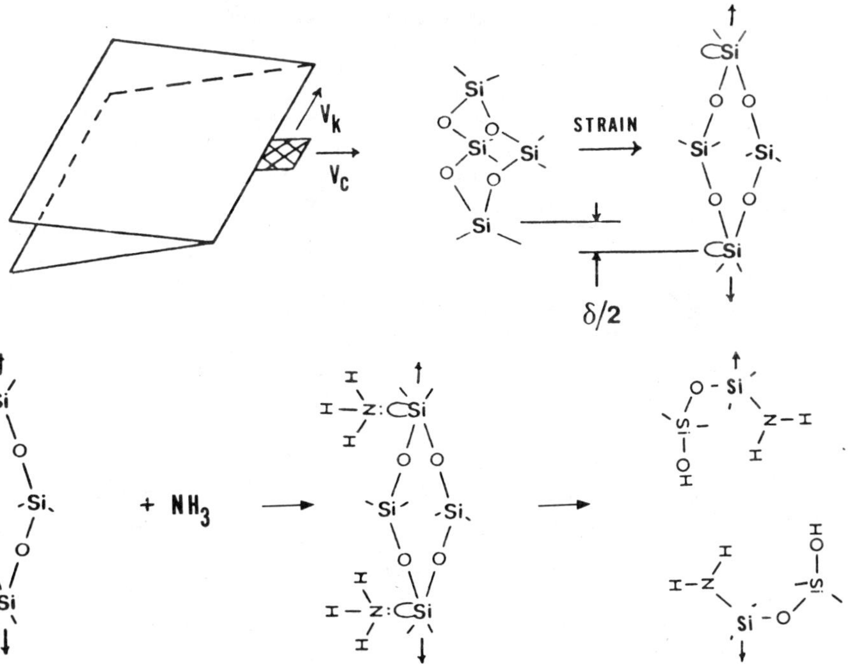

$$V = \frac{f^* \times f_A \times k}{N_A}$$

Fig. 4. Kink nucleation and growth model for crack growth in silica. Crack tip strain field converts a fraction of crack tip bonds (f^*) into chemically active sites. A fraction of those bonds (f_A) adsorb environmental species and react at rate (k) to yield macroscopic crack velocity (V). N_A is the density of bonds on the crack plane.

TABLE I. COMPARISON OF SLOW FRACTURE AND SURFACE ADSORPTION ON SILICA

Substance	Crack Growth	Surface Adsorption
H_2O	active*	active*
NH_3	active*	active*
CH_3OH	active*	active*
CH_4	inactive±	inactive±
O_2	inactive±	inactive±
CO	inactive±	inactive±
CH_3CN	inactive±	inactive±

*active - causes slow fracture; exhibits dissociative chemisorption
±inactive - doesn't cause slow fracture; exhibits physisorption

transferred to its neighbors increasing their probability of becoming chemically active. In this fashion the kinks move laterally to extend the macroscopic crack at the rate V. This atomistic model for stress corrosion crack growth predicts that the strain dependence of the crack velocity only involves the generation of kink sites and should depend on the properties of the silica network and not on the interaction with environmental species. The confirmation of this prediction is shown in Figure 5 where all of the active chemicals yield crack growth curves with the same slope. (Previous work by Wiederhorn[15], et. al., has shown that the slope relates directly to the strain dependence of the crack tip reaction). The kink nucleation and growth model also predicts that molecular adsorption of the attacking species is a precursor rate limiting step to crack growth. A recent experimental study by Fisk and Michalske[16] used data for thermal effects on the crack growth rate to confirm the importance of physisorption to the stress corrosion process.

The results presented above show that environmentally controlled bond rupture reactions can be modeled through a combination of theoretical and experimental surface reaction studies. The atomistic fracture model which we developed can be used to explain both the stress dependence and environmental dependence of measured crack growth behavior in vitreous silica. However, the bond rupture model does not explain the deviations from power law behavior observed in more complex glasses at low crack velocities. To explain such deviations, we need to consider reactions in addition to those which occur at the immediate crack tip.

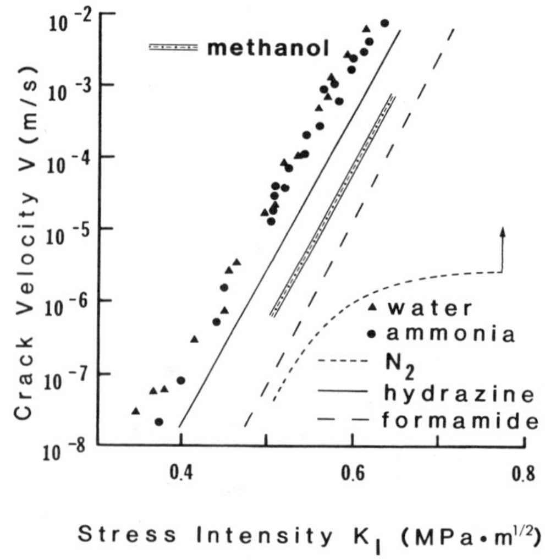

Fig. 5. V vs K_I diagram for vitreous silica showing that all surface active chemicals yield the same slope.

2) Alteration layer formation

For vitreous silica, reactions with water occur at the immediate glass surface, leading to congruent dissolution. However, for most other glasses, selective leaching via ion exchange or network hydrolysis, surface hydration, and network repolymerization reactions lead to the formation of surface alteration layers. The chemically altered layers have mechanical properties which are distinct from the bulk glass, changing the mechanics of fracture.

The formation of alteration layers on glass surfaces can create surface stresses which can influence slow crack growth. For binary alkali silicate glasses, tensile stresses in leached layers can become large enough to cause surface crazing[17]. The tensile stresses have been attributed to a volume contraction associated with repolymerization of silanol groups in the leached glass. Bunker, et. al.[17], have demonstrated that repolymerization occurs after the silanol groups are created during selective alkali leaching via ion exchange. Since rate of ion exchange increases as the solution pH decreases, the formation of tensile leached layers on glass surfaces occurs faster in acidic solutions than in neutral or basic environments.

The formation of a tensile layer on the fracture surface can contribute to a crack tip stress intensity which is independent of the applied stress. Borrowing from the analysis of McMeeking, et. al.[18], and Budiansky, et. al.[19], for phase transformation toughened brittle materials, we can treat the dealkalized layer as a transformed zone on the crack surfaces. The expression derived to determine the magnitude of stress intensity (K_{Ir}) due to the stress in the reaction layer (S_r) is:

$$K_{Ir} = -0.64\, S_r\, H^{1/2} \qquad (2)$$

where H is the reaction layer thickness. Since the stress in the dealkalized layer is tensile, the resulting stress intensity adds to the applied stress intensity K_{IA} to produce a crack tip stress intensity given by:

$$K_{Itip} = K_{Ir} + K_{IA} \qquad (3)$$

TABLE II. STRESS INTENSITY RESULTING FROM ALTERATION LAYER ON SODA-SILICA GLASS

V*(m/s)	d**(μm)	K_{Ir} (MPa·n$^{1/2}$)	K_{Ir}/K_{IA}
1.0×10^{-6}	0.063	0.007	0.017
1.0×10^{-7}	0.2	0.013	0.037
1.0×10^{-8}	0.63	0.024	0.08
1.0×10^{-9}	2.0	0.044	0.176

* V crack velocity
** d alteration layer thickness

Qualitatively, the generation of a stressed surface layer can be used to interpret the crack growth results obtained by Simmons and Freiman[3] for soda-silica glass (see Fig. 1). In water at a stress intensity of approx. 0.3 $MPam^{1/2}$ they observed a marked decrease in the slope of the crack growth curve. This means that the crack was moving faster than would be predicted by an extrapolation of the high stress intensity data. If a tensile layer was forming at these low stress intensities (and low crack velocities) it could add to the crack tip stress intensity to increase the measured crack velocity. This explanation for the low velocity behavior is consistent with the pH effect observed for crack growth in these glasses (see Fig. 6). Simmons and Freiman found that the velocity at which the plateau occurred increased with decreasing solution pH. Since the rate of dealkalization increases in more acidic solutions, it is expected that the effect of the tensile layer would be observed sooner at low pH. Simmons and Freiman argued that the low velocity plateau was a measure of the inherent corrosion rate of the silica network. However, their model is contradicted by data[20] which shows that the rate of silica release during alkali silicate glass dissolution decreases with decreasing solution pH. In addition, the surface stress model is able to explain the decreased plateau velocity obtained when salts were added to the aqueous solution. Figures 1 and 6 show that the plateau velocity decreases an order of magnitude with 1M $LiNO_3$ or $CsNO_3$ additions. This result agrees with work by McGrail[21] showing that salt additions decrease the rate of stressed layer formation.

Quantitative predictions based on the surface stress model require knowledge of the rate of silica rich layer formation. McGrail[21], et. al., have measured the rate of sodium release from a sodium tri-silicate composition at pH values from 2.3 to 10 and 22°C. Their results show that the alkali release follows a $t^{1/2}$ behavior. Using the measured alkali release rate at pH 7 and assuming that all of the alkali is removed from the surface layer, the thickness of that layer can be determined as a function of time. In order to calculate a reaction layer profile, we model the moving crack tip as though it were fixed in position and the crack walls were moving away. The distance away from the tip can be converted to a time for which that portion of the surface has been exposed to the aqueous solution. Figure 7 shows the silica rich layer generated as a function of

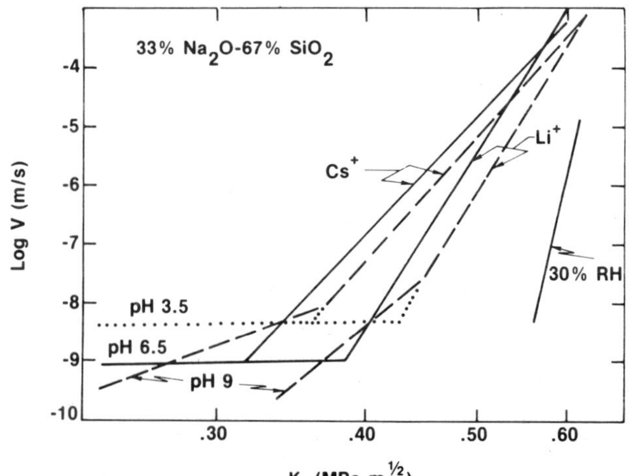

Fig. 6. V vs K_I diagram showing effect of solution pH on low velocity plateau in sodium silicate glass. (after Simmons and Freiman[3])

the crack velocity calculated for several different values of V. In addition to calculating the reaction layer profile, we have also determined the extent of layer formation in advance of the moving crack tip. This is accomplished by setting the first derivative of relation for reaction layer formation equal to the crack velocity of interest. It can be seen that crack velocities greater than or equal to 1.0×10^{-6} m/s yield a reaction layer on the crack wall near the tip which is less than one Si-O bond distance.

In order to calculate stress intensities resulting from these reaction layers, we need to know the magnitude of the stress generated. No experimental measurements for these reaction layer stresses exist so we have assumed that S_r is 70 MPa which is the value normally associated with the strength of glass. Taking H as 1/2 the deepest extent of our reaction zone as suggested by Tada[22] for such zone profile, we can determine the K_{Ir} as a function of crack velocity (see Table II). Table II also gives the ratio of K_{Ir} to the K_{IA} determined by extrapolating the high stress intensity portion of the crack growth curve. These results indicate that the effect of the tensile surface layer is negligible (< 5% in total stress intensity) for crack velocities above 1.0×10^{-7} m/s. Crack velocities calculated from K_{Ir} values in Table II are compared to the experimental crack growth curve in Fig. 8. (Calculated values of V were obtained by substituting K_{Itip} determined from Equation 3 into the power law relationship observed at higher stress intensities). Although these calculations are in reasonable agreement with the measured velocities, they do not represent a rigorous treatment since we do not presently have an experimental measure of the surface stress resulting from the corrosion process. However, the stressed surface model for crack growth in sodium silicate glass does predict the dependence of the plateau velocity on both pH and dissolved salt concentrations.

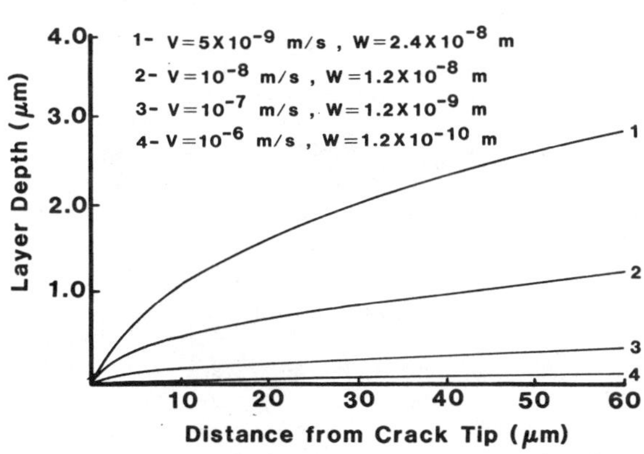

Fig. 7. Alteration layer depth calculated for fracture surface of sodium tri-silicate glass in D.I. water. The crack velocity is given by (V) and (W) is the extent of layer formation ahead of the crack tip for each velocity treated.

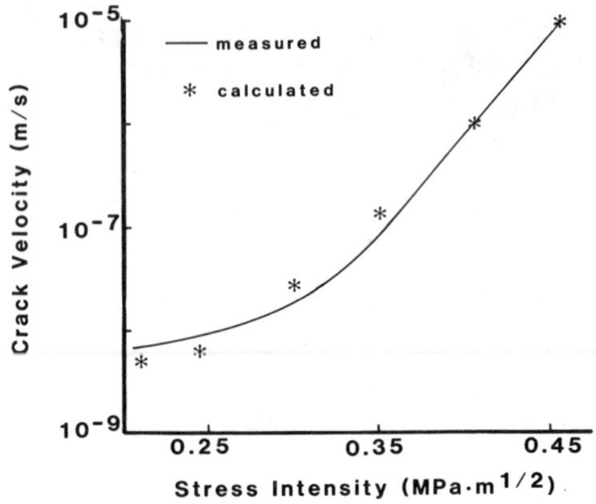

Fig. 8. V vs K_I diagram comparing experiment measurements for sodium tri-silicate glass tested in water[3] with calculated results derived from surface stress model.

Stresses generated in leached glass surfaces are not always tensile. Selective leaching of the sodium borate phase from phase separated sodium borosilicate glasses can form swollen, silica rich, leached layers which produce compressive surface stresses[23]. The equation for crack tip stresses used above for the case of tensile stresses can be used when compressive stresses are present. Calculated crack growth data for phase separated borosilicate glass (Fig. 9) indicate that compressive surface stresses could produce a fatigue limit in crack velocity data. For the phase separated borosilicate glass studied by Weiderhorn[24] ($80SiO_2$, $14B_2O_3$, $4Na_2O$ and $2Al_2O_3$, wt%) the calculated crack velocity curve for glass leached in acidic solutions (based on a corrosion rate of 1×10^{-8} m/s) agrees well with the experimental data. The effects of compressive surface stresses leading to a fatigue limit are apparent at velocities below 10^{-6} m/s. Deviations from power law behavior are observed at lower crack velocities for the same glass leached in neutral pH solutions, which is consistent with the slower rate of leached layer formation (6×10^{-10} m/s) expected for neutral pH's. Unfortunately, this analysis also suffers from a lack of surface stress measurements; we assumed 70 MPa of compressive stress for these calculations. Additionally, structural changes in alteration layers suggest another mechanism which can also contribute to a limit, as described below.

Phase separated sodium borosilicate glasses provide systems for studying how the introduction of a porous zone at the crack tip influences slow crack growth. When phase separated borosilicate glasses dissolve, the sodium borate-rich phase is selectively leached to produce a porous surface layer consisting of the silica rich phase[25,26]. If the glass composition

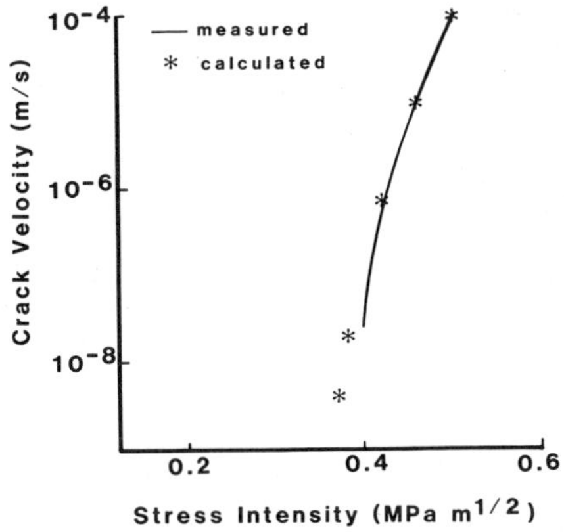

Fig. 9. V vs K_I diagram comparing experimental measurements for sodium borosilicate glass tested in acidic solution[24] with calculated resultes derived from surface stress model.

and degree of phase separation are known, the rate of alteration layer formation can be estimated based on extensive leaching data in the literature[25,26]. In all cases, leaching occurs more rapidly in acidic than in neutral pH solutions.

In crack growth experiments, if the leached layer grows faster than the crack, a porous zone is predicted to be produced in front of the growing crack. In essence, a porous layer acts as a low modulus cladding around the crack tip. This cladding effectively shields the embedded crack tip from the full extent of the applied stresses thereby reducing the crack tip stress intensity. As the layer formation rate approaches the crack growth rate, the crack tip will be partially shielded and the actual crack tip stress intensity will decrease. This stress intensity decrease will manifest itself as a velocity decrease so that the observed crack growth curve will steepen and tend toward a fatigue limit.

Simmons and Freiman[3] measured the crack growth behavior of a 7 Na_2O, 23 B_2O_3, 70 SiO_2 (mole %) glass that was heat treated to induce phase separation. The crack growth curves obtained at pH 7 (see Fig. 10) indicate that crack arrest occurs at stress intensities < 0.35 $MPam^{1/2}$. Following crack arrest it was very difficult to restart the crack without causing catastrophic failure. A later analysis of the fracture generated surfaces[27] showed that the aged crack had to repropagate through a porous zone that formed ahead of the original crack tip (see Fig. 11). These findings are consistent with a reduction in crack tip stress due to the formation of a porous layer surrounding the crack tip. In this case, the applied stress intensity necessary to repropagate the crack through the porous layer was apparently greater than K_{IC} for the virgin glass so that catastrophic

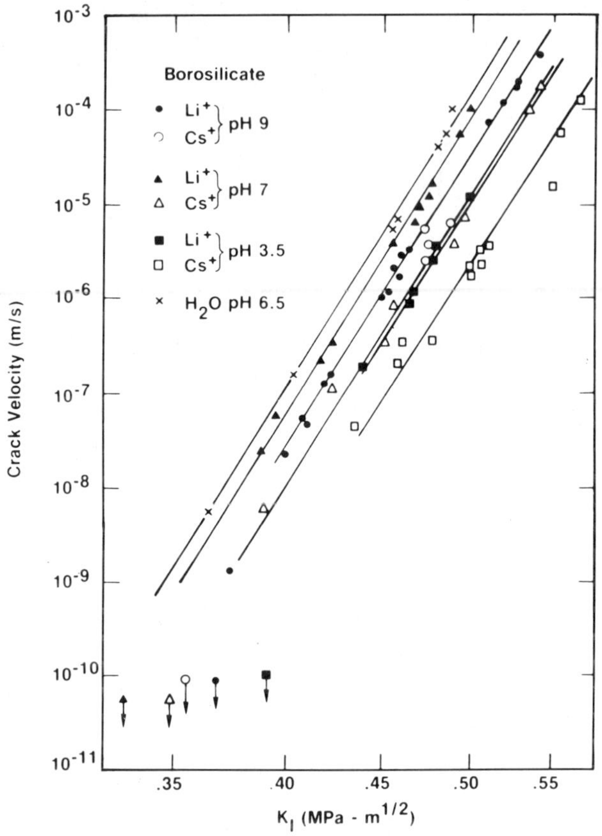

Fig. 10. V vs K_I diagram showing the onset of a fatigue limit in a phase separated sodium borosilicate glass. (after Simmons and Freiman[3])

failure was observed. The fracture rate for the onset of crack arrest ($< 1.0 \times 10^{-9}$ m/s) is in reasonable agreement with the measured rate of porous layer formation determined from corrosion experiments[26].

For the case of phase separated borosilicate glasses, the creation of a porous zone at the crack tip appears to dominate the low stress intensity behavior; however, a compressive surface layer may also contribute to the observed fatigue limit. Detailed measurements of alteration layer properties will be required to determine the relative importance of these crack growth mechanisms.

3) Surface deposition

Another type of chemical reaction which can alter the nature of a glass surface is precipitation from solution. Iler[28] has reviewed data for the solubility of silica in aqueous solution and shows that it depends strongly on particle size. With decreasing radius of curvature, the solubility increases whereas negative curvature decreases solubility (see Fig. 12). This effect is used to explain the ripening of silica particles in solutions of pH 9-10 and the aggregation of silica colloids by neck formation. In these processes, silica is preferentially dissolved from the more positively curved regions and deposited on the more negatively curved areas. Since the crack tip represents an area which has an extremely small negative radius of curvature, it is reasonable to expect that silica which is dissolved from the flat fracture surface will tend to deposit at the tip.

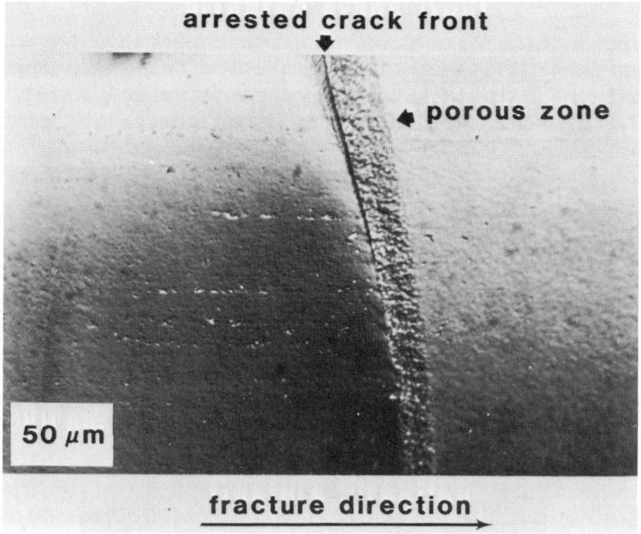

Fig. 11. Fracture surface showing crack arrest and subsequent repropagation of crack in phase separated sodium borosilicate glass tested in water. Following arrest, crack repropagated through porous region resulting from selective leaching by aqueous solution.

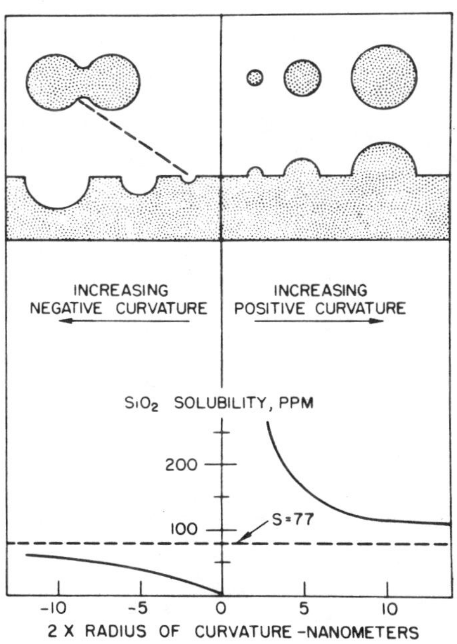

Fig. 12. Effect of radius of curvature on solubility of silica in water. Negative curvature decreases solubility. (after Iler[28])

The deposition of silica from solution can change the curvature of the crack tip. Inglis[29] used a continuum solution for the stress around an elliptical opening to approximate the effect of crack tip curvature on the crack tip stress field. In this continuum approach, the stress at the crack tip is given by:

$$S = 2\ S_a\ (C/P)^{1/2} \tag{4}$$

where S is the crack tip stress, S_a is the applied stress, C is the crack length and P is the radius of curvature of the crack tip. Assuming that some critical value of crack tip stress is required for growth, an increase in crack tip radius at constant crack length will necessitate a larger applied stress for fracture.

In an attempt to test this hypothesis, Ito and Tomozawa[30] measured the strength of precracked vitreous silica bars which had been soaked in water or saturated $Si(OH)_4$ for various times. Their results (see Fig. 13) showed that a strength increase was observed after soaking at 88°C. Although the strength increase was the same for water and $Si(OH)_4$ solution, the samples soaked in water were slower to achieve the full effect. The experimental results are consistent with a dissolution/reprecipitation process. The samples soaked in water were slower to strengthen because silica had to first dissolve from the crack walls before it could be precipitated at the tip. It was hypothesized that the limit in strengthening could be attributed to the decrease in deposition rate with increasing crack tip radius. In a more recent study, Bando, et. al.[31], used high-resolution electron microscopy to investigate silica deposition at crack tips. A thin glass film was precracked then soaked in 90°C water prior to examination. After soaking, a silica deposit was observed which changed the crack tip radius from 1.5 nm to 5.0 nm. This radius change could explain the strength effects reported above.

However, the significance to glass fracture of a dissolution/reprecipitation mechanism depends upon the nature of the silica deposit that is formed. The actual crack tip deposit must be rigid and capable of supporting stress, and not an incompletely cross linked gel material. To date, the strength of crack tip deposits has not been determined.

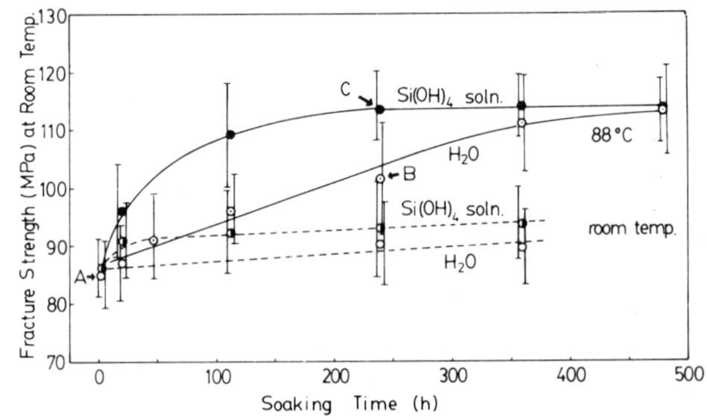

Fig. 13. Strength increase observed when abraded silica glass is soaked in either water or saturated $Si(OH)_4$ solution. (after Ito and Tomozawa[30])

4) Intermolecular surface forces

Studies on colloidal silica systems show that surface adsorbed species can produce intermolecular surface forces which may either attract or repel neighboring surfaces. These forces may arise from a variety of interactions including Van der Waals attractions, hydrogen bonding, electrostatic fields or network repolymerization. Although these forces act over very short distances (< 1.0 nm), they may contribute to the stresses at the crack tip where the crack wall separation approaches zero. If the surface forces are repulsive, they would tend to increase the crack opening and thereby increase the crack tip stress intensity. One may expect that such a condition could result in a slope decrease at low applied stress intensity where the crack walls are closer together. Alternatively, attractive surface forces tend to close the crack opening and decrease the crack tip stress intensity. In this case one would expect an increase in the slope of the V - K_I curve at low stress intensity followed by crack closure.

Crack closure in silicate glass has been reported by several investigators[32,33,34]. A recent study by Michalske and Fuller[35] attempted to relate the measured crack closure and repropagation energies to specific surface interaction mechanisms. In this study, samples of vitreous silica and soda-lime-silica glass were cracked part way in controlled humidity environments. After cracking, the applied stress was reduced until the crack surfaces healed together. At the onset of crack healing, the crack tip strain field (as observed in polarized light) moved back along the fracture indicating that the healed surfaces were supporting stress. At humidities of 20% or greater, cracks in vitreous silica and soda-lime-silica glass healed against an applied load with a closure energy of 0.15 J/M^2. It was shown that this closure energy can be predicted by the formation of hydrogen bonded linkages between surface adsorbed water molecules (Fig. 14a). This model predicts and experiments confirm that relative humidities < 15%, which correspond to submonolayer surface coverage of water, result in decreased closure energies. The decrease in closure energy corresponds to a decrease in the number of hydrogen bond linkages available. It was also shown that the repropagation energy for healed cracks in vitreous silica corresponded to the rupture energy of the hydrogen bonds. In the driest environments used (0.01% RH) healed cracks in soda-lime-silica glass required 1.7 ± 0.2 J/m^2 to reopen. This increased surface attraction was attributed to the formation of cationic bridges between deprotonated silanol surface groups and sodium ions present in the glass (see Fig. 14b). A prediction of the fracture energy for cationic bridges was made by analogy

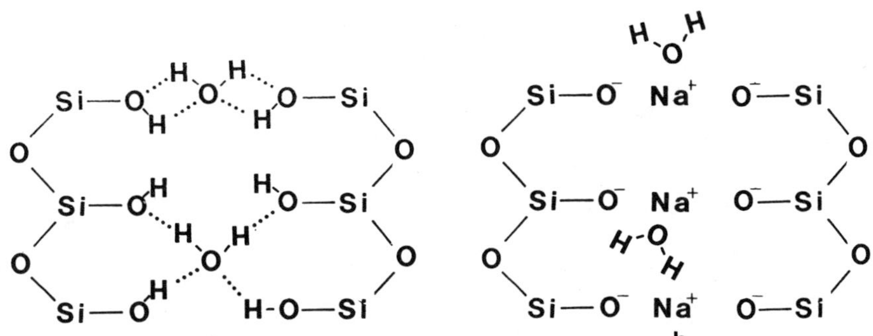

Fig. 14. Proposed mechanisms for crack healing in silicate glass; (a) hydrogen bonding through adsorbed water molecules on vitreous silica and (b) cationic bridging in soda-lime-silica glass.

to the cleavage energy for mica crystals and shown to be 1.75 J/m^2 in good agreement with the measured crack repropagation energies.

From a structural design viewpoint, the onset of crack closure is very significant since it marks the stress intensity below which no further crack extension can occur. The measurements discussed above show that cracks in vitreous silica or soda-lime-silica glass in humid air will not grow when the applied stress intensity is < 0.11 $MPam^{1/2}$. Unfortunately, this result is of little practical consequence since the crack velocity which is expected in the absence of surface forces is < 1.0 X 10^{-30} m/s. However, the healing force attributed to cationic bridging in the alkali containing glass was observed to eliminate the growth of cracks reloaded to stress intensities < 0.35 $MPam^{1/2}$. This increase in the crack healing energy could be important for systems which experience a cyclic loading history. In this case, the healing of cracks during unloading could retard crack extension when the load is increased.

The results reviewed here show that attractive interactions between crack surfaces can decrease the crack tip stress intensity and that the magnitude of these closure forces can be predicted from a knowledge of the species present on the glass surface. In the future it will be interesting to determine whether other types of surface interactions, e.g., double layer forces, can alter the crack tip stress intensity to produce crack arrest or healing. Specifically, it is important to know if repulsive forces can increase the crack tip stress intensity to produce a low velocity plateau as seen in the binary alkali silicate compositions.

FATIGUE LIMIT IN SODA-LIME-SILICA GLASS

In this final section we will show how the concepts developed in the previous sections can be used to interpret the fracture behavior of a complex commercial glass composition (soda-lime-silica plate glass). Crack growth data for soda-lime-silica glass fractured in water[36] (see Fig.1) displays an increased slope at low stress intensity indicating a fatigue limit. In addition, Michalske[2] determined that cracks in soda-lime-silica glass aged at low stress intensities in water required some incubation period before repropagating at higher stress intensity levels. As discussed in the previous sections, this type of fracture behavior can be a result of several different glass surface/aqueous solution interactions. We will attempt to determine which of those mechanisms best explains the data for this glass composition.

To determine if the slope increase and incubation effect can be attributed to alteration layer formation, we need to determine the character of the layer and its rate of growth. Lanford, et. al.[37], have used elemental depth profiles to determine that a hydrated alteration layer forms on the soda-lime-silica surface exposed to aqueous solutions. The data suggests that layer formation is a diffusion controlled process with an effective diffusion coefficient of 9 X 10^{-16} cm^2/sec. Tomozawa[38] has shown that the incorporation of molecular water into silicate glasses via steam hydration can produce a compressively stressed, low modulus surface layer. These results suggest that either a hydrated, low modulus crack tip zone or a compressively stressed surface behind the crack could give rise to the observed fracture behavior.

In order to produce a fatigue limit via the formation of an alteration layer at the immediate crack tip, the rate of layer growth needs to be fast enough to stay ahead of the advancing crack. Our analysis suggests that a significant hydrated zone (thicker than 0.4 nm) will not be produced for soda-lime-silica glass until the crack velocity is below 10^{-10} m/s, which is substantially below the velocity (10^{-7} m/s) at which deviations from power law behavior are observed. Although this analysis indicates that alteration layer formation does not explain the presence of a fatigue limit for soda-lime-silica glass, the analysis is based on reactions which occur in unstressed glass specimens. For fused silica, Tomozawa[38] has diffusion data which suggests that the rate of surface hydration can be accelerated in the presence of tensile stresses. Using Tomozawa's estimate for the stress dependence of surface hydration and estimates of the stresses encountered at the immediate crack tip, the rate of layer formation could be increased two orders of magnitude at a distance of 0.5 μm from the tip. Therefore, the stress dependence of hydration in soda-lime-silica glass must be measured before crack tip hydration can be positively ruled out.

In terms of a fatigue limit controlled by the creation of compressively stressed crack walls, our analysis suggests that the highest velocity at which surface stresses would begin to influence slow crack growth is approximately 10^{-9} m/s. (This again assumes a stress of 70 MPa in the alteration layer). Since the rate of layer formation along the crack walls is not influenced by the crack tip stress, this estimate suggests that compressive stresses do not produce the fatigue limit in soda-lime-silica glass.

We can illustrate the implications of crack tip deposition mechanism by examining both corrosion and crack growth data for soda-lime-silica glass. The source of silica deposited at the crack tip is the silica released into solution from all glass surfaces exposed to the solution. If we assume that all silica in solution has access to the crack tip, then the rate determining step for deposition is the rate at which silica becomes available for deposition via glass dissolution. Using the dissolution data of Douglas, et. al.[39], for soda-lime-silica glass, we estimate that significant silica deposits could form in cracks moving slower than 10^{-7} m/s, which agrees with the onset of the fatigue limit. Although the agreement with crack velocity data is good, further experiments will be required to determine if the fatigue limit is due to crack tip deposition or stress enhanced alteration layer formation. This example serves to illustrate the need to understand the rate of reactions on glass surfaces as well as the physical properties of the reaction products in order to properly interpret the crack growth behavior.

SUMMARY

In this paper we have shown how various chemical reactions that occur on glass surfaces in aqueous or humid environments can alter the mechanics of glass fracture. A review of glass corrosion results show that many different surface effects can be generated depending upon glass composition and environmental conditions. We have selected some specific surface reactions and attempted to show how the physical effects of these reactions can be incorporated into simple fracture mechanics models. The fracture mechanics models were then used to predict previously measured crack velocity effects. In most cases there is insufficient data to adequately test these concepts, but we did demonstrate how these chemical effects can be used to predict trends in the experimental fracture data for silicate glasses.

The chemical reactions on glass surfaces were divided into four separate categories and example glass systems were chosen to demonstrate both the surface reaction results and corresponding fracture behavior for each. The following summarizes the results for each category:

1) Bond rupture reactions - Reaction studies for surface defect sites on silica show that bond angle strain can greatly enhance the rate of Si-O bond rupture reactions. Surface reactions can be used to predict the reactivity of highly strained crack tip bonds in silica glass and identify

active stress corrosion species. In addition, rates of surface defect reactions can be used to predict the rate of stress corrosion crack growth in vitreous silica through the use of a kink nucleation and growth fracture model. The model describes power law crack growth behavior which is observed at high crack velocities.

2) Alteration layer formation - Aqueous corrosion reactions can alter the chemical composition and physical properties of glass surfaces. Binary alkali silicate glasses develop a surface tensile layer when exposed to aqueous environments. This tensile layer behind the crack tip can contribute to the crack tip stress and increase the rate of crack extension to produce a low velocity plateau in the crack growth curve. This model for the fracture behavior is consistent with the solution pH effects both for glass surface corrosion rates and crack velocity effects. Phase separated sodium-borosilicate glasses undergo selective leaching to produce a porous reaction layer. The porous layer around the crack tip reduces the stress which is transferred to the crack tip and therefore reduces the crack velocity. This model is consistent with fracture measurements which show an increase in the slope of the V - K_I curve at low stress intensity.

3) Surface deposition - Solubility studies for silica show that deposition from solution is favored at regions of high negative curvature. This suggests (and reports in the literature confirm) that silica in solution can deposit at crack tips to change the radius of curvature. The crack tip stress is expected to decrease with decreasing curvature to produce an increase in fracture resistance with silica deposition. This effect was used to explain strength effects observed for precracked silica glass samples aged in saturated $Si(OH)_4$ solutions.

4) Intermolecular surface forces - Silica colloids demonstrate that surface adsorbed species can produce either attractive or repulsive forces between neighboring surfaces. Attractive forces between surfaces are expected to decrease the crack tip stress and eventually lead to crack healing. Crack healing measurements for vitreous silica and soda-lime-silica glass can be explained by the formation of hydrogen bonded linkages of surface adsorbed water molecules. Further attractive forces can be seen when surface adsorbed cations form electrostatic bonds between anionic surface groups. Presently no fracture results exist which demonstrate the effect of surface repulsive forces.

These results suggest that many fracture effects observed in silicate glasses can be understood by combining chemical corrosion results with fracture mechanics formulations. In order to make more quantitative predictions of the fracture behavior based on this approach, it will be necessary to measure the physical properties of the reaction layers more carefully so that the fracture mechanics expressions can be evaluated properly.

REFERENCES

1. W. M. Wiederhorn, J. Am. Ceram. Soc. 56:227 (1973).
2. T. A. Michalske, pp. 277 in "Fracture Mechanics of Ceramics," Vol. 5, R. C. Bradt, A. G. Evans, D. P. H. Hasselman and F. F. Lange, eds., Plenum Press (1983).
3. C. J. Simmons and S. W. Frieman, J. Am. Ceram. Soc. 64:683 (1981).
4. T. A. Michalske and S. W. Freiman, J. Am. Ceram. Soc. 66:284 (1983).
5. H. Naono, R. Fujiwara and M. Yagi, J. Colloid Interface Sci. 76:74 (1980).
6. T. A. Michalske and B. C. Bunker, J. Appl. Phys. 56:2686 (1984).
7. B. H. W. S. DeJong and G. E. Brown, Jr., Geochim. Cosmochim. Acta 44:491 (1980).
8. J. A. Jossell and G. V. Gibbs, Am. Mineral 61:287 (1976).
9. M. D. Newton and G. V. Gibbs, Phys. Chem. Mineral 6:221 (1980).
10. G. V. Gibbs, Am. Mineral 67:421 (1982).

11. B. A. Morrow and L. A. Cody, J. Phys. Chem. 80:1995 (1976).
12. B. A. Morrow and L. A. Cody, J. Phys. Chem. 80:1998 (1976).
13. B. A. Morrow and L. A. Cody, J. Phys. Chem. 80:2761 (1976).
14. B. C. Bunker, T. A. Michalske, W. L. Smith and D. Haaland, to be published.
15. S. M. Wiederhorn, E. R. Fuller, Jr. and R. Thomson, Metal Sci. J. 14:450 (1980).
16. G. A. Fisk and T. A. Michalske, J. Appl. Phys. 58:2736 (1985).
17. B. C. Bunker, G. W. Arnold and E. K. Beauchamp, J. Non-Cryst. Solids 58:295 (1983).
18. R. M. McMeeking and A. G. Evans, J. Am. Ceram. Soc. 65:242 (1982).
19. B. Budiansky, J. W. Hutchinson and J. C. Lambropoulus, Harvard Univ. Report, Div. of Appl. Sci, MECH-25 (1982).
20. T. M. El-Shamy, J. Lewis, and R. W. Douglas, Glass Tech. 13:81 (1972).
21. B. P. McGrail, L. R. Pederson, and D. A. Peterson, submitted to Phys. Chem. Glasses.
22. P. C. Paris, H. Tada and G. R. Irwin, "The Stress Analysis Handbook," Del Research, St. Louis (1973).
24. S. M. Wiederhorn and H. Johnson, J. Am. Ceram. Soc. 56:192 (1973).
25. P. B. Adams and D. L. Evans, Mater. Sci. Res. 12:525 (1978).
26. B. C. Bunker, G. W. Arnold, D. E. Day, and P. J. Bray, submitted J. Non-Cryst. Solids.
27. C. J. Simmons, S. W. Freiman and T. A. Michalske, presented at Surfaces and Interfaces in Ceramic and Ceramic-Metal System, Univ. of California at Berkeley, July 28-Aug. 1, 1980.
28. R. K. Iler, "The Chemistry of Silica," Wiley, New York (1979).
29. C. E. Inglis, Trans. Inst. Naval Archit. 55:219 (1913).
30. S. Ito and M. Tomozawa, J. Am. Ceram. Soc. 65:368 (1982).
31. Y. Bando, S. Ito and M. Tomozawa, J. Am. Ceram. Soc. 67:C-36 (1984).
32. V. M. Finkel and L. A. Kutkin, Doyl. Akad. Nauk SBR. 143:90 (1962): Sov. Phys.-Dokl. 7:231 (1962).
33. S. M. Wiederhorn and P. R. Townsend, J. Am. Ceram. Soc. 53:486 (1970).
34. B. Stavrinidis and D. G. Holloway, Phys. Chem. of Glasses 24:19 (1983).
35. T. A. Michalske and E. R. Fuller, J. Am. Ceram. Soc. 68:586 (1985).
36. S. M. Wiederhorn and L. H. Bolz, J. Am. Ceram. Soc. 53:543 (1970).
37. W. A. Lanford, K. Davis, P. Lamarche, T. Laursen, R. Groleau, and R. H. Doremus, J. Non. Cryst. Solids 33:249 (1979).
38. M. Tomozawa, J. Non-Cryst. Solids 73:197 (1985).
39. M. A. Rana and R. W. Douglas, Phys. Chem. Glasses 2:196 (1961).

CONTRIBUTORS

Co-Chairmen

R. C. Bradt, Professor of Ceramic Science and Engineering and Head of the Department of Materials Science and Engineering, The Pennsylvania State University, University Park, Pennsylvania.

A. G. Evans, Professor of Materials Science, Department of Materials Science, University of California at Berkeley, Berkeley, California.

D. P. H. Hasselman, Whittemore Professor of Engineering, Department of Materials Engineering, Virginia Polytechnic Institute and State University, Blacksburg, Virginia.

F. F. Lange, Senior Research Scientist, Structural Ceramics Group, Rockwell International/Science Center, Thousand Oaks, California.

Session Chairpersons

R. C. Bradt, Department of Materials Science and Engineering, University of Washington, Seattle, Washington USA

F. F. Lange, Rockwell International, Thousand Oaks, California USA

P. F. Becher, Oak Ridge National Laboratory, Oak Ridge, Tennessee USA

D. Munz, University of Karlsruhe, Karlsruhe, FEDERAL REPUBLIC OF GERMANY

V. J. Tennery, Oak Ridge National Laboratory, Oak Ridge, Tennessee USA

N. Claussen, Max-Planck Institute, Stuttgart, FEDERAL REPUBLIC OF GERMANY

D. S. Wilkenson, McMaster University, Hamilton, Ontario CANADA

E. R. Fuller, National Bureau of Standards, Washington, DC USA

AUTHORS

H. Abe, Asahi Glass, Yokohama, 221 JAPAN

M. M. Abou el leil, United Technologies, East Hartford, Connecticut 26108 USA

J. E. Bailey, University of Surrey, Guildford ENGLAND

P. F. Becher, Metals and Ceramics Division, Oak Ridge National Laboratory, Oak Ridge, Tennessee 37831 USA

B. A. Bender, Naval Research Laboratory, Washington, DC 20375 USA

G. Bernhardt, Societe Europeenne Propulsion, Bordeaux, FRANCE

R. C. Bradt, Department of Materials Science and Engineering, University of Washington, FB 10, Seattle, Washington 98195 USA

T. K. Brog, Department of Materials and Metallurgical Engineering, University of Michigan, Ann Arbor, Michigan 48109 USA

B. C. Bunker, Sandia National Laboratories, Albuquerque, New Mexico 87185 USA

F. E. Buresch, Kernforschungsanlage, Juelich, FEDERAL REPUBLIC OF GERMANY

N. J. Bylina, Research and Development, Norton Company, Worcester, Massachusetts 01606 USA

F. A. Camaratta, United Technologies, East Hartford, Connecticut 01606 USA

R. Carlsson, Swedish Institute for Silicate Research, Box 5403, Goteborg, SWEDEN

E. Carlstrom, Swedish Institute for Silicate Research, Box 5403, Goteborg, SWEDEN

E. D. Case, Department of Materials Science and Engineering, University of California, Berkeley, California 94520 USA

K. S. Chan, Southwest Research Institute, San Antonio, Texas 78284 USA

S. Chanda, University of Surrey, Guildford ENGLAND

J. Chang, General Motors Corporation, Indianapolis, Indiana USA

P. O. Charreyron, Research and Development Division, Norton Company, Worcester, Massachusetts 01606 USA

C. P. Chen, Jet Propulsion Laboratory, Pasadena, California 91109 USA

E. P. Chen, Sandia National Laboratory, Albuquerque, New Mexico 87185 USA

J. L. Chermant, Caen University, Caen, FRANCE

L. Chuck, Fracture and Deformation Division, National Bureau of Standards, Gaithersburg, Maryland 20814 USA

V. M. Chushko, Institute for Strength, Kiev, UNION SOVIET SOCIALIST REPUBLICS

K. C. Chyung, Research and Development, Corning Glass Works, Corning, New York 14831

N. Claussen, Max-Planck Institute, Stuttgart, Federal Republic of Germany

D. R. Clarke, IBM Corporation, Yorktown Height, New York 10598 USA

R. R. Cook, Fracture and Deformation Division, National Bureau of Standards, Gaithersburg, Maryland 20814 USA

R. F. Cooper, Research and Development, Corning Glass Works, Corning New York 14831 USA

R. F. Davis, Department of Materials Science and Engineering, North Carolina State University, Raleigh, North Carolina 27695

J. U. Derby, Engineered Materials Division, Standard Oil of Ohio, Niagara Falls, New York 14302 USA

R. R. DiGenova, United Technologies, East Hartford, Connecticut 26108 USA

N. B. Eden, University of Surrey, Guildford, ENGLAND

A. E. Emery, Department of Mechanical Engineering, University of Washington, Seattle, Washington 98195 USA

A. G. Evans, Department of Materials Engineering, University of California, Berkeley, California 94520 USA

C. J. Fairbanks, Fracture and Deformation Division, National Bureau of Standards, Gaithersburg, Maryland 20814 USA

G. Fantozzi, I.N.S.A. Villeurbane, FRANCE

S. W. Freiman, Fracture and Deformation Division, National Bureau of Standards, Gaithersburg, Maryland 20814 USA

H. Fujita, Asahi Glass, Yokohama, 221 JAPAN

E. R. Fuller, Fracture and Deformation Division, National Bureau of Standards, Gaithersburg, Maryland 20814 USA

T. Garino, Department of Materials Science and Engineering, Massachusetts Institute of Technology, Cambridge, Massachusetts 02139 USA

M. G. Gee, National Physical Laboratory, Teddington, Middlesex, UNITED KINGDOM

B. S. Glaesman, Department of Mechanical Engineering, University of Massachusetts, Amherst, Massachusetts 01003 USA

M. Gomina, Caen University, Caen, FRANCE

R. S. Gordon, University of Utah, Salt Lake City, Utah 84112 USA

F. J. Gray, Department of Civil Engineering, University of British Columbia, Vancouver British Columbia, CANADA

D. J. Green, Department of Materials Science and Engineering, Pennsylvania State University, University Park, Pennsylvania 16802 USA

F. G. Hannoosh, Research and Development, Norton Company, Worcester, Massachusetts 01606 USA

D. P. H. Hasselman, Department of Materials Engineering, Virginia Polytechnic Institute and State University, Blacksburg, Virginia 24061 USA

T. Hirai, Tohoku University, Sendai, JAPAN

J. Hominy, Department of Ceramic Engineering, University of Illinois, Urbana, Illinois 61801 USA

S. Ichikawa, Kyoto University, Kyoto JAPAN

K. Jakus, Department of Mechanical Engineering, University of Massachusetts, Amherst, Massachusetts 01003 USA

H. Jinno, Kyoto University, Kyoto JAPAN

J. M. Jouin, University of Caen, Caen, FRANCE

M. Kawai, Asahi Glass, Yokohama, 221 JAPAN

P. Khandelwhal, General Motors Corporation, Indianapolis, Indiana USA

T. Kishi, University of Tokyo, Tokyo JAPAN

A. Kobayashi, Department of Mechanical Engineering, University of Washington, Seattle, Washington 98195 USA

B. B. Koepke, Honeywell Corporation, New Hope, Minnesota 55428 USA

S. Kohara, University of Tokyo, Tokyo JAPAN

S. P. Kovalov, Institute for Strength, Kiev, UNION SOVIET SOCIALIST REPUBLICS

A. S. Krausz, University of Ottawa, Ottawa, CANADA

B. A. Kuzmenko, Institute for Strength, Kiev, UNION SOVIET SOCIALIST REPUBLICS

J. C. Kyonka, Honeywell Corporation, New Hope, Minnesota 55428 USA

L. W. Lambert, Honeywell Corporation, Elk River, Minnesota 55428 USA

J. Lankford, Southwest Research Institute, San Antonio, Texas 78284 USA

S. Lawrence, Naval Research Laboratory, Washington, DC 20375 USA

B. R. Lawn, Fracture and Deformation Division, National Bureau of Standards, Gaithersburg, Maryland 20899 USA

M. H. Leipold, Jet Propulsion Laboratory, Pasadena, California 91109 USA

D. Lewis, III, Naval Research Laboratory, Washington, DC 20375 USA

B. M. Liaw, Department of Mechanical Engineering, University of Washington, Seattle, Washington 98195 USA

H. Y. Liu, Max-Planck Institute, Stuttgart, Federal Republic of Germany

J. Mace, Societe Europeene Propulsion, Bordeaux FRANCE

D. B. Marshall, Rockwell International, Thousand Oaks, California 91360 USA

Y. Matsuo, Tokyo Institute of Technology, Tokyo, JAPAN

D. Maugis, Lab Central des Ponts et Clausees, Paris, FRANCE

K. D. McHenry, Honeywell Corporation, New Hope, Minnesota 55428 USA

K. R. McKinney, Naval Research Laboratory, Washington, DC 20375 USA

J. J. Mecholsky, Pennsylvania State University, University Park, Pennsylvania 16802 USA

T. A. Michalski, Sandia National Laboratories, Albuquerque, New Mexico 87185 USA

W. Meyer, Kernforschungsanlage, Juelich, FEDERAL REPUBLIC OF GERMANY

M. Miyata, Kyoto University, Kyoto, JAPAN

A. Mocellin, Ecole Polytechnique de Lausanne, Lausanne, SWITZERLAND

H. Monji, Kyoto University, Kyoto, JAPAN

K. L. More, Department of Materials Science and Engineering, North Carolina State University, Raleigh, North Carolina 27695 USA

R. Morrell, National Physical Laboratory, Teddington, Middlesex, UNITED KINGDOM

D. Munz, University of Karlsruhe, Karlsruhe, FEDERAL REPUBLIC OF GERMANY

A. Nakahira, Tohoku University, Sendai, JAPAN

J. Nakayama, Asahi Glass, Yokohama, 221 JAPAN

P. S. Nicholson, Department of Metallurgical and Materials Engineering, McMaster University, Hamilton, Ontario CANADA

K. Niihara, Tohoku University, Sendai, JAPAN

T. Okada, Kyocera Corporation, Laguno Hills, California 92653 USA

G. Orange, I.N.S.A. Villeurbane, FRANCE

F. Osterstock, Caen University, Caen, FRANCE

R. A. Page, Southwest Research Institute, San Antonio, Texas 78284 USA

J. J. Petrovic, Los Alamos National Laboratory, Los Alamos, New Mexico 87545 USA

G. G. Pisarenko, Institute for Strength, Kiev, UNION SOVIET SOCIALIST REPUBLICS

G. G. Pisarenko, Institute for Strength, Kiev, UNION SOVIET SOCIALIST REPUBLICS

G. D. Quinn, Army Materials and Mechanics Research Center, Watertown, Massachusetts 02172 USA

J. E. Ritter, Department of Mechanical Engineering, University of Massachusetts, Amherst, Massachusetts 01003 USA

A. G. Robertson, Department of Metallurgy and Materials Engineering, McMaster University, Hamilton, Ontario CANADA

M. Sakai, Department of Materials Science and Engineering, University of Washington, Seattle, Washington USA

T. H. Service, Department of Mechanical Engineering, University of Massachusetts, Amherst, Massachusetts 01003 USA

S. G. Seshadri, Engineered Materials Division, Standard Oil of Ohio, Niagara Falls, New York 19302 USA.

D. L. Shelleman, Department of Materials Science and Engineering, Pennsylvania State University, University Park, Pennsylvania 16802 USA

S. K. Shetty, University of Utah, Salt Lake City, Utah 84112 USA

G. Sines, Department of Materials Engineering, University of California, Los Angeles, California 90024 USA

J. P. Singh, University of Utah, Salt Lake City, Utah 84112 USA

M. Srinivasan, Standard Oil of Ohio, Engineered Materials Division, Niagara Falls, New York 14302 USA

L. W. Stephens, CSIRO, Melbourne, AUSTRALIA

R. L. Stewart, Research and Development, Corning Glass Works, Corning, New York 14831

L. J. Storz, Sandia National Laboratories, Albuquerque, New Mexico 87185 USA

M. V. Swain, CSIRO, Melbourne, Victoria, AUSTRALIA

P. L. Swanson, National Bureau of Standards, Gaithersburg, Maryland 20899 USA

L. M. Taylor, Sandia National Laboratory, Albuquerque, New Mexico 87185 USA

M. P. Taylor, Research and Development, Corning Glass Works, Corning, New York 14831 USA

T. N. Tiegs, Metals and Ceramics Division, Oak Ridge National Laboratory, Oak Ridge, Tennessee 37831 USA

T. Y. Tien, Department of Materials and Metallurgical Engineering, University of Michigan, Ann Arbor, Michigan 48109 USA

T. Troczynski, Department of Metallurgical and Materials Engineering, McMaster University, Hamilton, Ontario, CANADA

T. Uchiyama, Tohoku University, Sendai, JAPAN

A. Venkateswaran, Department of Materials Engineering, Virginia Polytechnic Institute and State University, Blacksburg, Virginia 24061 USA

A. V. Virkar, University of Utah, Salt Lake City, Utah 84112 USA

S. Wakayama, University of Tokyo, Tokyo, JAPAN

J. S. Wallace, Department of Materials and Metallurgical Engineering, University of Michigan, Ann Arbor, Michigan 48109

S. M. Wiederhorn, Inorganic Materials Division, National Bureau of Standards, Gaithersburg, Maryland 20899 USA

D. S. Wilkenson, Department of Metallurgy and Materials Engineering, McMaster University, Hamilton, CANADA

C. C. Wu, Naval Research Laboratory, Washington, DC 20375 USA

J. Z. Zhe, Research Institute of Building Materials, Beijing, PEOPLE'S REPUBLIC OF CHINA

INDEX